# The ARRL Antenna Compendium

## Volume 3

**Editor**
Gerald L. (Jerry) Hall, K1TD

**Assistant Editors**
Joel P. Kleinman, N1BKE
Bob Schetgen, KU7G
Larry D. Wolfgang, WR1B
Jim Kearman, KR1S
Steve Ford, WB8IMY

**Production**
Jean Wilson
Joe Shea

**Technical Illustrations**
Dianna Roy
David Pingree, N1NAS

**Cover Design**
Sue Fagan

### About the Cover:

**Clockwise from top left:** Beginning on page 79, Steve Powlishen, K1FO, takes aim at the EME polarization-mismatch problem with a 432-MHz rear-mount Yagi array. The drawing shows the calculated H-plane pattern for 4 × 14-element Yagis stacked 49 inches apart.

Three antenna experimenters collaborated on a 160-meter antenna for small lots, "The Square-Four Receiving Array." It's described in the article beginning on page 33. This view is the calculated azimuth radiation pattern when the array is operated in a square configuration. (It can also operate as a diamond.)

The calculated three-dimensional pattern of a commercial antenna—the 20-meter Hy-Gain 205CA, 70 feet above real earth.

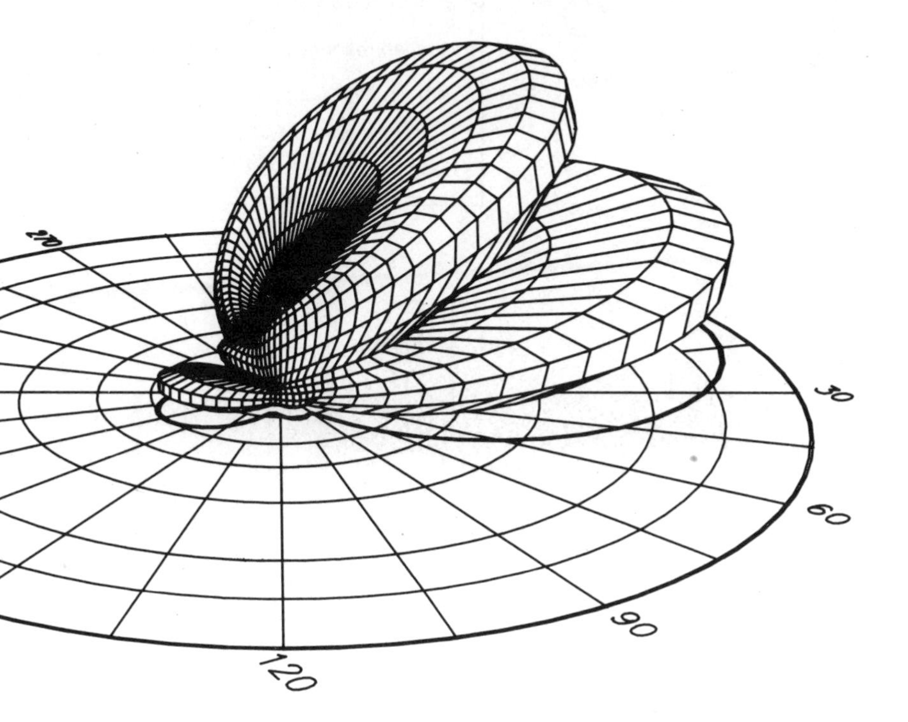

Printed in USA



ISBN: 0-87259-401-7

First Edition
Second Printing, 1993

# Contents

## VERTICALS, GROUND PLANES AND SLOPER ANTENNAS



## MINIATURE, TRAP AND MULTIBAND DIPOLES



## YAGI ANTENNAS

## LOOP ANTENNAS



## A QUAD ANTENNA



## LOG PERIODIC ARRAYS



## A MOBILE ANTENNA



## A DIRECTION FINDING SYSTEM



## CONTROLLED CURRENT DISTRIBUTION ANTENNAS



## OTHER ANTENNA TYPES



## COMPUTERIZED ANTENNA MODELING AND FEED-LINE CALCULATIONS

## GROUND EFFECTS



## TRANSMISSION LINES AND FEED METHODS



## IMPEDANCE MATCHING TECHNIQUES



## INSTALLATION



## RECEIVER OVERLOAD



## MEASUREMENTS AND TEST EQUIPMENT



## PROPAGATION



*Software to accompany this article is available for the IBM PC or compatible computers on a separate diskette.

# Foreword

This book contains 40 previously unpublished articles on antennas, transmission lines, computer modeling and other related subjects. Some, like Steve Cerwin's HF mobile antenna, are practical and straightforward. Others, like Frank Witt's discussion of off-center-fed multiband wire antennas, stress the theoretical bases for more-complex types of antennas.

Regardless of your level of expertise, you'll find much of interest in this book. There's something for everyone in Volume 3 of *The ARRL Antenna Compendium*!

Have an antenna-related project in mind? We'd appreciate the opportunity to review it for possible publication in Volume 4.

David Sumner, K1ZZ
Executive Vice President

Newington, Connecticut
December 1992

**Please note:** Throughout this book, notes and references appear at the end of each article.

# Diskette Availability

Programs for five papers in this book are available in IBM and IBM-compatible format on an optional diskette offered by the ARRL. (The papers with accompanying programs are denoted with asterisks in the table of contents.)

The diskette is copyrighted but not copy protected.

The program files have been supplied by authors of corresponding papers, and filenames have been assigned to resemble the paper titles. There are four BASIC programs and one spreadsheet file in Quattro and Lotus 1-2-3 WK1 format. The diskette is made available as a convenience to the purchaser. The ARRL has verified that the programs run properly, but does not warrant program operation or the results of any calculations.

Diskettes are available in both 5¼-inch (ARRL Order No. 4033) and 3½-inch (ARRL Order No. 4041) formats.

# The Triband Triangle

**By Robert J. Zavrel, Jr, W7SX**
**117 Locatelli Lane**
**Scotts Valley, CA 95066**

I have been interested in phased-array vertical antennas for many years. These antennas have a low radiation angle. The ability to steer these arrays electrically makes them even more attractive. Much of the experimental work has concentrated on the low-frequency bands, where practical Yagi antennas become a bit out of reach for most hams. At the peaks of sunspot cycles, phased verticals might be useful at the higher frequencies (14-30 MHz). The cost of towers, rotators, and triband trap Yagis can easily top several thousand dollars. Additional factors such as aesthetics, occupied real estate, and safety turn me off towers. Yagis and quads can be set on roof-mounted tripods and provide good results. Roof-mounted ground-plane antennas produce good DX results for a fraction of the cost of a roof-mounted Yagi or quad.

My goal for this project was to take advantage of the economic, mechanical, and performance characteristics of a roof-mounted ground-plane antenna. I wanted to achieve multiband operation and some gain with my antenna system, while keeping cost and complexity down. Performance comparable to a roof-mounted Yagi or quad should be possible at a substantial savings. Several design goals were defined:

1) A low-angle radiation pattern
2) Multiband operation (10, 15, and 20 meters)
3) Small size
4) Simple impedance matching
5) Low cost
6) Easy installation
7) Gain on all frequencies
8) Patterns steerable to cover 360°
9) Simple electrical steering from the shack
10) Elimination of parasitic effects

One of the striking features of articles written about phased vertical arrays is that they usually offer only monoband operation.[1] Many of these designs require extensive impedance matching and phasing systems for single-band operation. Building multiband, multidirectional matching and switching circuits could make it difficult to achieve my goals of providing simple impedance matching, low cost and easy installation.

Several of my goals seemed to indicate the possibility of using multiple phased ground-plane antennas. I decided to do some experimental work with these antenna systems. To achieve maximum H-plane gain, I chose a ⅝-λ antenna for the highest operating frequency (28 MHz). This should add about 3 dB to any directional gain achieved on 10 meters. Multiband operation seemed to mandate in-the-shack tuning of the array. I have an old Johnson Kilowatt Matchbox, so I decided to use that with the antenna system. By keeping the feed impedances high at the antennas and choosing my feed-line lengths properly, I would obtain a proper match with the tuner. For several months I experimented with a single antenna at different lengths and with various feed-line lengths. I also did extensive experiments with ground-radial systems. Each step involved on-the-air testing with DX stations.

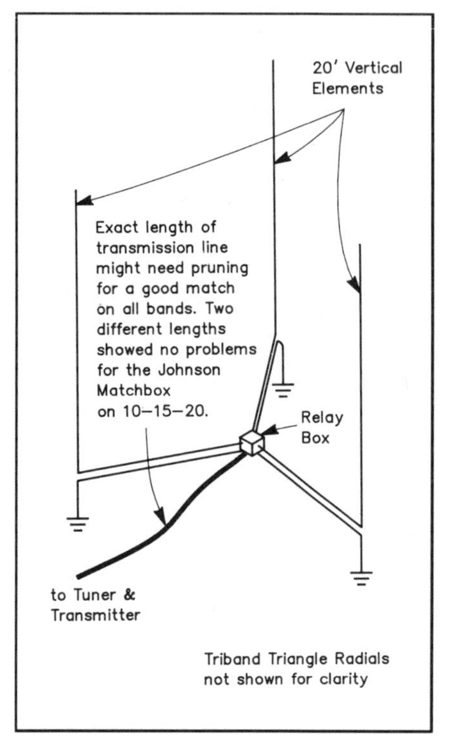


Fig 1—The Triband Triangle consists of three 20-foot-long elements, spaced 10 feet apart at the corners of an equilateral triangle. The radial system is shown in Fig 4.

### System Configuration

Multiple, identical ground-plane antennas seemed to work best, but I had to decide what configuration to use. Should I use two, three, or four elements? Simple switching and impedance matching required either 0 or 180° phasing. That is easily accomplished for multiband operation.

Two elements would provide gain, but achieving multiband operation with a steerable pattern covering 360° becomes a difficult trade-off.

Four elements at the corners of a square form an attractive arrangement. The "four-square array" provides four switchable unidirectional patterns with excellent front-to-back ratios. The switching network for this array is somewhat complex, however, violating one of my design goals.

Three elements arranged at the corners of an equilateral triangle provide a solution to all ten goals. Two of the three elements are selected and fed 180° out of phase, providing three switchable bidirectional patterns. Fig 1 shows how the antenna is laid out.

Each vertical element is 20 feet long, and the sides of the equilateral triangle are 10 feet. I chose a length slightly shorter than ⅝ λ at 28 MHz to avoid resonance at 21 MHz. The high self-resonant base im-

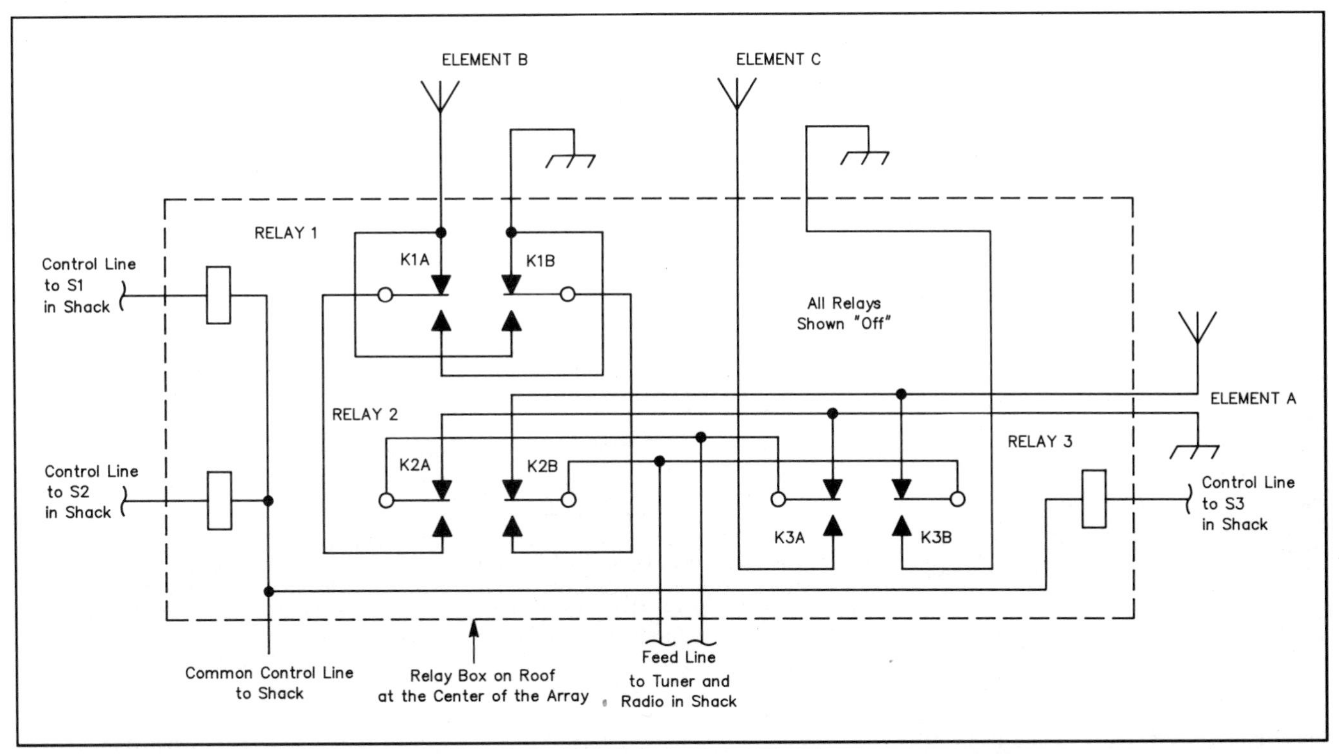


Fig 2—Three DPDT relays select the vertical elements and phasing to steer the Triband Triangle. A common wire can be used to all three relays. The switches are SPST types that connect each relay coil to the required control voltage supply.

pedance might make it difficult to feed the array at 21 MHz, so I used a 20-foot length. I chose 10-foot spacing to raise the feed-point impedance on 20 meters to an acceptable value while maintaining gain over the range from 14 to 28 MHz. One might call this array a "switchable ground-plane W8JK," but "Triband Triangle" is more succinct. The two active elements induce equal signals 180° out of phase in the unused third element. There is no current in the unused element, and no pattern distortion results.

### Feed-Line Considerations

I tried several feed-line configurations. Calculations from the Smith Chart, mutual impedances, and element self-impedances together yielded relatively low common feed-point impedances. Open-wire line was desirable to reduce losses, but the characteristic impedances are quite high. Then I discovered 300-Ω tubular TV twin-lead. Since the system is always fed out of phase, it can be treated as a balanced system. We can ignore coaxial cable, with its high losses. All transmission lines are 300-Ω tubular cable, which costs about 5 cents per foot. In earlier tests I tried RG-8 coax and 600-Ω open-wire line. Losses with the coaxial cable were high. With homemade open-wire line, the high characteristic impedance (600 Ω) made matching more difficult since the SWR was very high. Hams may want to pay more attention to 300-Ω tubular line! The optimum feed-line length to the transmitter depends on the characteristics of the tuner to be used in the shack, such as the range of impedances the tuner can match.

### Switching and Phasing Considerations

On a trip to a local surplus dealer, I secured some 120-V ac DPDT relays for $1.50 each (Struthers-Dunn type 425XBX). I use three DPDT relays to switch between the three possible antenna combinations. Fig 2 shows a wiring diagram. A plastic weatherproof box houses the relays, which are mounted on a ¼-inch-thick piece of plywood inside the box. A big advantage of using a plastic box is that all the RF terminals can be machine screws and nuts mounted through holes drilled in the box. Multiconductor control cable may be the most expensive part of the array! Three toggle switches in the shack control the relays. Fig 3 shows a truth table for the switch settings. One word of caution; if you use a common conductor for the relays it must be able to handle the total relay current. You may want to consider using 12-V relays if you don't like the idea of 120-V lines running to your antenna, but that requires heavier control cable and a power supply. In either case, remember to switch off all power to the relays before working on the box!

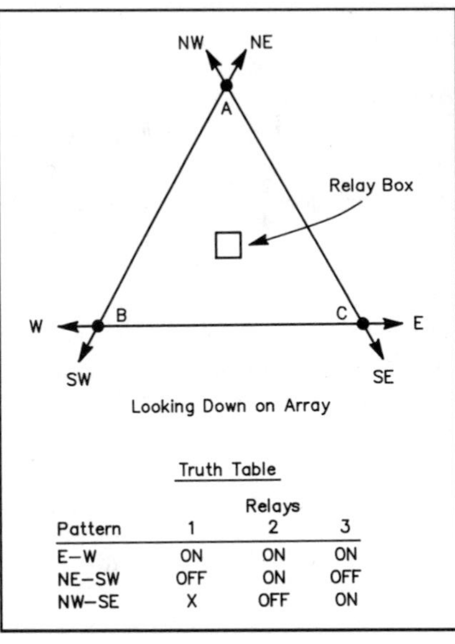


| Pattern | Relays 1 | 2 | 3 |
|---|---|---|---|
| E–W | ON | ON | ON |
| NE–SW | OFF | ON | OFF |
| NW–SE | X | OFF | ON |

Fig 3 —The possible directions you can steer the array. These directions assume elements B and C are aligned on an east-west line and element A is due north of the center point between them. The truth table shows switch positions to select each desired pattern direction; X = don't care.

Since phasing is always 180°, phasing switches are simply cross-wired relays.

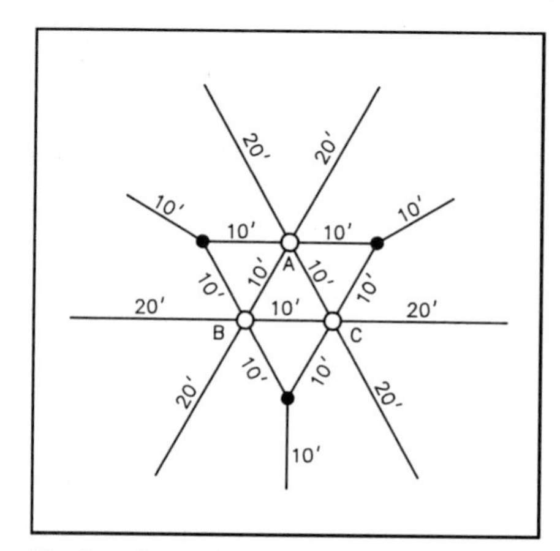


Fig 4 — A top view of the radial system used with the Triband Triangle. The radials should form a plane parallel to level ground or tilt downward from the array symmetrically.

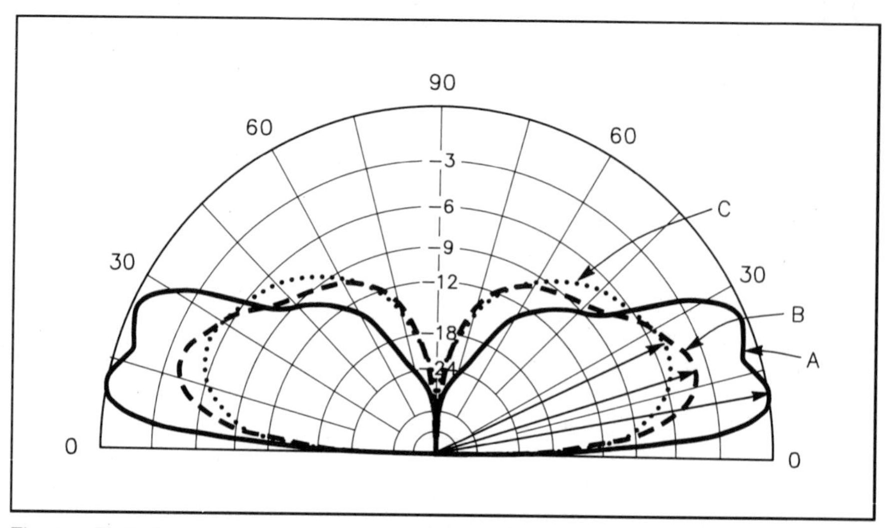


Fig 5 — ELNEC was used to compare the Triband Triangle (curve A) with a 20-foot-long vertical antenna (curve B) and a ¼-λ vertical (curve C) on 10 meters. This is the elevation plot for all three patterns. The graph is scaled to show the Triband Triangle maximum gain at 0 dB. This 0-dB reference is actually 5.34 dBi. The elevation angles of maximum response for the three antennas are 11, 18 and 26°.

This system is inherently broadbanded.

Although my antenna works on 10, 15 and 20 meters, this array should also work on 24 and 18 MHz. The array should also serve as a good SWL antenna, covering the bands between 14 and 28 MHz. A suitable antenna tuner in the shack is all that you need to tune the array to any frequency over the octave for transmitting.

### Antenna Elements

You can make the twenty-foot vertical elements from many different materials. Aluminum electrical conduit comes in 10-foot lengths, so you can use six of these, with couplings to thread them together. I used thin-wall tubing guyed with heavy-duty nylon twine. (I found tubing sections that telescoped together in progressively smaller sizes from 1¼ inches in diameter at the bottom to ½-inch diameter at the top.) I anchored the guy lines using small hooks screwed into my wooden shake-shingle roof. Guy lines should be insulated material. I paid $3 for all the twine required for the array. The hooks cost $5. Three ceramic stand-off insulators serve as base mounts for the 20-foot pipes. Small glass bottles will do nicely as base insulators if ceramic is not available. A good base insulator is important since the voltages here can be quite high in this array.

### Ground System

Any discussion of ground systems for vertical antennas brings up work on elevated verticals by Arch Doty (K8CFU) and Al Christman (KB8I).[2,3] Four slightly elevated radials are as effective as 120 buried radials. "Slightly elevated" can be 15 feet at 1.8 MHz. Many hams are experimenting with new ideas for radial systems. Previous "conventional wisdom" is collapsing under the weight of new ideas and models that seem to work quite well in practice.

I tried several ground systems in my experiments. I tried no ground, different grounds for each antenna, and a common ground. By far the most effective and electrically stable approach is a common ground. No violation of conventional wisdom here! A rule of thumb seems to be that a more extensive radial system is better. Some advantage in pattern symmetry seems to result when the radial lengths are equal to the vertical element height. Fig 4 shows an overhead view of the radial system.

### Other Configurations

This array could be extended to 20-foot spacing and 40-foot verticals for a 20, 30, and 40-meter array. Additional scaling should be possible for 40 and 80 or 80 and 160-meter operation. It may be possible to use the Triband Triangle in conjunction with elevated vertical techniques to provide multiband low-frequency systems. The variations seem endless, and this work may serve as a model for other low-band systems.

This array might work over 1.5 octaves. For example, three 33-foot verticals with 20-foot spacing could be used between 7 and 21 MHz. There are two areas of concern that should be noted for such an array. The shorter element lengths at 7 MHz will lower the feed-point impedance, making it more difficult to match. At 14 MHz, the lengths approach ½ λ, raising the input impedance to a high level. This may also make it more difficult to match the impedance. Array balance may also be more difficult to achieve. These potential problems are certainly not insurmountable.

### Gain Calculations

Antenna gain is often given relative to reference isotropic or dipole antennas. It seems more reasonable to use a single ¼-λ ground-plane antenna as the reference for the Triband Triangle. Be careful about comparing the Triband Triangle gain figures with horizontally polarized arrays such as Yagis and quads. Such comparisons may be misleading. Many factors influence E- and H-plane gain characteristics. These include the Brewster angle, nearby objects and antenna characteristics. Keep in mind that optimum azimuth gain with Yagis and quads is achievable only with high antennas.[4] That would violate my goals of small size, low cost and easy installation. It is interesting to compare the Fig 5A graph with Yagi-antenna patterns from Chapter 5 of *Yagi Antenna Design*.

Using a single ¼-λ ground-plane antenna as a reference, the favored-direction gain is

- 6.4 dB at 10 meters (3.4 dB phased-array gain + 3.0 dB gain from increased element length)
- 5.6 dB at 15 meters (3.6 dB phased-array gain + 2.0 dB gain from increased element length)
- 4.0 dB at 20 meters (3.8 dB phased-array gain + 0.2 dB gain from increased element length)

The higher gain figures at the higher frequencies result from the longer electrical length of the elements. The phased-array gain figures are taken from *The ARRL Antenna Book.*[5] [Computer gain calculations indicate the figures given above may be somewhat optimistic.—*Ed.*] Fig 5 shows elevation-plane radiation patterns on 10 meters, calculated using ELNEC, a method of moments analysis program. This plot compares the gain of a ¼-λ vertical, a 20-foot vertical and the Triband Triangle.

### Impedance Matching

Take care to make the system symmetrical. Antenna tuning should be consistent for all three directional settings. An exact match may be difficult to obtain, but a near-ideal compromise setting can be found for all three direction settings corresponding to the phone and CW bands. I obtained an SWR of 1.5:1 or better over the desired bandwidths for all three directions. This permits switching directions without requiring me to adjust the matching network. Of course with some "tweaking" of the tuner, you can optimize the match to squeeze out that last fraction of a dB.

Follow these guidelines for the best system symmetry.

1) The feed lines to each element should be identical. (Remember to count the inches inside the relay box!)

2) The three elements should be mounted as far away from other objects as possible (walls, trees, power lines, and the like).

3) The ground-radial system should be as symmetrical as possible. This means identical radial lengths laid out at identical angles around each element.

4) The vertical elements should be perfectly vertical.

### Conclusions

This antenna is relatively easy to build. Construction requires a minimum of time, money, and materials. The design is inherently broadband over at least one octave. It is advisable to erect a simple single-element ground-plane antenna to check your location for noise before considering the Triband Triangle. Vertical antennas are more susceptible to noise pickup than horizontally polarized antennas. A disadvantage of this array is the lack of front-to-back rejection when receiving, although I measured 20-dB side nulls on ten meters.[6] The DX performance-per-dollar ratio of this antenna system is hard to beat if you have a location that is relatively free from electrical noise. This array will not compare with high Yagis or quads, but will compete admirably against the more common roof-mounted Yagis and quads.

### Notes

[1]J. Hall, Ed, *The ARRL Antenna Book*, 15th ed (Newington: ARRL, 1988), Chapter 29, "Topical Bibliography on Antennas." See the "Phased Array" topic, p 29-17.

[2]A. C. Doty, Jr, J. A. Frey and H. J. Mills, "Efficient Ground Systems for Vertical Antennas," *QST*, Feb 1983, pp 20-25.

[3]A. Christman, "Elevated Vertical Antenna Systems," *QST*, August 1988, pp 35-42.

[4]J. Lawson, *Yagi Antenna Design* (Newington: ARRL, 1986), Chapter 5, "The Effects of Ground." See the "Antenna Performance Over Ground" and "Best Height" sections, pp 5-6 to 5-14.

[5]J. Hall, Ed, *The ARRL Antenna Book*, 16th ed (Newington: ARRL, 1991), pp 8-6 and 8-7. (Some gain figures in previous editions are in error.)

[6]Accurate gain measurements can be made with the receiver AGC "off," as long as the receiver is operating in its linear response region (usually signals less than S9). Simply measure the detected audio voltages and take the ratios.

### Other References

J. Devoldere, ON4UN, *Low Band DXing* (Newington: ARRL, 1988).

# Making Tower-Mounted Half Slopers Work for You

By Duane R. Sanderson, WØTID
3735 SE Stanley Rd
Tecumseh, KS 66542

Amateur Radio, like most things in life, tends to follow trends. The current trend in station equipment typically involves factory-built transceivers, amplifiers and antennas. Today's amateur equipment has advanced to a high level of performance and sophistication. In fact, we have nearly reached the point where average hams cannot create anything of similar quality on their own.

There are many amateurs, however, who still possess that inner desire to create at least a portion of their station layout with their own two hands.

With this idea in mind, the station antenna system provides a great opportunity for creative experimentation and discovery without the need for elaborate and expensive test equipment. This is especially true for various types of wire antennas. Sooner or later, however, most amateurs feel the need to erect a tower.

The typical project will often include a tower, a triband beam or monobanders, and additional antennas for VHF and other bands. I occasionally work amateurs on 10, 15, or 20 meters who have towers and beams. When the discussion shifts to the topic of working the lower bands, I usually hear responses such as, "I don't work the lower frequency bands. I don't have room for larger antennas," or, "Yeah, I tried some slopers on my tower, but I couldn't get a decent SWR. It was a waste of time."

Despite such a pessimistic assessment, I can assure you that it *is* possible to create an all-band antenna system, on *one* tower, in a backyard situation. The result can be an antenna array that performs very well without adding a great deal of time and expense to a basic tower project.

The motivation to begin my own project came one spring evening about a year ago, when a tornado devastated my QTH. It totally destroyed my home and my antennas, creating the need to literally rebuild everything from the ground up. Despite the bleak task that confronted me, I realized that I could use the opportunity to correct some of the shortcomings in my original antenna layout. The result is a beam and half-sloper combination that provides very satisfying results to this day.

### The Tower

Since I have been a ham for over forty years, enough time has passed to reduce my enthusiasm for climbing towers. Therefore, I chose a tower that permitted a ground-level work position. My requirements included

1) a tilt-over type design approximately 50 feet high with hand-crank operation,
2) multi-level guying capability,
3) a large poured-concrete base, and
4) a multiple ground-rod system with each rod driven deeply into the earth.

All of these requirements are fairly common considerations and were achieved with standard hardware construction techniques.

### Guy Wires and Overall Tower Resonance

I deviated from the standard practice of breaking up guy wires with insulators. Actually, I did not place insulators in any of the guy wires. All of the guys make direct metal-to-metal contact at their attachment points with the tower *and* with the grounded guy anchors. This is an important step in lowering the natural resonant frequency of the tower to its lowest possible point. By enlarging the electrical size and cross section as much as possible, the tower, the beam and the guy wires combine to form a large vertical mass with its own resonant frequency. This is a crucial point to keep in mind when considering the use of tower-mounted half slopers.

As you physically view any tower, you are looking at a metal structure projecting above the earth. The bottom is usually grounded, making it the low-impedance end. As you progress up the tower, the impedance increases significantly. If you are about to install a half sloper and you select an operating frequency that is close to the resonant frequency of the tower, you will quickly discover that you can't feed it! The impedance mismatch is so bad, the low coax impedance (52 ohms) can't even begin to transfer power to the tower/sloper combination. The impedance on the tower is just too high. Small wonder that so many amateurs have had so much difficulty with tower-mounted half slopers!

It is a very frustrating experience to search endlessly for the resonant frequency of the half sloper you have just put up, unable to find a point where your pruning will drop the SWR to an acceptable value. The grim fact is, a high feed-point impedance creates a large SWR reading that masks any efforts to prune the sloper for a resonant frequency.

Obviously, low tower resonance is a *must* for successful operation of half sloper antennas. The benefits of this approach should become clear as we progress through the tune-up and usage of the slopers.

### Other Advantages of Uninsulated Guy Wires

There are two other advantages of the uninsulated guy wire approach: It creates a large ground footprint which helps to dissipate direct lightning strikes, and it produces a large RF ground footprint which enhances the ground return of any antenna connected to the tower.

### A Word of Caution

The guy wires should be maintained in a reasonably taut condition. This insures that the metal-to-metal contact is constant, eliminating any noise that may result from loose connections. My installation is completely noise-free in this regard.

### Antenna System Goals

The following criteria were established for the antennas on my tower:

1) the provision to operate on *all* HF bands from one tower,

2) coax feeds that do not require coils or traps,

3) survivability in severe weather and high wind conditions,

4) broadband operation without the need for a tuner,

5) good performance day and night, and

6) no additional towers or supporting masts required.

This is a fairly demanding list! However, with the successful installation of half slopers, all of the requirements were met.

### Past Experience

Over the years, I have tried most of the HF antennas described in *The ARRL Antenna Book*. The sloper type antenna usually gives better performance than horizontal dipoles and has demonstrated good DX ability as well.

The shortcomings of a sloping dipole have been evident at my QTH in terms of wind and weather survival. The sloping dipole has a hanging coax connected to its center. Not only does the coax whip around in the wind, but it is also subject to the stresses of ice loading. The result has been the frustration of watching the coax break off in the middle of a winter storm when it's not possible to do anything about it until good weather returns! The half sloper does not have this problem.

I would like to say at this point that a collection of half slopers is not the only antenna arrangement that will provide all-band performance on a single tower. A center-fed Zepp with a tuner would accomplish the same for many amateurs. However, considering the six goals I listed earlier, the half sloper is the best choice.

Fig 1—The upper half of the tower with guy lines and sloper wires visible.

### Putting It All Together

A half sloper is a ¼-λ wire fed with 52-ohm coax at the top end where it is attached, with an insulator, to the top of the tower. The braid of the coax is connected to the tower and the center conductor is connected to the sloping wire.

The length of any half sloper is computed using the standard quarter-wave antenna formula ($\ell_{ft} = 234/f_{MHz}$). I cut my slopers several inches long so that I could prune them down to the desired resonant frequency. This is a lot easier than soldering small lengths of wire onto a short sloper! I used RG8X mini-foam coax to feed the slopers. While I prefer the smaller diameter and greater flexibility of RG8X, RG8 is obviously the more durable coax.

I used stainless-steel hose clamps to secure the coax braids to the tower. I soldered the center conductors of the coax directly to the sloper wires. Be sure to use a good quality moisture sealant on the exposed coax opening around the braid and center conductor insulation. Twelve-gauge solid copper house wiring was used for the sloper wires. All insulators were made from 1½-inch schedule 40 PVC pipe.

### Interaction...And A Bonus!

If you were to look down on my tower from a helicopter, it would look like a spoked wheel without a tire. The slopers and guy wires fan out around the entire tower (see Fig 1). The horizontal separation angles between adjacent slopers varies from 15° to 30°. The vertical slope of each wire is about 45° with the exception of the 160-meter antenna, which is long enough to require a much flatter angle. With this arrangement there is virtually no interaction between the slopers. There is one exception—that I use to my advantage!

It seems that most amateurs usually have to cope with the old problem of 75 versus 80 meters. After all, everyone wants an antenna that will work both ends of that band. A tuner could be used to alleviate the problem, but my goal was to create a layout that would *not* depend on a tuner to operate properly.

While the half slopers on my tower were fairly broad-banded, I shared the same 75- versus 80-meter dilemma. So, I put up *two* slopers, one for 75 and one for 80. The 75-meter sloper resonated at 3800 kHz and the 80-meter sloper resonated at 3580 kHz. Unfortunately, pruning either sloper had some effect on the other. I tried positioning them on opposite sides of the tower, but the interaction still existed. With a little experimentation I found that pruning the 80-meter antenna produced a greater effect on the 75-meter antenna than vice versa. That was when I made a discovery that turned out to be a real bonus.

With the 75-meter sloper connected to the transceiver, I *shorted* the 80-meter coax in the shack with a wire connected between the braid and the center conductor. As I did so, the 75-meter sloper jumped upward in resonant frequency to 3850 kHz. At the same time, the signals received from the east suddenly increased about 5 dB on the S meter.

My 75-meter sloper is on the east side of the tower. The 80-meter sloper is on the west side of the tower. The shorted 80-meter sloper became a *reflector* for the 75-meter antenna, effectively creating a two element vertical beam!

### Another Bonus

Believe it or not, you can feed two or more half slopers with the same coax line. It's not a complicated procedure. Just tie the top ends of the slopers to the same coax center conductor. As of this writing, I have not tried to feed more than two slopers with one coax. Two slopers fed by one coax are best suited for a common feed when they are positioned next to each other.

I advise against trying to feed both the 75- and 80-meter slopers with the same coax. I tried it and soon realized that I was going through an endless and unsuccessful attempt to unscramble the interaction.

### Interaction With Guy Wires

The guy wires appear to be neutral as far as any noticeable disturbance to nearby slopers. The guy wires at the top of the tower slope downward at about 25°. A sloper trailing away from the tower at 45° has a considerable distance developing between itself and the guy as the wires progress downward. This permits a sloper to be positioned within inches of a guy at the top of the tower with no noticeable effect. The bottom ends of all my slopers are tied off past the end insulators to metal fence posts driven into the soil. The locations of the metal posts were chosen to provide the direction, spacing and 45° slope angle mentioned above.

### Planning for the Worst

The coaxial feed lines are taped to the tower legs at one-foot intervals all the way up the tower. This is my own standard for securing coax since I don't want to see *anything* moving in a 50-mi/h wind.

Midwest ice storms can take down all but the strongest of antennas. Even so, ice-loaded sloping wires are mechanically more durable and will survive better than horizontal wires. The strain load on the tower is generally downward and balanced with a reduced likelihood of tower damage from excessive sideload stress.

### Tune-Up

Without exception, every one of my half slopers was resonated to the spot in the band I wanted, and produced a 1:1 SWR. A small amount of pruning was necessary on most bands to locate the optimum resonant frequency. The only serious deviation from the $1/4$-λ formula was on the 160-meter band. The 126-foot formula length proved to be too long for resonance at 1850 kHz. Since it was so near to the earth's surface, I suspected that the low fractional wavelength on the vertical position of this sloper was effectively lowering the resonant frequency. I had to prune off several feet to bring it on frequency.

### SWR Measurements

All half slopers will produce a 1:1 SWR at their resonant frequencies. The bandwidths vary, with the 160, 80 and 75-meter slopers covering at least half of the band with a 1.5:1 SWR. The 40-meter sloper covers 200 kHz of the band with a 1.5:1 SWR. The higher frequency slopers cover their entire bands with a 1.5:1 SWR.

### How Well Do They Work?

How well *do* they work? Very well—both day and night!

Half slopers are more or less vertically polarized, which is excellent for low-angle nighttime propagation. Some propagation experts suggest that vertically polarized signals are tilted off the vertical plane when they travel near the earth's surface. This tilt condition would help explain why all types of slopers perform so well. There also appears to be a very small amount of directivity in the direction of the slope.

I have made comparisons at my station between an 80-meter full-wave horizontal loop and a sloper. Daytime performance slightly favors the loop. At night, the sloper will be generally one to two S units better.

### Band Coverage

As of this writing, I have erected tower-mounted half slopers for the following bands: 160, 80, 75, 40, 30 and 17 meters.

The 75- and 40-meter slopers are fed by the same coax, as are the 80- and 17-meter slopers. The 40-meter sloper has a low SWR on 15 meters and operates as a $3/4$-λ antenna on that band. I usually use my triband beam on 15 meters, so the choice is optional.

I have left one band out: 12 meters. In my case, 12 meters happens to be a band of low interest. However, it's just a matter of connecting one more wire to the array and choosing an existing sloper as its mate so that both can be fed from the same cable. I will probably connect it to the 80/17-meter pair, feeding three slopers with one coax.

### Interaction from the Tribander Beam

Rotation of the tribander beam on the top of the tower does not have a significant effect on any of the slopers. I would expect a full-size 20-meter Yagi to produce some noticeable effects. Certainly a 40-meter Yagi would create a large overhang of overall tower mass that would influence the radiation angle significantly.

Other half-sloper users have reported that changes in sloper directivity become very noticeable when big Yagis are rotated. The difference between the ends of the Yagi elements hanging over the slopers versus the broadside portions of the elements in the same position appears to be the mechanism that varies performance. It can be readily seen that the angle of radiation would be modified in this case. Investigation of this phenomenon will probably be a future adventure at my QTH.

### Conclusion

I have been using half slopers in the windy and turbulent midwest for eight years now. No other antenna has demonstrated their combined survivability and performance. It is a good feeling to look out the window on the morning after a big storm and see that the sloper farm is still in one piece. Give these half slopers a try on your tower. If you can keep the tornadoes away, you'll have the same satisfying view from your window too!

# The Double Cross Vertical Antenna

**By Robert Wilson, AL7KK**
**Box 110955**
**Anchorage, AK 99511**

Antennas have been part of my business for years and I have placed my designs at large stations in a number of countries. (Money is no object with the big commercial arrays.) Even though I design antennas for a living, I still like to play with them at home. Part of the fun of antenna construction is producing a high quality, efficient antenna that is truly low in cost.

One snowy Alaskan afternoon I was inspired to design a home-brew antenna project. Simplicity was the primary consideration. It had to be something I could build with just a couple of coax connectors, a few coat hangers and my trusty soldering iron. It also had to be theoretically feasible and efficient. Despite my best intentions, my antennas have a tendency to grow like rabbits. Within a short period of time my living room was filled with paper and wire!

### The Double Cross Design

Where would antenna designers be without computers? After a bit of work at my keyboard I soon developed the "X" antenna. As I examined the design I noticed a unique property of the "X": It could be mounted on a metal pole with almost no interaction. (The vertical center line between elements was also a null line.) Stacking the antenna seemed natural, and after looking at the result (see Fig 1) the name "Double Cross" seemed natural too!

This antenna is a variation on a vertical dipole. Imagine two V-shaped wires, one opening upward and one downward. Now imagine that both V wires are connected to a coax cable at their apexes. The coax shield is soldered to one V and the coax center wire to the other V.

By adjusting the angle of the V to 70°, the antenna can be made to resonate at the desired frequency and that angle will give the best bandwidth. The angle does not have to be extremely precise. Angles from about 60 to 110 degrees do little to change the feed impedance. (The impedance remains about 30 to 35 Ω over these angles.) The mid-line between the V wires is a null line. This permits the support pole and coax feeder harness to be mounted in the center line area with no problems.

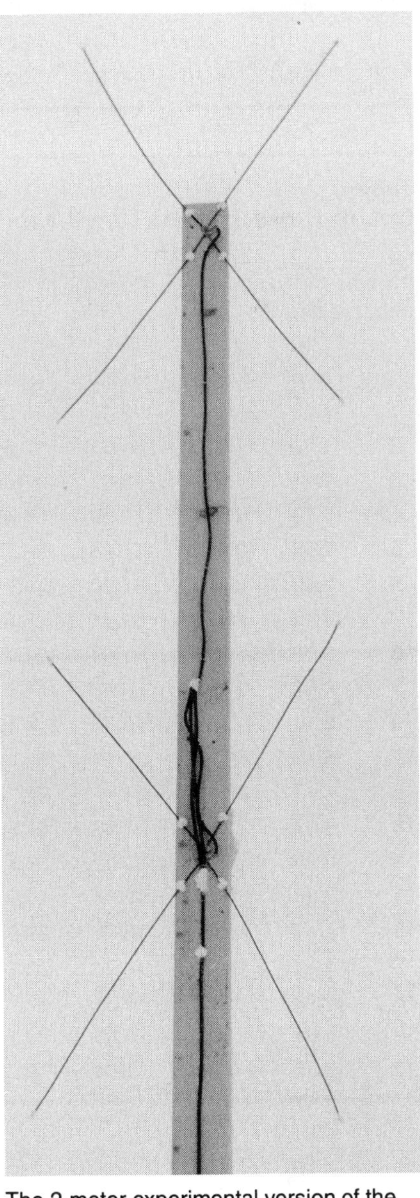

The 2-meter experimental version of the Double Cross antenna made with no. 10 wire, coax, silicone glue and a 2 × 4 board.

The angle of the V apex is set easily by adjusting the tip-to-tip spacing of the V elements. On paper and in actual practice it seems to be reasonably noncritical. For example, a tip-to-tip error of ±5% seems to make little difference in the operation of the antenna.

It also occurred to me that grounding the topmost elements on the antenna would offer lightning protection for the receiver. Turning the top dipole upside down and mounting the hot element of the next lower dipole *upward* also seemed to be a neat symmetrical method to balance the antenna currents. Feeding the dipoles 180° out of phase is all that is required for both protection and balance. This can be easily accomplished by adjusting the length of the feed harness (see Fig 2).

Old hams told me that using two wires in a V would not give a circular pattern because it was not balanced. This "old ham's tale" is simply not true; the pattern of the Double Cross is only 0.5 dB out of round. For the sake of a little irregularity there isn't a compelling reason to increase the complexity by adding more wires. To prove that more complex designs were not necessary, I constructed a three-dimensional version of the antenna and found it was indeed very difficult to handle. After a thorough examination of the calculations and construction of the flat "X" antenna, I proved that a two-dimensional design was entirely sufficient.

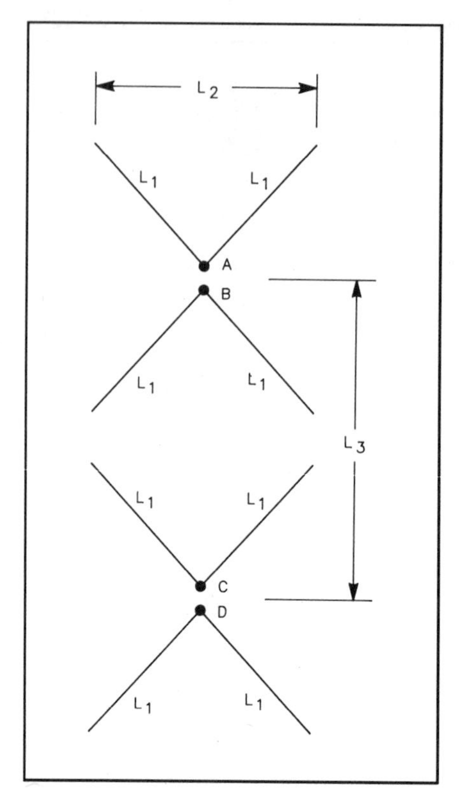


Fig 1—Double Cross antenna design.

After I built the first coat-hanger "X" antenna I also discovered that the measured bandwidth was better than expected. The element diameter was increased to ½ inch on the 2-meter theoretical model and this allowed the antenna to cover more than the full band.

Table 1 provides element lengths, tip-to-tip spacings, element diameters and phasing-line information for most ham bands. Table 2 offers the same information for 1 to 30 MHz in arbitrarily numbered bands. Incidentally, using wire for the elements is quite acceptable even though tubing is indicated. For ham operation a sufficient bandwidth can be obtained by using thin wire elements.

A single stacked "X" dipole could be strung between two trees for low-frequency operation. Attempting the Double Cross stacking arrangement on 80 or 160 meters would be unreasonable, but a single unstacked "X" would perform as an excellent wide-band vertical on these bands.

It may be a good idea to employ a matching transformer between the 50-Ω coax and the antenna. Such an RF transformer can be made by using a high quality, high frequency powdered core with a cross section of at least ½ by ¼ inch for 150 watts. For the HF bands the 50-Ω primary should be 10 turns and the 30-Ω secondary should be 8 turns. I prefer to use 18 gauge wire with Teflon insulation, wrapping the first winding and then interlacing the second. I also like to tie down the ends with fishing line and coat the transformer with a heavy layer of clear silicone glue. A properly constructed transformer should last up to 50 years—if it doesn't take a direct lightning strike!

**Table 1**
**Double Cross Antenna Lengths for Amateur Bands**

All dimensions are in feet. "Bandwidth" shows the lower and upper 2:1-SWR frequencies, MHz.

| *Band* | *Freq.* | *L1 Element* | *L2 Tip-tip* | *L3 Spacing* | *L4 Coax 1* | *L5 Coax 2* | *L6 Diam.* | *Bandwidth* |
|---|---|---|---|---|---|---|---|---|
| 160 | 1.90 | 111.05 | 127.38 | 323.68 | 256.36 | 85.45 | 3.04 | 1.7—2.1 |
| 80 | 3.70 | 57.03 | 65.41 | 166.22 | 131.64 | 43.88 | 1.56 | 3.3—4.1 |
| 75 | 3.90 | 54.10 | 62.06 | 157.69 | 124.89 | 41.63 | 1.48 | 3.5—4.3 |
| 40 | 7.15 | 29.51 | 33.85 | 86.01 | 68.12 | 22.71 | 0.81 | 6.4—7.9 |
| 30 | 10.13 | 20.84 | 23.90 | 60.74 | 48.11 | 16.04 | 0.57 | 9.1—11.1 |
| 20 | 14.18 | 14.89 | 17.07 | 43.39 | 34.36 | 11.45 | 0.41 | 12.8—15.6 |
| 17 | 18.11 | 11.65 | 13.36 | 33.96 | 26.90 | 8.97 | 0.32 | 16.3—19.9 |
| 15 | 21.23 | 9.94 | 11.40 | 28.98 | 22.95 | 7.65 | 0.27 | 19.1—23.3 |
| 12 | 24.93 | 8.46 | 9.71 | 24.67 | 19.54 | 6.51 | 0.23 | 22.4—27.4 |
| 10 | 28.50 | 7.40 | 8.49 | 21.58 | 17.09 | 5.70 | 0.20 | 25.7—31.4 |
| 6 | 50.10 | 4.21 | 4.83 | 12.28 | 9.72 | 3.24 | 0.12 | 45.1—55.1 |
| 2 | 146.00 | 1.45 | 1.66 | 4.21 | 3.34 | 1.11 | 0.04 | 131.4—160.6 |

**Table 2**
**Double Cross Antenna Lengths for Various SWL Bands**

All dimensions are in feet. "Bandwidth" shows the lower and upper 2:1-SWR frequencies, MHz

| *Band* | *Freq.* | *L1 Element* | *L2 Tip-tip* | *L3 Spacing* | *L4 Coax 1* | *L5 Coax 2* | *L6 Diam.* | *Bandwidth* |
|---|---|---|---|---|---|---|---|---|
| 1 | 1.00 | 211.00 | 242.02 | 615.00 | 487.08 | 162.36 | 5.77 | 0.9—1.1 |
| 2 | 1.20 | 175.83 | 201.68 | 512.50 | 405.90 | 135.30 | 4.81 | 1.1—1.3 |
| 3 | 1.44 | 146.53 | 168.07 | 427.08 | 338.25 | 112.75 | 4.01 | 1.3—1.6 |
| 4 | 1.73 | 121.97 | 139.89 | 355.49 | 281.55 | 93.85 | 3.34 | 1.6—1.9 |
| 5 | 2.07 | 101.93 | 116.92 | 297.10 | 235.30 | 78.43 | 2.79 | 1.9—2.3 |
| 6 | 2.49 | 84.74 | 97.20 | 246.99 | 195.61 | 65.20 | 2.32 | 2.2—2.7 |
| 7 | 2.99 | 70.57 | 80.94 | 205.69 | 162.90 | 54.30 | 1.93 | 2.7—3.3 |
| 8 | 3.58 | 58.94 | 67.60 | 171.79 | 136.06 | 45.35 | 1.61 | 3.2—3.9 |
| 9 | 4.30 | 49.07 | 56.28 | 143.02 | 113.27 | 37.76 | 1.34 | 3.9—4.7 |
| 10 | 5.16 | 40.89 | 46.90 | 119.19 | 94.40 | 31.47 | 1.12 | 4.6—5.7 |
| 11 | 6.19 | 34.09 | 39.10 | 99.35 | 78.69 | 26.23 | 0.93 | 5.6—6.8 |
| 12 | 7.43 | 28.40 | 32.57 | 82.77 | 65.56 | 21.85 | 0.78 | 6.7—8.2 |
| 13 | 8.92 | 23.65 | 27.13 | 68.95 | 54.61 | 18.20 | 0.65 | 8.0—9.8 |
| 14 | 10.70 | 19.72 | 22.62 | 57.48 | 45.52 | 15.17 | 0.54 | 9.6—11.8 |
| 15 | 12.84 | 16.43 | 18.85 | 47.90 | 37.93 | 12.64 | 0.45 | 11.6—14.1 |
| 16 | 15.41 | 13.69 | 15.71 | 39.91 | 31.61 | 10.54 | 0.37 | 13.9—17.0 |
| 17 | 18.49 | 11.41 | 13.09 | 33.26 | 26.34 | 8.78 | 0.31 | 16.6—20.3 |
| 18 | 22.19 | 9.51 | 10.91 | 27.72 | 21.95 | 7.32 | 0.26 | 20.0—24.4 |
| 19 | 26.62 | 7.93 | 9.09 | 23.10 | 18.30 | 6.10 | 0.22 | 24.0—29.3 |
| 20 | 31.95 | 6.60 | 7.57 | 19.25 | 15.25 | 5.08 | 0.18 | 28.8—35.1 |

### Design Calculations

A simple four-function calculator can be used to calculate the antenna dimensions.

1) The lengths of each leg of the V elements (L1):

$$L1 = \frac{64.2 \text{ meters}}{f_{MHz}} = \frac{211 \text{ feet}}{f_{MHz}}$$

2) The tip-to-tip distance (L2) of the open end of the V element gives a 70° apex angle:

$L2 = 1.147 \times L1$ (results in either feet or meters, according to the original L1 values)

3) The separation of two "X" elements is ⅝ λ or L3:

$$L3 = \frac{187.0 \text{ meters}}{f_{MHz}} = \frac{615 \text{ feet}}{f_{MHz}}$$

4) L4 represents the length of the 50-Ω solid polyethylene dielectric coax (velocity factor 0.66) from the **upper** dipole to the summing junction where upper and lower dipoles are connected:

$$L4 = \frac{224.4 \text{ meters} \times 0.66}{f_{MHz}} = \frac{738 \text{ feet} \times 0.66}{f_{MHz}}$$

5) L5 represents the length of 50-Ω solid polyethylene dielectric coax from the **lower** dipole to the summing junction mentioned above, *and* the length of the two parallel 70-Ω coaxial cables used for impedance matching:

$$L5 = \frac{75 \text{ meters} \times 0.66}{f_{MHz}} = \frac{246 \text{ feet} \times 0.66}{f_{MHz}}$$

6) L6 represents the diameter of the V legs required for the indicated bandwidth. However, ordinary wire also works well for practical ham antennas:

$$L6 = \frac{1.759 \text{ meters}}{f_{MHz}} = \frac{69.2 \text{ in.}}{f_{MHz}} = \frac{5.77 \text{ ft}}{f_{MHz}}$$

### Double Cross Construction

Construction of the first "X" antenna was accomplished with coat-hanger wire soldered to a coax connector. The coax line was connected and the free end was pulled through a ½-inch support pipe for testing. The results were excellent. The SWR was low and the bandwidth was exactly as calculated.

The next step was construction of a stacked 2-meter Double Cross antenna with a complete coaxial feed harness. This was done using no. 10 Copperweld wire salvaged from an open-wire telephone system. It was screwed to a 2 × 4 board and glued in place with silicone adhesive. There is no question that this was a minimum cost antenna! Once again, everything worked fine and the SWR was 1.1:1 over the entire 2-meter band.

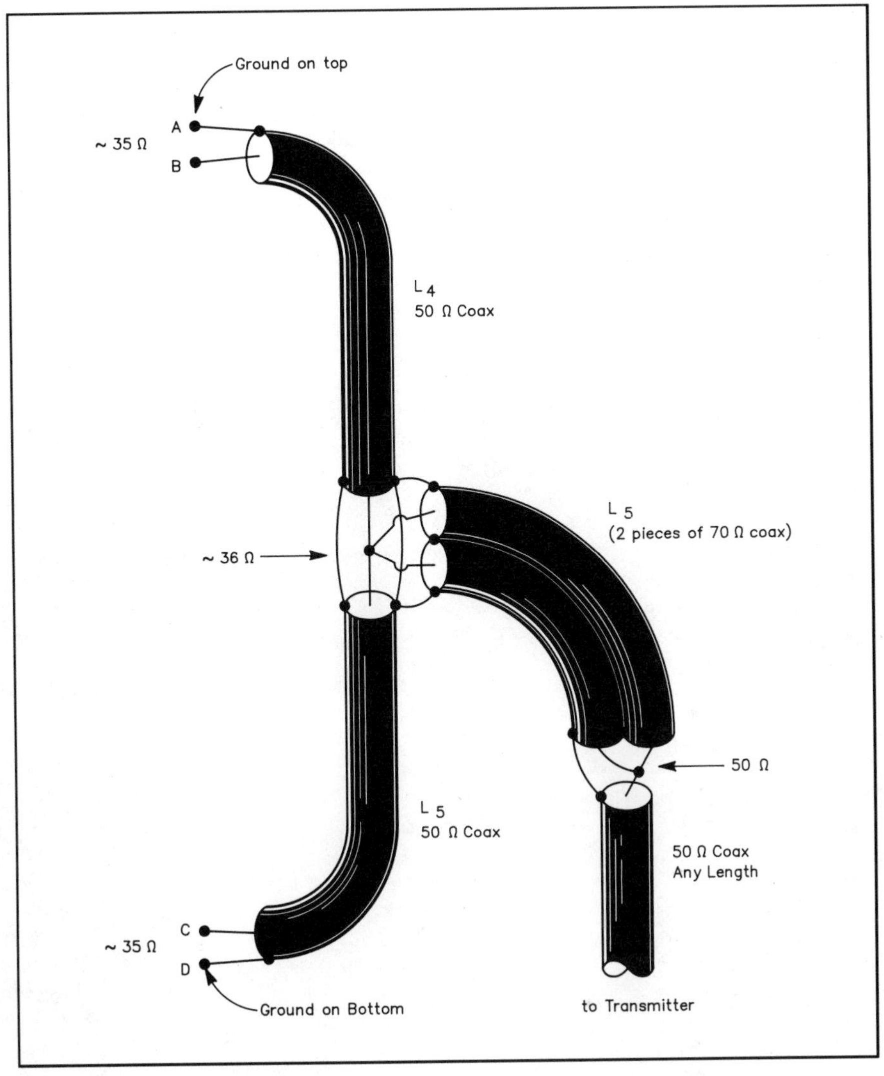


Fig 2—Double Cross coax harness.

The stacking harness shown in Fig 2 is designed for a 180° phase reversal. The shield of the ¾-λ section of 50-Ω coax goes to the top V. This is length L4. The shield of the ¼-λ 50-Ω coax section goes to the bottom V. This is length L5.

The two 50-Ω phasing lines are soldered together at a common point. The impedance at this point becomes about 36 Ω because there are two 50-Ω lines in parallel.

Matching is easy with a "Q" section, a ¼-λ series matching section made from 42-Ω coax. Two pieces of solid polyethylene coax are cut to length L5 and soldered in parallel (shield to shield and center to center) to make the "Q" section. I used two pieces of 70-Ω coax and achieved an excellent match, but it is possible to substitute one length of 95-Ω coax if you need to improve the match further. Examples of 95-Ω coax are RG-180B or RG-195A. Alternatively, it is possible to build short pieces of 42-Ω coax from brass hobby tubing. The inner conductor needs to be 0.217 inch or about 7/32 inch, and the outer conductor's inner diameter should be 0.5 inch. However, I recommend starting with two parallel 70-Ω coax cables first.

All harness connections should be kept short. I prefer to connect all center conductors first. Make one last check against Fig 2 to be sure that everything is correctly connected. Then smooth the joints and wrap them tightly with Teflon plumber's tape. Now connect the shields, taking care not to melt the polyethylene insulation. A wet cloth will quickly cool the joint after soldering. Let it cool for several minutes because the polyethylene core cools much more slowly than copper. The joint can then be dried and coated with silicone glue to make it waterproof. If you insist on a neater appearance, slide some shrink tubing over the joint before the silicone hardens and shrink it. Wipe off the extra silicone for a first-class job.

The final step is to use a VOM and make a resistance check of the antenna with the coax in place. The path from the center conductor of the coax to the shield should exhibit an infinite resistance (open circuit). The path from the center conductor to the two inside V elements should show a short. The resistance from the coaxial shield to the outside V elements should also indicate a short.

Route all coax straight down the middle line of the antenna. Secure it in place so that wind, ice and time will not change its location.

These Double Cross stacked dipoles have given me the extra low-angle gain necessary for improved 2-meter coverage. I keep thinking about how well a long dipole stack would perform on the UHF bands and how nicely a 20-meter Double Cross could work—if I could only get two tall trees to grow in my muskeg swamp!

# Ground Planes, Radial Systems, and Asymmetric Dipoles

By L. A. Moxon, B Sc, C Eng, MIEE, G6XN
Gorse Hill, Tilford Road
Hindhead, Surrey GU26 6SJ England

Several years have elapsed since I first drew attention to the advantages of replacing λ/4 radials by much shorter ones sharing a common loading inductance.[1] This greatly reduces the space occupied without adding significantly to the losses, and through being "shrunk" in this way a "ground plane antenna" (GPA) becomes ideally suitable as a parasitic element for use in close-spaced beams (which can be fully assembled, tested, and used at or near ground level). This also overcomes another problem: that of ensuring equal current amplitudes and correct phasing when resonant wires of high Q (and subject to finite tolerances) are connected in parallel.

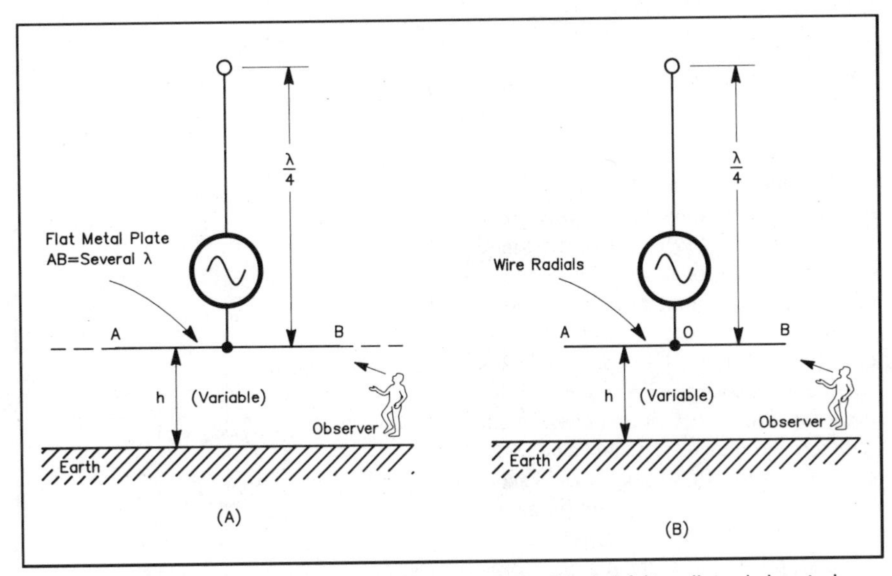


Fig 1—A comparison of ideal and real GPAs. At A, the ideal. A λ/4 radiator is located above a solid, conductive plane that extends several wavelengths in all directions. The radiating element is not visible, electrically, from the shadow of the plane. B shows a typical GPA installation; the λ/4 radiator is located above several λ/4 wire or tubing radials. The antenna is electrically visible from all points on the ground.

Unfortunately this topic has remained clouded by controversy. It is obvious that a true ground plane could hardly continue functioning after "shrinking" in such fashion. Consider such a ground plane: a flat metal sheet extending for a sufficient distance from the antenna base, as in Fig 1A. The antenna would then have a radiation resistance, $R_r$, of 36 Ω independent of its height above true ground (from which it would be shielded by the ground plane). When a ground plane is replaced by a set of λ/4 radial wires (Fig 1B), no such shield exists. We have an antenna with two poles. One (the radials, collectively) is constructed in a way that prevents it from radiating. The other is a short wire that (like any other such wire) produces a symmetrical cos θ radiation pattern that is not influenced by the presence of the radials. Since the radiating element is only half the length of a λ/2 dipole, it follows that the GPA $R_r$ is equal to that of a dipole divided by 4, not 2, other things being equal. Therefore, it has a value of 18 Ω (remote from ground influences) plus a small correction to recognize the 0.4-dB gain difference between short and λ/2 dipoles; this brings it up to about 19.5 Ω. As height decreases, the free-space value of R fluctuates in the normal manner (as a consequence of ground reflections) and rises to the usual value of 36 Ω at zero height. This is shown in Fig 2, which was derived from mutual-impedance data (assuming perfect ground and treating the antenna with its image as a collinear pair). Despite approximations and although the heights plotted are those of the feed point, this is almost identical with results obtained by Fletcher, VK2BBF, on the basis of rigorous mathematical analysis.[2] As Fletcher points out, the height at which the impedance changes to its "elevated" value is remarkably low. Except at very low heights (where errors appear to cancel), the "center of gravity" of the current distribution more accurately represents the true height, and the graph has been drawn accordingly. (To obtain the mutual resistances, 0.08 λ was subtracted from the heights used.) Illustrating this, mutual resistance is zero and $R_r$ is therefore equal to its free-space value of 19.5 Ω for separations of 0.7 and 1.2 λ between sources (corresponding to radial heights of 0.27 and 0.52 λ).

Recognition of the GPA as an asymmetrical dipole is at odds with much "conventional wisdom." It requires both convincing proof and answers to a number of questions before the opportunities can be fully grasped. For example, it is well known that significant inductive loading

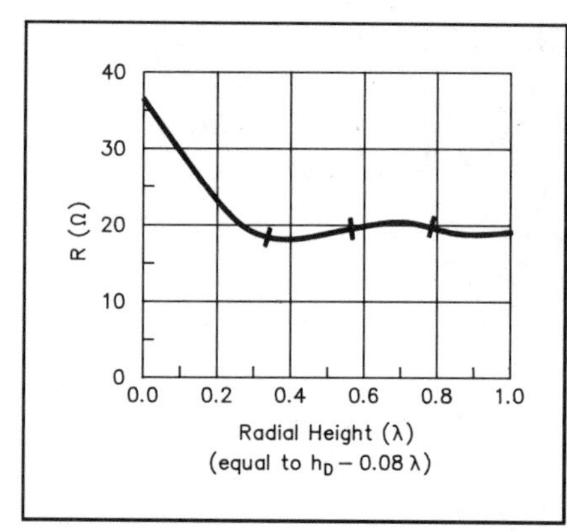


Fig 2—Radiation resistance, $R_r$, versus feed-point height for a λ/4 GPA over perfect ground. The calculation was performed for a pair of point sources separated by $2h_D$, where $h_D$ is the height measured to the center of current distribution. The marked points are those for which $R_r$ *should* equal its free-space value of 19.5 Ω (see text), as derived by rigorous mathematical analysis. Note the agreement between this chart and that in Ref 1.

seriously degrades the efficiency and bandwidth of dipole antennas, so why should GPAs be exempt? What are the best methods, the limitations, and the penalties? What are the ground influences and the implications for feeder systems? Are these questions interrelated, or can they be considered independently? Are there implications for other types of antennas? Armed with a few of the answers and a belief that the others were within easy reach, I set out in search of a complete set of guidelines, encountering a number of setbacks and not a few surprises before the main objectives were achieved.

Since the radials have no screening or reflecting properties and form no part of the radiating system, they can be reduced in length without any direct influence on efficiency. Nonetheless, we must take due account of losses in any loading components needed to maintain resonance, and the radials must not be hindered in discharging their proper functions: providing the only return path for the antenna currents and anchoring the coaxial cable shield firmly at ground potential.

As stated in the text of Ref 1, initial experience with such arrangements was encouraging (including their use as parasitic elements), though at that time a number of questions relating to (1) basic limitations and (2) the influence on bandwidth had not been fully resolved. Further, there appeared to be no reliable information on ground losses and none at all on how these might be affected by reducing the length of radials.

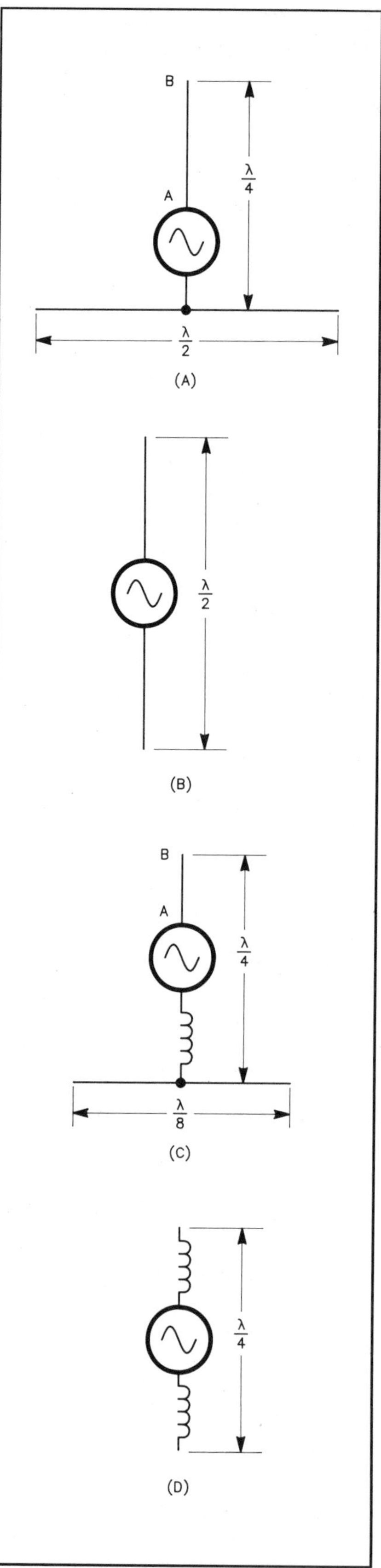


Fig 3—Loading of radials. A and B show a full-size GPA antenna and a vertically oriented full-size dipole, respectively. C and D show similar antennas that have been reduced in size by inductive loading.

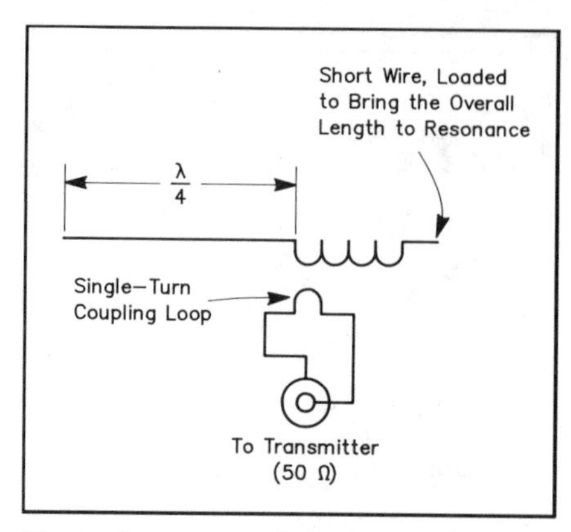


Fig 4—An asymmetric horizontal dipole used to verify current distribution with the equivalent of very short radials.

This was the starting point for what eventually became a protracted investigation covering all aspects of asymmetric dipoles, including an extension of short-radial principles. Most of the observations can be explained on the basis of equivalent circuits and relatively well-established behavior of horizontal wires close to earth. There were three major topics, as follow.

*Radial lengths*: Predictably, radial lengths can be varied between extremely wide limits (from λ/4 to λ/20) with no significant effect on SWR or field strength, and only about 50% reduction of bandwidth. Less obviously, the radial height below which losses from currents in the ground become appreciable is extremely low and independent of radial length. There were no discernible losses in parasitic elements. Attention was also directed to the need for nonresonant radial wires, in order to avoid unequal current amplitudes and phases.

*End-fed λ/2 dipoles:* The results indicated that half-wave dipoles can be end fed with coaxial cable, without losses, feed-line radiation or bandwidth restrictions characteristic of other methods of end feeding. Bandwidth has proved to be readily calculable, but other problems can arise and have been identified.

*Resonance of loaded horizontal antennas*: Other aspects include the discovery of a close link between height and resonant frequency of loaded horizontal wires at low heights. The link has been analyzed and used to determine the height above which ground losses can be neglected.

## Radial Loading

Fig 3A shows a so-called GPA in free space with only two radials. Fig 3B shows a λ/2 dipole with an $R_r$ of 73 Ω; its radiation pattern approximates a cos θ curve. Since

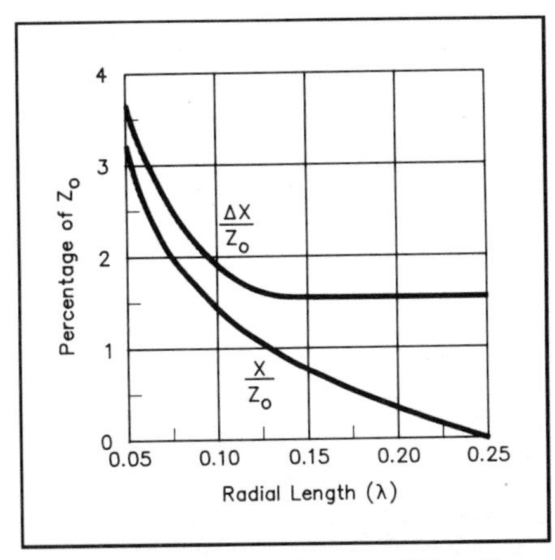


Fig 5—Variation of $X/Z_0$ and $\Delta X/Z_0$ with radial length, where $Z_0$ is the approximate characteristic impedance of a single radial or the antenna.

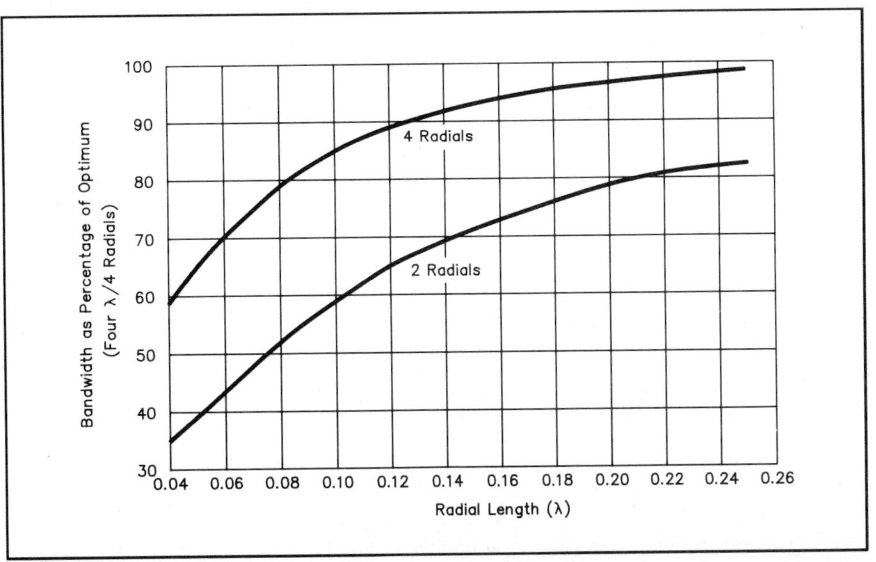


Fig 6—Variation of bandwidth with number and length of radials. A set of four λ/4 radials is taken as standard. All conductors have the same diameter.

nearly all radiation from the antenna in Fig 3A comes from wire AB, the pattern also approximates cos θ; *the field strength at any point in space is the same for both antennas*. Because the length is halved in Fig 3A, $R_r$ is divided by 4, not by 2 as commonly assumed. This *implies* an initial disadvantage in terms of bandwidth and efficiency, **but** there is one other very important and rather less obvious difference between the two antennas. Perhaps it can be best appreciated by trying to fit each antenna into a small space as illustrated in Figs 3C and D. For the dipole, $R_r$ drops sharply to about 7 Ω; whereas the GPA radiator (AB) is unchanged, and $R_r$ remains at some value between about 18 and 36 Ω, depending on its height above ground. In principle, the shortening process can continue until the resistance of the loading coil becomes comparable with $R_r$, which could mean a radial length of about 0.015 λ at 14 MHz.

Though this option is emphatically discouraged, it is less strange than it appears, and correct operation was nearly achieved with a single helically wound radial only slightly longer than 0.015 λ. Adjustments, though, were much too critical. I suspect that the coax shield was making an "unauthorized" contribution to the ground system. Lengths of λ/8 to λ/12 are recommended, and considerably shorter radials (λ/16 to λ/20) have been used successfully as follows.

When a pair of λ/16 radials were substituted for λ/4 radials, antenna current appeared to increase marginally. Nonetheless, the signal induced in a nearby vertical antenna dropped by about 2 dB. What caused this? The radial currents appeared to be more than half of the antenna current. Although there was no apparent current on the coax shield (because the power level was too low) such current was later found to be the cause of the trouble. Because the direction of shield current flow is opposite to that in the antenna, field strengths are no longer directly related to the current in the radiator, and the feed-point impedance is augmented by additional losses so that *complete suppression of shield currents is critical* (as described later).

Prior to the shield-current discovery, a nonsinusoidal current distribution was thought to be a plausible explanation (in view of the very small size of one antenna pole). This was difficult to test because most of the vertical element was out of reach. After a precarious balancing act failed to produce a conclusive result, a horizontal model of the antenna was set up as shown in Fig 4. This restored confidence in short radials and eventually provided the pattern for more efficient methods of end feeding horizontal antennas.

Placement of the feed point is extremely critical, in proof of my contention elsewhere that asymmetric feeding of antennas is viable if, but only if, resonance is established in both directions from the feed point (see Ref 1, p 42). Otherwise, the feeder inevitably becomes part of the radiating system.

Another experiment at 14 MHz used two radials, with lengths varying from λ/4 to λ/16 while observing SWR, bandwidth and the current induced in another vertical antenna at a range of some 50 meters (164 ft). There was little change apart from decreases in bandwidth (in line with that shown Fig 6) and a just-perceptible drop in field strength at each end of the range. The drops were caused by radiation from the λ/4 radials at one extreme and losses in the loading coil at the other. It is interesting to note that radiation from radials decreases very rapidly with their length. This results in part from their smaller "current × length" product, but the main cause is closer spacing of the antiphased sources.

**Estimation of Bandwidth**

The following calculations assume three or five conductors (a radiator, and two or four radials) of equal diameter. Impedances (and their variations with frequency) were read from the outer scale of a Smith Chart and plotted in Fig 5 (which is valid for any value of characteristic impedance, $Z_0$). For each 1% of detuning, the radiator generates a reactance $2\pi Z_0/400$ Ω (to which must be added $\Delta X_r/n$ for the radials and $0.01 X_r/n$ for the loading coil). From the Smith Chart, we also find that when $R_r$ is matched, the SWR rises to 2 when the reactance equals 0.7 R, so the percentage half-bandwidth is given by

$$\frac{0.7 \times R}{2\pi Z_0 + \dfrac{\Delta X_r + 0.01\, X_r}{n}} \qquad \text{(Eq 1)}$$

Based on a set of four λ/4 radials as standard, Fig 6 shows the extent to which bandwidth is degraded by decreasing the number and/or length of radials. Note that a set of four λ/10 radials yields the same bandwidth as a pair of λ/4 radials. For a $Z_0$ of 500 Ω) and an $R_r$ of 36 Ω, the standard bandwidth is 5.4% and the worst case included in Fig 6 (a pair of λ/25 radials) is

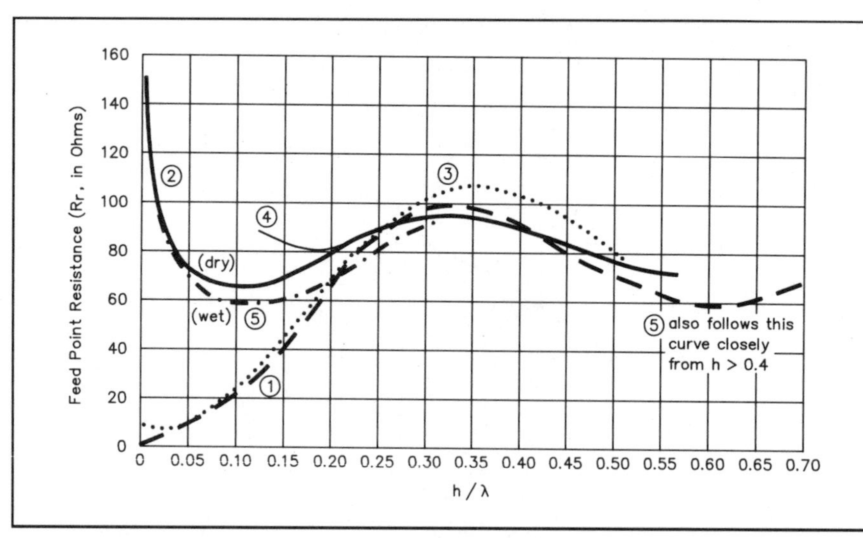


Fig 7—Comparison between perfect ground (Ref 5) and measured results for typical ground (Ref 4) showing the steep rise in feed-point resistance of horizontal dipoles at very low heights and the effect of a ground screen (mesh = 0.0003 λ), curves 1, 2 and 3, respectively. Calculations for dielectric ground (curve 4, k = 5) show good agreement with curve 2, except as indicated (when h < 0.2 λ). Curve 3 checks closely with theory putting k = ∞. Curve 5 illustrates the effect of wet weather. Note that the difference between curves 1 and 2 below h = λ/4 consists of ground losses.

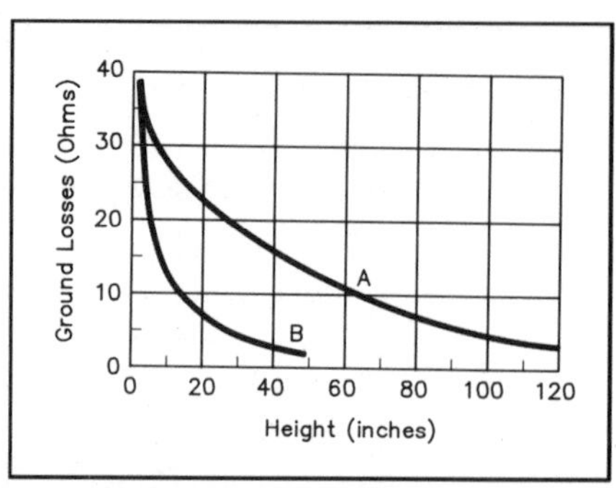


Fig 8—Variation of ground losses with radial height. Curve A is a crude estimate based on dividing the values of curve 2 in Fig 7 by four. Curve B was plotted from rough measurements at 14 MHz (with various radial lengths). The measured values were obtained by subtracting $R_r$ (36 Ω in this case) from the total resistance at the feed point. Note the convergence of the two curves towards a similar high value at zero height.

1.9%, which is more than enough to cover the 20-m phone band. So, faced with the previously mentioned problems of critical adjustments and feed-line radiation, bandwidth is probably the least of one's worries if the reduction of radial length is carried to its ultimate limit consisting probably of a pair of short whiskers loaded by a very large coil.

## Ground Losses

When a λ/2 horizontal antenna is moved down towards ground from a height of λ/4, $R_r$ at first closely follows the curve for perfect ground, but it departs (increasing) from the curve at about 0.17 λ. At ground level, $R_r$ reaches about twice the free-space value, as illustrated in Fig 7. That figure is based on Ref 3, which mentions good agreement between measurements at 50 MHz and theory based on the assumption of dielectric ground, but questions the validity of this assumption at lower frequencies (particularly below 7 MHz and at very low heights), though the findings seem to be in good accord with Ref 4 and my own observations.

I have been unable to find similar results with respect to vertical antennas. Nonetheless, two observations point strongly to the conclusion that there are no significant ground losses associated with vertical antennas as such, only in ground connections and radial systems consisting in the main of horizontal wires, which may be analyzed in the same way as horizontal antennas:

1) Failure to observe any ground loss when a GPA is excited parasitically.

2) A horizontal λ/2 dipole within about an inch of the ground was found to have an $R_r$ of 140 Ω, which would be expected to translate into a value of 35 Ω if its center is used as a ground connection. That is what happened. The variation with height is plotted as the lower curve in Fig 8. Comparing this with the upper curve, we can see that the disappearance of ground loss with increasing height is much more rapid than with horizontal dipoles. In conjunction with Fig 11 (described later), this is convincing evidence that ground effects (other than as described by Fig 2) disappear completely at radial heights exceeding about λ/16—a result entirely to be expected.

To explain this, it is helpful to start by considering a λ/4 transmission line in free space, Fig 9. If this line is unfolded to make a dipole, its characteristic impedance increases to a value given by[5]

$$Z_0 = 276 \log\left(\frac{2\ell}{a}\right) - 120 \qquad \text{(Eq 2)}$$

Except for the acquisition of $R_r$, it remains otherwise unchanged. Like any other λ/4 resonator, it can be approximated by an inductive reactance, $X_L$, equal to $Z_0$ and tuned by the capacitance between the conductors. We can therefore construct the equivalent circuit shown in Fig 10 (minus C2 and C3 for the moment).

As the line approaches ground level, it reaches a point where the capacitance from each pole to ground provides an easier current path than that between the poles. The antenna starts to resemble a pair of series-connected unbalanced lines. This is where C2 and C3 take over from C1. The impedance of each line (see Ref 5, p 174) is given by

$$Z_0 = 138 \log\left(\frac{2h}{a}\right) \qquad \text{(Eq 3)}$$

In terms of the equivalent circuit, this is equal to $X_L/2$. For a given conductor of radius "a" near ground level, $Z_0$ is dependent solely on height. (Remote from ground influences, $Z_0$ depends only on length.) As an example, let us take a 20-m dipole for which $2\ell$= 10 m and assume, in anticipation of Fig 11, a height of 1 m. Typical values of "a" range from 0.5 mm for an "invisible" wire dipole to 10 mm for a typical rotatable dipole. Corresponding values of $X_L$ (given by the free-space formula) range from 1067 to 708 Ω and (from the low-height formula) 911 to 635 Ω. These last two figures are extremely height dependent.

This has interesting practical consequences as demonstrated by the lower (experimental) curve in Fig 11. That curve illustrates the above behavior and reveals a method to determine the presence of ground losses with no instrumentation other than a dip meter. As a dipole is shortened and resonance is restored by inductive loading, the required inductance is proportional to the $Z_0$ of the dipole. Where ground losses are present, $Z_0$ is height dependent and antenna resonant frequency varies with height. (The height-dependent formula considers the return path through ground. If it ceases to apply, it is reasonable to conclude the absence of ground losses.)

Note from Fig 11 that nearly all fre-

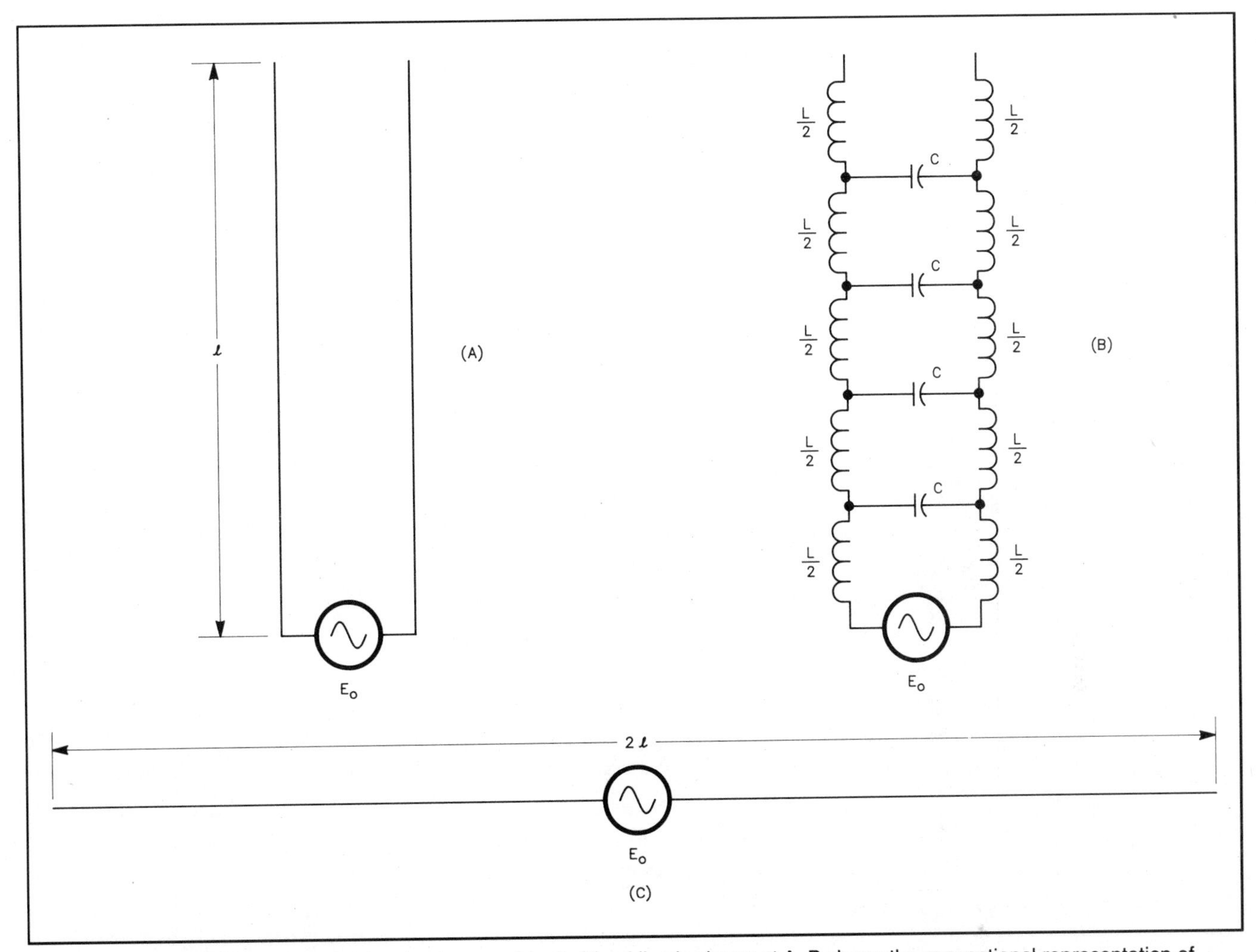


Fig 9—Derivations of equivalent circuits of antennas. A typical feed line is shown at A. B shows the conventional representation of such a line in terms of lumped constants. C shows the same line opened out to form a dipole.

quency change in a 28-MHz λ/4 loaded horizontal wire takes place at heights below two feet. Rough tests have indicated that this figure is proportional to wavelength but independent of radial length.

It is well known that the resonant frequency at very low heights is influenced by the dielectric properties of the ground, but this effect is very much smaller than the one illustrated by the curves antenna height. An example is shown as a single point in Fig 11. You can see that, in the case of the unsymmetrical dipole (Fig 10B), reducing radial length must decrease C1 and increase $Z_0$.

The top curve in Fig 8 was a first attempt to relate the ground losses of radial systems to the relatively well-established behavior of horizontal dipoles. The dipoles were "converted" into radials by using their centers as ground connections. From inspection and for a given efficiency, lower heights are acceptable in the case of radials, but this is pessimistic. It ignores the field cancellation that prevents radiation, and can be expected to result in a considerable reduction of ground loss (except at zero height, as already indicated). This is demonstrated dramatically by the lower curve (which is based on measurements of feed-point impedance) even when allowance is made for inaccuracies.

Inaccuracies inspired a search for a "more accurate" method of measurement. This took the form of parasitic excitation from a remote antenna, observing the effect on current from the insertion of known resistances. It was here that matters got really interesting: *The expected height dependence was nonexistent.* This was in line with observations dating from many years earlier and hitherto unexplained. It leads to the conclusion that for *parasitic* excitation of a GPA, there are no significant ground losses even at a height of only an inch or so. In retrospect, this is less surprising: In a base-fed element at ground level, the antenna current must return to the source via the ground, whereas the signal injection of a parasitic element is distributed along its length.

Though intriguing, the practical value of this discovery is limited. From the evidence of Figs 8 and 11, the radial heights at which ground losses occur are very low. (They are more suitable for a system of trip-wires than antennas compatible with domestic harmony!) It also appears that ground loss, particularly with short radials, should be readily avoidable with the help of a small earth mat that covers the area of ground subject to fields generated by the radials. (See Refs 3 and 4).

In this context, an interesting point has emerged that tends to reconcile the findings of myself and others: There is no significant advantage in increasing the number of radials from two to three or four, despite advice from some quarters (such as Ref 6) to use very large numbers (up to a hundred or more at low heights). As height above a ground screen approaches zero, the radials can be perceived as merging into it. There appears to be no reason why the screen itself should not be constructed as a

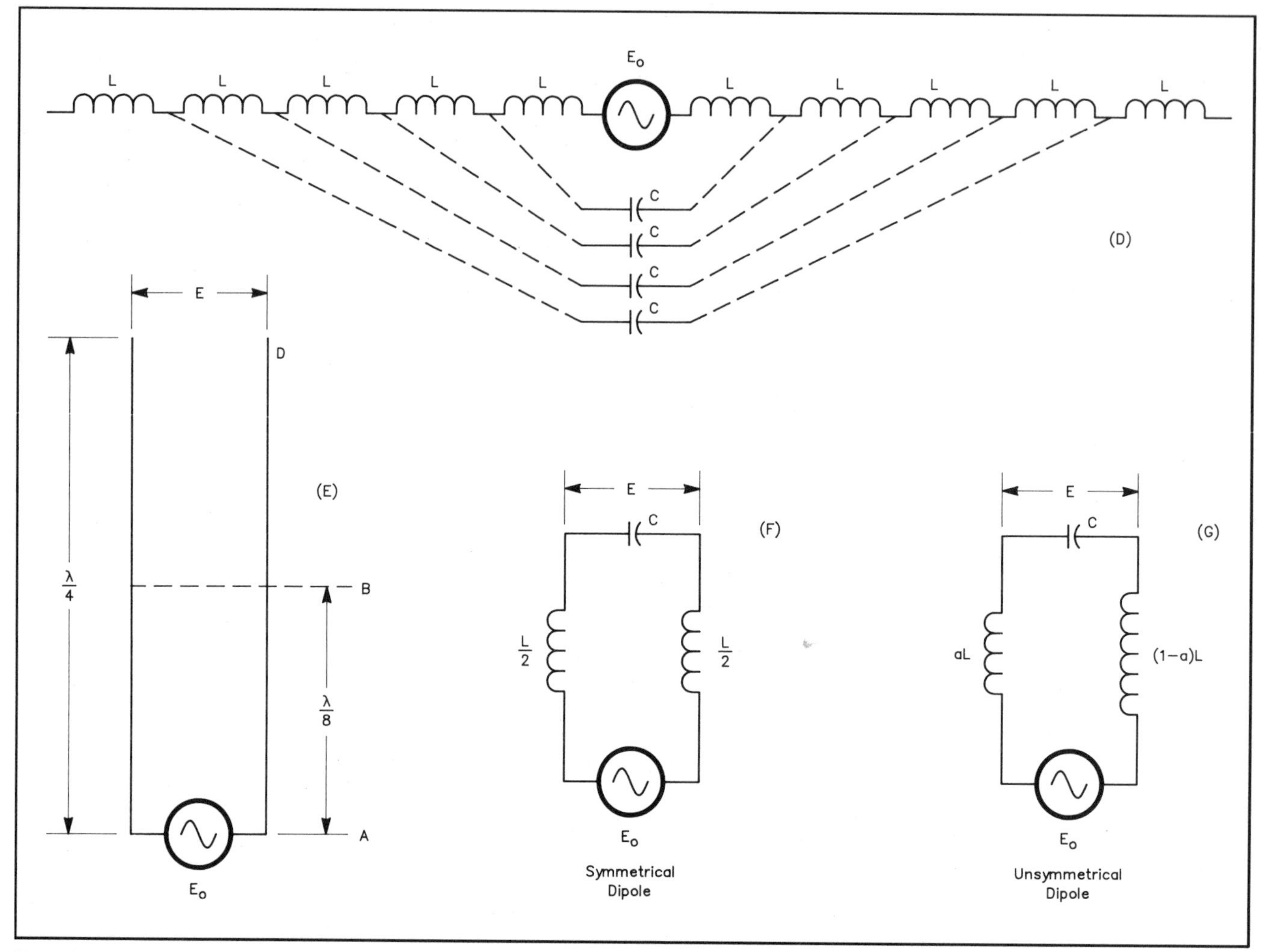


Fig 9 (continued)—Derivations of equivalent circuits of antennas. D is the equivalent circuit of Fig 9C (see previous page, which differs from B only to the extent that L and $Z_0$ are larger, C is smaller, and radiation resistance (not shown) has been acquired. E, F and G illustrate simpler versions of the equivalent circuit for the special case of $\ell = \lambda/4$. The inner half of the line is replaced by a single inductance for which $\omega L = 1/\omega C = Z_0$. The voltage ratio $E/E_0$ is given in all three cases by $Z_0/R$ or $\omega L/R$, where R is the internal impedance of the generator or, in the case of an antenna, the radiation resistance $R_r$.

set of radial wires, provided that mesh-size requirements are met. Assuming a radius of λ/4 and a mesh size of 0.01 λ (which, from Ref 3, is sufficient to give a result almost identical with curve 3 of Fig 7) the required number of radials is 79. The maximum useful mesh size is about 0.07 λ. With less than 10 radials, no significant improvement can be expected. Nevertheless, Fig 8 shows a simpler solution, at least for the higher frequencies: Increase the height slightly (so the radials cease to function also as trip-wires). Demands for even greater heights are impossible to reconcile with well-established properties of wires close to ground (as already discussed). They may be attributable to feed-line radiation.

Alternative feed possibilities include link coupling into loading coils with no direct connection. Though used successfully, this merely alters the problem without removing it. I strongly recommend that short vertical radiators at low height should be rearranged as symmetrical end-loaded dipoles. Very successful beams constructed on this basis have been described by me (Ref 1, p 193) and by GØGSF/ZS6BKW (Ref 7), who used the MININEC computer program to take the guesswork out of the design. After adding the experience recorded in these references to the observations with parasitic elements recorded above, this appears to be a simple and effective way of eliminating ground losses.

## Other Unsymmetrical Dipoles

Unsymmetrical dipoles along the lines of Fig 4, or possibly with two "radials," are one solution to the problem of end feeding horizontal dipoles, half-wave verticals and the like with coax. The necessity to ensure resonance in both directions from the feed point (so as to maintain the coax shield at ground potential) cannot be too strongly emphasized. To this end, a radiator length of λ/4, as thus far assumed, is a step in the right direction, since with the further assumption of overall resonance the required condition should be fully met.

Shortened radial procedures can be applied to other radiator lengths, horizontal as well as vertical, subject to suitable tuning. Methods of loading short radiators have received extensive coverage, and the reader should find little difficulty in relating them to the use of shortened radials.

Arrangements such as in Fig 12 appear to be new, however, and could be of some importance as they extend the idea of "end-feeding with coax" to λ/2 elements (see Ref 8). Fig 13, prepared from a Smith Chart, shows how this can be achieved by moving inwards from one end a short distance, Δλ, to find a resistive impedance suitable for

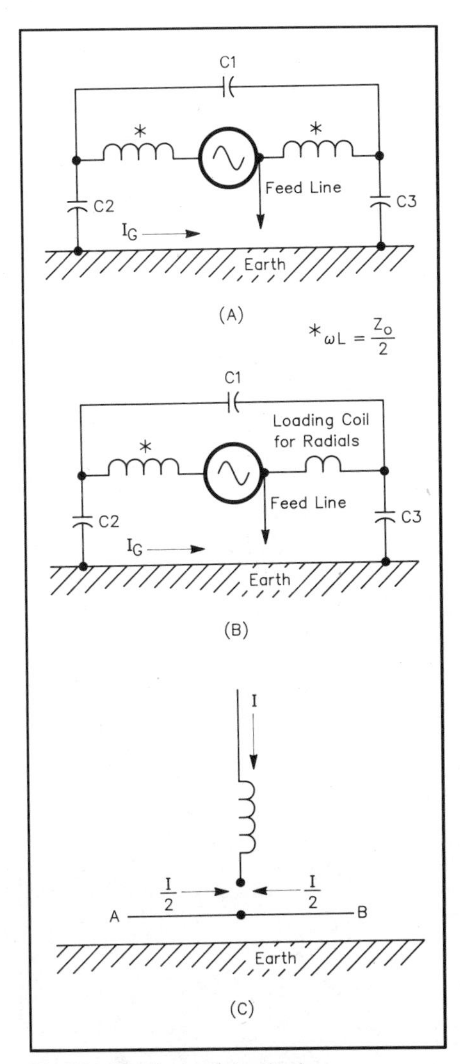


Fig 10—Equivalent circuits for symmetrical (A) and unsymmetrical (B) dipoles. The feed-line shield will radiate unless maintained close to ground potential. C shows why the ground current, $I_G$, might be reasonably expected to disappear in the case of a ground-plane antenna with parasitic excitation or if the transmitter were to be incorporated in the antenna.

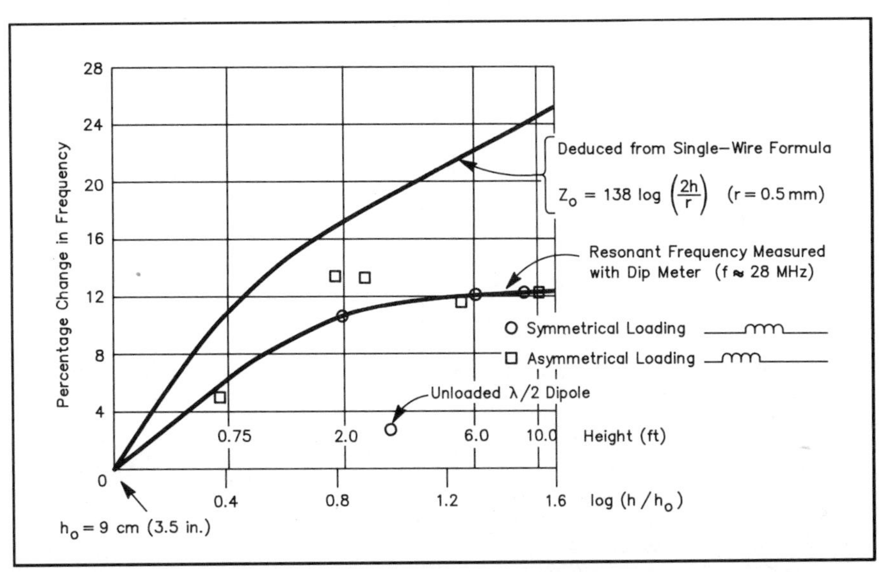


Fig 11—Resonant frequency versus height of an inductively loaded λ/4 horizontal dipole.

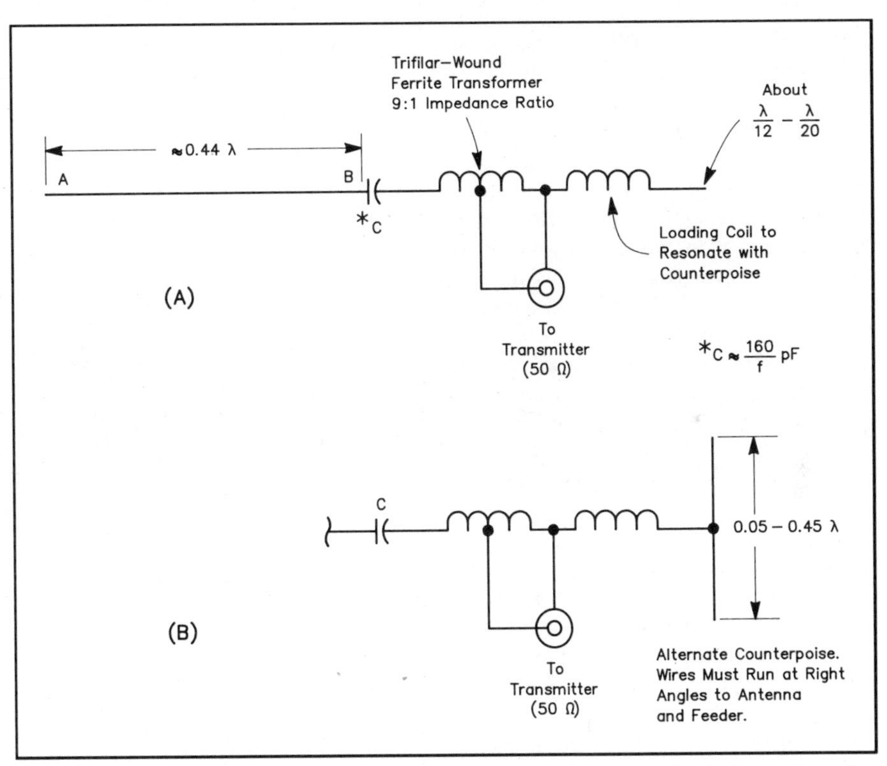


Fig 12—A is a dipole that is end fed with coaxial feed line. The length AB depends on R and $Z_0$ (see Fig 13). With a 9:1 transformer, options include single-wire feed line (typical $Z_0$ of 500 Ω giving an SWR of about 1.1:1). A 4:1 transformer may be used in place of the 9:1 unit. Then AB must be reduced to 0.405 λ.

matching to coax with the help of a suitable transformer. The relatively large value of series inductance is easy to tune out by a small capacitor, as illustrated in Fig 12. This technique is readily adaptable for other types of radiator such as the inverted GPA, Fig 14A, or "Maypole" beam elements, Fig 14B (see Ref 1, p 190, and Ref 9).

For Maypoles, the use of short Windom-like inserts overcomes the problem of high RF voltages that normally make it difficult to switch the ends of λ/2 elements. A counterpoise or radial system is essential but, because of the relatively large values of resistance, even less demanding than in the case of the GPA.

Fig 13 is plotted for two values of matching impedance; 500 Ω is suitable for single-wire feeders, though the method of Fig 12 is much to be preferred. The design of the transformer for this value of impedance posed problems initially, but these were overcome for monoband operation by using trifilar windings (which were self resonated by varying the amount of core insertion). In Fig 12, the bandwidth was virtually identical to that for a conventional center-fed dipole. The broad bandwidth typical of widely spaced parallel wires, Fig 14B, was fully evident in marked contrast to narrow bandwidths previously imposed by the use of Zepp feeders.

## Equalization of Currents in Radial Systems

The almost universal use, hitherto, of resonant radial systems raises issues that have been left till last in view of inherent

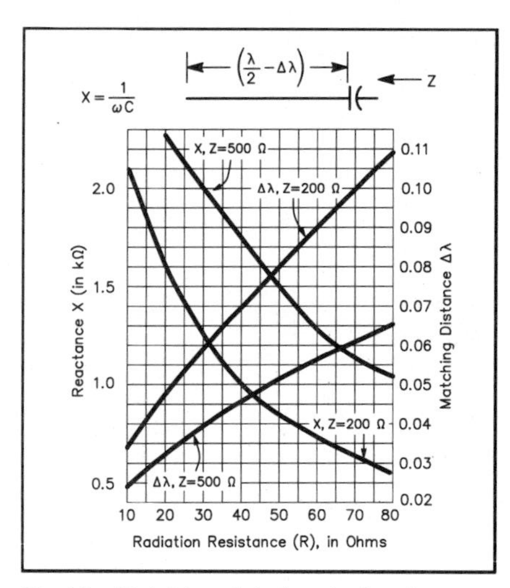


Fig 13—Matching data for single wires that are end fed with coaxial cable.

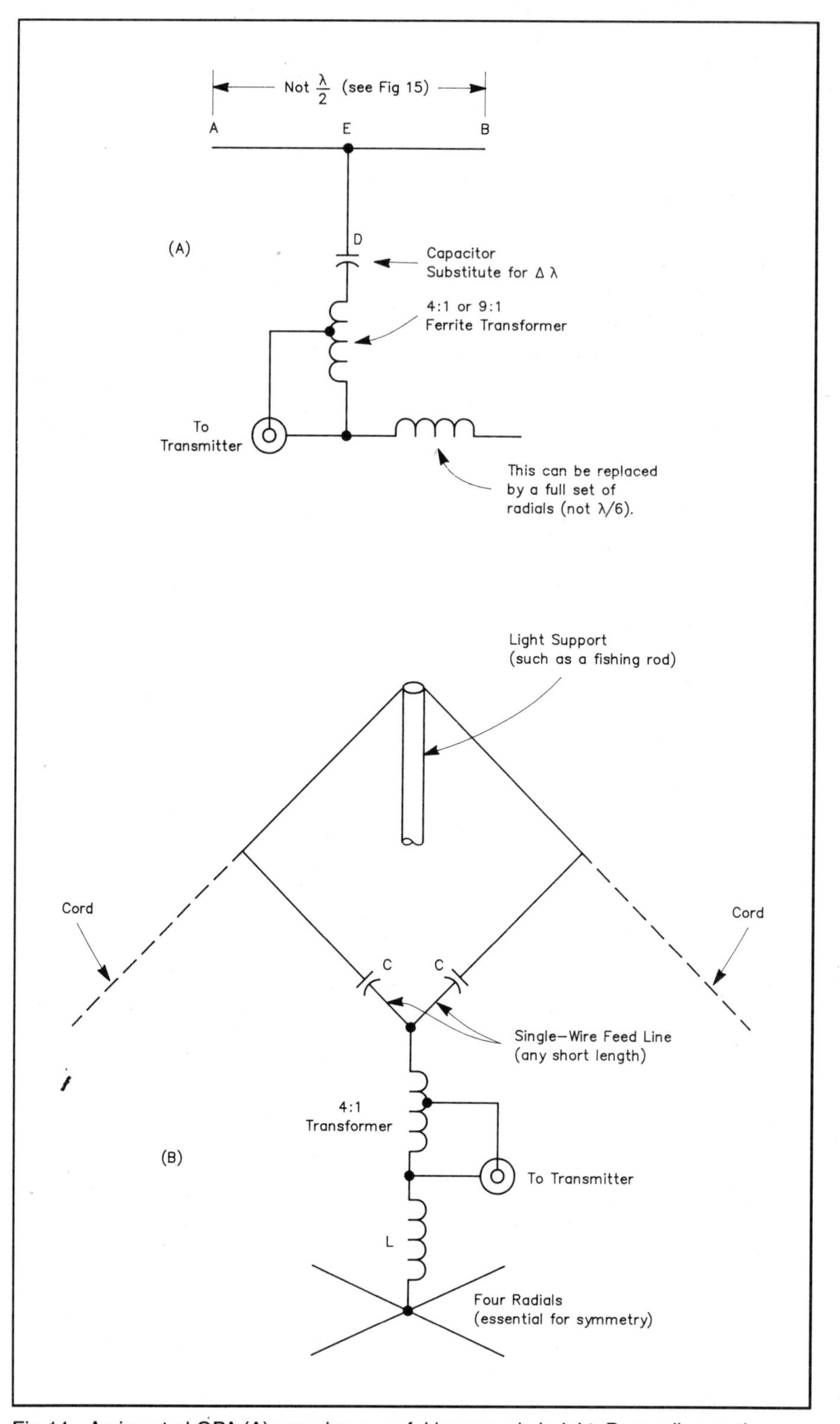


Fig 14—An inverted GPA (A) can give a useful increase in height. Depending on the length, AB, of the flat top, R and therefore both C and Δλ can vary between wide limits. For design assistance see Fig 13. Length ED, loaded by AB, is short of resonance by Δλ, as defined in Fig 13. At B, the Maypole array uses four wires (λ/2) that hang from a single pole and are switched in adjacent pairs; the sketch shows the driven element only. The reflector (tuned by 60-100 pF in series, for 14 MHz) uses the same ground plane.

difficulties and lack of direct bearing on other matters. Away from ground influences, the Q of the radial system is likely to be extremely high. The significance of this is apparent from Fig 15, which contrasts the way that phase varies along perfectly matched and completely mismatched lines. In the absence of losses (which are never specified), this implies phase reversals over a frequency range corresponding to the radial-length tolerances (which are always finite). This problem is not confined to radial systems. I have found (to my cost) that it tends to crop up (heavily disguised and with devastating effect) when least expected unless due respect is paid to the following axiom: **If loss-free resonant conductors connected in parallel are required to carry equal currents, they must not be connected solely at points of maximum current.**

The significance of this in the context of resonant radials is hard to estimate. Any imbalance results in radiation from radials and from any feed system, hence degrading the Q and tending to remove the cause. After experiencing problems of equalization even at ground level, and given the ease with which such problems can be avoided (by detuning the individual radials and restoring resonance by means of a reactance common to all of them), it seems clear: A radial length of λ/4, far from being mandatory, is the one length to be avoided!

### Practical Aspects

Because vertical antennas are inherently unsymmetrical, special care is needed for the prevention of feeder radiation. Extreme shortening of radials aggravates the problem, but within the limits recommended here, measures such as those advised in Ref 4 (p 26-6) usually suffice.

Linear traps (see Fig 16) have been found very effective, but the user must be alert to a problem that can make matters worse: The traps can establish a "match" to the coax shield. This should be checked with a sensitive probe at two or more

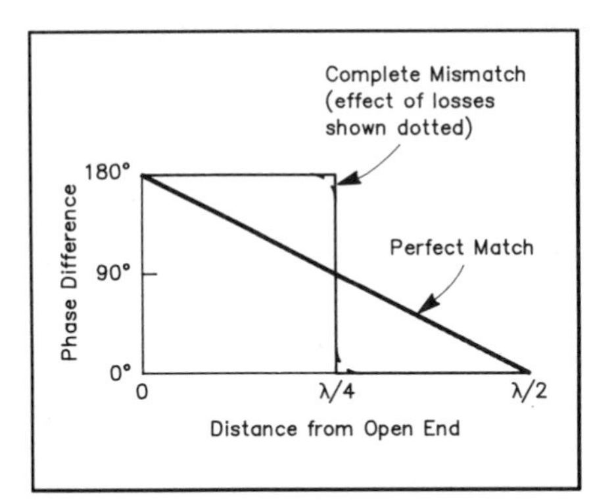


Fig 15—Variation of current phase along transmission lines.

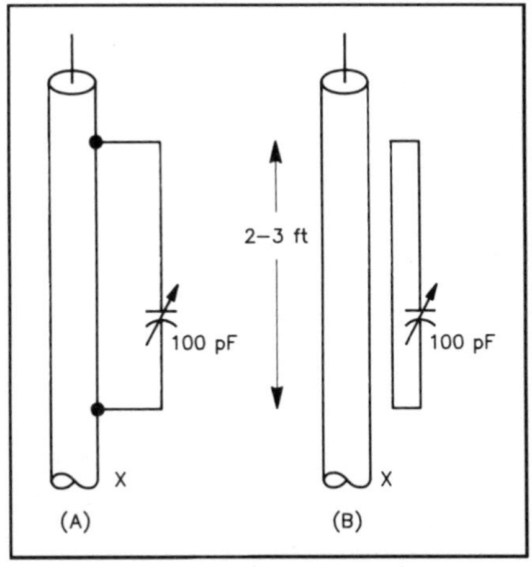


Fig 16—Traps for eliminating antenna current on coaxial cable shields. The values and dimensions shown are for 14 MHz, but they are not critical. Current should be zero at points marked X, but it may be large within the loops. At A, a tuned wire is connected directly to the shield to form a loop with the shield as part of the loop. At B, a complete tuned loop is taped to the cable jacket. The loop couples to the shield through the jacket, and jacket damage is avoided.

points. Minimum current on the coax shield is the best indicator for correct tuning of the radials. Adjust the radiator for minimum SWR, and repeat the process if necessary. Tune radials by altering either their length or the inductance of the loading coil. With this method of loading, there is no need for accurate equalization of lengths.

In some cases, a single short radial (at right angles to the radiator) may be acceptable. Since the inner ends of loaded radials are at high RF potential, it is essential to ensure that they are well insulated. Existing GPAs can be checked for current equality with a probe consisting of a small loop connected to a rectifier and meter; guard against the possibility of current reversals by checking at more than one frequency.

Since (as indicated above) radial lengths can be varied between wide limits, tuning by alteration of the inductance should be feasible over a range up to two octaves, but this has not yet been tried. For multiband operation, buried grounds have obvious attractions if some extra loss is acceptable. The amount of loss should be easy to determine by comparative field-strength measurements using a temporary pair of radials.

It is implicit from the decrease in $R_r$ illustrated in Fig 2 that an increase of base height to about λ/4 should result in a "height gain" of some 2 to 3 dB at low angles. An additional benefit may accrue from the clearing of obstructions.

Beams based on ground-plane elements with short loaded radials have been described elsewhere (Ref 1, p 190). Particular advantages of this type of element include (1) portability, (2) easy access, and (3) the ability to optimize coupling between reflectors and driven elements (in order to achieve deep nulls by equalization of currents; see Ref 9).

The above height gains apply equally to beams. The Maypole array (also featured in the references) tends to be more attractive, particularly if the problems of directional switching are resolved in accordance with the possibilities suggested by Fig 14B: The extra height gain is obtainable without the feed point becoming inaccessible.

### References

[1]L. Moxon, G6XN, *HF Antennas for All Locations* (Potters Bar, Herts: RSGB, 1982 and 1986), pp 44, 104, 156, 186-196.

[2]G. Fletcher, VK2BBF, "The Feed Impedance of an Elevated Vertical Antenna," *Amateur Radio* (Australia), Aug-Oct 1984.

[3]R. Proctor, "Input Impedance of Horizontal Dipole Aerials at Low Heights Above the Ground," Part 3, *Proceedings of the IEE*, May 1950.

[4]15th ed., *The ARRL Antenna Book* (Newington: ARRL, 1988), p 3-11.

[5]F. E. Terman, *Radio Engineers' Handbook*, 1st ed. (New York, London: McGraw-Hill Book Co, 1943) p 864.

[6]W. Orr, W6SAI, *Antenna Handbook* (Wilton, CT: Radio Publications, Inc, 1980), p 93.

[7]B. Austin, GØGSF/ZS6BKW, *Radio Communication*, Sep 1989, p 44.

[8]"Technical Topics," *Radio Communication*, May 1991.

[9]L. Moxon, "Two-Element HF Beams," *Ham Radio*, May 1987.

# Phased Arrays for the Low Bands

**By Al Christman, KB8I**
**EE Department, Grove City College**
**100 Campus Drive**
**Grove City, PA 16127-2104**

During a conversation with Dr. Jim Breakall, WA3FET, who teaches Electrical Engineering at Penn State, I learned we had both been modeling phased arrays of inverted-V antennas for the 40-meter band. We ended up comparing notes. Jim mentioned that Floyd Koontz, WA2WVL, had found a way to get excellent front-to-back ratio from a three-element driven array by optimizing the phase angles of the feed currents.

Jim, who lives in central Pennsylvania, wanted to put up a 40-meter wire beam firing northeast and southwest. He had planned to construct a Yagi using three inverted-V elements, with the center element driven. The outer two elements would be switchable parasitic elements capable of functioning either as reflectors or directors. Now, however, the promise of terrific front-to-back ratios, which seemed to be available from an all-driven array, had caused him to change his plans.

I had recently obtained a copy of ELNEC, an antenna-modeling program for PC-compatible computers,[1] and this news prompted me to investigate further. The results are contained in this article.

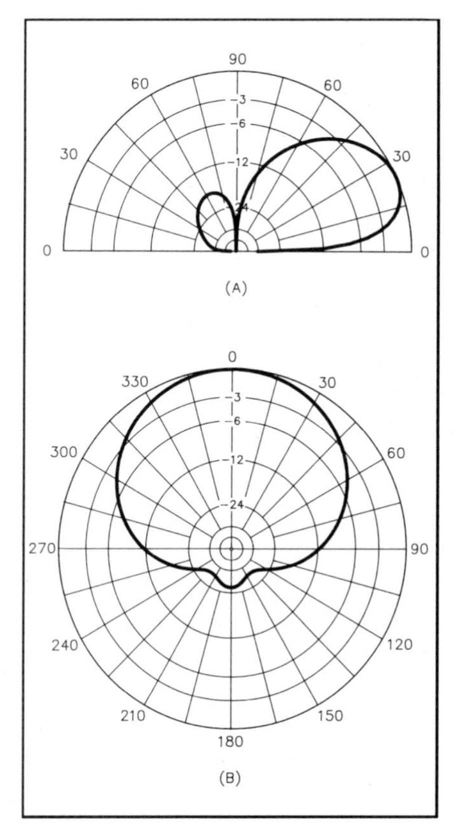


Fig 1—Radiation patterns for the four-square array of vertical monopoles, using ¼-λ spacing and 90° current phasing. At A, the elevation-plane pattern; at B, the azimuthal plane at a 24° elevation angle. The 0-dB reference is 5.4 dBi.

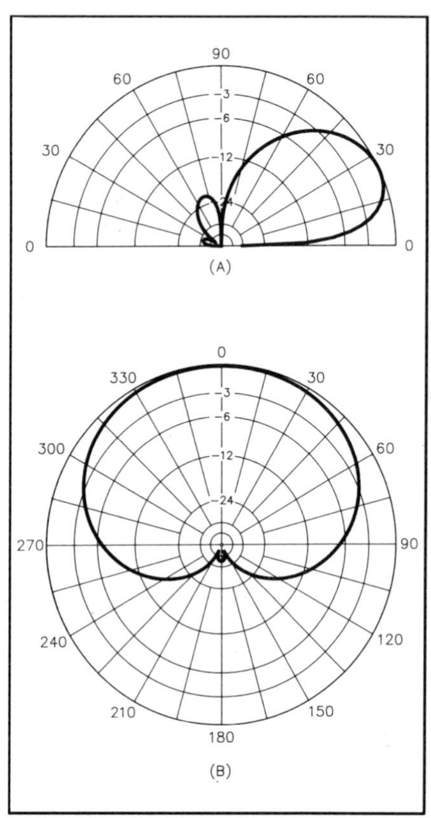


Fig 2—Radiation patterns for the 3-in-line array of vertical monopoles, using ¼-λ spacing and 90° current phasing. At A, the elevation plane; at B, the azimuthal plane at a 25° elevation angle. The 0-dB reference is 4.1 dBi.

## Background

A two-element phased array of vertical monopoles can generate a number of different and useful radiation patterns that provide azimuthal directivity, such as the familiar bidirectional broadside and endfire configurations, or the well-known and very popular cardioid pattern, with its pronounced rear null. With the proper current phasing, a major lobe of radiation can be placed in any compass direction, although there are often unwanted sidelobes of significant magnitude.

The widely used four-square array[2] provides excellent coverage in any of four quadrants, with a broad frontal lobe having an azimuthal beamwidth of about 100°, and rear rejection of 25 dB or better. See Fig 1. (All plots in this article were calculated with the antenna located over "real" earth having a conductivity of 5 mS/m and a dielectric constant of 13.) Forrest Gehrke, K2BT, described the excellent front-to-back ratio obtainable from an in-line array of three elements, using ¼-λ spacing and feed currents of $1\,/\underline{90°}$, $2\,/\underline{0°}$ and $1\,/\underline{90°}$ (Fig 2).[3]

Note the improvement in front-to-back ratio with the 3-in-line array as compared to the four-square (Figs 1B and 2B). How-

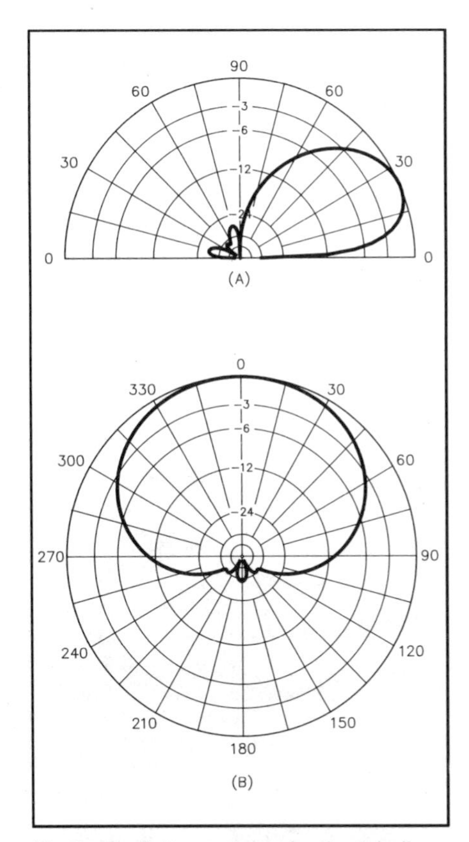


Fig 3—Radiation patterns for the 3-in-line array of vertical monopoles, using ¼-λ spacing and improved current phasing. At A, the elevation plane; at B, the azimuthal plane at a 24° elevation angle. The 0-dB reference is 5.0 dBi.

Fig 4—Radiation patterns for the Cross or five-square array of vertical monopoles, using ¼-λ spacing and improved current phasing. At A, the elevation plane; at B, the azimuthal plane at a 24° elevation angle. The 0-dB reference is 5.0 dBi.

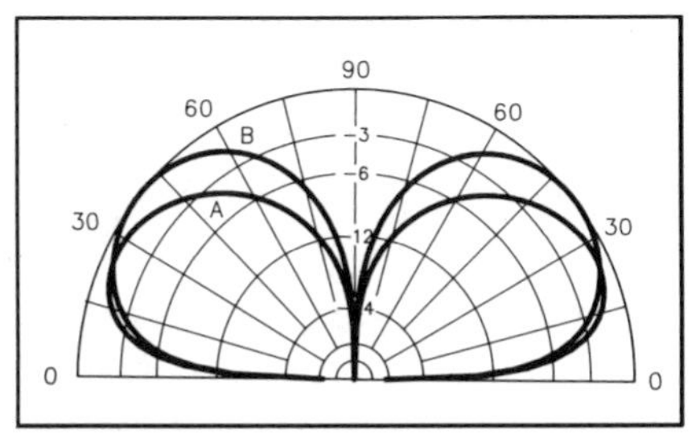


Fig 5—Elevation-plane radiation patterns for the Cross or five-square array of vertical monopoles with ¼-λ spacing when used in the omnidirectional mode. The 0-dB reference is 0.8 dBi. Pattern A results from driving only the center element, and pattern B results when all four outer elements are driven in phase. Unused elements are open-circuited at their bases in either case.

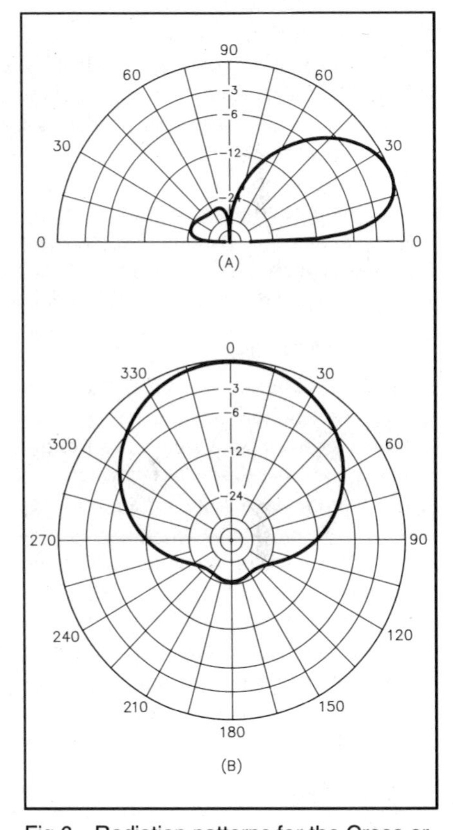


Fig 6—Radiation patterns for the Cross or five-square array of vertical monopoles using ⅛-λ spacing and improved current phasing. At A, the elevation plane; at B, the azimuthal plane at a 22° elevation angle. The 0-dB reference is 5.5 dBi.

ever, the 3-in-line array has lower forward gain and a broader azimuthal beamwidth.

## Changing the Current Phasing

Fig 3 shows what is possible with a 3-in-line array when the current phase angles are "adjusted." Rejection off the back is much improved, and the gain has increased by about 1 dB over the standard version. This was achieved on 40 meters (7.15 MHz) using 34.2-foot vertical elements spaced 34.4 feet apart (¼ λ), with base currents of $1\,\angle 117°$, $2\,\angle 0°$, and $1\,\angle -116°$. This antenna is almost as good as the four-square in terms of forward gain, and its front-to-back ratio is far superior.

Analysis with ELNEC indicates that two of these modified 3-in-line arrays can be combined in the form of a cross with the center element driven at all times. The two unused outrigger elements are open circuited at their feed points to force their base currents to zero. As a result, these floating 90° elements are then essentially detuned and are almost invisible electrically. Fig 4 shows the radiation patterns, which are very similar to those in Fig 3. This "Cross" or "five-square" array may be attractive to low-band operators with sufficient space for a larger antenna having better rejection off the back.

Omnidirectional radiation may be obtained from the Cross by using either of two different feed methods. All four of the outrigger elements can be open circuited and only the center monopole driven, or the center vertical can be open circuited and all four of the outer elements driven in phase, which yields some additional gain and better coverage of high takeoff angles. Both patterns are plotted in Fig 5.

For the three active elements of the Cross, ELNEC gives input-impedance values of $Z_1 = 9.8 + j0.4\ \Omega$, $Z_2 = 27.5 + j23.5\ \Omega$, and $Z_3 = 45.6 + j82.5\ \Omega$. MININEC and its derivative programs, including ELNEC, calculate input impedances based upon "perfect earth" beneath the antenna, so these magnitude values may not be absolutely accurate.

The Cross can also be constructed using an element spacing of ⅛ λ (17.2 ft), and these patterns are shown in Fig 6. The closer spacing increases mutual coupling; the input impedances calculated by ELNEC are $Z_1 = 1.9 - j3.4\ \Omega$, $Z_2 = 11.6 + j17.4\ \Omega$, and $Z_3 = -22.9 + j33.2\ \Omega$. For the ⅛-λ element spacing, currents of $1\,\angle 146°$, $2\,\angle 0°$ and $1\,\angle -149°$ were used.

Notice that, for an element spacing of only ⅛ λ, the front-to-back ratio deteriorates considerably when two of the 3-in-line arrays are combined to form a cross, dropping from 31.4 dB to only 24.7 dB. In contrast, for an element spacing of

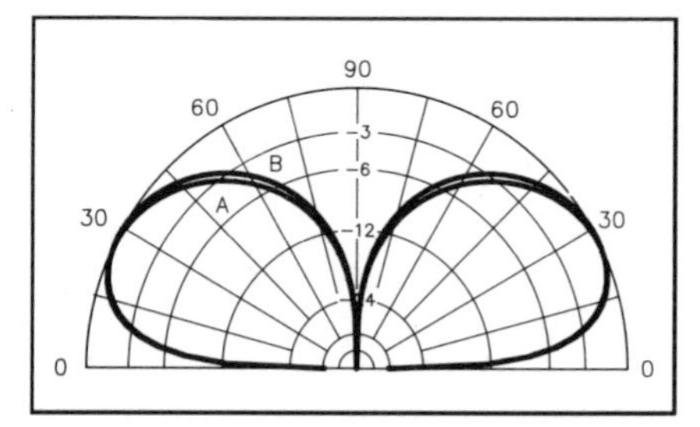


Fig 7—Elevation-plane radiation patterns for the Cross or five-square array of vertical monopoles with ⅛-λ spacing, when used in the omnidirectional mode. The 0-dB reference is 0.1 dBi. Pattern A results from driving only the center element, and pattern B when all four outer elements are driven in phase (bases of unused elements open).

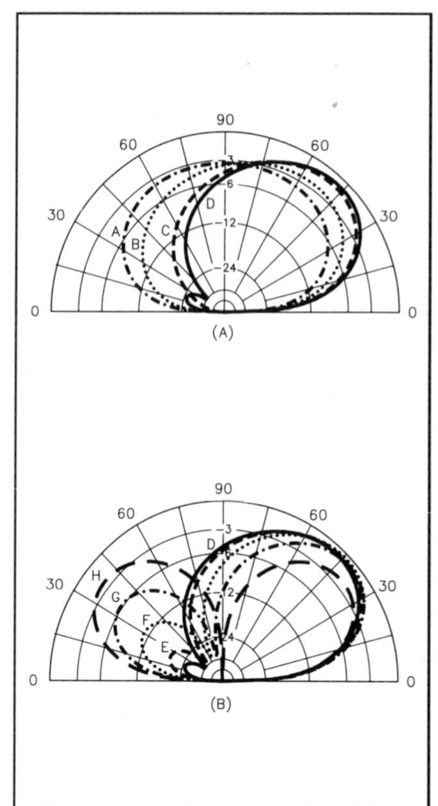


Fig 8—Elevation-plane radiation patterns for a two-element array of inverted Vs with an apex height of 36 feet, using ⅛-λ spacing and equal-amplitude currents. The 0-dB reference is 8.7 dBi. The current phase angles for the patterns are A: 0°, B: −90°, C: −135°, D: −145°, E: −150°, F: −160°, G: −170°, and H: −180°.

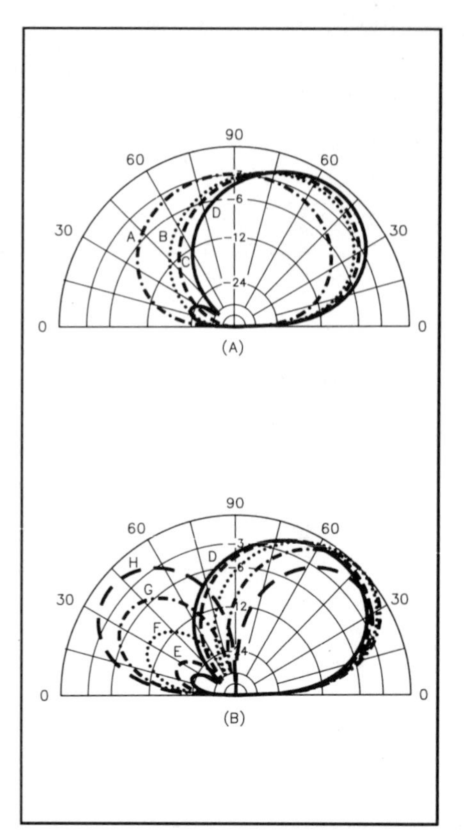


Fig 9—Elevation-plane radiation patterns for a two-element array of inverted Vs with an apex height of 36 feet, using ¼-λ spacing and equal-amplitude currents. The 0-dB reference is 8.7 dBi. The current phase angles for the patterns are A: 0°, B: −75°, C: −90°, D: −110°, E: −120°, F: −140°, G: −160°, and H: −180°.

¼ λ, the front-to-back ratio decreases just 1 dB (from 33.8 dB to 32.8 dB) when the two extra vertical elements are added. If the extra room is available, it appears that ¼-λ spacing is definitely superior.

Omnidirectional low-angle radiation can be obtained either by feeding the center element alone, or by feeding all of the outer elements in phase, as discussed previously. Fig 7 displays the results for both cases.

### Elevated Antennas

MININEC-based programs don't adequately model horizontal wires near the earth (in terms of wavelength), so I didn't model any of these arrays as elevated verticals with elevated radials. I am told that at least one elevated four-square is presently being used on 80 meters with good results, and I know Paul Bittner, WØAIH is very pleased with his single full-size elevated ¼-λ vertical on 160. Based on these reports, I assume the vertical-monopole arrays described in this article will work well when properly phased and elevated above ground.

To my knowledge, the only disadvantage of elevated vertical arrays is that the null at the zenith, which is present in all ground-mounted vertical antennas, disappears when the array is elevated. [This is caused by horizontally polarized radiation from the radial wires.—Ed] The exception is when all elements of the elevated array have symmetrically disposed radials and identical currents.

### Arrays of Inverted-V Elements

I modeled several 2-element phased arrays of inverted Vs on 40 meters (7.15 MHz), using spacings of either ⅛ λ (17.2 ft) or ¼ λ (34.4 ft). In each case, the apex of the V was at 36 feet (36-foot pushup masts are available from Radio Shack). The legs were 33.65 feet long, and the included angle between the legs was 120°. The results of this analysis, for ⅛-λ spacing, are summarized in Fig 8, while the plots for ¼-λ spacing are shown in Fig 9.

Figs 8 and 9 reveal that in-phase currents produce a major lobe at the zenith, while antiphase currents yield two equal lobes pointing in opposite directions. For equal-amplitude currents, adjusting the phase to some intermediate angle between 0° and −180° produces a rear null. The null can be placed at any desired elevation angle, depending upon the amount of phase shift between the two currents. There is, however, always a significant amount of radiation off the back of the array. The null is rather deep and is very sharp. It attenuates signals only over a narrow range of arrival angles.

Performance is much better if a three-element driven array is used. Fig 10 displays the radiation patterns for an array with ¼-λ (34.4 ft) spacing and currents of $1\angle 117°$, $2\angle 0°$ and $1\angle -116°$. The driving-point impedances given by ELNEC are $Z_1 = 34.1 - j37.6\ \Omega$, $Z_2 = 45.3 + j22.4\ \Omega$, and $Z_3 = 0.7 + j110.8\ \Omega$.

An element spacing of ⅛ λ (17.2 ft) can also be used with three radiators (Fig 11). The driving-point currents are $1\angle 146°$, $2\angle 0°$ and $1\angle -149°$, yielding impedances of $Z_1 = 13.7 - j45.4\ \Omega$, $Z_2 = 15.9 + j2.3\ \Omega$, and $Z_3 = -45.7 - j1.6\ \Omega$ at the antenna inputs.

It is possible to feed just the center element of these inverted-V arrays (or both of the outer elements, in phase) to produce high-angle radiation in all compass directions. The plots for ¼- and ⅛-λ spacing are shown in Figs 12 and 13 respectively.

Examining the various radiation-pattern plots shows that the two-element inverted-V antenna arrays have somewhat less forward gain than the 3-element versions (7.4 to 8.8 dBi versus 9.5 to 9.8 dBi), and it appears that the extra radiator improves front-to-back ratio considerably.

### Comparisons

For comparison, the plots in Fig 14 show the elevation-plane radiation patterns

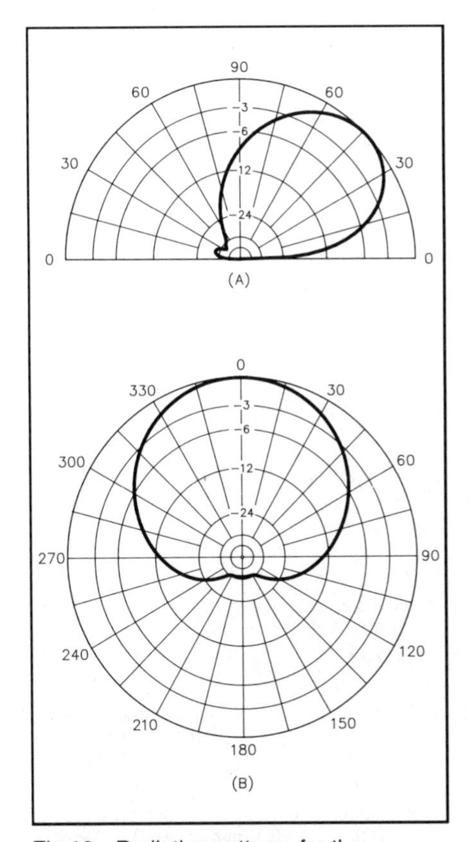


Fig 10—Radiation patterns for the 3-in-line array of inverted V antennas with an apex height of 36 feet, using 1/4-λ spacing and improved current phasing. At A, the elevation plane; at B, the azimuthal plane at a 43° elevation angle. The 0-dB reference is 9.8 dBi.

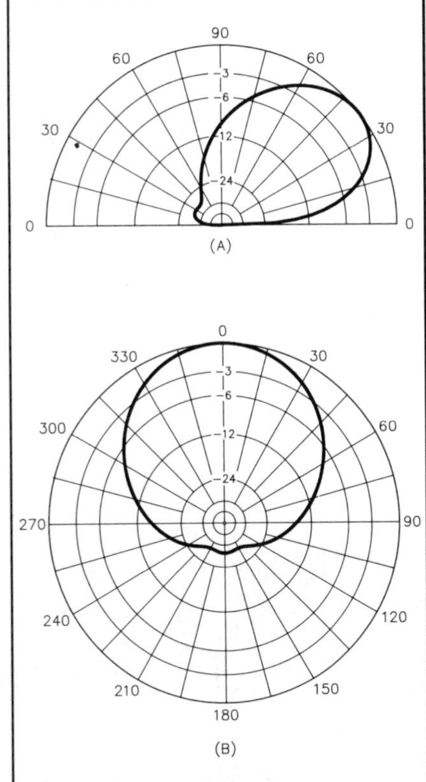


Fig 11—Radiation patterns for the 3-in-line array of inverted Vs with an apex height of 36 feet, using 1/8-λ spacing and improved current phasing. At A, the elevation plane; at B, the azimuthal plane at a 39° elevation angle. The 0-dB reference is 9.5 dBi.

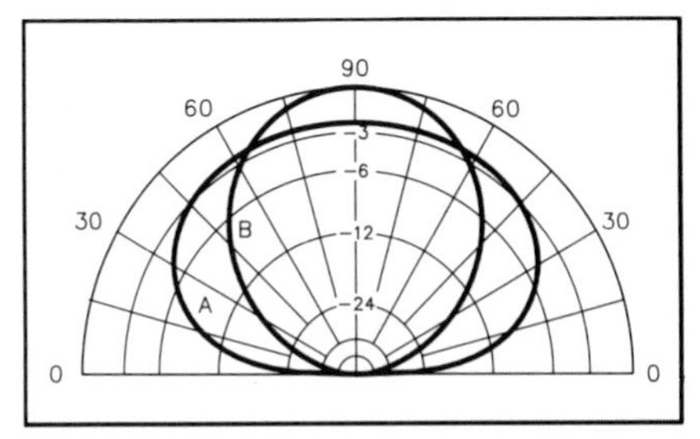


Fig 12—Elevation-plane patterns for the 3-in-line array of inverted Vs with an apex height of 36 feet, using 1/4-λ spacing and operating in the omnidirectional mode. The 0-dB reference is 7.8 dBi. Pattern A results from driving only the center element, pattern B from driving all four outer elements in phase (unused elements opened at the base).

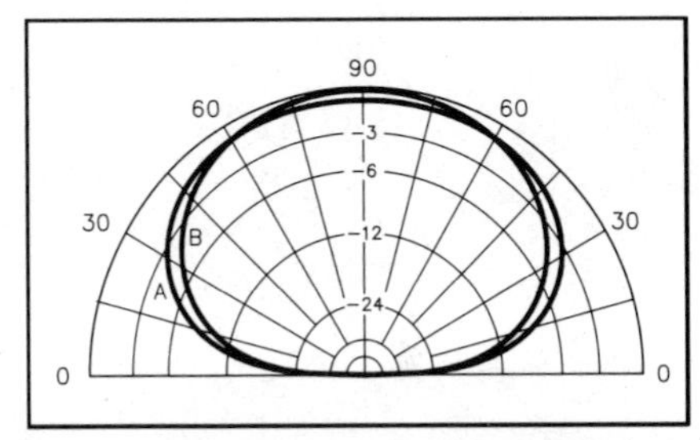


Fig 13—Elevation-plane patterns for the 3-in-line array of inverted Vs with 1/8-λ spacing, when used in the omnidirectional mode. The 0-dB reference is 6.2 dBi. Pattern A results from driving only the center element; pattern B results when all four outer elements are driven in phase (unused elements opened at the base).

of a 3-in-line vertical-monopole array and a three-element inverted-V array. Both cases use 1/4-λ spacing. As expected, the verticals hold the edge at very low takeoff angles, while the inverted Vs are superior everywhere else, with peak values of forward gain as much as 10 dB higher at some angles.

From another perspective, one could say that the (ground-mounted) vertical array rejects all high-angle signals, whether they come toward the front, back, or sides of the antenna. If you want the ability to select wave polarization (horizontal or vertical), or to have a choice of elevation angles, you may decide to build both types of array.

### Arrays of Arrays

For additional gain, two (or more) of these three-element arrays may be constructed side by side and fed in phase. The spacing between subarrays should be 0.5 to 0.625 λ. Figs 15 and 16 display the radiation patterns for six-element arrays of vertical monopoles and inverted Vs, using a spacing between subarrays of 1/2 λ; in both cases, the spacing between individual elements within a subarray is 1/4 λ.

### Other Bands

Any of these antennas can be used on 75/80 or 160 meters, by simply scaling all dimensions by an appropriate factor (equal to 7.15 MHz divided by the desired frequency). Changing the element spacing at a given frequency within a particular band, however, requires an adjustment in the current phasing. The height of the masts may be varied at will, and the elements can be erected as horizontal dipoles rather than inverted Vs.

Remember that the current phase angles given here are probably not optimum. Other combinations of element spacings and driving-point currents may well produce better results. No attempt was made to alter the current amplitudes. For that reason, it is likely that improvements in forward gain and front-to-back ratio can be made by adjusting both the amplitude and the phase of the feed-point currents. I presently have no software to facilitate this process.

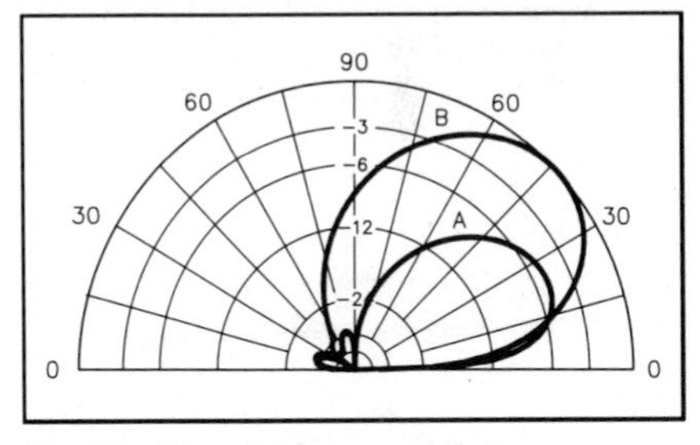


Fig 14—Elevation-plane radiation patterns for two types of 3-in-line arrays with 1/4-λ spacing and improved current phasing. Pattern A is for an array of vertical monopoles, pattern B for an array of inverted Vs. The 0-dB reference is 9.8 dBi

### Network Design

I have not discussed the design or construction of passive networks required to achieve a specific set of current magnitudes and phases. This topic has been covered extensively in the amateur literature, with excellent articles by Gehrke,[4], Lewallen[5] and others,[6] as well as a book by John Devoldere, ON4UN,[7] which provides detailed analysis of many array types.

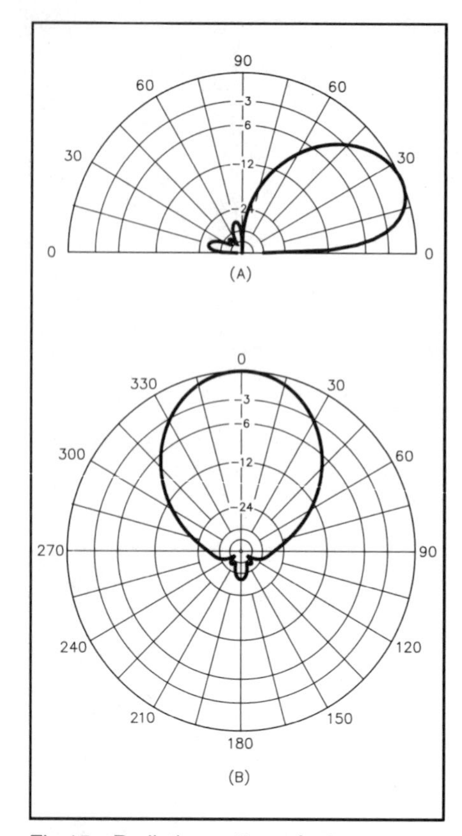


Fig 15—Radiation patterns for two 3-in-line arrays of vertical monopoles with ¼-λ spacing and improved current phasing. These arrays are placed side by side, ½ λ apart, and fed in phase. At A, the elevation plane; at B, the azimuthal plane at a 24° elevation angle. The 0-dB reference is 7.7 dBi.

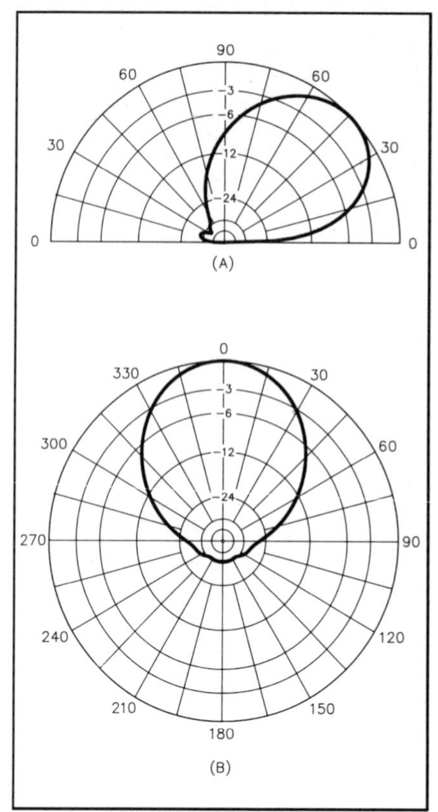


Fig 16—Radiation patterns for two 3-in-line arrays of inverted Vs with ¼-λ spacing and improved current phasing. These arrays are placed side by side, ½ λ apart, and fed in phase. At A, the elevation plane; at B, the azimuthal plane at a 43° elevation angle. The 0-dB reference is 11.6 dBi.

## Conclusion

I hope this report serves as food for thought for those with an interest in antenna design and construction. Perhaps some enterprising ham will manufacture and sell a "phase box," which will allow us to experiment more easily with three- or four-element driven arrays. An accurate and inexpensive antenna impedance bridge would also be helpful.

## Notes

[1]ELNEC, a program based on MININEC3 and permitting use of true current sources, is available commercially from Roy Lewallen, W7EL, PO Box 6658, Beaverton, OR 97007.

[2]D. W. Atchley, Jr, "Switchable 4-Element 80-Meter Phased Array, *QST*, Mar 1965, pp 48-52, and "Updating Phased Array Technology," *QST*, Aug 1978, pp 22-25.

[3]F. Gehrke, "Vertical Phased Arrays," *Ham Radio*, May, Jun, Jul, Oct and Dec 1983, and May 1984.

[4]See Note 3.

[5]R. Lewallen, W7EL, "The Simplest Phased-Array Feed System...That Works," *ARRL Antenna Compendium, Volume 2* (Newington: ARRL, 1989).

[6]See Chapter 8, *The ARRL Antenna Book*, 15th or 16th eds (Newington: ARRL, 1988 or 1991).

[7]J. Devoldere, ON4UN, *Low-Band DXing* (Newington: ARRL, 1987).

# The Square-Four Receiving Array

By Gary R. Nichols, KD9SV
4100 Fahlsing Rd
Woodburn, IN 46797

John C. Goller, K9UWA
4836 Ranch Rd
Leo, IN 46765

Roy W. Lewallen, W7EL
5470 SW 152 Ave
Beaverton, OR 97007

This article presents a new and innovative solution to low-band reception on a small lot without beverage receiving antennas. The system was conceived by John and built by Gary, with a lot of technical assistance from Roy. John came up with the original idea of a miniature receiving antenna system, one that would have good front-to-back ratio and be switchable in directions.

With the inception of antenna modeling programs for PCs, it became possible to look at many antenna systems, vary all of the parameters, and in a few minutes see what the design does and what the radiation patterns look like. All the modeling for this antenna was done using ELNEC.[1] As low-band enthusiasts, John and Gary were looking for better, quieter receiving antennas that would fit on an average size lot and still hear well enough to work DX on 160 meters. The system described here has consistently heard as well as or better than an 800-foot beverage antenna in the directions that have a clear shot toward the horizon.

The basic system consists of four self-supporting electrical half-wave vertical dipoles for 160 meters, center fed, and they are only twenty feet high. Fig 1 shows the antenna system. The elements are used as a nonresonant vertical dipole array, center loaded so they exhibit a $-jX$ or capacitive input, and then are artificially tuned to zero reactance at the feed point so the phasing will work properly. The elements are very close spaced, less than ⅛ λ, and are fed both as a diamond and square-four configuration, which enables switching the array in eight directions.

At this point let us emphasize that this array is very difficult to build and phase correctly, and should not be attempted without the means of proper test equipment for measuring input reactances and impedances, as well as a dual-trace oscilloscope to set up the phase delays and currents.

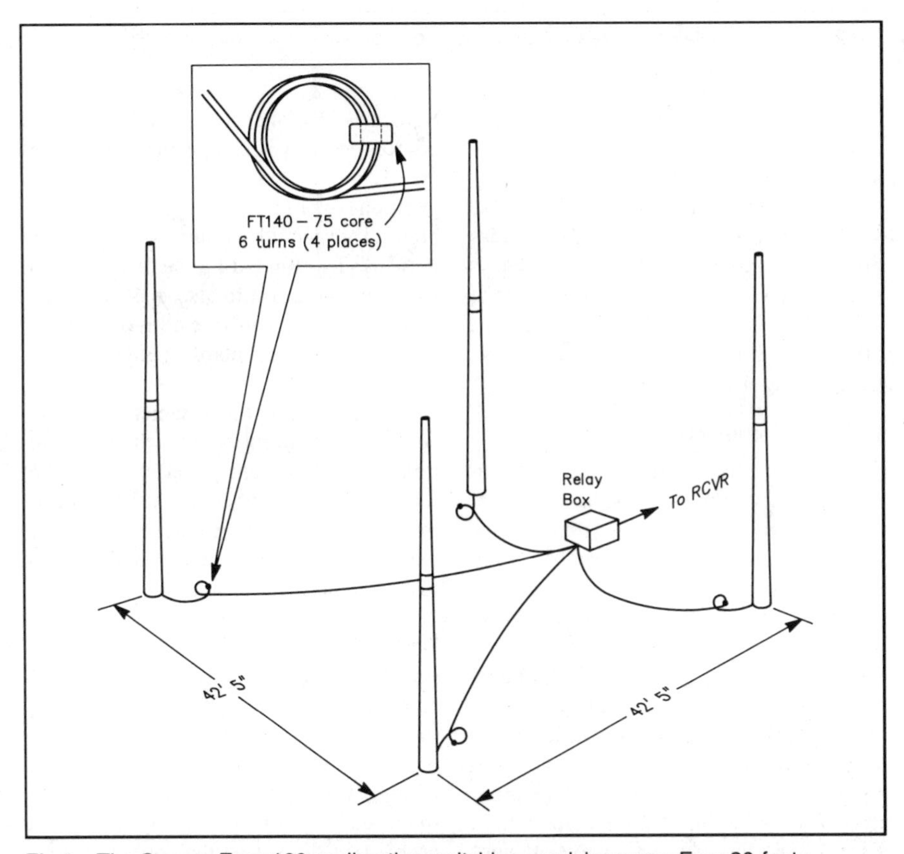


Fig 1—The Square-Four 160-m direction-switching receiving array. Four 20-foot elements are constructed of telescoping aluminum tubing. The coaxial feeder enters the base of each element and exits at the center feed point. Ferrite-core chokes are used at the base for decoupling, shown in the inset.

### Physical Description

The elements are made of telescoping sizes of 6061-T6 aluminum. The elements are placed in a square, 42 ft 5 in. on a side, or 30 ft from center for a diagonal of 60 feet. The feed line goes up the inside of each element and is decoupled at the base of the element with a high permeability ferrite toroid. The elements are insulated at the center with a fiberglass rod. A 4-in. ferrite rod is used for a center-loading coil of about 6800 Ω reactance, Fig 2. The feed line is link coupled to the loading coil for impedance matching, and a small toroid coil is placed in series at the feed point to tune out the capacitive reactance. Gary used 93-Ω coax, although any coax could

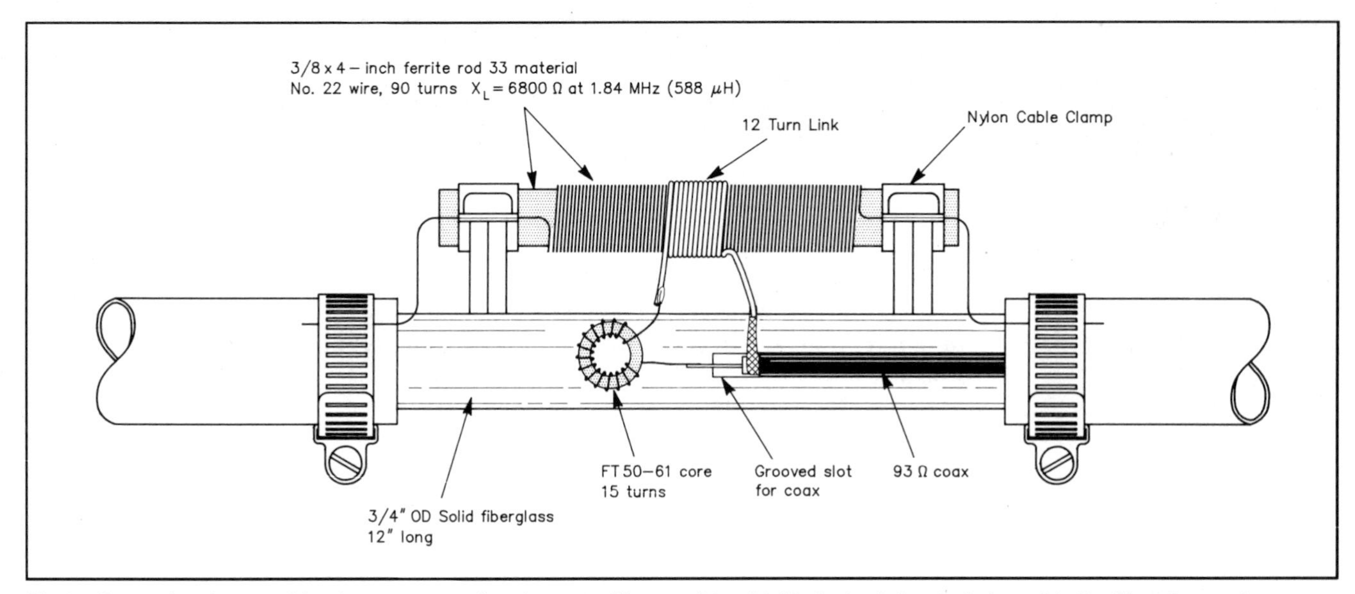


Fig 2—Center insulator and feed arrangement for elements. The small toroidal inductor is located alongside the fiberglass rod.

be used and the number of turns on the link adjusted to match the characteristic impedance of the coax.

Direction switching is done with twelve small 12-V dc DPDT relays, and both sides of the coax are switched. The phase delays are L-C networks and provide delays of 155° and 310°. The 310° delay is built as a 50°-lead high-pass network. See Figs 3 (on the next page) and 4.

### Builders Comments

This antenna system, being nonresonant, has an extremely wide bandwidth of about 75 kHz for a 25-dB front-to-back ratio. Because the element length is very small compared to a full-size dipole, the mutual coupling normally associated with a phased array is very near nonexistent. Gary could look at one element through a half-wavelength of coax with the vector impedance meter and see very little change when the other elements were laid down on the ground. This lack of interaction is probably the only reason it is possible to build an array of this size and make it work properly.

### User Comments

Calculated azimuth radiation patterns are shown in Fig 5. Gary used this array through the winter and found it to compare very closely with his 800-foot beverages. Close-in stations with signals coming in at high angles are greatly attenuated with this system, as there is no high-angle response in the antenna patterns. This can be a blessing when trying to work DX, or a curse if you are trying to work stateside; it all depends on your receiving requirements. Gary found however, that he used the Square-Four more often than the diamond configuration, since the square has about 10 dB more gain. Probably four switch positions would be adequate instead of eight. This would simplify the switching networks by eliminating the diamond configuration, and would also make the array much less phase sensitive. Plus or minus 5° still gives an acceptable pattern on the Square-Four.

A year of using this antenna indicates it is a good system for people who have limited space and want an effective 160-m receiving array. Probably other systems of this type will be coming from many authors who use computer modeling, and we have just scratched the surface.

### Comments from Roy

Several factors always must be considered in the design of a phased array. Among these, the effects of mutual coupling are perhaps the most important. Mutual coupling causes the element feed-point impedances to change, complicating the problem of delivering the correct currents to the elements. These effects are covered in detail in *The ARRL Antenna Book*.[2]

A close look at this antenna provides insight into some of the factors that need to be considered. It also reveals some of the factors playing a major role in the antenna's operation that aren't immediately obvious.

The first factor to consider is the effect of mutual coupling. Let's see how it effects this antenna. If we were to take a single element, without the feed system, mount it one foot above the ground, and look at the impedance at its center, we would see about $0.5 - j4900\ \Omega$—a very low radiation resistance in series with a large capacitive reactance. This is entirely characteristic of a short dipole. If we then were to mount four elements and feed them with currents as done in this array, the feed-point impedances would become

$-1.3 - j4900\ \Omega$ for the leading element
$1.2 - j4900\ \Omega$ for the lagging element
$0.06 - j4900\ \Omega$ for the middle elements

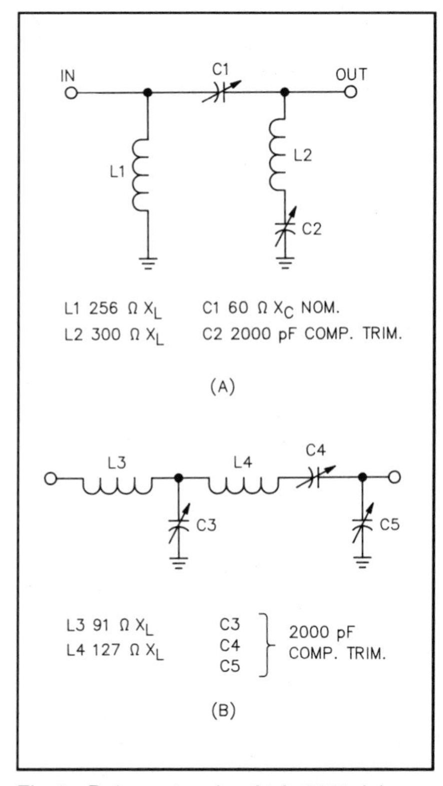


Fig 4—Delay networks. At A, 310° delay (actually a 50° lead). At B, 155° delay.

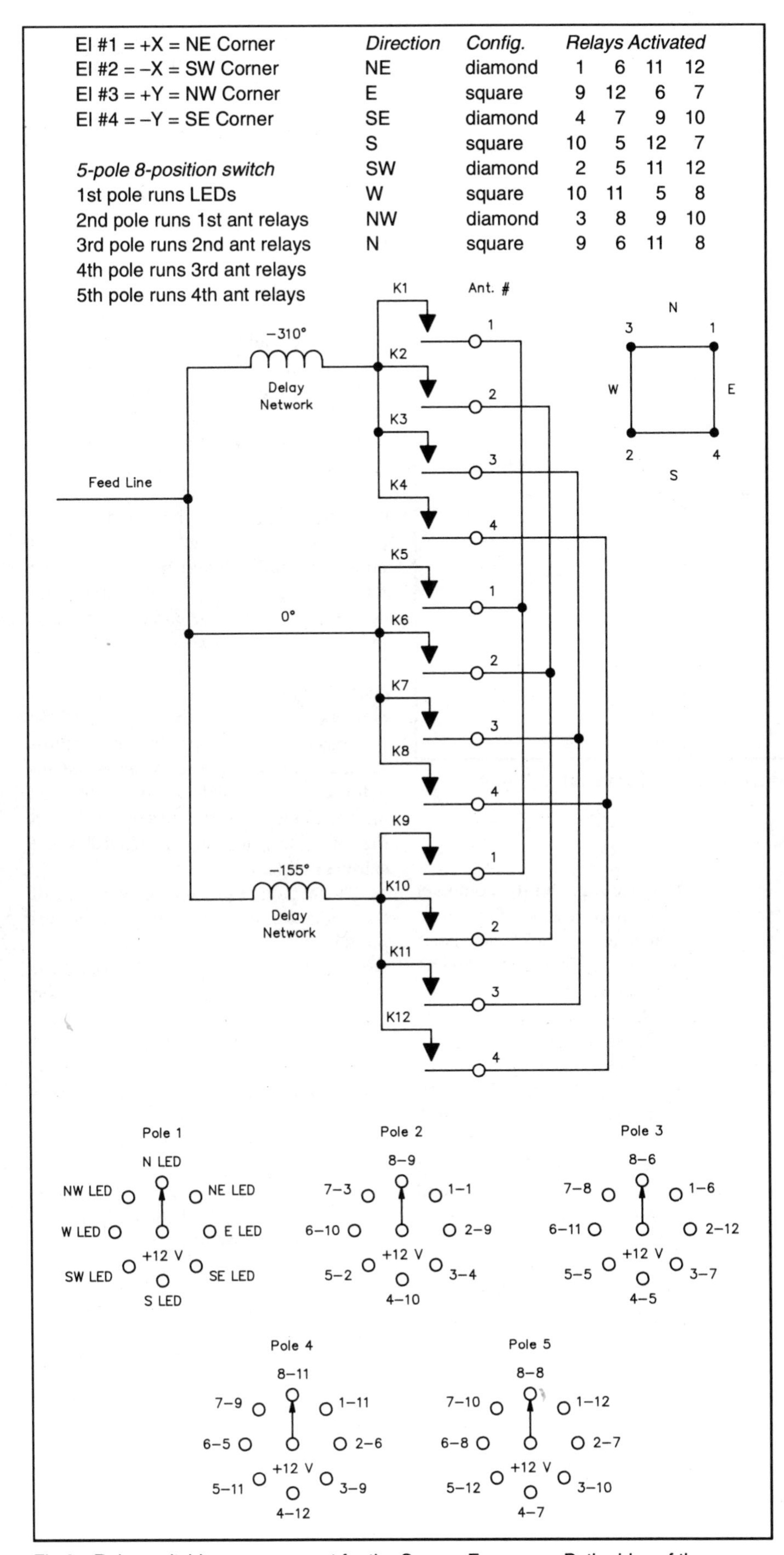


Fig 3—Relay switching arrangement for the Square-Four array. Both sides of the coaxial feeders are switched, shield and center conductor. Only one side is shown here for simplicity. Twelve 12-V dc DPDT relays and a 5-pole 8-position switch are used.

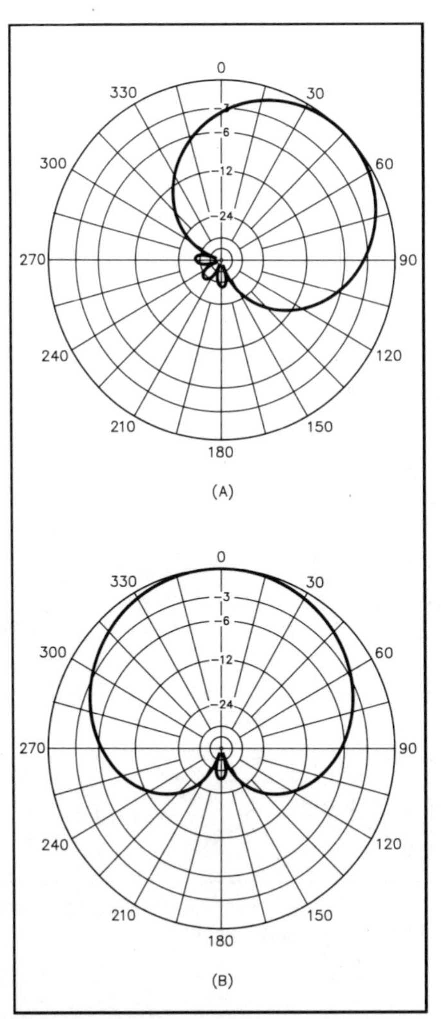

Fig 5—Radiation patterns of the Four-Square array. At A, the pattern when switched to operate as a diamond, and at B the pattern when operated as a square. The calculated gain of the square arrangement exceeds that of the diamond by approximately 9.6 dB.

(The negative resistance of the first element isn't a misprint. This can, and does, occur in phased arrays. It simply means that, if power were applied to the array, the negative-resistance element delivers power back to the feed system. It gets this power through coupling to the other elements.) If we were to design a lossless matching system for this array, the different element resistances would have to be accounted for. But, of course, we can't design a matching system with no loss. And, since this array is to be used for HF receiving only, there's no real need to keep losses low anyway.

The equivalent circuit of a single element is shown in Fig 6. The center loading/matching transformer provides a relatively large amount of loss. The ferrite-core winding's measured reactance of 6800 Ω and Q of 32 translate to an equiva-

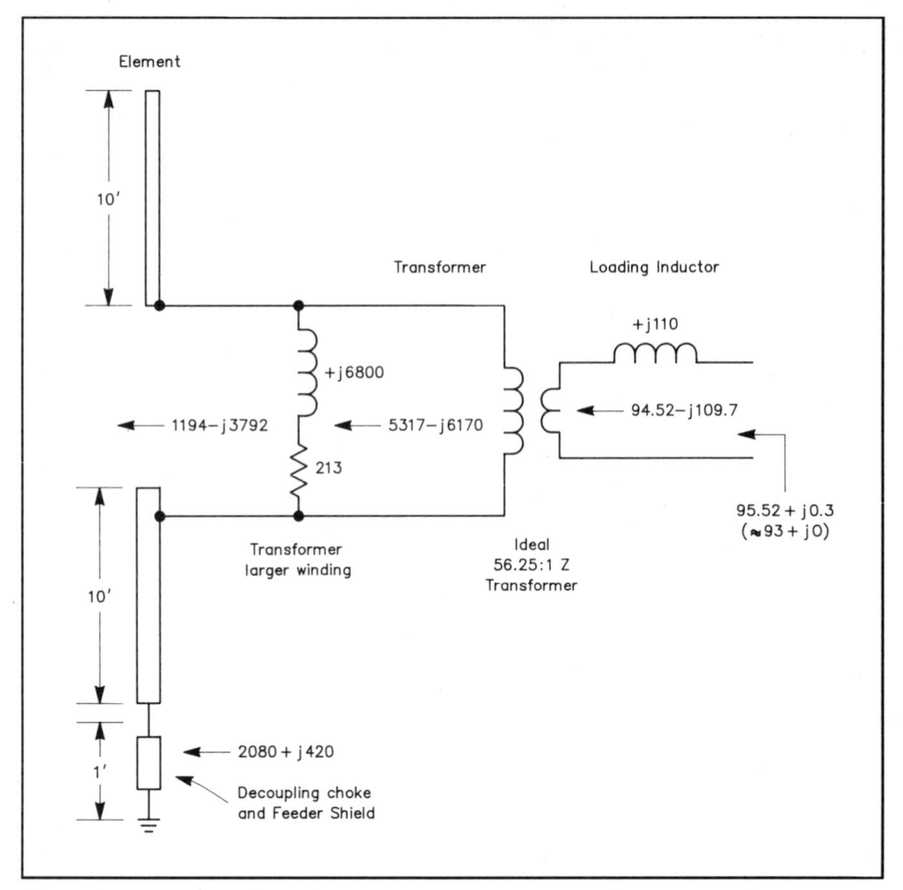


Fig 6—The equivalent circuit of a single element of the Four-Square array. Resistances and reactances are in ohms.

lent series resistance of 213 Ω. Although the transformer impedance appears in parallel with that of the element, the effect of the loss is to raise the combined resistance of the element and transformer by 1400 Ω(!), and to make it very insensitive to the actual element resistance. This illustrates an important point in the design of phased arrays: In the presence of a large amount of loss, the effects of mutual coupling can be disregarded.

Before doing any further analysis of the feed method, it is necessary to look more closely at the decoupling of the feed line. In this design, the feed line exits the element at its end, rather than more conventionally at its center. Without any decoupling, this causes the outside of the coax to become an extension of the element, with substantial current flow along the coax. Experience at W7EL has shown that conventional decoupling methods, such as ferrite cores or chokes, are totally inadequate for reducing this flow to an acceptable level. This has been confirmed by computer modeling, and the computer was put to work on the decoupling system used in this antenna. The choke impedance is known to be about 1000 + *j*0 Ω, but an unknown is the impedance looking along the outside of the feed line that is lying on the ground. The latter impedance is dependent on the feed-line length and the characteristics of the ground, so it is difficult to predict. Therefore, a "backward" approach was used. Various load impedances were connected to the bottom of the element until the computer model showed the known center impedance. The resulting load was 2080 + *j*420 Ω. Fairly large changes to the load impedance made relatively small changes to the feed-point impedance, so the result is only approximate. However, it is known to be at least 1000 Ω, and probably not more than 2000, so the result is in the ballpark.

As suspected, the impedance of the ferrite choke is inadequate to decouple the coax feed line. The impedance of an element (without the transformer) is about 1190 – *j*3790 Ω with the "decoupled" coax coming from the end—quite a change from 0.5 – *j*4900 without it! The increased resistance represents increased loss; indeed, the computer shows another 30 dB loss in the decoupling choke. How well has the current on the feed line been reduced? The computer analysis shows the current on the outside of the feed line to be ¾ the current flowing at the center of the dipole! Even so, the impedance at the center of the element is significantly different with and without the choke.

Considering the vastly changed element impedance, the effects of mutual coupling should be reinvestigated. However, again due to losses, the elements maintain nearly the same impedances when fed together as each does when isolated.

When designing the elements for phased arrays, shunt networks at the feed points generally should be avoided. The reason is as follows. Feed systems are designed to deliver the correct currents to the feed points. Since elements generally have different feed-point impedances because of mutual coupling, identical shunt networks will divert differing proportions of the current from each element, disturbing the desired current balance. However, in this array the impedances of all elements are nearly the same, because of losses, and the shunt transformer impedance won't cause a problem.

The method of having the feed line exit the element at the end is the most serious problem to anyone trying to duplicate this antenna. With this method, the outside of the feed line becomes a very significant part of the antenna system. Ground characteristics and placement of the feed line will alter the element feed-point impedances, but correction can be made to some degree by readjusting the phasing networks. More serious is the possibility of the fields from the feed lines contributing to the pattern, making deep nulls impossible. If duplication is attempted, careful attention must be paid to feed-line placement.

**Notes**

[1]All computer modeling was done with ELNEC, an analysis program based on MININEC and commercially available from Roy Lewallen, W7EL, PO Box 6658, Beaverton, OR 97007. (The ARRL in no way warrants this offer.)

[2]G. Hall, Ed., 15th or 16th eds., *The ARRL Antenna Book* (Newington: ARRL, 1988 or 1991), Chapter 8.

# Phased Verticals with Continuous Phase Control

By Peter H. Anderson, KZ3K
Department of Electrical Engineering
Morgan State University
Cold Spring Lane & Hillen Rd
Baltimore MD 21239

This paper describes a simple arrangement for phasing two verticals. The arrangement utilizes a simple symmetrical pi network to provide a continuous phase delay and avoid the inconvenience of switching transmission-line sections.

The continuous phasing arrangement may be used on several bands. In fact, I have used this circuit to great advantage on 80, 40 and 20 meters. By changing the angle, or "steering" the array, I have been able to consistently vary signals by as much as 20 dB. The required parts are inexpensive and readily available. Construction is straightforward and simple.

I have two Butternut verticals spaced about 60 feet apart. They were erected without plans for future phasing and were tuned to different frequencies on the 80-meter band. The possibility of phasing them was posed to my undergraduate electrical engineering design class. They seized upon the idea and soon developed a workable system. When we put the circuitry on the air, we were pleasantly surprised with the results.

Most phasing arrangements are presented on the premise that the user can feed the antennas in phase. In my installation, the relative lengths of the coaxial runs is unknown. In addition, one of the verticals has a 160-meter coil and the other has a 30-meter coil. As a result, I have no idea of the relative electrical lengths from the shack to the antennas, which of course will vary from band to band. Under the circumstances it seemed as though the benefits of phasing were clearly reserved for amateurs who possessed two identical antennas, lots of coaxial cable, and a great deal of leisure time.

Fortunately for the rest of us, this reasoning is flawed. By adding a 0° to 90° variable phase lag to either of the two antennas, and by reversing the phase as necessary, it is possible to adjust the relative phase of one antenna to the other by a

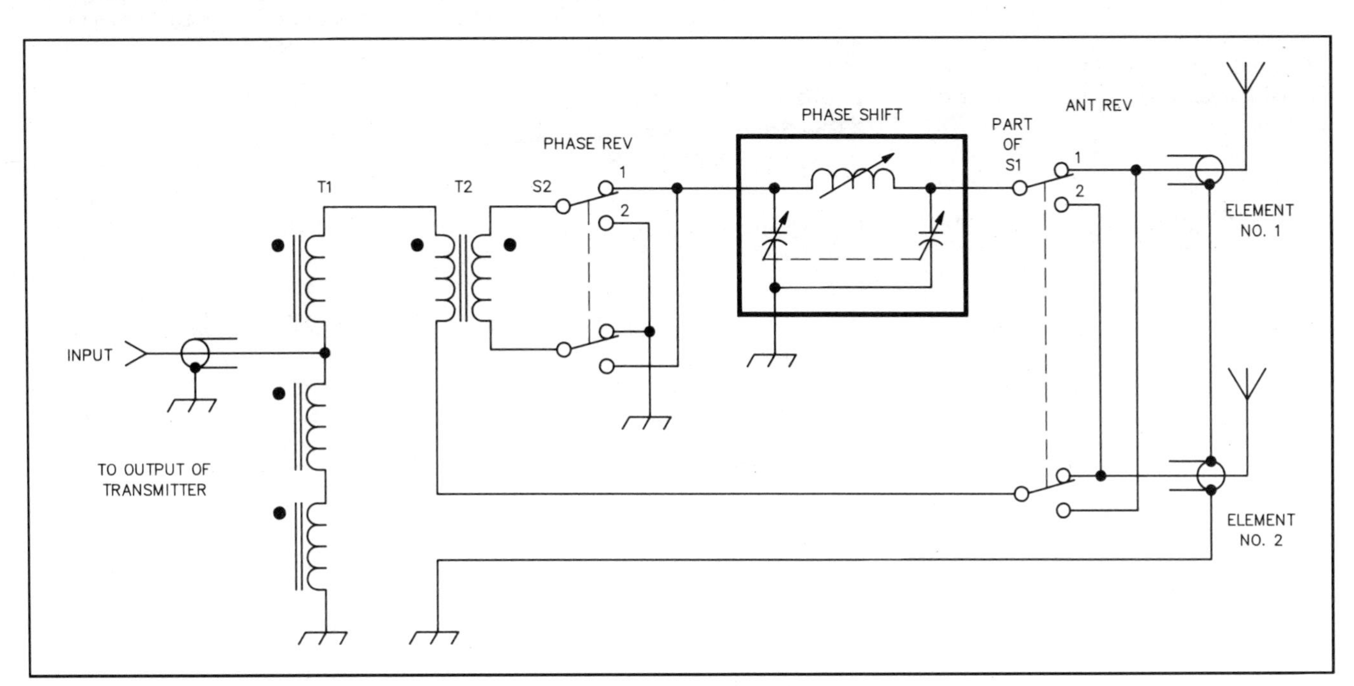


Fig 1—An abbreviated circuit diagram. The phase of the input to the phase shift network may be inverted using DPDT toggle switch S2. S1, used to interchange the antennas, is part of a larger function switch discussed in the text.

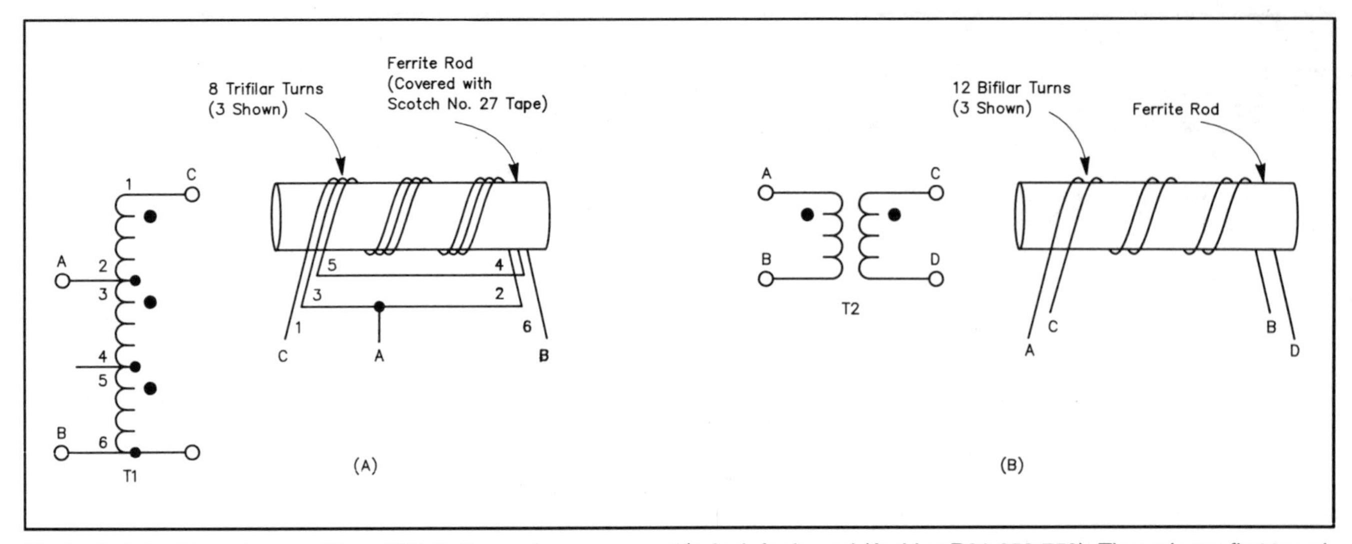


Fig 2—Details of transformers T1 and T2. Both transformers use a 7½-inch ferrite rod (Amidon R61-050-750). The rods are first taped using Scotch no. 27 glass cloth electrical tape. The no. 14 AWG thermaleze wire is available from Amidon. T1, shown at A, consists of eight trifilar turns connected as illustrated. The three conductors should be tightly bound together, either by twisting or by using small pieces of tape. (The length of each conductor is about 18 inches.) To prevent the rod from moving within the coil, small pieces of tape or epoxy may be used. T2, at B, consists of 12 bifilar turns which are similarly wound on a ferrite rod. The length of each conductor is about 24 inches. Amidon offers a special kit consisting of the two rods, 10 feet of wire and a 66-foot roll of Scotch no. 27 tape. When contacting Amidon, refer to this article by title and author. (See note 5.)

full 360°. The fact that the actual phase lag is unknown is unimportant. With the system described in this article, the user will be free to vary the phase so as to effectively feed the antennas broadside, end fed toward element no. 1, or end fed toward element no. 2. By using the phase circuitry to "peak" the antenna system for maximum received signal strength, the user will optimize the transmitted signal pattern as well.

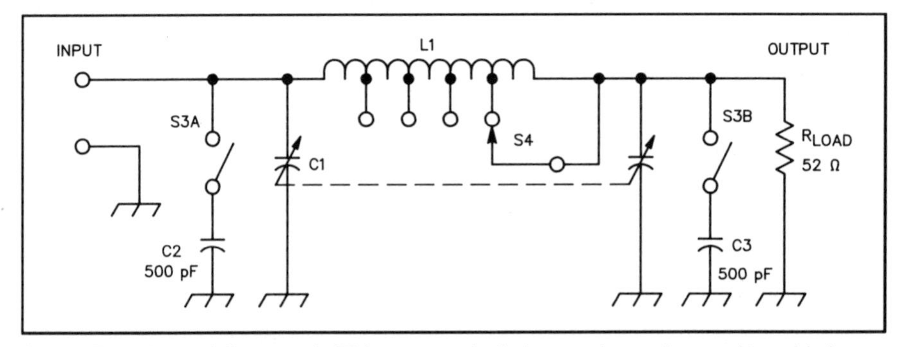


Fig 3—The phase-shift network. This symmetrical pi network may be used to add phase delay while maintaining an input impedance that is close to the 52-Ω load resistance. Using S3, C2 and C3 may be switched in to provide a 0- to 1000-pF range for use on 80 meters.

C1—Dual ganged 0-500 pF air variable capacitor.

C2, C3—500-pF doorknob type capacitor.

L1—See Fig 6 for construction details.

S3—DPST toggle.

S4—Single pole rotary switch with multiple positions to switch in the desired inductance.

### Overall Configuration

An abbreviated circuit diagram is illustrated in Fig 1. Transformer T2 "floats" one of the antennas, permitting the antennas to be fed in series. Therefore, assuming that both antennas are at 52 Ω, the impedance of the series pair is 104 Ω. A simple 2:3 turns-ratio transformer (T1) transforms this impedance to 4/9 × 104 or 46 Ω. Use of these broadband transformers permits the same circuitry to be used on many different frequencies. The construction details of T1 and T2 are presented in Fig 2.

A symmetrical pi network of the type illustrated in Fig 3 is used to provide a phase shift while maintaining the 52-Ω input impedance. Note that the capacitors on either side of the inductor are equal.

Fig 4 may be used to determine the values of C (on each side of the inductor) and L to provide any phase in the range of 0° to 90°. For example, to obtain a phase shift of 45° at 7.1 MHz, the L and C values are determined to be slightly more than 0.8 µH and 175 pF. Note that Fig 4 was developed for a frequency of 7.1 MHz and it is reasonably accurate for the entire 40-meter band. However, you can use Fig 4 for any frequency by simply scaling the values.

For example, assume that the frequency is 3.905 MHz and 45° of phase lag is desired.

$$L = 7.1/3.9 \times 0.8 = 1.45\ \mu H$$

$$C = 7.1/3.9 \times 175 = 320\ pF$$

At f = 14.2 MHz, you would simply halve the values at 7.1 MHz; L = 0.4 µH and C = 87 pF.

Note that you could construct such a network for use on 40 meters by using a roller inductor, variable between 0 and 1.18 µH, and a dual-ganged capacitor having a range of 0 to 425 pF. With the appropriate settings, any phase in the range of 0 to 90° could be obtained. By scaling the values of L and C as shown in Table 1, the network could be used on any amateur band.

I started the project with a roller inductor and a dual-ganged 0-500 pF variable capacitor (with the capability of switching in an extra 500 pF for 0-1000 pF on 80 meters). However, roller inductors are a considerable expense for many amateurs. To make matters worse, I found the whole

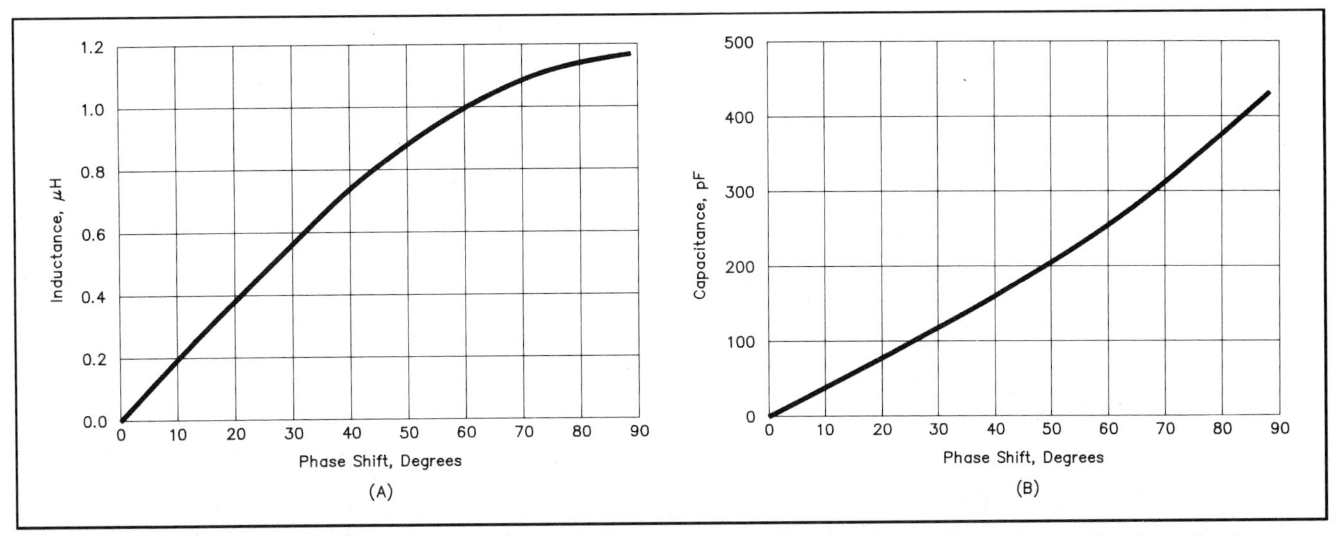


Fig 4—Phase shift versus inductance and capacitance at 7.1 MHz, for a load of 52 Ω. These plots may be used to determine the values of L and C in a symmetrical pi network to provide any phase delay in the range of 0 to 90°. The L and C values may be scaled for frequencies other than 7.1 MHz. (See text.)

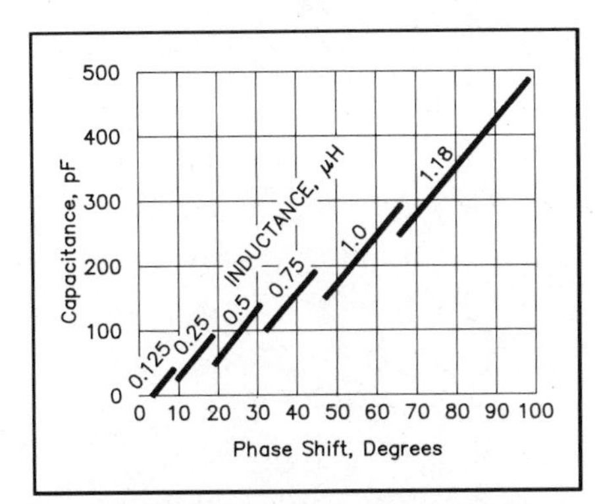


Fig 5—Capacitance versus phase for switched inductance at 7.1 MHz, for a load of 52 Ω. The phase may be adjusted by varying capacitor C over a limited range with an SWR of better than 1.2:1. For example, with an L of 1.18 µH, the phase delay may be varied from 65 to 98° by adjusting C in the range of 250 to 500 pF.

arrangement to be too confusing to use on all bands.

With this deficiency in mind, I began to explore a switched-inductor arrangement. I found that with a fixed L, the C could be varied within a limited range, achieving a variable phase while maintaining an SWR of less than 1.2. This is summarized in Table 2 and is presented graphically in Fig 5. Note that all values are at 7.1 MHz.

The results are significant. By using a small switched inductor, a dual-ganged 0-500 pF variable capacitor and a very simple six-position switch, any phase in the range of 0 to 90° can be obtained. There are no cumbersome transmission-line sections to insert! Further, the pi network values can be easily scaled to provide similar phase shifts on either 80 or 20 meters (see Tables 3 and 4).

**Table 1**
**L and C Network Values for Various Amateur Bands**

| *Freq (MHz)* | *L (µH)* | *C (pF)* |
|---|---|---|
| 3.55 | 0 – 2.4 | 0 – 950 |
| 7.1 | 0 – 1.2 | 0 – 425 |
| 10.1 | 0 – 0.85 | 0 – 300 |
| 14.2 | 0 – 0.6 | 0 – 210 |
| 18.1 | 0 – 0.5 | 0 – 166 |
| 21.3 | 0 – 0.4 | 0 – 140 |

**Table 2**
**Phase Shifts at 7.1 MHz with Fixed Inductance**

| *Fixed L (µH)* | *C range (pF)* | *Phase Range (degrees)* |
|---|---|---|
| 0.12 | 0 – 50 | 3.1 – 9.5 |
| 0.25 | 25 – 100 | 9.5 – 19 |
| 0.50 | 50 – 150 | 19 – 32 |
| 0.75 | 100 – 200 | 32 – 46 |
| 1.0 | 150 – 300 | 46 – 65 |
| 1.18 | 250 – 500 | 65 – 98 |

**Table 3**
**Phase Shifts at 3.5 MHz with Fixed Inductance**

| *Fixed L (µH)* | *C range (pF)* | *Phase Range (degrees)* |
|---|---|---|
| 0.25 | 0 – 100 | 3.1 – 9.5 |
| 0.50 | 50 – 200 | 9.5 – 19 |
| 0.75 | 100 – 300 | 19 – 32 |
| 1.0 | 200 – 400 | 32 – 46 |
| 1.18 | 250 – 600 | 46 – 65 |
| 2.36 | 500 – 1000 | 65 – 98 |

**Table 4**
**Phase Shifts at 14.2 MHz with Fixed Inductance**

| *Fixed L (µH)* | *C range (pF)* | *Phase Range (degrees)* |
|---|---|---|
| 0.06 | 0 – 25 | 3.1 – 9.5 |
| 0.12 | 12 – 50 | 9.5 – 19 |
| 0.25 | 25 – 75 | 19 – 32 |
| 0.37 | 50 – 100 | 32 – 46 |
| 0.50 | 75 – 150 | 46 – 65 |
| 0.60 | 125 – 250 | 65 – 98 |

You could develop a switched inductor network for any or all of the three bands. One simple technique I used was to develop a single tapped inductor (actually three inductors in series) as illustrated in Fig 6. This arrangement provides taps at 0.062, 0.125, 0.25, 0.37, 0.5, 0.6, 0.75, 1.0, 1.18 and 2.36 µH, and may be used on all three bands. (If you desire three separate tapped coils, one for each band, the construction data can be easily adapted.)

**Operation**

Refer to the detailed schematic shown in Fig 7. S1 (multiposition) permits the user to feed *only* element no. 1 or no. 2. These switch positions also allow the user to assure that both antennas present the same impedance to the phasing network. Two other positions on S1 insert the phase-shift network in element no. 1 or no. 2. These are the primary operating positions and are labeled PHASE 1 and PHASE 2. An additional position (TUNE) is provided to connect the input (through the phasing network) to a dummy load. This is especially helpful when tuning the phasing network. Another position connects the input directly to a dummy load.

After the impedance to both antennas has been verified to be roughly equal and the transmitter has been loaded using the dummy load, the phasing network can be adjusted. With S1 in the TUNE position, a particular L value is selected and C is adjusted for minimum SWR. By repeating this procedure at various L values, you can quickly build a table of the C values corresponding to the L setting for each band. You will also note that there is a range of

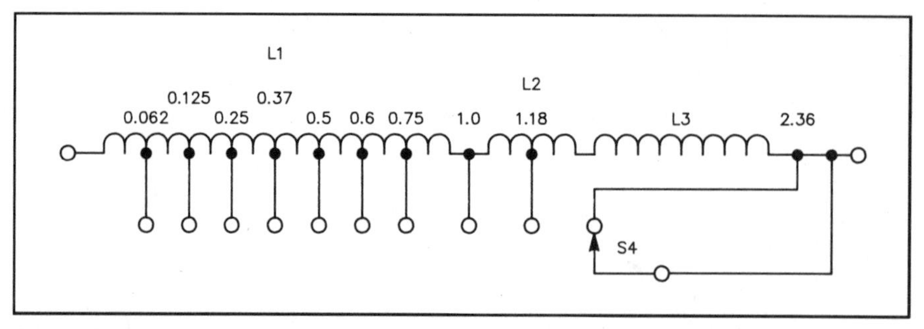


Fig 6—Construction of the tapped phase shift inductor. L1 is tapped at 1.5 turns (0.062 $\mu$H), 2.25 turns (0.125 $\mu$H), 3.75 turns (0.25 $\mu$H), 5 turns (0.37 $\mu$H), 6.5 turns (0.5 $\mu$H), 7.5 turns (0.6 $\mu$H) and 9 turns (0.75 $\mu$H). The 1.0 $\mu$H tap is at the junction of L1 and L2. L2 is tapped at 4.25 turns (1.18 $\mu$H). The combination of L1, L2 and L3 provides the 2.36 $\mu$H tap. Note that in using a single switched inductor, the lower five settings are used for 20 meters. All but the highest tap may be used on 40 meters and all settings may be used on 80 meters. Some of the phase settings will overlap on 40 and 80 meters.

L1, L3—Barker & Williamson 804T, 1 inch diameter, 4 turns per inch, length 3 inches. (See note 6.)

L2—Barker & Williamson 604T, ¾ inch diameter, 4 turns per inch, length 2 inches.

S4—Centralab PA-2001, 1 pole, 2 to 12 position rotary switch (Newark). (See note 7.)

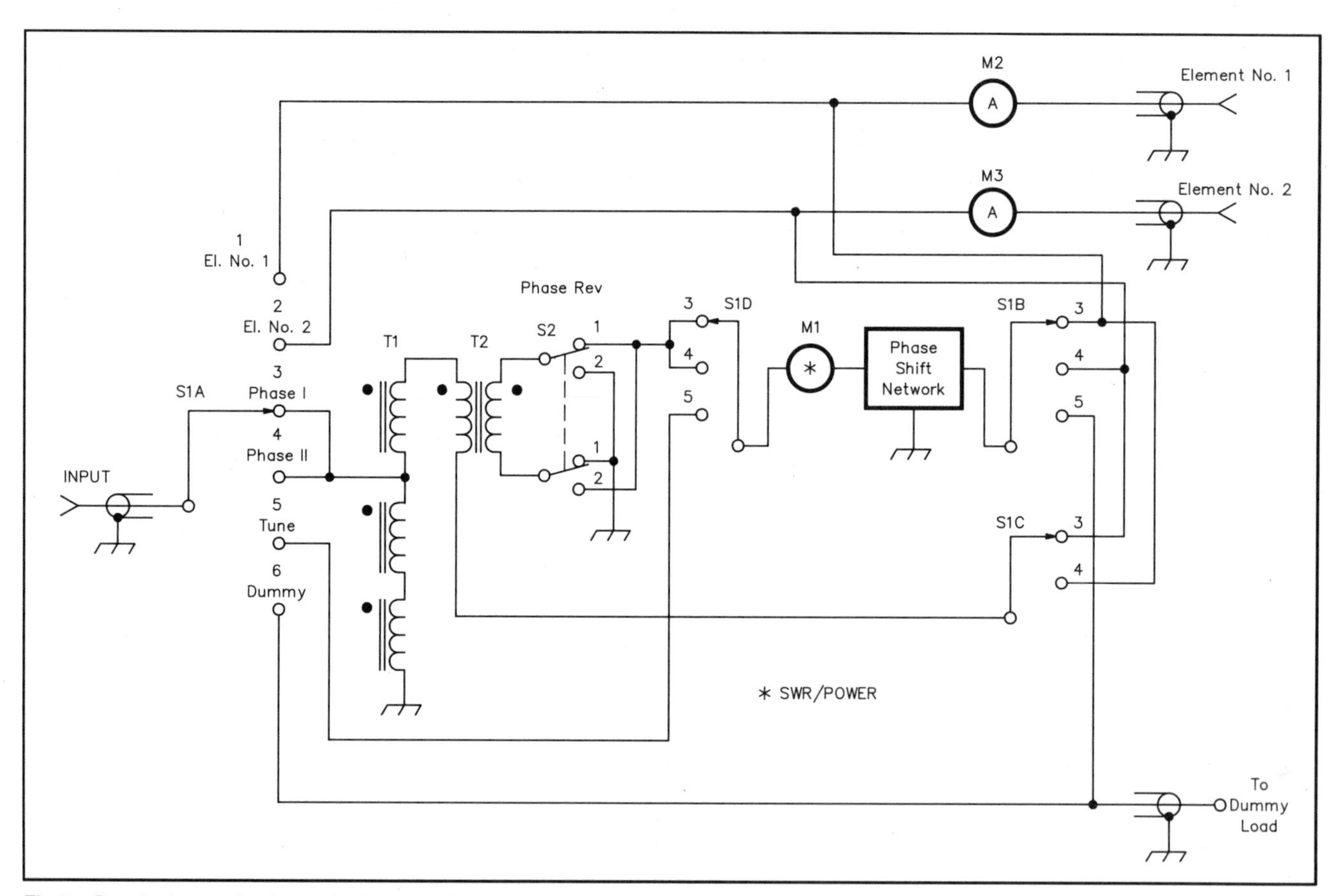


Fig 7—Detailed overall schematic diagram. See text for details on switching capabilities.

M1—SWR/power meter; see text.

M2, M3—RF ammeters; see text.

S1—Centralab PA-2013, 4 pole, 2 to 12 position rotary switch (Newark).

S2—DPDT toggle.

S3—DPST toggle.

Phase-shift network—See Figs 3 and 6.

T1, T2—See text and Fig 2.

C which gives you a low SWR. This range should be substantial for the highest L setting associated with each band, providing the ability to continuously modify the phase for a particular L setting.

With the table in hand, turn on the receiver and listen. Find a distant signal (not sky wave) and set the PHASE REV switch for whichever setting provides the weakest signal. Now switch through the various L settings, carefully adjusting capacitor C until the signal is at its absolute weakest point. Throw the PHASE REV switch to the opposite position and you should find that the signal has increased to its strongest possible level.

If you wish to work Scandinavia from the US with a beam, you might adjust the rotator to 25°, depending on where you live. This is one type of degree setting. It is important to understand that in using the phase network arrangement, you can vary the phase of one antenna by a full 360° relative to the other and, in doing so, dramatically change the directivity of the array. However, 25° of phase lag is not the same as the bearing to Scandinavia. Instead, you "steer" the array by changing the phase lag and experimentally determining the best settings to hear Scandinavia. Be sure to log the settings for future reference.

This may sound very complex, but it is not. I hunt counties on all three bands and consult a simple table for the settings that favor New England, Florida, Texas and the Northwest from my QTH here in Maryland. (Note that the settings vary from band to band.) I can switch from one band to another and be fully operational within 30 seconds.

I continuously monitor the SWR and power while utilizing the network circuitry. This fact is not emphasized in Fig 7. I repackaged a Heath HM-102 SWR/power meter into the unit (shown as M1 on the schematic) to provide this capability. Two RF ammeters (M2 and M3) were also used to continuously monitor power to each antenna. These are luxuries; you should be able to live with a single SWR and power meter between the rig and this circuit, provided you are careful when adjusting the C setting to agree with the L setting. Use of the S1 TUNE position is a great aid in quickly performing this adjustment.

### Components

Note that switches S1 and S4 are relatively expensive if purchased new. You should be able to find all the switches, the variable capacitor and the 500-pF doorknob capacitors at a hamfest.

### 160-Meter Considerations

I do not have two 160-meter verticals and, therefore, I was not able to test the network on that band. The phase-shift network may be simply scaled by a factor of two from Table 3, the 3.5-MHz table. Note that a dual 0-2000 pF variable capacitor and an inductance of 4.72 µH are required.

The limiting factors are transformers T1 and T2. It is important that the inductive reactance associated with a winding be greater than the terminating resistance by a factor greater than five. Thus, the self-inductance must be greater than 22 µH (250 Ω at 1.8 MHz). This translates to 21 or more bifilar or trifilar windings on the Amidon R61-050-750 rods, or a total of over 60 conductor turns for T1. If this guideline is observed, the phasing arrangement should work on the top band. I would greatly appreciate any feedback on efforts to use this design on 160 meters.

### Notes and References


[1]B. Alexander, "Steerable Array for the Low Bands," *ARRL Antenna Compendium, Volume 2* (Newington: ARRL, 1989), pp 10-15.

[2]P. H. Anderson, "Impedance Matching Transformers and Ladder Line," *Ham Radio*, May, 1989.

[3]L. A. Moxon, *HF Antennas for All Locations* (Potters Bar, Herts: RSGB, 1982 and 1986).

[4]The author has a number of C language (Borland Turbo C) and spreadsheets (Borland Quattro Pro) which were used to analyze the pi network and to calculate the physical characteristics of the inductors. They are available on a 5¼-inch, 360k disk in MS/DOS format. Please include $2.00 to cover postage and handling. (The ARRL in no way warrants this offer.)

[5]Amidon Associates, PO Box 956, Torrance, CA 90508.

[6]Barker and Williamson, 10 Canal St, Bristol, PA 19007.

[7]Newark Electronics, Administrative Offices, 4801 North Ravenwood St, Chicago, IL 60101.

# The High Profile Beam— A High Performance Vertical Antenna with No Radials

**By Ed Suominen, NM7T**
**16529 28th Avenue West, Apt A**
**Lynnwood, WA 98037**

Vertical phased arrays are excellent DX performers for hams interested in specific areas of the world. Low-angle radiation comes naturally to a vertically polarized antenna over an expanse of very good ground or near sea water, even when ground-mounted. With proper phasing and spacing, excellent gain and a clean directive pattern can be achieved. Although the beam is fixed, it is bidirectional. Therefore, amateurs must choose an orientation that covers their areas of interest.

Such arrays are not without problems. Two quarter-wave verticals each require their own structural supports and effective ground systems. Hundreds of feet of radials are required in order to avoid ground losses, a problem made worse by the lowered radiation resistance of the phased elements. These radial systems and the typical ⅝-λ spacing take up a lot of backyard!

Enter the High Profile Beam, Fig 1. By joining the phased verticals into one antenna and feed point, we get all of the advantages of two elements with the simplicity of a single antenna.

### The High Profile Beam as an End-Fire Array

The usual method of feeding two verticals is in a broadside array, with equal phases and spaced about ½ λ apart. This provides gain and directivity because the wave fronts *add* when viewed from the side (they're in phase), but *cancel* from the end. The wave from one element has reversed by the time it gets to the other and very little radiation takes place from the end of the array.

The same gain and directive pattern can be obtained by feeding the elements with opposite phases and spacing them very close together (about ⅛ λ). The operation of the antenna in this configuration is a bit more complex.

In this case, the wave fronts cancel when viewed from the side because they are of opposite phases. When viewed from the end, however, the same delay effect occurs. The wave from one element has begun to reverse again when it reaches the other. The cancellation is not complete and some radiation takes place.

Although the antenna is experiencing cancellation, it still needs to get rid of the energy being fed to it. It responds by lowering its radiation resistance, which increases the current in the elements. Since current is what actually causes radiation,

**Verifying the Directive Pattern of a Fixed Beam**

Verifying the directive pattern of a rotary beam is easy; the signal strength from a fixed source is measured while the beam is rotated. With a fixed beam such as the High Profile Beam, signals from many sources must be compared to get an idea of its pattern. This presented a problem for me. My original wire vertical was too close to the beam and parasitically coupled to it. The result was poor, but equal, performance from both antennas. To see what the High Profile Beam could really do, I needed to remove the wire vertical and compare the beam to something else.

A second receiver with a small omnidirectional antenna was used to provide a comparison. It was placed on a table to the right of my station receiver and I sat between them. This created a stereo effect, with signals from the beam heard on my left side and signals from the small omni antenna audible to my right. The results were dramatic!

While listening to DX QSOs, the station in Europe or Western Asia was heard from the left side (the beam), with only a faint signal coming from the other receiver. When the stateside station responded from off the side of the beam, the signal was always heard from the right (the small omni antenna). This stereo effect was especially entertaining during pileups with the DX station on the beam. The hordes of stateside suitors were heard from the center and right, heightening the effect of their signals coming in from all different directions. When the DX station responded, however, the signal was always heard clearly coming from the left, plucked out of the crowd by the beam.

This technique has convinced me that the High Profile Beam has considerable gain, low-angle radiation, and a clean directive pattern.—*Ed Suominen, NM7T*

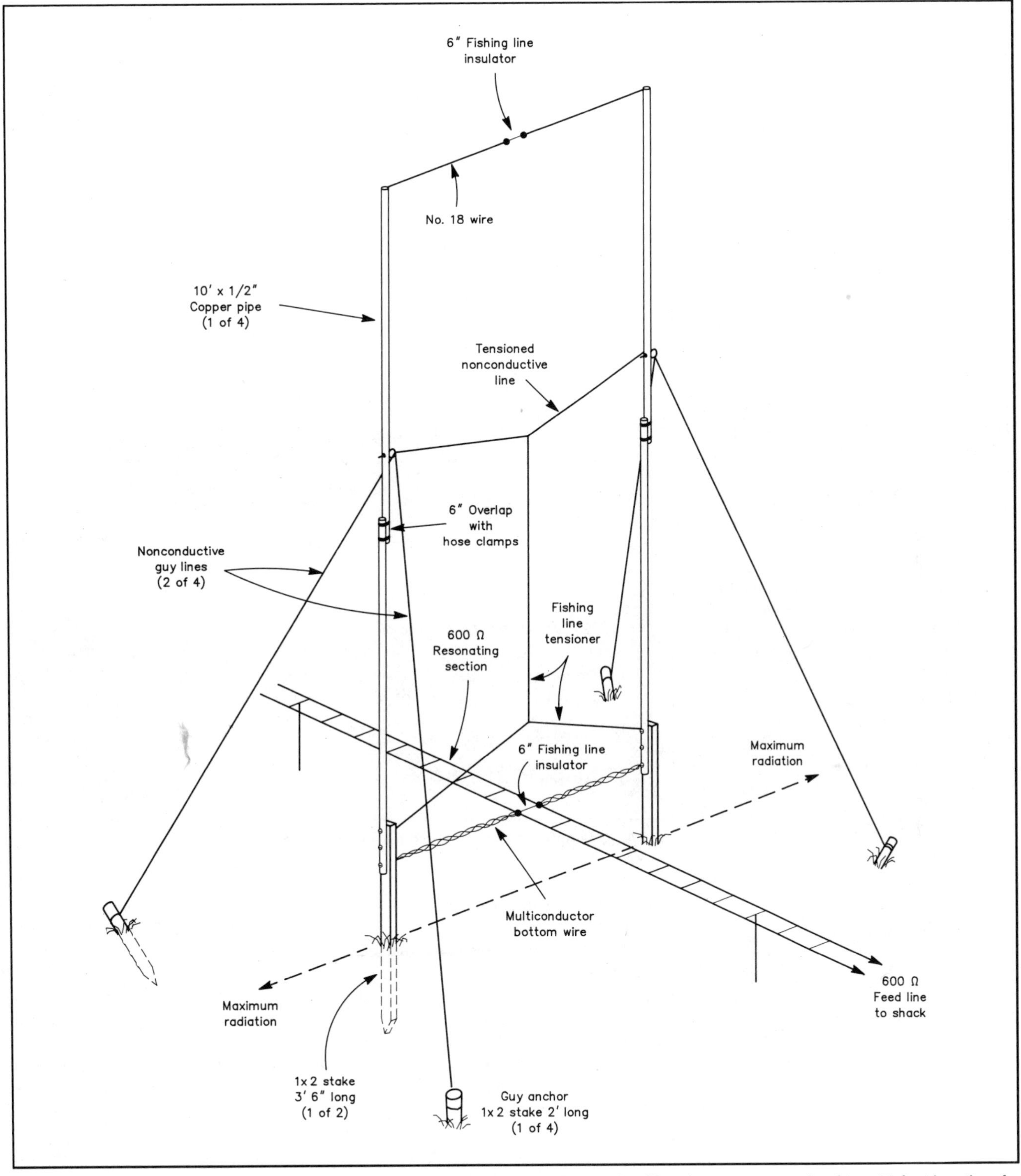


Fig 1—The High Profile Beam: a two element end-fire array with a single feed point. The elements are made of two 10-foot lengths of ½-inch copper pipe, supported by wood ground stakes and nonconductive guy lines. All descriptions given are for the 20-meter version built by the author, but the measurements can be scaled for other bands as well.

the partial cancellation is overcome. In fact, because of the complete cancellation from the side, radiation from the end is actually greater than from an individual element. This gives the array an end-fire pattern, and as much gain as a broadside array.[1]

The end-fire array is not often used with quarter-wave verticals because of its extremely low radiation resistance. (Ground losses become very significant.) If we can take ground out of the circuit entirely, however, we don't need to worry about its loss resistance. The only remaining loss is caused by the conductors, which can be very efficient if made of tubing.

How can we remove ground from a vertical antenna? Earth ground is only used as an infinite sink of electrons to allow current to flow and create a closed circuit. In a balanced antenna such as a dipole, a closed circuit is formed by opposite currents flowing in each leg. The High Profile Beam is a balanced antenna with the opposite currents in each vertical leg forming a 180° phased end-fire array.

### Two Half Waves Out of Phase

The balanced antenna that is folded to make this array could have been a dipole, but there are serious problems involved. Let's take a look at the antenna current curve.

$I = \sin(\ell)$

Radio waves are sine waves, and when they zip along a wire prior to launch they form a sine current curve. This means that the amount of current flowing in an antenna half-wave element (nearly all antennas are made of portions or groups of half-wave elements) is zero at the end and maximum in the center. With this simple formula we can tell a great deal about which parts of an antenna are really important for radiation.

By using a bit of calculus, we can measure the area under this antenna current curve and find out how much current there is in a certain part of an antenna. Since only current causes radiation, this will provide a general indication of how much work that part does.

Fig 2A shows us that a substantial amount of the total current is flowing in the bottom horizontal leg of the folded dipole (only a foot high). The radiation from this current is almost completely canceled by the lossy earth ground. Whatever radiation does occur will fill the nulls in the end-fire pattern.

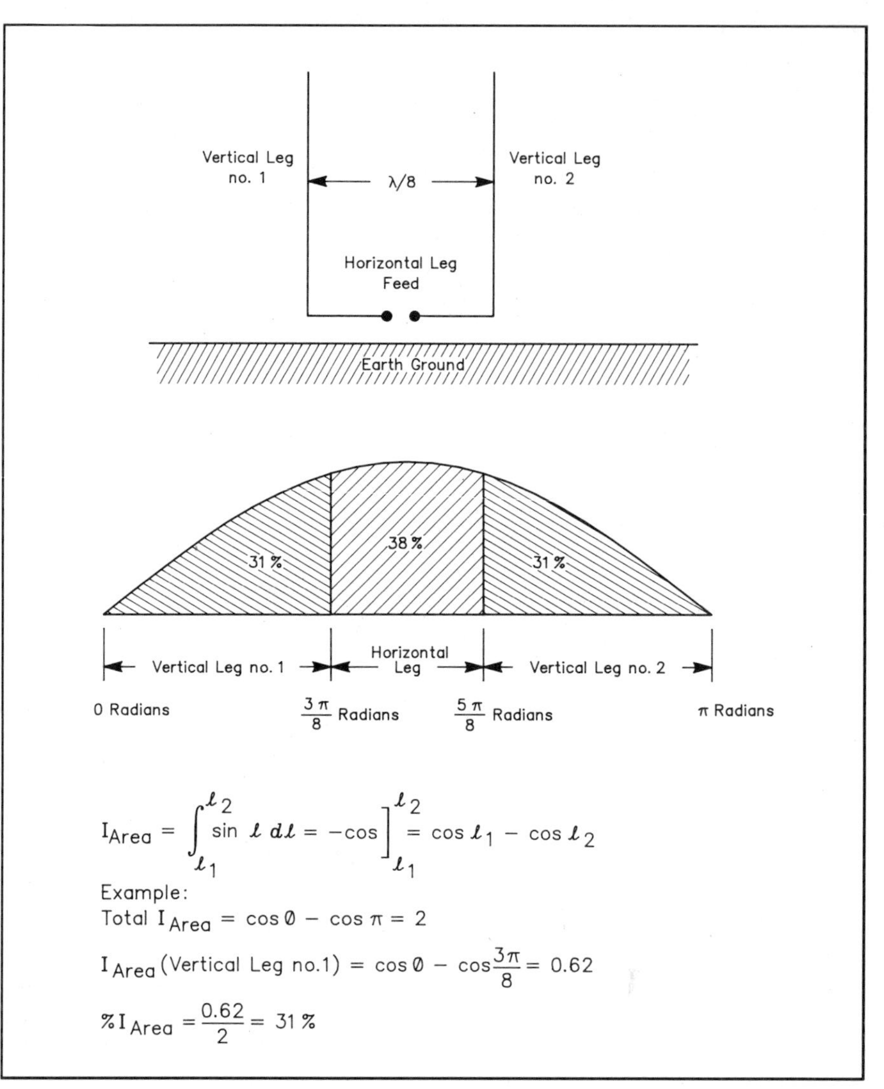


Fig 2A—The antenna current curve for a folded dipole.

Instead of using a single half wave, we'll use two, as shown in Fig 2B. This gives two current maxima nearly halfway up the vertical elements, where they belong. The amount of undesirable radiation in the bottom leg is now much less. Most of what is radiated is healthy, vertically polarized RF!

The use of two half-wave elements has another advantage. Since the current maxima are farther from ground, low radiation angles will be favored even more. A small amount of additional gain can be expected from this improved vertical pattern.[2]

### The Antenna as a Leaky Transmission Line Section

The High Profile Beam has just been described as a folded, balanced antenna working as an end-fire array. There is, however, an entirely different way of visualizing the antenna which will help to explain a few of its characteristics. Let's take a look at the antenna as a leaky transmission line section.

The most familiar transmission line section is the shorted quarter wave. An infinite impedance will appear at the feed point when the line is shorted a quarter wavelength away. This results from the applied wave reflecting from the short with opposite phase, and reversing phase again in the transit time to and from the short. This gives the reflected wave the same phase as the applied wave, and they oppose each other to give an infinite impedance. This principle works so well with low-loss lines that VHF insulators are sometimes made of metal in the form of quarter-wave sections.[3]

If the line is made a half-wavelength long, an *impedance repeater* is formed. An infinite impedance at one end will look like a short in the center (after traveling ¼ λ) and will repeat to an infinite impedance at the end. Here the wave travels 180° and reflects with no reversal from the open end, then travels another 180° on the return trip. Once again, the result is equal applied and reflected waves, and infinite impedance at the feed.

In both of these cases, a perfect transmission line with no loss was assumed. With loss, however, the reflected wave will not completely oppose the applied wave and the impedance will not be infinite. As the loss increases the reflected wave and feed impedance will decrease.

This can be illustrated with a half-wave section of open-wire line. Here the loss results mainly from radiation as the conductors are spaced farther apart. The wave fronts from the opposing conductors do not cancel completely and the line begins to act like an end-fire array. More power is used in the line and more current will be drawn at the feed point.

If this wide spacing is taken to extremes, the very high impedance at the feed point will be brought down to a level that can be matched and fed with real power.

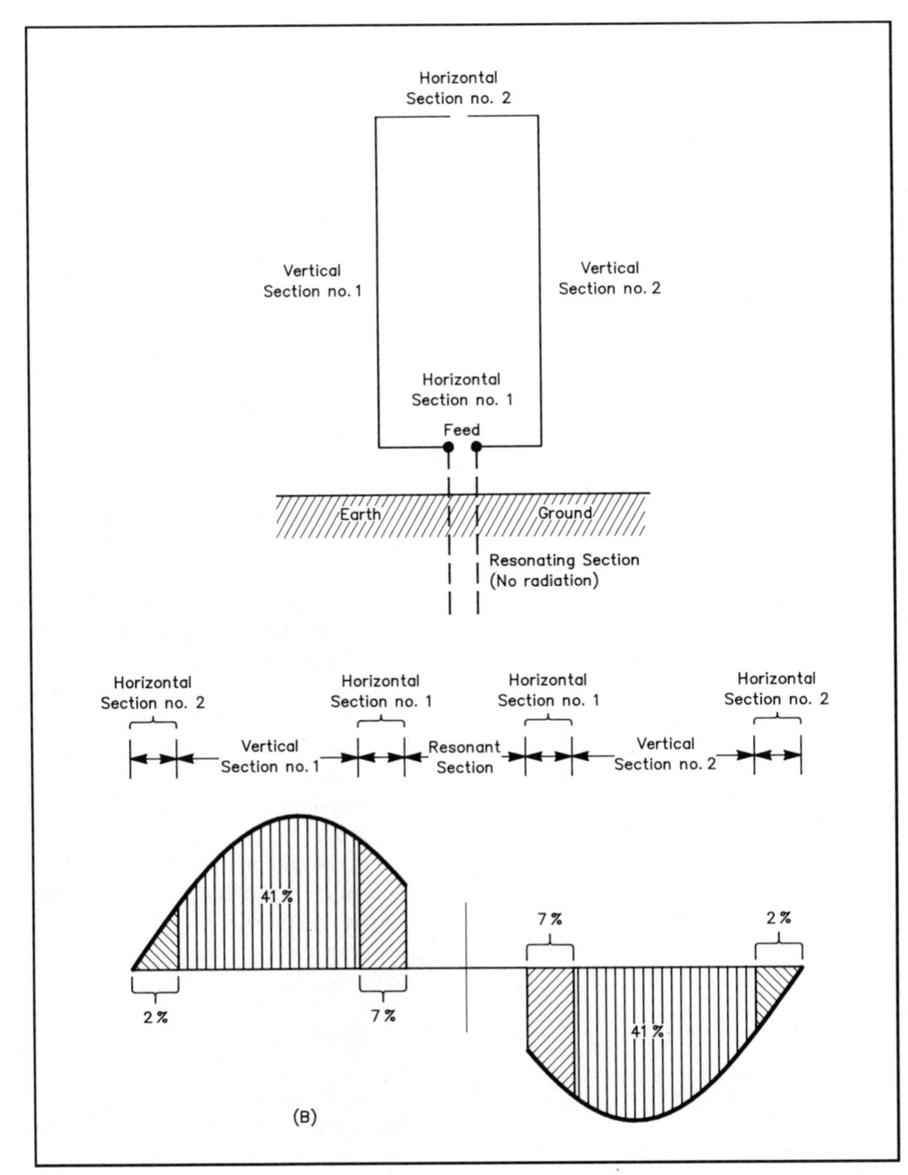


Fig 2B—The antenna current curve for the High Profile Beam. Percentages shown are general indicators of radiation from each antenna section and do not take into account coupling between the sections and ground.

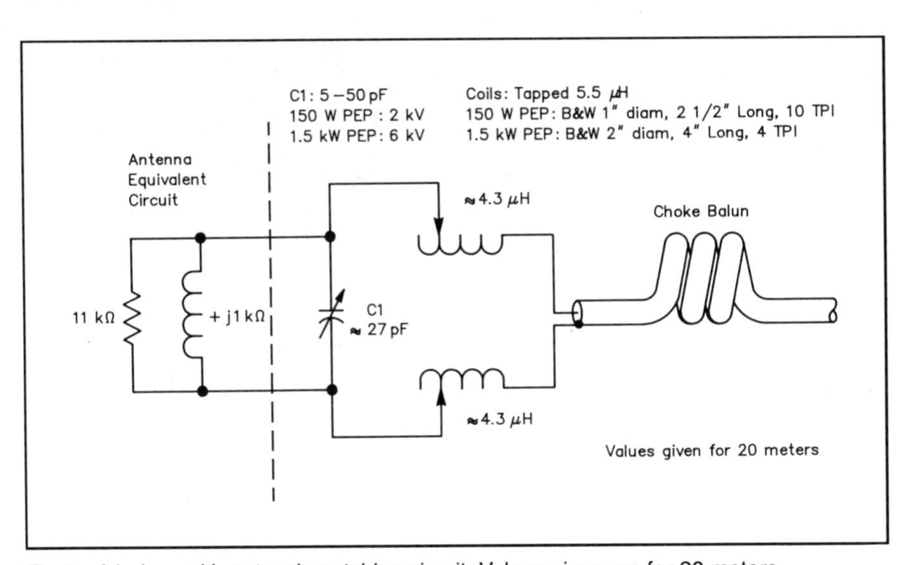


Fig 3—A balanced L-network matching circuit. Values given are for 20 meters.

All of that power will "leak" from the section as end-fire radiation. This leaky half-wave section, with conductors stretched apart and mostly bent vertically, becomes an efficient radiator—the High Profile Beam.

Although its feed impedance is not the theoretical infinity of an open half-wave section, the antenna still acts much like one and its impedance will remain quite high, around 11,000 Ω. The ⅛-λ conductor spacing is narrow enough to retain much of the wave applied to the feed and create a reflection that largely opposes it. Properly feeding the antenna requires matching the high impedance to overcome this reflected wave.

To put it another way, the wave fed to the antenna needs to reflect back and forth quite a few times before it all leaks out. This doesn't mean that the antenna is inefficient, but it does result in high impedance and another drawback: high Q.

Q is the ratio of reactance, or "tuning" part of a circuit, over the resistance, or "working" part. Higher Q gives narrower bandwidths, since there is more tuning for less work. With an antenna that acts like a tuned transmission line section, high Q is an unfortunate fact of life. The High Profile Beam has a Q of nearly 60, calculated from a measured 1.5:1 SWR bandwidth of 100 kHz on 20 meters.

### Feeding and Tuning

Since the antenna is not quite a full wavelength, it is resonated with a 3⁄32-λ open-ended section of 600-Ω ladder line attached to the feed. This completes the sine curve by allowing the voltage maxima to move out to the end of the section. In a matching network, this section may be replaced by additional shunt capacitance.

The best way to feed the antenna's somewhat awkward 11,000-Ω impedance is with 600-800 Ω ladder line to a good balanced tuner.[4] This allows coverage of the entire band and is the easiest system to tune.

The SWR on a 600-Ω feed line will be nearly 20:1, so its length should be a multiple of a half wavelength to preserve the nonreactive high impedance at the tuner terminals. With this high SWR, even ladder line will develop loss and should be limited to 1 or 2 half-wavelengths (34 feet or 68 feet at 14.1 MHz).

With this system, operation on other bands near the design frequency (such as 17 meters and 15 meters for a 20-meter antenna) should be possible, although the nonresonant length of the feed line and antenna will present a complex impedance

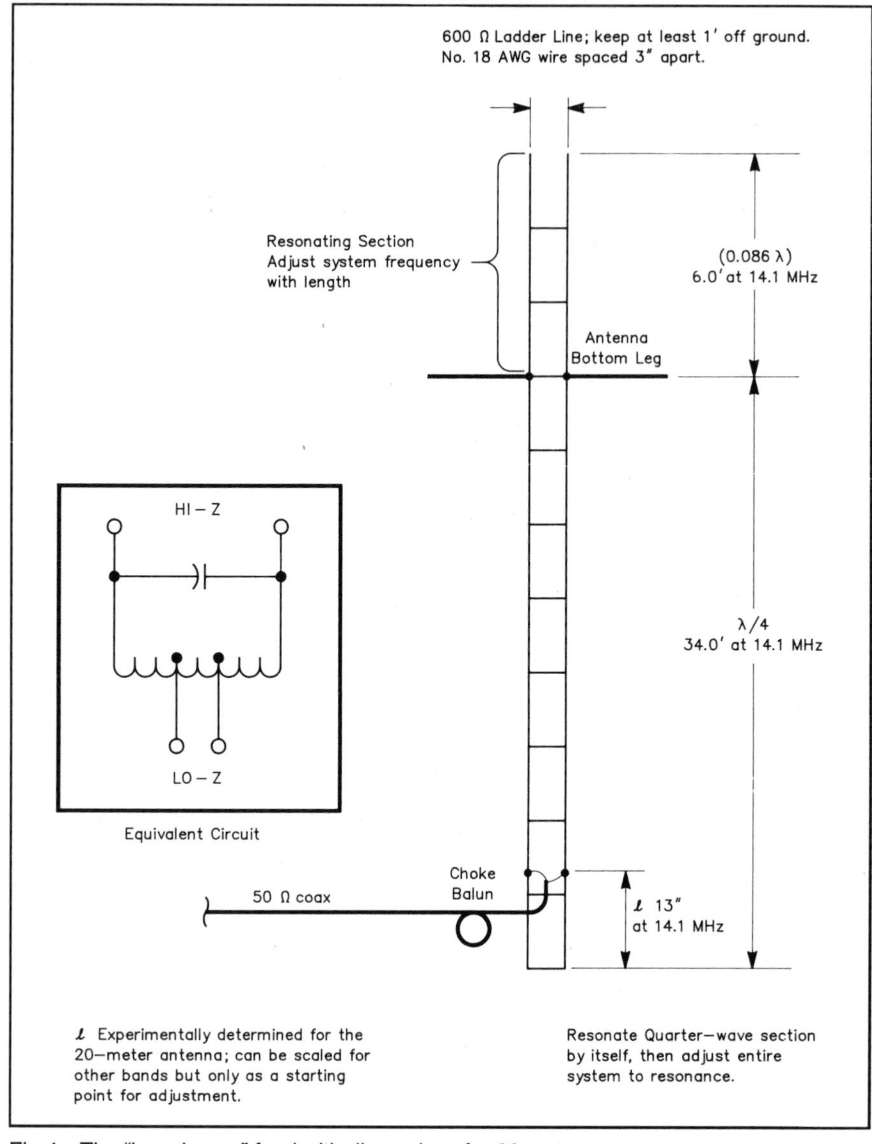


Fig 4—The "bare bones" feed with dimensions for 20 meters.

to the tuner. This may result in a more difficult, less efficient match than that obtained at the design frequency.

If the tuner to be used has a balun on its output (as is the case with most commercially available models), it will *not* efficiently match very high impedances.[5] A ¼-λ transformer section of 600-800 Ω ladder line will bring the impedance down very close to 52 Ω, enabling almost any tuner to do the job. The ¼-λ section should be connected between the antenna and the 52-Ω coaxial feed line to the tuner. Ten feet of the coax should be coiled up at the connection to the ladder line to form a choke balun. Although this feed will work only on a single band, the loss from SWR on the coaxial feed line will remain insignificant across that entire band.

With either of these two methods, the resonating section may be made part of the ladder line leading to the tuner, lengthening it by 3⁄32 λ. This would, for example, make the ¼-λ transformer 11⁄32 λ long.

If no tuner is available, an impedance match needs to be made at the antenna feed. Two methods were tested, and worked equally well.

The balanced L network shown in Fig 3 eliminates the need for ladder line. The resonating section is absorbed into the shunt capacitance. A servo motor on the capacitor would allow easier coverage of the entire band.

The "bare bones" feed method uses a shorted ¼-λ section of 600-Ω ladder line as a resonant transformer. The open end of the section is connected to the antennas, and the coaxial feed line (with a choke

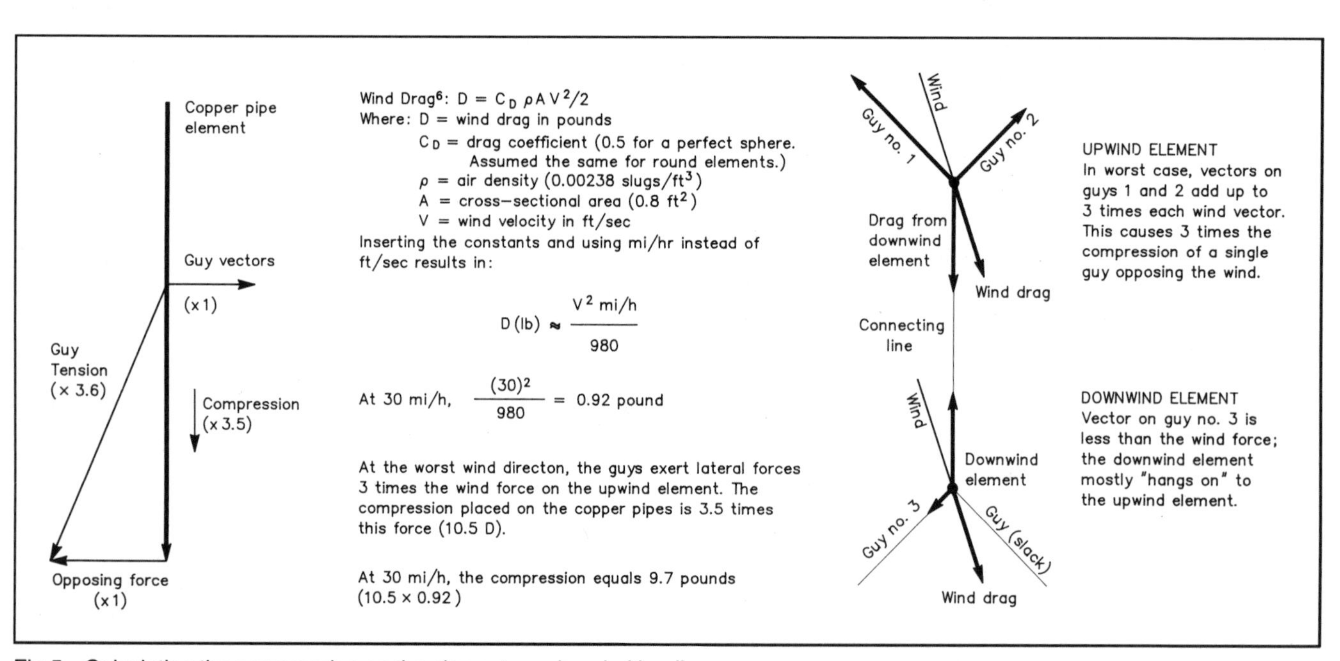


Fig 5—Calculating the compression on the elements under wind loading.

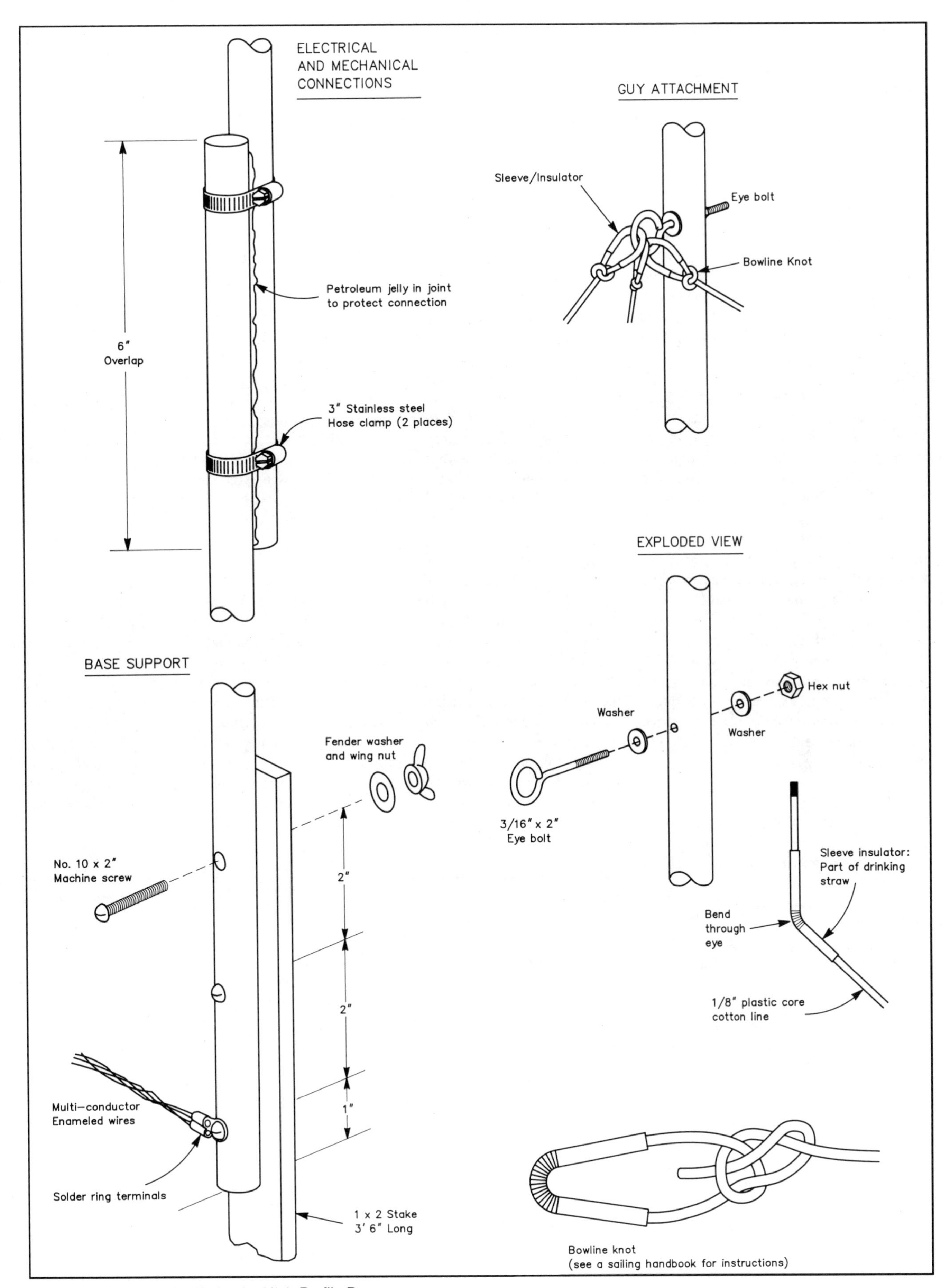


Fig 6—Construction details for the High Profile Beam.

balun) is tapped to a point near the shorted end for a match to 50 Ω. Although the resonating section could be made part of the ¼-λ section, it is preferable to lead it from the antenna feed point in the opposite direction to allow easier adjustment of resonance.

**Mechanical Design**

The vertical elements have two guy lines each and a connecting line between them. The use of a guyed support system eliminates the severe stresses which would otherwise be encountered under wind loading. Instead, a simple compressive force is placed on the elements. This allows the use of light copper pipe and simple 1 × 2 wood stakes for the ground mount.

Calculations of wind loading[6] and the vectors associated with the guy lines showed the compression on the upwind element (which bears most of the burden for both elements) to be about 10 lb in a 30-mi/h wind. This force increases with the square of the wind speed, to about 20 lb at 45 mi/h and nearly 40 lb at 60 mi/h. Although the maximum allowable compression was not known, 40 lb would certainly seem to be the most a thin copper pipe could handle before bending severely. This would put the maximum wind speed at 60 mi/h. Observation of the antenna in strong winds shows that this is a safe wind speed to use as an upper limit. During 30-mi/h gusts, hardly any bending or swinging was noticed in the elements, although a good deal of tension could be felt in the upwind guy lines.

**Construction Details**

This construction procedure was planned for simplicity. The materials needed for construction are listed in Table 1. All materials except the fishing line can be purchased at a hardware store and no special tools are required. The elements, ground mounting stakes and guy anchor stakes are identical and can be worked on together in assembly-line fashion.

The first step in construction is to drill all of the holes. Drill three 11⁄16-inch holes at distances of 1 inch, 3 inches, and 5 inches from one end of two pipes. These will become the bottom sections of the elements. Take the other two pipes (the top sections), and drill a single 3⁄16-inch hole exactly 2 feet from one end of each. (This will be the bottom end of the top section.) Fasten the eye bolt and its hardware here, as shown in Fig 6.

Drill a single hole in each guy anchor stake, 2 inches from the top. Fasten an eye bolt through each stake using a fender washer on the opposite side.

Only one hole is drilled in the ground mounting stakes at this time, since all three would probably not line up with their counterparts in the copper pipe. Drill the center hole 5 inches from the top of the stakes.

The guy anchor stakes should receive at least two coats of varnish. The ground mounts are also used as insulators and at least three coats are recommended to keep the wood absolutely dry. Marine spar varnish allows for some flexing of the wood and is probably the best for the job.

The elements are formed from two 10-foot lengths of ½-inch copper pipe, jointed at a 6-inch overlap with a pair of hose clamps. The pipe surfaces which will make electrical contact in the overlap should be finely sanded and cleaned just before clamping them together. Petroleum jelly should then be applied to the area around the connection, followed by a wrapping of duct tape. (This treatment should also be given to all fasteners used on the antenna to prevent corrosion.) Before fully tightening the hose clamps, sight down the end of the pipes to ensure that they are straight with respect to each other. Also make sure that the eye bolts and mounting bolts will point in the same direction. Five lengths of nonconductive line and two lengths of wire attach to the elements. The four guy lines can be cut with little regard for accuracy, but the other measurements are more critical.

The top section of 18-gauge wire needs to be just a bit longer than the distance between the elements to keep it from bending their tops inward. Make a 6-inch fishing line insulator, then cut two wires and attach them to the insulator. The total length of the top section must be exactly 9 feet, with both wires of equal length. Attach the ends of the wire to the top ends of the elements by bending an inch of wire under a hose clamp. With all electrical connections, take the same precautions of sanding and cleaning the contact surfaces, and covering them with petroleum jelly and duct tape.

The bottom wire section is best measured at this point but should not be attached until the antenna is raised. It consists of two (or more) strands of insulated wire twisted together to form a conductor of lower RF resistance than a single larger wire. (There is considerable RF current flowing in the bottom section, and skin effect limits the conductivity of a single wire.) Measure 25 feet of the wire and double it back on itself. Tie the loose ends down, and insert the bend at the other end of the doubled wire into the chuck of a drill. Now use the drill to twist the pair until it looks stable. Cut the twisted pair into two lengths of 8 feet, 2 inches. This allows 4 inches for an adjustable fishing-line insulator which will be used to apply tension to the bottom section. The ends of the section will be attached to the bottom through-bolts on the elements with ring terminals. This will be a high- current electrical connection, so prepare it accordingly.

The mechanical connection between the elements is made with a length of the same nonconductive line used in the guys. Accurate measurement is important here, as this line maintains tension on the entire guying system. The line should measure exactly 8 feet, 9 inches from the eye bolt when the ends are tied. This allows an inch

**Table 1**
**List of Materials for the High-profile Beam**

Four lengths of ½-inch × 10-foot copper pipe
Two 1 × 2 stakes, 3 ½ feet long (suitable for pounding)
Four 1 × 2 stakes, 2 feet long (suitable for pounding)
70 feet of light nonconductive line (⅛ inch plastic-core cotton line is recommended)
35 feet of fishing line
30 feet of insulated copper wire (#18 AWG enamel is recommended)
Two 3⁄16-inch × 2-inch eye bolts with washers and hex nuts
Six #10 machine screws with fender washers and wing nuts
Four 3⁄16 inch × 1½-inch eye bolts with fender washers and hex nuts (for guy anchors)
Four 2-inch stainless-steel hose clamps
Two ⅝-inch stainless-steel hose clamps
Two #10 solder ring terminals
Six flexible drinking straws
Plenty of petroleum jelly and duct tape
Marine spar varnish (½ pint can)
Materials for whatever feed system is to be used

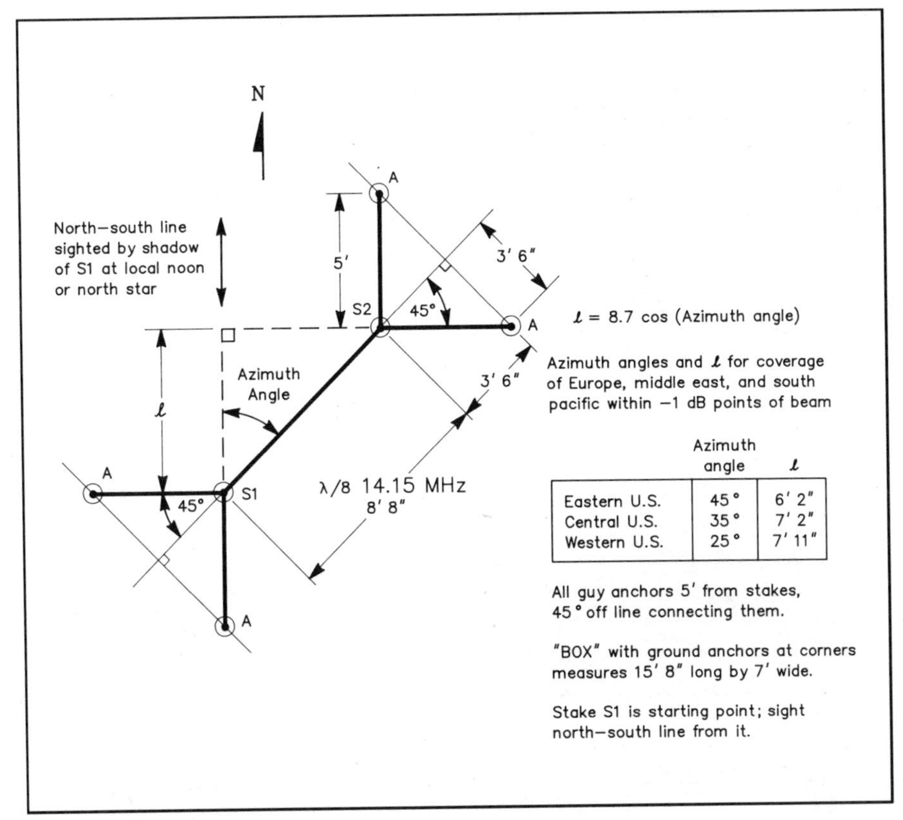


| | Azimuth angle | $\ell$ |
|---|---|---|
| Eastern U.S. | 45° | 6' 2" |
| Central U.S. | 35° | 7' 2" |
| Western U.S. | 25° | 7' 11" |

Fig 7—Staking out the antenna.

of slack, which is taken up by the fishing-line tensioner shown in Fig 1. The four guy lines should be cut to lengths of 15 feet and tied to the eye bolts.

To reduce the effects of weather, the connecting line and guys should have some insulation from the elements. Fig 6 illustrates the use of a piece of flexible drinking straw on the loop of a bowline knot.

Now that the elements are ready, the supporting stakes should be driven into the ground in the layout shown in Fig 7. The southwest element mounting stake is the starting point; the other is located 8 feet, 8 inches from it, in the direction of maximum radiation. Locate the guy anchors by finding a reference point 3 feet, 6 inches from each mounting stake on the line connecting them, and running a perpendicular line from that point. Drive in the guy anchors on this line, 3 feet, 6 inches from the reference point.

Before raising the elements, loosely tie down the guy lines. The elements must be walked up at the same time and this will require two sets of hands. Once the elements are vertical, they should be fastened to the mounting stakes by slipping a bolt through the center hole on the bottom of each. Make sure that the eye bolts point toward each other and lead the guys away from either side. Then both partners should walk away from the elements, holding the guy lines to keep them steady. Tighten the guys one at a time until the elements are vertical. The whole system is tensioned by tightening the fishing line on the connecting line.

Complete the job by drilling the top and bottom holes in the mounting stakes and fastening the rest of the through-bolts. Then attach the bottom wire section and whatever feed system is to be used.

### On-the-Air Performance

At my QTH, the antenna performs as an end-fire array should, with a noticeable improvement of the European signals it's pointed at. The signal reports are consistent with the relative power of the stations giving them, indicating that the antenna is radiating most of the power fed to it. The high measured Q is also a good sign of radiation efficiency.

The real indicator of this antenna's performance as a beam is on the *received* signals, where a dramatic improvement is noticed. Signals barely audible on an omnidirectional antenna are clearly copied on the High Profile Beam. Strong QRM off the side of the beam is cut down by a few S units, allowing the very enjoyable experience of hearing European and stateside stations with comparable signal strengths much of the time.

### Acknowledgments

Special thanks go to Stan Oehmen, K7BZ, who introduced a curious teenager to Amateur Radio and gave years of patient instruction on the mysteries of RF and antennas. Without his service as an "Elmer," this article would not have been possible. I would also like to thank Neil Hill, K7NH and Mark Smythe, WA7URJ, for loaning their equipment for the project.

### Notes

[1] G. L. Hall, *The ARRL Antenna Book*, 14th ed. (Newington, CT: ARRL, 1987), p 6-5.

[2] *The ARRL Antenna Book*, 14th ed., p 2-23.

[3] *The ARRL Antenna Book*, 14th ed., p 15-25. Shows an example of this in the standard-gain antenna.

[4] For an excellent article describing what makes a "good balanced tuner," and why most aren't, see R. L. Measures, "A *Balanced* Balanced Antenna Tuner," *QST*, February 1990, pp 28-32.

[5] See Note 4.

[6] D. Halliday and R. Resnick, *Fundamentals of Physics*, 3rd ed. (New York, NY: John Wiley and Sons, 1988), p 109.

# A Miniature Ground-Independent Dipole for 40 through 10 Meters

**By Jack Kuecken, KE2QJ**
**2 Round Trail**
**Pittsford, NY 14534**

You may not have the space or the privilege of erecting a full sized antenna. Or, it may be impossible to get a good RF ground at your location. A second- or third-story apartment, fiberglass boat, camper or recreational vehicle, are typical examples. Some hams use mobile whip antennas in such locations. This solution often makes equipment cabinets "hot" with RF, causing annoying RF tingles from the mike, and RFI problems. My miniature dipole solves these problems.

One way to avoid RF in the shack is to use a balanced antenna, like an ordinary dipole. If the dipole is fed with coax and a balun is used at the feed point, the feed line will not be "hot." However, the impedance of a dipole considerably smaller than a half-wave is so high that it is difficult to construct an adequate balun.

Note that, as the antenna becomes shorter, its radiation resistance falls but its capacitive reactance rises. To make the balun reactance five times that of the antenna, an antenna with $-j1250\ \Omega$ of reactance requires a balun reactance of $+j6250\ \Omega$. It is almost impossible to obtain a reactance greater than $+j2500\ \Omega$ from any inductor because the large stray capacitance in the winding resonates with the inductance.[1] If the balun is resonated to obtain the required reactance, its resonant bandwidth is too narrow to be practical.

The impedance of electrically small antennas is difficult to match. A 0.1-$\lambda$ dipole has an impedance of about $12 - j1250\ \Omega$. When used with 50-$\Omega$ feed line, the resulting SWR would be 2609:1! Even if a Transmatch is used in the shack, even a slightly lossy feed line would excessively

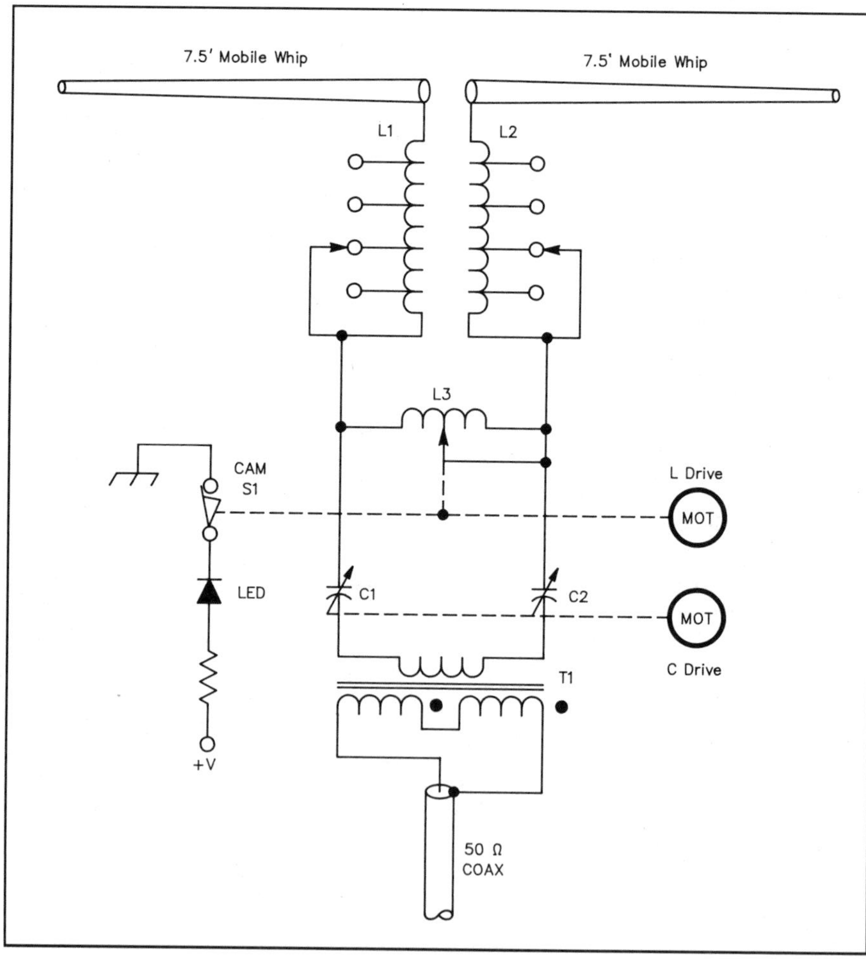


Fig 1—Schematic diagram of the remotely tuned miniature dipole.

C1, C2—30- to 290-pF variable, 0.050-in. spacing, ball-bearing shafts.
L1, L2—28 turns, no. 14 enameled, on 1.5-inch diameter white PVC pipe (10 turns/in.). See text for tap information.
L3—Roller inductor, 16 turns, 10 µH (E. F. Johnson 229-201).
S1—Lever-actuated switch (turns counting).
T1—15 trifilar turns, no. 18 Teflon-insulated hookup wire, on three FT-114-61 cores (µ=125).

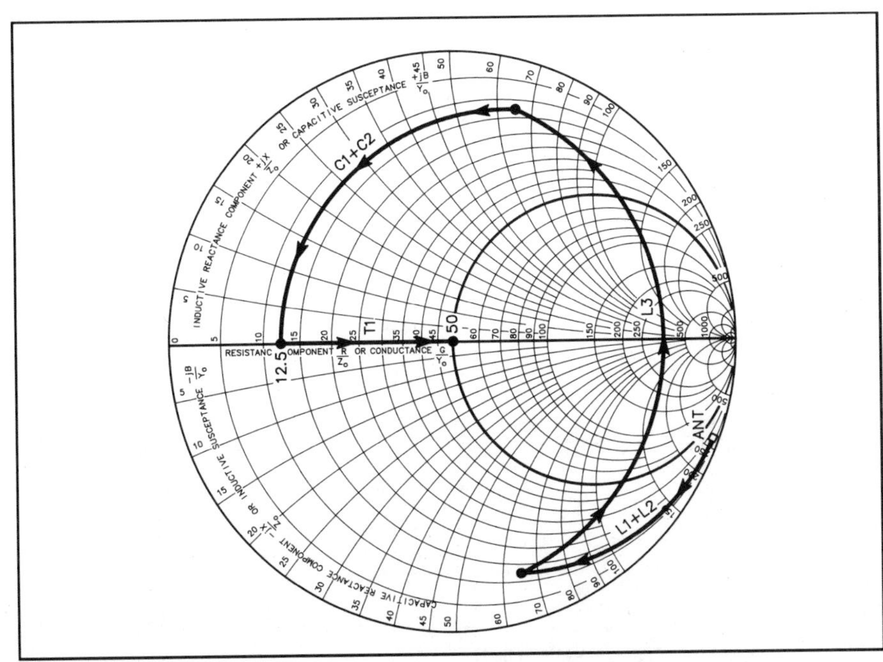


Fig 2—Smith Chart depiction of impedance matching to the miniature dipole.

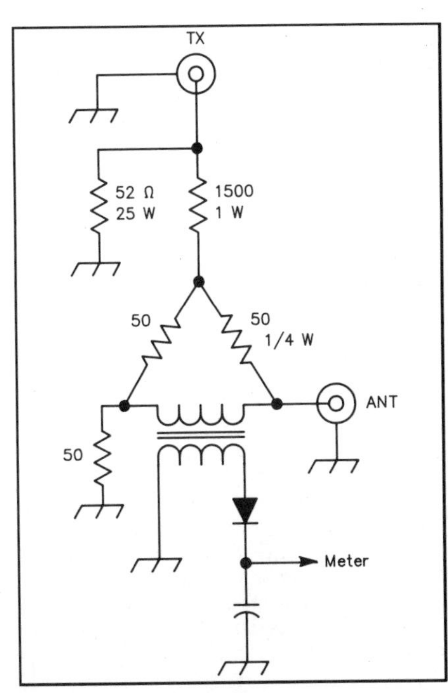


Fig 3—Fixed Wheatstone bridge used for adjusting the antenna at reduced power levels.

reduce radiated power. To get any efficiency at all from such an antenna, you have to match the impedance at the antenna itself.

For example, if you insert a coil of *j*1250 Ω reactance at the antenna, the SWR will be reduced to about 4:1. While this value is still too high for coax line, it is considerably better than the previous example. Of course, with a real inductor, you expect a finite Q and some losses. Suppose we can achieve a Q of 208 in the coil. Now the loss resistance of the coil is 6 Ω and the overall system is 66% efficient, assuming no other losses.

Some mobile antennas and many 160-meter antennas are end loaded, giving a substantial lower section with high current. Such end loading is more efficient than base loading if the coils are of equal Q.[2] However, the presence of a large loading coil near the end of the radiator poses certain physical problems, particularly if the antenna is to be tunable over a wide range of frequencies. In some military applications the antenna must be tunable from 2 to 30 MHz. These applications use a simple whip, tuned with an antenna coupler. Naturally, the tuner must be remotely adjusted or automatically adjust itself.

### A Practical Mini Dipole

The basic diagram for the miniature dipole is shown in Fig 1. The radiating elements are two 7.5-ft stainless-steel whips. You can use a fiberglass whip of similar size, but don't use a helically loaded CB whip. Note that coupler symmetry is maintained right to the output transformer.

L1 and L2 represent two base-loading coils. They are not coupled magnetically. The inductances are

| | |
|---|---|
| Full | 14.4 µH |
| Tap 1 | 12.4 |
| Tap 2 | 8.8 |
| Tap 3 | 4.9 |
| Tap 4 | 1.7 |

C1 and C2 are 30 to 290 pF transmitting capacitors with about 0.050-in. spacing. The capacitors should be capable of continuous rotation (more than 360°). The capacitors should be identical and should be locked together at the same capacitance setting.

The capacitors I used are from the BC-458A "Command Set" transmitter (C67). They were used as a fixed padder in the final amplifier tank circuit. Their shafts have ball bearings, although the shafts were locked in place at the factory. L3 could actually be the antenna loading coil from the same set (L5). I had already used L5 in a mobile antenna coupler. T1 is a 12.5- to 50-Ω transformer.

L1 and L2 take some of the curse off the antenna at the low end of the coupler band. They greatly reduce the maximum value of L3 required at the bottom of the band. Shunt coil L3 transforms the impedance to 12.5 + *j*XXX Ω and the series capacitors C1 and C2 cancel the *j*XXX term (ie, they restore the power factor). T1 transforms the impedance up to 50 + *j*0. This matching path is shown schematically on the Smith Chart (Fig 2).

I put taps on L1 and L2 so I could remotely switch them for band changing. In practice however, I found the antenna can be tuned from about 6 through 30 MHz with tap 2 set at 8.8 µH. L3 is a roller inductor. Fig 3 is the schematic of a Wheatstone bridge I use to tune the dipole at low power levels.

### Motor Drives

I used small, permanent-magnet dc motors with a speed-reducing gear head in this and other remote-tuning couplers. Output shaft speed should be about 20 r/min, and the motor should run equally well in both directions. You reverse drive direction by reversing the power supply polarity. New motors from Globe or Pittman in a full MIL-SPEC temperature range are quite expensive. Surplus dealers however, may have suitable motors for a few dollars apiece. Remember, the motors must run outdoors in whatever weather your area experiences. Automotive-type motors used for electric windows and antennas will usually operate over a wide temperature range.

Fig 4 shows how I drove the roller inductor (L3). S1 and S2 are microswitches installed at opposite ends of L3. When the roller wheel opens one of the switches, the motor stops. D1 and D2 permit reverse-polarity current to flow around the open switch so the coil can turn in the opposite

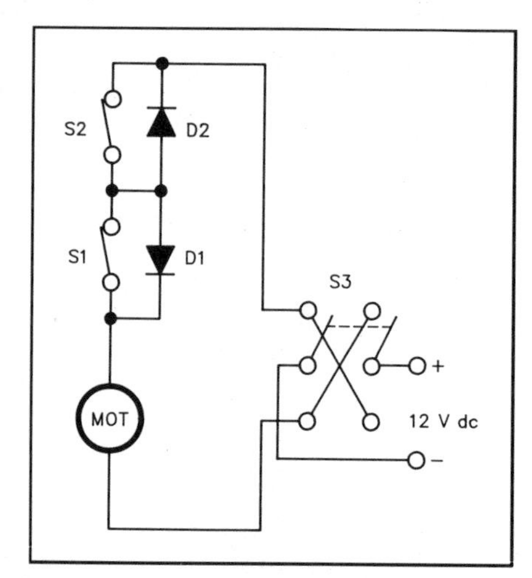

Fig 4—Schematic diagram of the roller-inductor drive-motor circuit. S1 and S2 are limit switches. Select D1 and D2 for the current required by the motor. S3 is a DPDT toggle. The motor is a 12-V dc reversible gear motor.

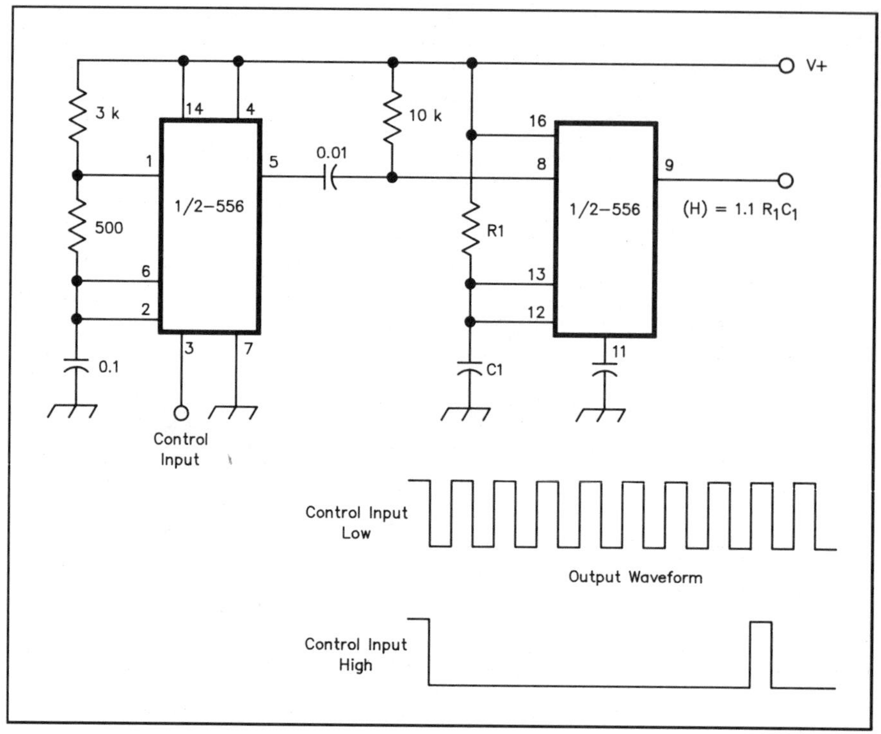

Fig 5—Schematic diagram of the simple pulse-width motor-speed-control circuit. Select R1 and C1 to suit the motor used.

direction. Both ends of L3 are above RF ground, so the coil shaft must be insulated from the motor and the switches must be insulated from the roller.

For the capacitor drive, a motor speed adequate for the searching is too fast for fine tuning. The circuit of Fig 5 slows the motor so that it can be "tweaked" into position. I don't recommend a series resistor, as it doesn't allow sufficient starting or running torque, especially in cold weather. I have had much more success and satisfaction using this full-voltage pulse-width-modulated circuit. The first half of the 556 sets the pulse rate, and the second half sets the pulse width. Starting torque is superior with this circuit, since the full voltage is available to overcome the voltage drop in the motor brushes.

If the capacitors you use have a shaft on one end only (like the ones I used), join them with insulated couplings and use a drive gear on a shaft between them. The motor will then need a drive gear to mesh with the gear on the capacitors. Note that the capacitors have RF voltage on both their rotors and stators, although the voltage is much lower at the 12.5-Ω end than at the L3 end. When setting up the drive, make sure both capacitors are set to the same capacitance, to maintain circuit balance.

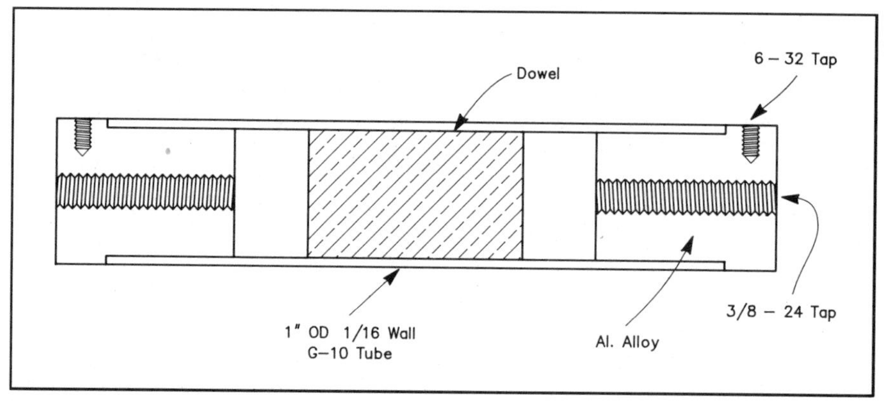

Fig 6—The dipole center insulator.

### Construction of the Dipole

Fig 6 shows the dipole center insulator, which I made from a piece of thin-wall G-10 tubing. I turned down a maple dowel to fit and epoxied it in place, to prevent crushing the tubing. Then I turned down a pair of aluminum plugs threaded ⅜-24 to fit the ends, and epoxied them in place. The ⅜-24 thread is used on many mobile whips. Two pairs of maple trunnions clamp the insulator to the top of the case.

A pair of G-10 feedthroughs are glued in the back of the case, with a downward slope. The antenna connecting wires are routed through them from the top of the coils, and up to the aluminum plugs into which the radiators screw. A drip loop is arranged in each wire.

The case itself is half-inch weatherproof plywood, with a door opening on the broad face. Build a doorjamb inside the box, on the sides and top. The door overlaps the bottom so that any water inside the box runs out alongside the door. While the seal isn't perfect, I have never found moisture inside the box. I gave the entire box a few coats of white latex house paint.

The coupler parts are mounted on a large Plexiglas plate, ¼-in. thick. The plate is secured in the box with four screws. For servicing, the entire works can be removed.

The horizontal whips droop about 5 in. at the tips. The droop has little effect on performance, and has the advantage that rainwater runs off the tips, rather than to the center and across the insulator.

RF enters through a close-fit hole around the PL-259. The control cable fits through a notch in the doorjamb. Make sure no significant holes are open to the outside; otherwise you'll find wasps the next time you open the box!

The box has a NURAIL fitting on the back to clamp it to a 2-in. aluminum mast, which is hinged at the bottom. The mast is

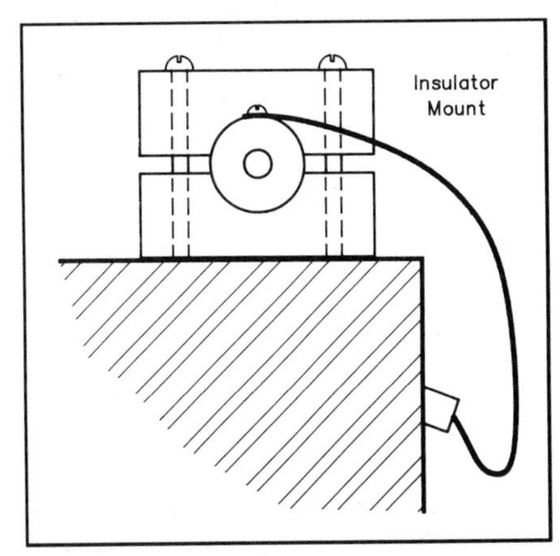


Fig 7—Mounting the center insulator to the box containing the tuning capacitors and inductors.

about 20 feet long. With the hinge and the box extension, the dipoles wind up about 24 ft above ground.

### Tuning Up

The cam on the inductor shaft actuates a turns counter. The LED flashes once per revolution. The tune-up procedure follows.

1) Run L3 to the low-inductance end stop.

2) Turn on the capacitor drive motor so that C1 and C2 continuously rotate.

3) Jog L3 out, one turn at a time, until you see the SWR start to drop. There will be two dips per capacitor revolution.

4) Stop the capacitor motor from searching and jog the capacitors for minimum SWR.

5) Fine tune L3 and the capacitor until the SWR is 1:1.

From this description you can see your transmitter will be exposed to a very high SWR during tuneup. For this reason, I use the circuit of Fig 3 for tune-up. The fixed Wheatstone bridge detects a matched condition on the antenna. The detector output is proportional to the voltage-reflection coefficient, and goes to zero at $50 + j0\ \Omega$. The padder isolates the transmitter and attenuates the signal to the antenna by about 30 dB. This action protects your transmitter, and the relatively quiet tuning will be appreciated by other hams.

I constructed my dummy load from 2-watt composition resistors in series-parallel. Once the antenna is tuned, the pad is bypassed by a switch not shown in the diagram.

### Results

I have used the antenna for some time on 40 through 10 meters. I use it most often on 40 and 20. I have pumped as much as 1.5 kW PEP into it with no signs of arcing.

Compared with a 36-foot vertical with nine buried radials, the dipole is much "less hot." When I installed the vertical, I had to put ferrite cores on the control cables and the feed coax, to keep RF off the microphone and other equipment. No such treatment was required with the dipole.

The miniature dipole should have a slight disadvantage in radiation efficiency, compared to the vertical. On 40 meters, efficiency probably does not exceed 66% or –1.8 dB. Nevertheless, in several months of operation I found that either antenna can be better on a given path at a given time. I do find the dipole is a considerably quieter receiving antenna than the vertical, however (typically 5 to 6 dB), particularly in the presence of locally generated noise.

All in all, I am quite pleased with the performance of this antenna. I recommend it to anyone with space or grounding limitations.

### Notes

[1]J. A. Kuecken, *Antennas and Transmission Lines* (Indianapolis: Howard W. Sams), Ch 24, "Reactive Elements and Impedance Limits."

[2]G. L. Hall, Ed., *The ARRL Antenna Book* (Newington: ARRL, 1982).

# An 80/40/17-Meter Super-Trap Dipole

By Albert C. Buxton, W8NX
2225 Woodpark Rd
Akron, OH 44333

Many hams now realize what a great band 17 meters is for worldwide DX. But getting on the air with an antenna better than a simple dipole without the cost and complexity of a rotary type is difficult. One of the better low-cost options which can be home-brewed is a third-harmonic dipole. It has four major radiation lobes providing a bow-tie radiation pattern and a pair of minor lobes. It has better low-angle radiation than a dipole and a small gain. It also has the low receiving noise level of horizontal polarization. For hams in the US this antenna, when installed running east/west, gives good coverage of Europe, Russia, Africa, South Pacific, Japan and South America with its major lobes. Its minor lobes pick up Asia, India and the polar regions. The length of such an antenna is about 80 feet, about 25 or 30 feet shorter than a typical 80/40 trap dipole. Would it be possible to build an 80/40 trap dipole with very high L/C ratio traps so it could be shortened enough to also work in a third-harmonic mode on 17 meters? This article shows one way to do that.

Fig 1 shows the 80/40/17 trap-dipole configuration. It is recommended that the antenna be fed with a 1:1 current balun and 75-Ω feed line. The radiating elements are no. 12 or no. 14 stranded copper wire. The antenna resonates very closely to the design frequencies of 3.75, 7.15 and 18.1 MHz. However, the antenna has about 125 Ω input impedance on 17 meters and the SWR may shift the apparent resonant frequency at the feed point, depending on feed-line length. SWR is no problem at the design frequencies on 40 and 80 meters, but the bandwidths are somewhat limited as for all trap dipoles. You may wish to use an antenna tuner to extend the bandwidths, especially on 80 meters. Note the overall length is about 95 feet, some 15 feet longer than a true 3⁄2-λ 17-meter antenna. This is because the traps behave like capacitors on 17 meters and their reactance must be offset by additional length of the dipole. However, the radiation pattern is not significantly altered by the added length, except for a beneficial small increase in gain and a slight outward shift in the direction of the major lobes. No new lobes are added.

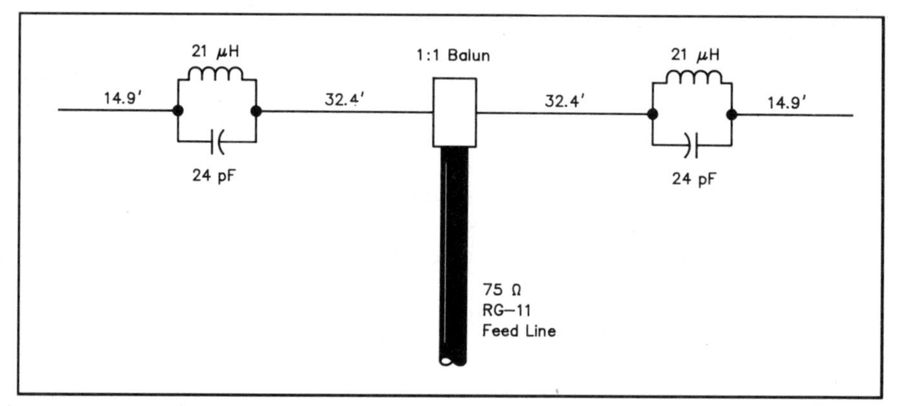


Fig 1—The trap dipole for 80, 40 and 17-meter coverage. See text and Figs 2 and 3 for Super-Trap construction.

### Super-Traps

Fig 2 shows the trap before wrapping it in tape for weatherproofing. The trap consists of 7.0 turns of RG-58 coax cable (Belden 8240) and 5.1 overlay turns of the inner core of RG-58. The "inner core" is simply RG-58 with the rubber sheath and shield braid stripped away. The windings are on a PVC form, 2¼ inches outside diameter by 2⅞ inches long, available at plumbing suppliers. Note the black electrical tape to hold the windings in place before adding the weatherproofing cover of tape.

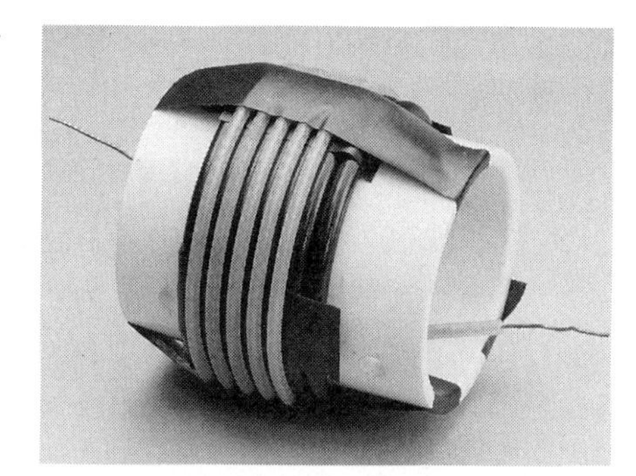

Fig 2—The Super-Trap assembly before wrapping it in tape for weatherproofing. All connections are made inside the form.

Fig 3 shows details of the trap interior with the feedback connections of the windings. Crimp-style connectors are used to eliminate the tough job of soldering in the interior of the trap. The input terminal is the center conductor of the coax on the left, marked A, and the output terminal is the end of the overlay winding on the right, B. Dip-meter tests indicate a trap resonant frequency of 7.15 MHz as it sets on the lab table free from adjacent equipment. The

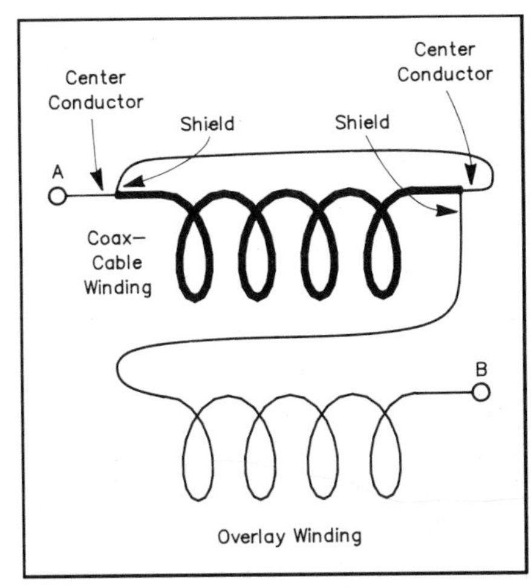


Fig 3—Details of the Super-Trap interior with the feedback connections of the windings. The end marked A is connected to the inner portion of the dipole.

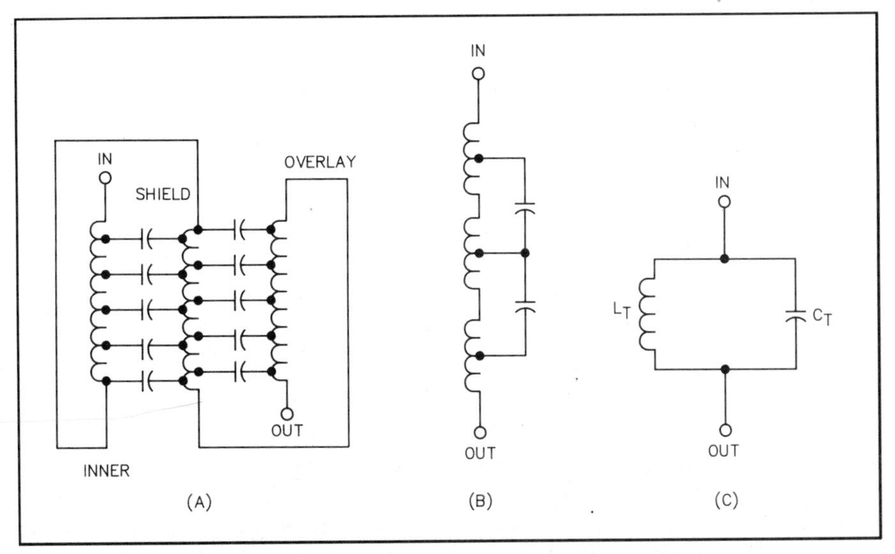


Fig 4—Development of the trap equivalent circuit.

pigtails of the trap lower its frequency by about 35 kHz, but their contribution is subsumed into the lengths of the radiating elements.

Studies on the Smith Chart indicated that a pair of 40-meter, high L/C traps were required with about 21 µH inductance and 24 pF capacitance. The conventional approach of using traps at both 17 and 40 meters to make an 80/40/17 trap dipole would destroy the third-harmonic radiation pattern desired on 17 meters. The aforementioned values of L and C correspond to a trap characteristic reactance of about 930 Ω, much higher than the usual form of coax-cable trap can supply. A design gimmick was needed. Of course a fixed capacitor and a single-layer air-core coil will do the trick, and some may opt to do just that. But I wanted the more weatherproof capability of coax-cable traps. So an attempt was made to modify the usual coax-trap design to increase its inductance and reduce its capacitance. Various gimmicks were tried before success was achieved. As sometimes happens, serendipity happily occurred.

The gimmick consisted of winding an additional layer of turns of coaxial-cable core around the coax trap in the same direction as the coax-cable turns, and connecting it in series with the turns of the coax portion of the trap. The core turns were laid in the grooves created by the coax winding. The number of turns on the added winding is limited to about 1 less than the number of turns on the coax if one sticks to the grooves formed by the coax. Additional inductance was provided alright, but significant capacitance between the new overlay winding and the outer shield winding of the coax was also provided. The capacitance between the RG-58 shield and the overlay coax core is 15 pF/ft, as measured on an accurate impedance bridge. This compares with the normal coax-cable capacitance of 28.5 pF/ft for RG-58 and 21 pF/ft for RG-59.

Use of ordinary insulated hookup wire for the overlay winding is not recommended. The region between the two layers of windings is under high dielectric stress, the same as the interior of coax cable, and requires a good low-loss dielectric for low trap losses. The best practical low-loss overlay winding therefore is made from the core of a length of coax cable. But where a different overlay wire is to be used with unknown capacitance to the shield of the coax, the unknown capacitance may be determined experimentally. Just make a trap of convenient configuration using the same type coax and overlay wire you expect to use in the final trap construction and measure the resulting trap frequency. The accompanying computer program may then be used repeatedly with trial values of capacitance per foot until the correct value is found.

Fig 4 shows the successive versions of the equivalent circuit leading to the final equivalent circuit of the enhanced trap. Determining the proper number of turns of the windings without excessive cut and try is dependent on having an accurate equivalent circuit.

## The Computer Program

Program 1 presents the computer program and equations which calculate the configuration of the traps upon entering the necessary design choices. The program is written in GWBASIC 3.2 for use directly on IBM compatibles. But it may be easily modified for use on other types since it is relatively simple. (The program is available on diskette; see **Diskette Availability** on a page at the front of this book.)

Definitions of terms appear in remarks from lines 260 to 480. Equations in lines 620 and 670 give the total inductance of the trap and total equivalent capacitance, respectively, referred to the trap terminals. These two equations also show the unexpected serendipitous benefit from the overlay winding: a significant reduction of the coax trap capacitance referred to the trap terminals because of a change in the capacitance transformation formula.

Note that the capacitance from the coax when referred to the trap terminals is now divided by the total trap inductance, which has been greatly increased over that of the conventional coax trap. The increase comes not just from the added self-inductance of the new overlay winding, but also from the added mutual inductances reflected into both the inner winding and the shield winding of the coax. Instead of the coaxial-cable distributed capacitance being transformed by the factor of 1:4 for the usual form of coax trap, it is now transformed by a factor approaching 1:9 when a tightly coupled, full-sized overlay winding is used. The cross-coupled mutual inductances, MEQ1 and MEQ2, are both negative in sign and though small in magnitude further enhance the L/C ratio of the traps. These beneficial effects also apply in referring the distributed capacitance of the overlay winding to the trap terminals. These factors are so significant they about double the characteristic trap reactance that can be attained with coax-cable traps, and justify calling the enhanced traps Super-Traps.

Lines 560 and 570 give the formulas for the mutual coupling coefficients between the overlay winding and the coax-cable windings. These coefficients are essentially the ratios of the enclosing volumes of the coax windings to the enclosing volume of the overlay winding. Certain approximations are present which bear further evaluation. Perhaps some readers having a good feel for mutual coupling coefficients would like to examine the problem. However, the results so far indicate adequate accuracy for most purposes. The overall accuracy for the total trap inductance and capacitance referred to the trap terminals is about 4% as confirmed by a General Radio 1650A impedance bridge. Correspondence is welcome on the problem. The mutual coupling coefficient between the two coax windings is taken as unity because the shield totally encloses the inner conductor of the coax.

Line 510 gives the equation for the self-inductances of the coax windings. Line 540 gives the equation for the self-inductance of the overlay winding. These equations take into account the outside diameter of the coil form, the outside diameter of the coax, and the outside diameter of the wire including the insulation of the overlay winding.

Other trap configurations using different diameter forms, coax, and overlay conductors may be created with the computer program. In using the program, a word of caution is in order. Not all of your attempted designs may be physically obtainable. If you have called for a physically impossible design, your computer will indicate an illegal function call in line 680. Two remedies are available: (1) Use a smaller coil form diameter or (2) use a greater turn deficiency for the overlay windings. (This second choice is not as desirable as the first.)

Conventional trap dipoles without a specified harmonic operating frequency may be built with significant reduction in the length beyond the trap if the very high L/C ratio traps afforded by the overlay winding technique are employed. However, to build this 80/40/17-meter antenna, the values of L and C of 21 µH and 24 pF, respectively, must be closely approximated.

It is convenient to prune the trap to exactly the right frequency by removing small fractions of a turn at a time. This pruning feature is very useful but in this application it must not be overdone lest the proper L/C ratio not be maintained. However, in many other trap applications high values of L/C are required but not a specified L/C ratio. In these cases more freedom to prune exists. Pruning requires entering a design frequency into the computer somewhat lower than the desired trap frequency and pruning the overlay turns until you get exactly the frequency you want. Of course it is best to do as much of the cut and try aspects of design as you can on the computer.

**Program 1**
**Listing for BASIC Program SUPERTRP.BAS**

This program is available on a companion diskette; see **Diskette Availability** on an early page of this book.

```
100 CLS:REM THIS PROGRAM DESIGNS COAX-CABLE SUPER-TRAPS WITH ADDED SERIES INDUCTANCE OVERLAYING
       THE OUTSIDE OF THE TRAP ."
110 REM  WRITTEN BY A C BUXTON , W8NX , 2225 WOODPARK ROAD , AKRON , OHIO ,44313
120 REM TEL 1-216-836-3854
130 INPUT "ENTER THE RG DESIGNATOR OF THE COAX FOR THE TRAPS ";RG$
140 INPUT "ENTER THE DESIRED TRAP FREQUENCY , MHZ";FT0
150 INPUT "ENTER THE DIAMETER OF THE TRAP COIL FORM , INCHES";D2
160 PRINT "NOTE : THE OUTSIDE DIAMETER OF RG58U IS .193 INCHES AND THE CAPACITANCE IS      28.5
       PF/FT "
170 INPUT "ENTER THE OUTSIDE DIAMETER OF THE COAX CABLE ,INCHES";TC
180 INPUT "ENTER THE CAPACITANCE OF THE COAX CABLE , PF/FT ";C0
190 PRINT "NOTE : THE OUTER DIAMETER OF THE RG58U CORE IS .119 INCHES "
200 INPUT "ENTER THE OUTER DIAMETER OF THE OVERLAY INDUCTOR , INCHES ";TC3
210 PRINT "NOTE: THE CAPACITANCE OF RG58U SHIELD TO RG58U CORE OVERLAY WINDING IS 15 PF /FT "
220 INPUT "ENTER THE CAPACITANCE PER FOOT OF RG58U SHIELD TO THE OVERLAY WINDING , PF/FT";CI
230 PRINT "NOTE : THE TURN DEFICIENCY OF THE OVERLAY WINDING IS THE NUMBER OF TURNS OF COAX MINUS
       THE NUMBER OF TURNS OF THE OVERLAY WINDING ,TYPICALLY .85 TO 3 TURNS ,NOT LESS THAN .85"
240 INPUT "ENTER THE TURNS DEFICIENCY OF THE OVERLAY WINDING ";TDEF
250 PI=3.141593:N2=10
260 REM INDUCTANCES ARE IN MICROHENRY AND CAPACITANCES ARE IN PICOFARAD .
270 REM LI=SELF INDUCTANCE OF INNER WINDING OF THE COAX .
280 REM LS=SELF INDUCTANCE OF SHIELD WINDING OF THE COAX .
290 REM LO=SELF INDUCTANCE OF OVERLAY WINDING .
300 REM D2=DIAMETER OF FORM , INCHES .
310 REM D3=DIAMETER OF OVERLAY WINDING , INCHES .
320 REM TC=THICKNESS OF THE COAX-CABLE , INCHES .
330 REM TC3=THICKNESS OF THE OVERLAY WINDING , INCHES .
340 REM N2=NO OF TURNS OF COAX WINDINGS .
350 REM N3=NO OF TURNS OF OVERLAY WINDING .
360 REM TDEF=N2-N3=TURNS DEFICIENCY OF THE OVERLAY WINDING .
370 REM KMIS , THE MUTUAL COUPLING COEFFICIENT BETWEEN THE INNER COAX WINDING AND THE  SHIELD IS
       UNITY .
```

```
380 REM KMOS=MUTUAL COUPLING COEFFICIENT BETWEEN THE OVERLAY WINDING AND THE COAX SHIELD WINDING
     .
390 REM C0= CAPACITY OF THE COAX CABLE , PF/FOOT .
400 REM C2=TOTAL CAPACITY BETWEEN THE INNER AND OUTER COAX WINDINGS BEFORE REFERRAL TO THE TRAP
     TERMINALS .
410 REM CI=CAPACITY , PF/FT , BETWEEN THE COAX SHIELD AND THE OVERLAY WINDING .
420 REM CIP=TOTAL CAPACITY BETWEEN THE OVERLAY WINDING AND THE COAX SHIELD BEFORE     REFERRAL TO
     THE TRAP TERMINALS .
430 REM MIS=MUTUAL INDUCTANCE BETWEEN THE INNER AND SHIELD WINDING OF THE COAX .
440 REM MOS=MUTUAL INDUCTANCE BETWEEN THE OVERLAY AND THE COAX SHIELD WINDINGS .
450 REM MIO=MUTUAL INDUCTANCE BETWEEN THE INNER COAX AND THE OVERLAY WINDINGS .
460 REM LH=LONGITUDINAL LENGTH OF THE COAX WINDING , INCHES .
470 REM LH3=LONGITUDINAL LENGTH OF THE OVERLAY WINDING , INCHES .
480 REM LT=TOTAL TRAP INDUCTANCE ; CT=TOTAL TRAP CAPACITANCE REFERRED TO THE    TRAP TERMINALS IN
     PARALLEL WITH LT .
490 N2=12
500 FOR W=1 TO 10
510 LI=(D2+TC)^2*N2^2/(18*(D2+TC)+40*N2*TC):LS=LI:N3=N2-TDEF
520 X=SQR((TC/2+TC3/2)^2-(TC/2)^2)
530 D3=D2+TC+2*X
540 LO=(D3/2)^2*N3^2/(9*D3/2+10*N3*TC)
550 KMIS=1
560 KMOS=N3/N2*((D2+TC+2*X)/(D2+2*TC))^2
570 KMOI=((D2+TC)/D3)^2*N3/N2
580 C2=C0*PI*N2*(D2+TC)/12+C0*(N2*TC+1)/12
590 CIP=CI*(PI*N3*D3)/12
600 MIS=LI:MOS=SQR(LS*LO)*KMOS:MIO=SQR(LI*LO)*KMOI
610 LH=N2*TC:LH3=N3*TC
620 LT=LI+LS+LO+2*(MIO+MIS+MOS)
630 NUM=((1+KMOS)^2*(LI+LS)*(LS+LO)/64)*1E-12
640 DENOM1=((LS/4+LO/4)*.000001)-1/(2*PI*FT0)^2/CIP
650 DENOM2=((LI/4+LS/4)*.000001)-1/(2*PI*FT0)^2/C2
660 MEQ1=NUM/DENOM1*1000000!:MEQ2=NUM/DENOM2*1000000!
670 CT=C2*(LI+MEQ1)/LT+CIP*(LI/4+LO/4+2*KMOS*SQR(LS*LO/16)+MEQ2)/LT
680 F0=1000/2/PI/SQR(LT*CT)
690 N2=N2*(1-(FT0-F0)/FT0)
700 NEXT W
710 CL=(PI*N2*(D2+TC))/12
720 CL3=PI*N3*D3/12
730 Z0=1000*SQR(LT/CT)
740 CLS
750 PRINT TAB(10);"SUMMARY OF SUPER-TRAP SPECIFICATIONS"
760 PRINT
770 PRINT "RG DESIGNATOR OF THE COAX-CABLE ";RG$
780 PRINT "SUPER-TRAP RESONANT FREQUENCY ";FT0;" MHZ"
790 PRINT "DIAMETER OF TRAP FORM ";D2;" INCHES"
800 PRINT "NUMBER OF CLOSE-SPACED COAX-CABLE TURNS ON THE TRAP ";N2
810 PRINT "NUMBER OF TURNS OF THE OVERLAY WINDING ";N3
820 PRINT "TURN DEFICIENCY OF THE OVERLAY WINDING RELATIVE TO COAX TURNS";TDEF
830 PRINT "COMBINED TOTAL INDUCTANCE ";LT;"MICROHENRY"
840 PRINT "TRAP TOTAL CAPACITANCE ";CT;" PICOFARAD"
850 PRINT "LENGTH OF COAX CABLE PER TRAP ";CL;" FEET"
860 PRINT "LENGTH OF OVERLAY WIRE PER TRAP ";CL3;" FEET"
870 PRINT "COAX CABLE CAPACITANCE ";C0;" PICOFARAD/FOOT"
880 PRINT "OVERLAY WIRE CAPACITANCE TO SHIELD OF COAX ";CI;"PICOFARAD/FOOT"
890 PRINT "LONGITUDINAL LENGTH OF COAX WINDINGS ON THE TRAP ";LH;" INCHES"
900 PRINT "LONGITUDINAL LENGTH OF THE OVERLAY WINDING ,";LH3;"INCHES"
910 PRINT "OUTSIDE DIAMETER OF THE COAX-CABLE ";TC;" INCHES"
920 PRINT "OUTSIDE DIAMETER OF THE OVERLAY CONDUCTOR ";TC3;" INCHES"
930 PRINT "TRAP CHARACTERISTIC REACTANCE IS ";Z0;" OHMS"
940 PRINT "MUTUAL COUPLING COEFFICIENTS,";"   KMIS=1";"   KMOS=";KMOS;"   KMOI=";KMOI
8990 END
9000 SAVE"SUPERTRP"
9010 END
```

# Trap Dipoles with Specified Harmonic Operation

By Albert C. Buxton, W8NX
2225 Woodpark Rd
Akron, OH 44333

This article presents a computer program for designing trap dipoles with a single pair of traps to provide at least 3-band coverage. It features a trap dipole covering the bands where the need for trap dipoles is greatest—160, 80 and 40 meters. The main purpose of the program is to give hams a tool for doing more home-brewing and getting more fun out of ham radio. It seems that hams nowadays are not getting as much fun out of their hobby as they used to because the opportunities for building are greatly reduced. Nobody can build a rig at home on a competitive basis with the commercially available transceivers. The same must be said for many of the instruments and other "black boxes" that we use. But antennas still must be tailored to fit available space, and here trap dipoles have a contribution to make. So pick your three favorite operating frequencies and construct a trap dipole that will give you a lot of fun as well as pride and satisfaction.

Fig 1 presents the layout and the important nomenclature in the case of the featured antenna. The antenna functions as a fundamental electrical half-wave dipole on both the 160- and 80-meter bands. On 40 meters it acts as a 3rd-harmonic antenna, so I have designated it the 160/80/40x3 trap dipole. Fortuitously, it also has useful coverage of 12 and 10 meters, although the input resistance is rather high and requires the use of an antenna tuner on these bands. The antenna has about 1 dB of gain on 40 meters. It is compatible with a feed arrangement using a 1:1 current balun and a 75-Ω coax line. It also may be built as a half-dipole or monopole with the missing half replaced by a good ground connection. In the latter case no balun is required and the antenna is best fed with a 50-Ω coax line.

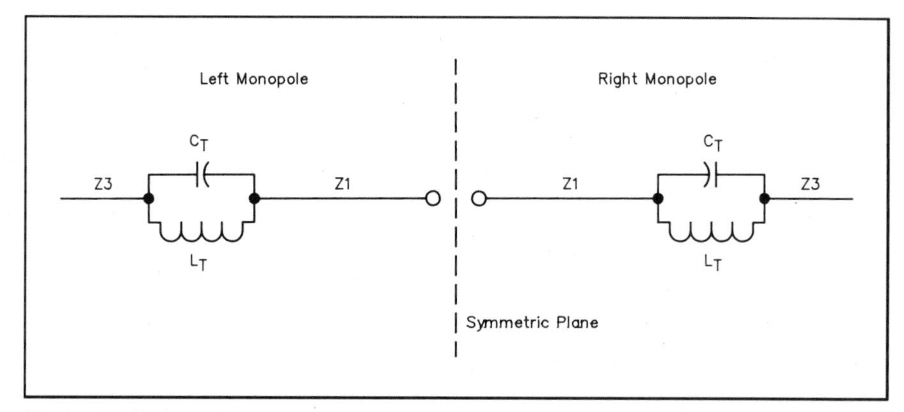


Fig 1—160/80/40x3 harmonic trap dipole.
$f_{HAR}$—7.2 MHz; also 24.7x11, 28.6x13 MHz.
$f_L$—1.825 MHz.
$f_M$—3.715 MHz.
$C_T$—97 pF.
$L_T$—18.8 µH.
Z1—62.2 feet.
Z3—48.0 feet.

Because of the size limitations of my lot, I have built and used the antenna as an inverted-L half-dipole using a good ground connection. The ground consists of a cold-water pipe, the aluminum siding of my house, and a 51-foot tower with a triband beam on top, all brought to a common electrical junction with short lengths of no. 4 copper wire. A good ground connection is essential because of radiation efficiency, as well as to prevent detuning the antenna by ground-connection reactance. The low frequency, $f_L$, is 1.825 MHz, near the low end of 160 meters where the activity seems to be greatest. The medium frequency, $f_M$, is 3.715 MHz, slightly favoring the CW end of the 80-meter band. The harmonic frequency is 7.2 MHz, toward the phone portion of 40 meters. The inboard section, Z1, is 62.2 feet long. The outboard section, Z3, is 48.0 feet long. The antenna performs closely to expectations. No pruning of the segment lengths or fine tuning beyond the initial design of the traps was required.

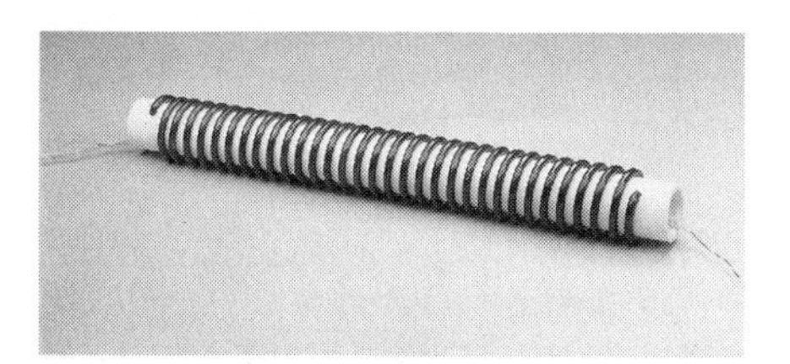

Fig 2—Coax-cable trap, wound on 1 inch PVC pipe. The coax is RG-58.

### Coax-Cable Traps

Fig 2 shows the coaxial-cable trap used in the featured antenna. The trap is tuned to the medium frequency, 3.715 MHz. The trap is wound on a nominal 1-inch PVC pipe, available at a plumbing supply house. The outer diameter of the pipe is 1.32 inches, which is its diameter as a coil form.

Check the outer diameter because not all PVC pipe may be alike. The trap has 32 turns of RG-58 coax with a center-to-center winding length of 13.5 inches. The turns are not closely compacted but have a gap of 0.23 inch between turns. Don't worry about the gap. Just drill the holes in the PVC form with a center-to-center spacing of 13.5 inches and maintain as nearly uniform turn spacing as you can.

To review coax-cable traps, one end of the center conductor of the coax is fed back through the inside of the coil form and connected to the shield conductor at the other end. See Fig 3. The connection inside the form may be made with a good crimp type connector if you don't like to use the soldering iron. The remaining two ends of the coax cable constitute the terminals of the trap, the shield at one end and the center conductor at the other. When placing the trap in the antenna, connect the coax center conductor to the inboard Z1 section of the monopole and the shield to the outboard Z3 section. If these terminals are inadvertently reversed the antenna will be slightly detuned, but no great harm is done.

It is good to check the resonant frequency of the trap with a dip meter if you have one to ensure a slipup hasn't occurred somewhere. The frequency should be within 50 kHz of 3.715 MHz as it sits on the laboratory bench free and clear of metal objects. For those with access to an R-L-C impedance bridge or a Q meter, the trap inductance should be close to 18.8 μH, and the capacitance should be close to 97 pF. The Q of the trap is 130, not as high as one of optimum dimensions but very adequate to ensure negligible trap losses.

**Table 1**
**Summary of Three-Band Trap Dipole Designs**

Z1 and Z3 are of no. 14 wire. Frequencies are in MHz.

| $f_L$ | $f_M$ | $f_{HAR}$ | *Order* | *Z1, feet* | *Z3, feet* | $L_T$ *μH* | $C_T$ *pF* |
|---|---|---|---|---|---|---|---|
| 1.85 | 3.75 | 7.15 | 3 | 62.1 | 48.4 | 17.2 | 104.4 |
| 1.85 | 3.75 | 10.1 | 3 | 61.3 | 25.8 | 63.5 | 28.4 |
| 1.85 | 3.75 | 14.175 | 5 | 61.4 | 27.8 | 57.5 | 31.3 |
| 1.85 | 3.75 | 14.175 | 7 | 62.0 | 58.7 | 5.7 | 318 |
| 1.85 | 3.75 | 18.1 | 7 | 61.6 | 34.2 | 41.0 | 43.8 |
| 3.75 | 7.15 | 14.175 | 3 | 32.1 | 22.9 | 7.0 | 71.1 |
| 3.75 | 7.15 | 18.1 | 3 | 32.3 | 14.9 | 20.3 | 24.4 |
| 3.75 | 7.15 | 21.3 | 5 | 32.2 | 24.4 | 5.4 | 92.9 |
| 3.75 | 7.15 | 24.9 | 5 | 31.9 | 16.5 | 16.6 | 29.7 |
| 3.75 | 7.15 | 28.5 | 5 | 32.4 | 12.7 | 26.4 | 18.8 |
| 3.75 | 7.15 | 28.5 | 7 | 32.4 | 27.7 | 2.0 | 250 |
| 7.15 | 14.175 | 28.5 | 3 | 16.0 | 11.8 | 4.2 | 29.9 |

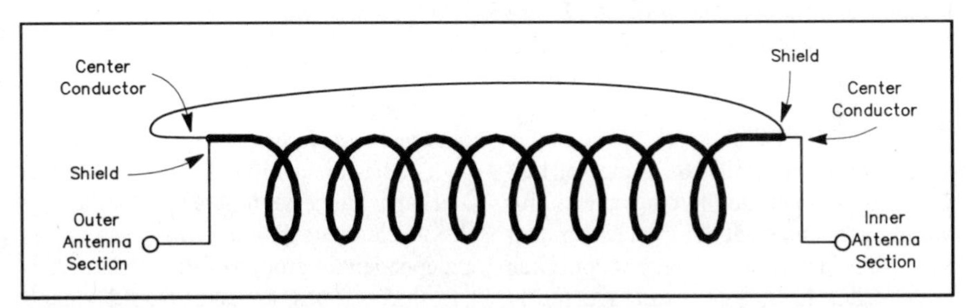

Fig 3—Showing the connections for the coax-cable trap of Fig 2. The center-conductor to shield connection is made inside the tubular winding form.

## Trap-Dipole Design

The usual method of trap-dipole design does not realize the full potential of such an antenna to provide multiple operating frequencies. It neglects one of the significant degrees of freedom available to the designer, namely the L/C ratio of the traps. For dipoles employing a single pair of traps with the correct L/C ratio as well as the correct LC product, one desired harmonic operating frequency may be specified in addition to the usual pair of upper and lower operating frequencies. Further, because of the harmonic relationships of the amateur bands, additional useful harmonic operating frequencies may sometimes be obtained although only one harmonic may be accurately specified. Also, with the wide availability of good antenna tuners (or Transmatches), it is acceptable practice to use an antenna on a higher "harmonic," even though the true harmonic frequency of the antenna may lie somewhat outside the amateur band. Experimenters in the past, of course, have proceeded along these lines with the results documented in many ARRL publications. However, much of this work was done prior to the 1979 WARC bands and does not reflect the full potential of trap dipoles to provide multi-band operation.

The computer program, HARTRPDI.BAS, is listed as Program 1. It is rigorously based on transmission-line theory including transmission losses due to radiation. (The program is available on diskette; see **Diskette Availability** on a page at the front of this book.) The program can be used to design trap dipoles with a single pair of traps that provide at least 3-band operation. Low impedance, current-fed input is provided at the antenna terminals at all operating design frequencies. The antenna is compatible with 75-Ω balanced feed or with unbalanced 75-Ω coaxial feed through a 1:1 current balun. Half-dipole or monopole versions may be implemented if a good ground connection is available to replace the missing half and if a change is made to 50-Ω coaxial-cable feed line.

The validity of calculation is such that most of the usual cut-and-try aspects of trap-dipole design are eliminated. The program is written in GWBASIC 3.2 but may be easily converted to other forms of BASIC. Design choices must be entered including wire diameter, the usual pair of operating frequencies, the harmonic operating frequency, the order of the harmonic, whether the trap is of coaxial-cable type or open-air inductor with a fixed capacitor, and nominal estimated dimensions and Q of the traps. The program determines the required lengths of the dipole segments, and the L and C of the traps. If desired it also calculates input impedance over any requested operating bands. Trap losses may be determined from the input impedance calculations. Operating bandwidths are indirectly inferred.

Table 1 summarizes some of the many possible 3-band trap dipole designs. It shows the resonant frequencies within the bands, segment lengths, and trap L and C for traps of nominal dimensions and Q. I have built and tested many of those of 3rd and 5th harmonic order with little need for pruning from the computer values. Since the various antenna configurations tend to be scale models of one another, a test for one of a given harmonic order is an approximate test for all of the same harmonic order. Thus, those of 7th and 9th harmonic order have not yet been tested. But little pruning is expected to be required for these antennas since even the 11th and 13th harmonics of the featured antenna are as calculated.

Table 2 gives a summary of some designs where more than three bands of operation are attained because of the fortuitous harmonic relationships of the various amateur bands. The second antenna in this table is the featured antenna. At the extra harmonic frequencies, assistance from an antenna tuner is required to cancel reactance because resonance may occur somewhat outside the amateur bands. But current feed is provided on all frequencies, compatible with 75-Ω feed line, and SWR values in the line do not get ruinously high at modest power levels.

**Table 2**
**Summary of Multiband Trap Dipoles**

Z1 and Z3 are of no. 14 wire. The second antenna listed below is the antenna featured in this article.

| *Freq (MHz), XOrder* | *Z1, Feet* | *Z3, Feet* | *$L_T$, μH* | *$C_T$, pF* |
|---|---|---|---|---|
| 1.85, 3.75, 18.1x9<br>Also 10.1x5, 14.1x7 | 62.1 | 59.2 | 5.2 | 345 |
| 1.825, 3.715, 7.20x3<br>Also 24.7x11, 28.6x13 | 62.2 | 48.0 | 18.8 | 97 |
| 3.9, 7.10, 14.025x3<br>Also 21.95x5, 31x7 | 32.5 | 22.2 | 5.0 | 100 |
| 3.55, 7.15, 18.05x3<br>Also 25.4x5 | 32.3 | 16.1 | 23.4 | 21.1 |

### The Program Itself

The program limits the choice of harmonic order operation to odd values. This assures low-impedance current input rather than high-impedance voltage input. If an even-order harmonic is inadvertently specified, the program will still give a design solution corresponding to the closest odd-harmonic operating condition. This is because the slope of the input reactance function is positive in the region of odd-harmonic operation and negative around even harmonics. The iterative design routine used in the program converges only for odd-harmonic operation. Also, physical realizability constrains the order of the odd harmonic chosen. For instance the order of the harmonic chosen must be such that

$$f_L < f_{HAR/ORDER} < f_M$$

Thus, if $f_L$ and $f_M$ are 3.75 MHz and 7.15 MHz, respectively, and $f_{HAR}$ is 18.1 MHz then either a 3rd or 5th harmonic design is realizable but a 7th harmonic design is unrealizable. Further, if you wish to maintain equal percentage bandwidths in the $f_L$ and $f_M$ bands then ideally $f_{HAR/ORDER}$ should be reasonably close to the mean value of $f_L$ and $f_M$. Also, the ratio $f_M/f_L$ seems to be practically limited to the following range.

$$1.4 < f_M/f_L < 7$$

If unrealizability constraints are ignored in your selection of input values to the program, then nonsense results may be obtained and starting over is required.

The basic equations for the inductance and capacitance per unit length for the wire dipole elements give the free space values independent of ground effects and interfering objects. These are taken from Terman's *Radio Engineers' Handbook*.[1] The transmission-line equations are rigorous and take into account losses due to radiation. Velocity of propagation along the dipole elements is taken as that of free space. Increased electrical length effects, typically 5% or so, associated with the fringing of the electrical field at the ends of the antenna are treated as the effects of equivalent end capacitors to ground. Both the series and shunt effects of the traps inserted in the transmission line are accounted for. The series effects are the primary effects and are those of the parallel impedance of the lumped L and C of the traps. The shunt effects are very small and are due to the stray distributed capacitance of the trap windings to free space. These are not to be confused with the capacitance between adjacent turns of the trap, which are subsumed into the ordinary C of the trap and slightly alter the trap resonant frequency. For coax-cable traps the stray capacitance to free space is approximately given by

$$C_{st} = 0.6\sqrt{\text{diameter} \times \text{length}}$$

where
diameter and length of the traps are in inches
$C_{st}$ is in pF

It is assumed that the inboard end of the coax-cable trap is the free end of the center conductor of the cable and the outboard end is the free end of the shield of the cable. It is also assumed that traps constructed of open air solenoid (single-layer) inductors and fixed capacitors have 50% higher stray capacitance than coax-cable traps. Traps typically have stray shunt capacitance in the range of 1 to 4 pF. Small unpredictable differences in stray shunt capacitance cause only negligible variations in resonant operating frequencies. Nominal dimensions of the traps were assumed based primarily on open-air inductor traps with fixed capacitors. If metallic enclosed cylindrical traps are used then different, but maybe negligibly different, stray shunt capacitance applies. At the frequency $f_M$, where the trap is anti-resonant, $C_{st}$ combines with the end capacitance of the inner dipole to act as a shunt to ground at the input side of the trap, the outboard length Z3 being divorced electrically by the high impedance of the trap. At all other frequencies one-half of $C_{st}$ is assumed to act as a shunt to ground at each end of the trap.

The program may be used to evaluate trap losses based on the assumption that all trap losses are chargeable to resistance losses of the inductor and none to dielectric losses or leakage in the capacitor. Input resistance may first be calculated assuming very high Q traps with zero loss. Input resistance may then be recalculated using the expected practical Q for the traps. The increase in input resistance is chargeable to trap losses. For instance the 80/40/17x3 trap dipole has an input resistance of

67.61 Ω on 40 meters when trap Q is 1 million. When traps with Q values of 100 are employed, the input resistance increases to 73.29 Ω and trap losses are 0.35 dB. Surprisingly, trap losses are even higher on 80 meters, 0.64 dB, because nearly all the current goes through the inductor where all the losses are. The losses are entirely negligible on 17 meters where nearly all the current goes through the trap capacitor. Encouragingly, careful laboratory measurements of Q values of coax cable traps made of RG-58 show values in excess of 100, up to as high as 150. But in general I believe that trap losses are somewhat of a false bogeyman and that traps with Q values greater than 100 have trap losses less than 1 dB.

The program may be used to perform trade-off studies of the bandwidths of trap dipoles using the computed input impedance versus frequency. Trap dipoles, of course, have narrower bandwidths in their individual bands than a separate dipole for each band. Studies show that trap dipoles have increasing loss of bandwidth in their low-frequency band as the L/C ratio of the traps is increased, corresponding to reduced length of the outboard segment of the dipole. Conversely, minimum reduction in bandwidth in the medium-frequency band is attained with very high L/C ratio traps. The effect on the bandwidth of the harmonic band may also be evaluated; the lower the trap L/C ratio, the lower the loss of bandwidth. Also, the higher the order of the harmonic, the lower the loss of bandwidth at the harmonic frequency.

If you like to experiment with trap dipoles, this program should be of real help in their design. Good luck with your home-built trap dipole jobs.

### Note

[1] F. E. Terman, *Radio Engineers' Handbook*, 1st ed. (New York, London: McGraw-Hill Book Co, 1943).

### Program 1
### Listing for BASIC Program HARTRPDI.BAS

This program is available on a companion diskette; see **Diskette Availability** on an early page of this book.

```
100 CLS
110 INPUT "DO YOU WANT THE PROGRAM PREAMBLE ?   ANS Y/N ";P$
120 IF P$="N" THEN 330
130 REM THIS PROGRAM IS FILED UNDER THE NAME HARTRPDI
140 PRINT "THIS PROGRAM DESIGNS AND CALCULATES THE PERFORMANCE OF A TRAP DIPOLE ANTENNA .ITWAS
      CREATED BY A.C. BUXTON , W8NX ,2225 WOODPARK RD , AKRON , OHIO  44313 ."
150 PRINT "TELEPHONE  1-216-836 3854"
160 PRINT "THE BASIC APPROACH TO THE TRAP DIPOLE ANTENNA DESIGN IS TO APPLY  TRANSMISION   LINE
      THEORY INCLUDING THE EFFECT OF  LOSSES DUE TO RADIATION .
170 PRINT "THE TRAPS ARE PARALLEL INDUCTOR/CAPACITORS TUNED TO THE MEDIUM  FREQUENCY , IN  THE
      USUAL MANNER ."
180 PRINT "AT THE  LOW OPERATING FREQUENCY THE TRAPS SUPPLY THE CORRECT AMOUNT OF
      INDUCTIVEREACTANCE TO BRING THE ANTENNA INTO AN EQUIVALENT ELECTRICAL HALF WAVELENGTH
      CONDITION ."
190 PRINT "AT THE MEDIUM OPERATING FREQUENCY THE TRAPS ARE ANTI-RESONANT , EFFECTIVELY
      PREVENTING THE OUTSIDE PORTION OF THE ANTENNA FROM RADIATING ."
200 PRINT "THUS THE INNER ELEMENTS OF THE ANTENNA  FORM AN ELECTRICAL HALF  WAVELENGTH AT  THE
      MEDIUM FREQUENCY ."
210 PRINT "AT THE HIGH HARMONIC OPERATING FREQUENCY THE TRAPS HAVE  A SMALL BUT
      SIGNIFICANTCAPACITIVE REACTANCE , SOMEWHAT REDUCING THE ELECTRICAL LENGTH OF THE ANTENNA ,
      THEREBY REQUIRING INCREASED PHYSICAL LENGTH ."
220 INPUT "PRESS RETURN TO CONTINUE ";KY$
230 PRINT "THE LENGTHS OF THE DIPOLE ELEMENTS GIVEN HERE ASSUME NEGLIGIBLY SHORT LEADS  TO THE
      TRAPS AND INSULATORS ,I.E. IF THE LEAD LENGTHS TO THE FEEDPOINT  AND TRAPS  ARE UNDULY
      LONG , THE ELEMENTS MUST BE SHORTENED ."
240 PRINT "THE VELOCITY OF PROPAGATION ALONG THE TRANSMISSION LINE IS TAKEN AS THAT OF FREESPACE
      ."
250 PRINT "THE LENGTHENING EFFECT DUE TO FRINGING OF THE ELECTRICAL FIELD AT THE ENDS OF   THE
      ANTENNA IS HANDLED AS AN EQUIVALENT END CAPACITIVE REACTANCE ."
260 PRINT "THE EFFECTS OF LOSSES AND NARROW-BANDING DUE TO THE LOADING TRAPS SHOW UP IN THEINPUT
      IMPEDANCE ."
270 PRINT "DESIGN CALCULATIONS REQUIRE AN ITERATIVE ROUTINE WHICH MUST BE CHECKED FOR
      CONVERGENCE . FAILURE TO CONVERGE IMPLIES YOU HAVE ASKED FOR A PHYSICALLY
      UNREALIZEABLE DESIGN . POOR CONVERGENCE IMPLIES CRITICAL OPERATION ."
280 INPUT "PRESS RETURN TO CONTINUE ";KY$
290 PRINT:PRINT "THE NOMENCLATURE FOR THE DIPOLE IS IN THE FOLLOWING FIGURE :":FOR I=1 TO 2500
      :NEXT I
300 GOSUB 2230:REM GRAPHICS SUBROUTINE
310 CLS
320 INPUT "ENTER TODAY'S DATE ";D$
330 INPUT "ENTER THE LOWEST OPERATING FREQ IN MHZ ";FL
```

```
34Ø INPUT "ENTER THE MEDIUM OPERATING FREQ IN MHZ ";FM
35Ø INPUT "ENTER THE HARMONIC OPERATING FREQ IN MHZ ";FHAR
36Ø INPUT "ENTER THE ORDER OF THE ODD HARMONIC FREQ ";ORDER
37Ø INPUT "ENTER THE DIAMETER OF THE WIRE RADIATING ELEMENTS IN INCHES ";DW
38Ø INPUT "WILL YOU BE USING COAX TRAPS OR OPEN INDUCTOR-CAPACITOR TRAPS ? ENTER C FOR COAX TYPES
      OR O FOR OPEN TYPES ";TT$
39Ø INPUT "ENTER AN ESTIMATE OF THE OUTSIDE DIAMETER OF THE TRAP ";DT
4ØØ INPUT "ENTER AN ESTIMATE OF THE LENGTH OF THE TRAP ,INCHES ";LH
41Ø INPUT "ENTER AN ESTIMATE OF THE Q OF THE TRAP ,";QTRAP
42Ø PRINT
43Ø FPSEUDO=FHAR:REM FPSEUDO IS THE HARMONIC FREQ THE ANTENNA WOULD HAVE IF THE TRAP WERE REMOVED
    .
44Ø FOR AB=1 TO 1Ø:REM THE DESIGN GOES THROUGH 1Ø ITERATIONS
45Ø PI=3.141592654#:KA=1:FMHZ=FHAR:REM INITAL ASSUMED VALUE FOR DIPOLE LENGTH FATOR IS 1 , THE
      SUBROUTINE ITERATES TO THE CORRECT VALUE .
46Ø GOSUB 215Ø:REM KA,ZØ,XCE,CE SUBROUTINE , KA=LENGTH FACTOR , CE=EQUIVALENT END CAPACITANCE ,
      XCE= EQUIVALENT END CAPCATIVE REACTANCE , ZO= CHARATERSITIC IMPEDANC OF LAMDA/2 WIRE
      SECTION.
47Ø ZT=983.57*(ORDER-1+KA)/4/FPSEUDO:FMHZ=FM:REM ZT=TOTAL LENGTH OF MONOPOLE ANTENNA
48Ø GOSUB 215Ø:REM KA,ZØ,XCE,CE SUBROUTINE
49Ø CS=.6*SQR(LH*DT):IF TT$="O" THEN CS=.9*SQR(LH*DT):ZØ=1ØØØ*SQR(L/C):REM CS IS THE STRAY SHUNT
      CAPACITY OF THE TRAP TO FREE SPACE . LH=TRAP LENGTH . DT=TRAP DIAMETER ,INCHES
5ØØ XCE=-ZØ*TAN(KA*PI/2):CE=-1ØØØØØØ!/2/PI/FM/XCE:REM XCE IS THE REACTANCE OF THE END CAPACITANCE
51Ø CSP=CS+CE-.18*SQR(LH*DT):XCSP=-1ØØØØØØ!/2/PI/FM/CSP:REM CSP IS THE COMBINED EFFECT OF THE
      TRAP STRAY CAPACITY TO FREE SPACE AND THE END CAPACITANCE DUE TO FRINGING OF THE
      ELECTRICAL FIELD . IT ASSUMES A 9 PERCENT NEGATIVE MUTUAL COUPLING COEFFICIENT .
52Ø KAP=ATN(-XCSP/ZØ)*2/PI:REM KAP IS THE TOTAL LENGTH FACTOR OF THE DIPOLE AT FM .
53Ø Z1=983.57*KAP/4/FM:Z3=ZT-Z1:FMHZ=FL:REM Z1= LENGTH OF INNER SEGMENT AND Z3 =LENGTH OF OUTER
      SEGMENT
54Ø PRINT "CHECK THE DESIGN CALCULATIONS FOR CONVERGENCE . "
55Ø GOSUB 215Ø:REM KA,ZØ,XCE,CE SUBROUTINE
56Ø XCEL=XCE:ZØL=ZØ:REM XCEL =END READCTANCE AND ZØL=CHARACTERISTIC IMPEDANCE BOTH AT FREQ FL
57Ø FFUND=983.57/4*.95/(Z1+Z3):REM FFUND IS THE FUNDAMENTAL FREQUENCY THE
58Ø NORLGTH=2*(Z1+Z3)*FL/983.57:RADRES=(27.Ø4*LOG(NORLGTH)+89.72):REM NORLGTH IS THE NOIRMALIZED
      PHYSICAL LENGTH OF THE ANTENNA . RADRES IS THE RADIATION RESISTANCE OF THE ANTENNA
      REFERRED TO THE CURRENT MAXIMUMS .
59Ø L=Z3:KV=1:ZØ=ZØL:FMHZ=FL:RL=.ØØ1:XL=XCEL:A=PI*RADRES/8/ZØL/(Z1+Z3)
6ØØ LAMBDA=983.57*KV/FMHZ:B=2*PI/LAMBDA:REM LAMBDA IS THE WAVELENGTH IN FEET.
61Ø GOSUB 2ØØØ:REM TRANSMISSION LINE SUBROUTINE
62Ø RINOUTL=RIN:XINOUTL=XIN:XCS=-1ØØØØØØ!/2/PI/FMHZ/CS
63Ø R7=.ØØØ1:X7=XCS*2:R8=RINOUTL:X8=XINOUTL
64Ø GOSUB 255Ø:REM PARALLEL IMPEDANCE SUBROUTINE
65Ø RINOUTLP=R9:XINOUTLP=X9
66Ø L=Z1:ZØ=ZØL:FMHZ=FL:RL=37.5:XL=.ØØØ1:A=-PI*RADRES/8/ZØL/(Z1+Z3)
67Ø LAMBDA=983.57*KV/FMHZ:B=2*PI/LAMBDA
68Ø GOSUB 2ØØØ:REM TRANSMISSION LINE SUBROUTINE
69Ø RININL=RIN:XININL=XIN
7ØØ R7=.ØØØ1:X7=2*XCS:R8=RININL:X8=XININL
71Ø GOSUB 255Ø:REM PARALLEL IMPEDANCE SUBROUTINE
72Ø RININLP=R9:XININLP=X9
73Ø XTRAPLP=-XININLP-XINOUTLP
74Ø LT=XTRAPLP*(1-(FL/FM)^2)/2/PI/FL:REM LT= TRAP INDUCTANCE ,UH AND CT =TRAP CAPACITY ,PF .
75Ø CT=1ØØØØØØ!/4/PI^2/FM^2/LT
76Ø FGEN=FHAR:FMHZ=FGEN
77Ø GOSUB 215Ø:REM KA,ZØ,XCE,CE SUBROUTINE
78Ø NORLGTH=2*(Z1+Z3)*FMHZ/983.57:RADRES=(27.Ø4*LOG(NORLGTH)+89.72)
79Ø RL=.ØØØ1:XL=XCE:L=Z3:A=PI*RADRES/8/ZØL/(Z1+Z3)*1.34:LAMBDA=983.57/FMHZ*KV:B=2*PI/LAMBDA
8ØØ GOSUB 2ØØØ:REM TRANSMISSION LINE SUBROUTINE
81Ø R7=RIN:X7=XIN:R8=.ØØØ1:X8=1ØØØØØØ!/2/PI/FMHZ/(-CS/2)
82Ø GOSUB 255Ø:REM PARALLEL IMPEDANCE SUBROUTINE
83Ø XLT=2*PI*FGEN*LT:XCT=-1ØØØØØØ!/2/PI/FGEN/CT:RTRAP=2*PI*FGEN*LT/QTRAP
84Ø MTN=SQR((XLT*XCT)^2+(RTRAP*XCT)^2):OTN=ATN(-RTRAP/XLT)
85Ø MTD=SQR(RTRAP^2+(XLT+XCT)^2):OTD=ATN((XLT+XCT)/RTRAP)
86Ø MT=MTN/MTD:OT=OTN-OTD
87Ø RP=MT*COS(OT):XP=MT*SIN(OT)
88Ø RLP=RP+R9:XLP=XP+X9
```

```
890 R7=RLP:X7=XLP:R8=.0001:X8=1000000!/2/PI/FMHZ/(-CS/2)
900 GOSUB 2550:REM PARALLEL IMPEDANCE SUBROUTINE
910 RL=R9:XL=X9:L=Z1
920 GOSUB 2000:REM TRANSMISSION LINE SUBROUTINE
930 FPSEUDO=FPSEUDO+8.999999E-03*XIN/RIN*FHAR
940 PRINT TAB(1);"LT=";LT;"UH";TAB(20);"CT=";CT;"PF";TAB(40);"Z1=";Z1;"FT";TAB(60);"Z3=";Z3;"FT"
950 PRINT
960 NEXT AB
970 INPUT "DO YOU WISH IMPEDANCE CALCULATIONS AT ANY OPERATING BANDS OF FREQUENCIES , ANS Y/N
       ";B$
980 IF B$="N" THEN 1780
990 INPUT "ENTER THE CENTER FREQ OF THE BAND 0F FREQUENCIES FOR WHICH YOU WISH  CALCULATIONOF
       INPUT IMPEDANCE ,MEGAHERTZ ";FCEN
1000 FOR I=1 TO 500:NEXT I:CLS
1010 PRINT "                  TRAP DIPOLE INPUT IMPEDANCE     " :PRINT
1020 PRINT TAB(3);"FREQ., MHZ.";TAB(20);"INPUT RES., OHMS"TAB(40);"INPUT REACT., OHMS ":PRINT
1030
       FGEN=FCEN:FMHZ=FGEN:NORLGTH=2*(Z1+Z3)*FGEN/983.57:RADRES=(27.04*LOG(NORLGTH)+89.72):A=PI*RADRES/8/60
1040 IF FGEN>1.2*FM THEN 1070
1050 IF FGEN<.8*FM THEN 1070
1060 NORLGTH=2*Z1*FGEN/983.57:RADRES=(27.04*LOG(NORLGTH)+89.72):A=PI*RADRES/8/600/Z1
1070 FOR I=1 TO 5
1080 GOSUB 2150:REM KA,Z0,XCE,CE SUBROUTINE
1090 IF FGEN>1.2*FM THEN 1120
1100 IF FGEN<.8*FM THEN 1120
1110 U=A:A=9.999999E-21
1120 RL=.0001:XL=XCE:L=Z3:FMHZ=FGEN:LAMBDA=983.57/FMHZ*KV:B=2*PI/LAMBDA
1130 GOSUB 2000:REM TRANSMISSION LINE ROUTINE
1140 CS=.6*SQR(LH*DT):IF TT$="O" THEN CS=.9*SQR(LH*DT)
1150 IF FGEN<.8*FM THEN 1180
1160  IF FGEN>1.2*FM THEN 1180
1170 CS=.0001
1180 R7=RIN:X7=XIN:R8=.0001:X8=1000000!/2/PI/FMHZ/(-CS/2)
1190 GOSUB 2550:REM PARALLEL IMPEDANCE SUBROUTINE
1200 CS=.6*SQR(LH*DT):IF TT$="O" THEN CS=.9*(LH*DT)^.5
1210 IF FGEN<.8*FM THEN 1240
1220 IF FGEN>1.2*FM THEN 1240
1230 CS=2*CSP
1240 XLT=2*PI*FGEN*LT:XCT=-1000000!/2/PI/FGEN/CT
1250 RTRAP=.001
1260 MTN=SQR((XLT*XCT)^2+(RTRAP*XCT)^2):OTN=ATN(-RTRAP/XLT)
1270 MTD=SQR(RTRAP^2+(XLT+XCT)^2):OTD=ATN((XLT+XCT)/RTRAP)
1280 MT=MTN/MTD:OT=OTN-OTD
1290 RP=MT*COS(OT):XP=MT*SIN(OT)
1300 RLP=RP+R9:XLP=XP+X9
1310 R7=RLP:X7=XLP:R8=.0001:X8=1000000!/2/PI/FMHZ/(-CS/2)
1320 GOSUB 2550:REM PARALLEL IMPEDANCE SUBROUTINE
1330 IF FGEN<.8*FM THEN 1360
1340 IF FGEN>1.2*FM THEN 1360
1350 A=U
1360 RL=R9:XL=X9:L=Z1
1370 GOSUB 2000:REM TRANSMISSION LINE SUBROUTINE
1380 A=A*RADRES/2/RIN:AP=A
1390 NEXT I
1400 FOR J=1 TO 15
1410 FGEN=FCEN+(J-8)*.01*FCEN
1420 FMHZ=FGEN
1430 GOSUB 2150:REM KA,Z0,XCE,CE SUBROUTINE
1440 A=AP*(FGEN/FCEN)^.5
1450 IF FGEN<.8*FM THEN 1480
1460 IF FGEN>1.2*FM THEN 1480
1470 U=A:A=1E-10
1480 RL=.0001:XL=XCE:L=Z3:FMHZ=FGEN:LAMBDA=983.57/FMHZ*KV:B=2*PI/LAMBDA
1490 GOSUB 2000:REM TRANSMISSION LINE ROUTINE
1500 CS=.6*SQR(LH*DT)
1510 IF FGEN<.8*FM THEN 1540
1520 IF FGEN>1.2*FM THEN 1540
```

```
1530 CS=.0001
1540 R7=RIN:X7=XIN:R8=.0001:X8=1000000!/2/PI/FMHZ/(-CS/2)
1550 GOSUB 2550:REM PARALLEL IMPEDANCE SUBROUTINE
1560 CS=.6*SQR(LH*DT):IF TT$="O" THEN CS=.9*SQR(LH*DT)
1570 IF FGEN<.8*FM THEN 1600
1580 IF FGEN>1.2*FM THEN 1600
1590 CS=2*CSP
1600 XLT=2*PI*FGEN*LT:XCT=-1000000!/2/PI/FGEN/CT
1610 RTRAP=2*PI*FGEN*LT/QTRAP
1620 MTN=SQR((XLT*XCT)^2+(RTRAP*XCT)^2):OTN=ATN(-RTRAP/XLT)
1630 MTD=SQR(RTRAP^2+(XLT+XCT)^2):OTD=ATN((XLT+XCT)/RTRAP)
1640 MT=MTN/MTD:OT=OTN-OTD
1650 RP=MT*COS(OT):XP=MT*SIN(OT)
1660 RLP=RP+R9:XLP=XP+X9
1670 R7=RLP:X7=XLP:R8=.00001:X8=1000000!/2/PI/FMHZ/(-CS/2)
1680 GOSUB 2550:REM PARALLEL IMPEDANCE SUBROUTINE
1690 IF FGEN<.8*FM THEN 1720
1700 IF FGEN>1.2*FM THEN 1720
1710 A=U
1720 RL=R9:XL=X9:L=Z1
1730 GOSUB 2000:REM TRANSMISSION LINE SUBROUTINE
1740 PRINT TAB(3);FGEN;TAB(22);2*RIN;TAB(42);2*XIN
1750 NEXT J
1760 INPUT "DO YOU WISH IMPEDANCE CALCULATIONS OVER ANOTHER BAND OF FREQS ?  ANS Y/N ";B$
1770 IF B$="Y" THEN 990
1780 CLS:PRINT TAB(23);"SUMMARY OF SPECIFICATIONS"
1790 PRINT TAB(23);"  HARMONIC TRAP DIPOLE "
1800 PRINT TAB(28);D$:PRINT
1810 PRINT "LOW OPERATING FREQ ";FL;"MHZ"
1820 PRINT "MEDIUM OPERATING FREQ ";FM;"MHZ"
1830 PRINT "HARMONIC OPERATING FREQ ";FHAR;"MHZ"
1840 PRINT "ORDER OF HARMONIC ";ORDER
1850 PRINT "LENGTH OF INNER ELEMENT OF MONOPOLE ";Z1;" FEET"
1860 PRINT "LENGTH OF OUTER EXTENSION BEYOND TRAP ";Z3;" FEET"
1870 PRINT "OVERALL LENGTH OF DIPOLE ANTENNA ";2*(Z1+Z3);" FEET"
1880 PRINT "DIAMETER OF RADIATING ELEMENTS ";DW;" INCHES"
1890 PRINT "TRAP RESONANT FREQUENCY ";FM;" MHZ"
1900 PRINT "TRAP INDUCTANCE ";LT;" MICROHENRY"
1910 PRINT "TRAP CAPACITANCE ";CT;" PICOFARAD"
1920 PRINT "TRAP CHARACTERISTIC REACTANCE ";1000*(LT/CT)^.5;"OHMS"
1930 PRINT "ESTIMATED OUTSIDE DIAMETER OF THE TRAP ";DT;"INCHES"
1940 PRINT "ESTIMATED LONGITUDINAL LENGTH OF THE TRAP WINDINGS ";LH;"INCHES"
1950 PRINT "ESTIMATED Q OF THE TRAP ";QTRAP
1960 PRINT "ESTIMATED TRAP STRAY CAPACITY TO INFINITE ";.6*SQR(LH*DT);"PF"
1970 L=.00508*(Z1+Z3)*12*(LOG(4*12*(Z1+Z3)/DW)-1)
1980 C=7.36*(Z1+Z3)/(LOG(2*12*(Z1+Z3)/DW)/LOG(10)-.5)
1990 END
2000 US=(EXP(A*L)-EXP(-A*L))/2*COS(B*L)
2010 VS=(EXP(A*L)+EXP(-A*L))/2*SIN(B*L)
2020 UC=(EXP(A*L)+EXP(-A*L))/2*COS(B*L)
2030 VC=(EXP(A*L)-EXP(-A*L))/2*SIN(B*L)
2040 C=US-XL*VC/Z0+RL*UC/Z0
2050 D=VS+RL*VC/Z0+XL*UC/Z0
2060 E=UC+RL*US/Z0-XL*VS/Z0
2070 F=VC+RL*VS/Z0+XL*US/Z0
2080 MNN=SQR(C^2+D^2):THETAN=ATN(D/C)
2090 IF C<0 THEN THETAN=THETAN+PI
2100 MND=SQR(E^2+F^2):THETAD=ATN(F/E)
2110 IF E<0 THEN THETAD=THETAD+PI
2120 M=MNN/MND:THETA=THETAN-THETAD
2130 RIN=Z0*COS(THETA)*M:XIN=Z0*SIN(THETA)*M
2140 RETURN
2150 FOR W=1 TO 10
2160 ZX=983.57/FMHZ/4*KA
2170 L=.00508*ZX*12*(LOG(4*12*ZX/DW)-1)
2180 C=7.36*ZX/(LOG(2*12*ZX/DW)/LOG(10)-.5)
2190 KA=(ZX/SQR(L*C)*1000/983.57)
```

```
2200 NEXT W
2210 Z0=1000*SQR(L/C):XCE=-Z0*TAN(KA*PI/2):CE=-1000000!/2/PI/FMHZ/XCE
2220 RETURN
2230 CLS
2240 FOR J=6 TO 18
2250 LOCATE 10,J:PRINT CHR$(61)
2260 NEXT J
2270 FOR J=23 TO 38
2280 LOCATE 10,J:PRINT CHR$(61)
2290 NEXT J
2300 FOR J=44 TO 59
2310 LOCATE 10,J:PRINT CHR$(61)
2320 NEXT J
2330 FOR J=64 TO 76
2340 LOCATE 10,J:PRINT CHR$(61)
2350 NEXT J
2360 LOCATE 10,39:PRINT CHR$(111)
2370 LOCATE 10,19:PRINT "TRAP":LOCATE 10,60:PRINT "TRAP"
2380 LOCATE 11,18:PRINT "LT,UH":LOCATE 12,18:PRINT "CT,PFD"
2390 LOCATE 11,59:PRINT "LT,UH":LOCATE 12, 59:PRINT "CT,PFD"
2400 LOCATE 9,11:PRINT "Z3,FT":LOCATE 9,67:PRINT "Z3,FT"
2410 LOCATE 9,27:PRINT "Z1,FT":LOCATE 9,49:PRINT "Z1,FT"
2420 LOCATE 10,43:PRINT CHR$(111)
2430 LOCATE 5,15:PRINT "LEFT MONOPOLE":LOCATE 5,56:PRINT "RIGHT MONOPOLE"
2440 LOCATE 6,17:PRINT "FL,FM,FHAR":LOCATE 6,58:PRINT "FL,FM,FHAR"
2450 FOR J=3 TO 17
2460 LOCATE J,41:PRINT CHR$(124)
2470 NEXT J
2480 LOCATE 17,42:PRINT "ZERO POTENTIAL PLANE "
2490 LOCATE 18,4:PRINT "ELEMENT DIAMETER ,DW , INCHES"
2500 LOCATE 21,22:PRINT "NOMENCLATURE , HARMONIC TRAP-DIPOLE ANTENNA"
2510 PRINT " ":PRINT " "
2520 INPUT "PRESS ENTER WHEN READY TO MOVE ON ";N$
2530 CLS
2540 RETURN
2550 MN=SQR((R7*R8-X7*X8)^2+(X8*R7+X7*R8)^2)
2560 THETAN=ATN((X7*R8+X8*R7)/(R7*R8-X7*X8))
2570 IF (R7*R8-X7*X8)<-.001 THEN THETAN=THETAN+PI
2580 MD=SQR((R7+R8)^2+(X7+X8)^2)
2590 THETAD=ATN((X7+X8)/(R7+R8))
2600 THETA=THETAN-THETAD:M=MN/MD
2610 R9=M*COS(THETA):X9=M*SIN(THETA)
2620 RETURN
8990 END
9000 SAVE"HARTRPDI"
9010 END
```

# How to Design Off-Center-Fed Multiband Wire Antennas Using that Invisible Transformer in the Sky

**By Frank Witt, AI1H**
**20 Chatham Rd**
**Andover, MA 01810**

This article describes the design procedure for harmonic wire antennas where the choice of feed point is optimized to provide maximum match bandwidth on a single band. These antennas can provide multiband operation without the use of an antenna tuner. The term "harmonic wire antennas" has been used to describe antennas whose length is a multiple of a half-wavelength. See Chapter 2 of *The ARRL Antenna Book*, 16th Edition, for a discussion of the harmonic wire antenna.[1] Since the term "Windom Antenna" is currently misused to describe any wire antenna with off-center feed, the class of antennas I address here might be called a type of Windom antenna.[2,3] A more accurate and descriptive name for this antenna class is the off-center-fed (OCF) antenna, since the antenna is useful even when a resonant condition does not exist.

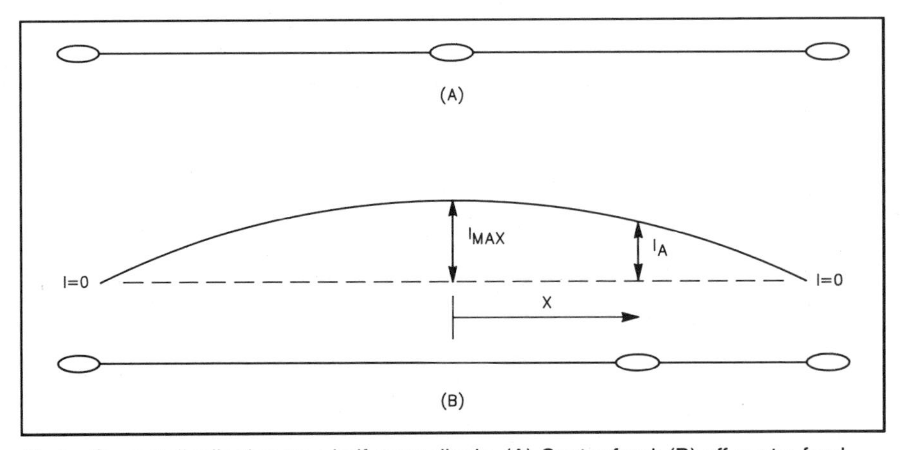


Fig 1—Current distribution on a half-wave dipole. (A) Center feed; (B) off-center feed.

I was drawn to this topic by the impressive claims I've read in articles and advertisements about OCF antennas. In particular, the large match bandwidth did not fit my understanding of the theory. Further, the theory does not support claims that a 6:1 transformer/balun is superior to a 4:1 transformer/balun when the feed line is 50-Ω coax. Reports of large changes in the SWR characteristic when feed-line length is changed, of RF in the shack and of high incidence of TVI suggests that substantial feed-line radiation can exist with an OCF antenna.

The result of my investigation is that the OCF antenna is a useful addition to a ham's arsenal, but only if fed properly. I designed an antenna to fit my needs. I operate mostly on 80 m (around 3.8 MHz), and I wanted to build an antenna that would also cover 160 m without an antenna tuner. (The built-in antenna tuner of my Kenwood TS-930S transceiver does not cover 160 m.) If I could get a low SWR on the HF bands, I would consider that a bonus. Out of the project came enough understanding so the same principles could be applied to other applications.

First I describe the "invisible transformer" concept. Then I show some MININEC simulation results that support the principle. I show how to maximize match bandwidth while taking into account the losses in the transmission line and matching transformer. Along the way I discuss how this type of antenna is prone to feed-line radiation, and what can be done to minimize its influence on the antenna performance. Last, I show how to design a multiband OCF antenna.

After the bulk of the material of this paper was prepared, I became aware of an interesting 1954 related paper by William R. Wrigley, W4UCW.[4] He reported on an in-depth analysis of the harmonic wire antenna. His major goal was to explore the possibility of designing practical antennas with low SWR on several HF bands. He concluded that such antennas were not very practical, primarily because of limited match bandwidth and because of potential feed-line radiation. He did not report on any experimental work. The results reported in my paper show that useful multiband

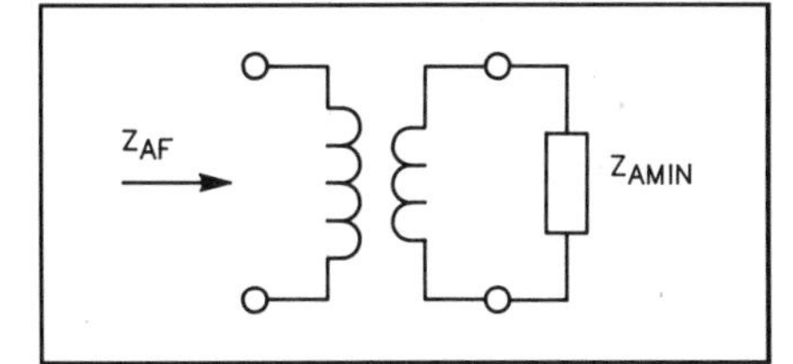


Fig 2—Equivalent circuit of the invisible transformer. Across its secondary is always connected the antenna impedance at the current maximum point.

designs are possible and that there are simple ways to overcome the feed-line radiation limitation. Because we now have the three WARC bands, the chances for a good match on several bands have increased since W4UCW did his analysis.

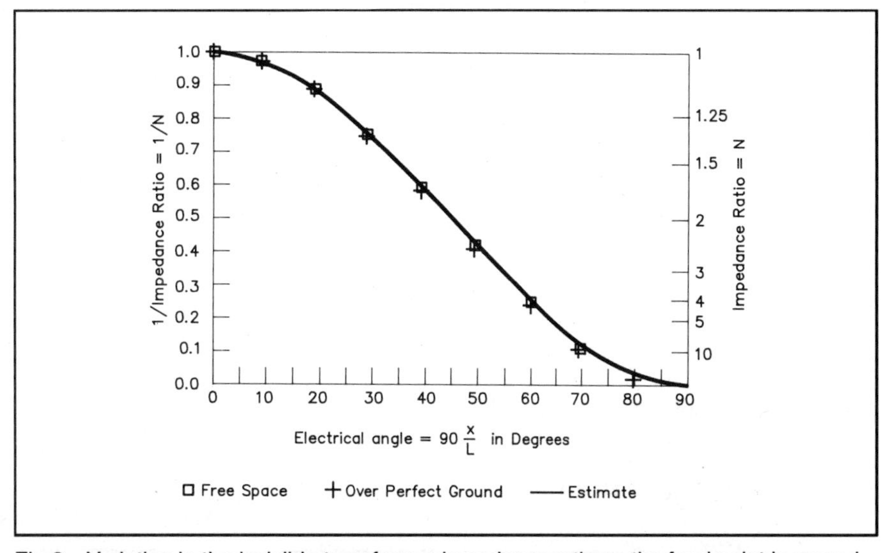


Fig 3—Variation in the invisible transformer impedance ratio as the feed point is moved off center. The solid curve is derived from Eq 2, while the data points are derived from MININEC simulations. (Free space or 60 feet above perfect ground, no. 14 AWG wire, one-half wavelength on 160 m.) For convenience in plotting, the reciprocal of N is plotted against the distance expressed as an electrical angle.

## The Invisible Transformer

It turns out that many antennas contain an invisible transformer that may provide a convenient adjustable impedance transformation mechanism. Here we will consider antennas operating near resonance, that is, near frequencies at which the feed-point impedance is purely resistive. To understand the principle, look first at the half-wave dipole shown in Fig 1. The amplitude of the current along the antenna is maximum at the center and is near zero at the ends. It can be approximated by a cosine function as follows.

$$I_A = I_{MAX} \cos\left(90 \frac{x}{L}\right)$$

where

$I_A$ = current at distance x from center

$I_{MAX}$ = current at the center of the antenna

L = a quarter wavelength of the standing wave measured along the antenna.

$90 \frac{x}{L}$ = distance expressed as an electrical angle in degrees

The more common feed point is the center of the antenna, as shown in Fig 1A. Feed-point impedance is lowest at that point. When the antenna feed point is moved off-center, Fig 1B, the same current distribution exists. The result is that the impedance at the new feed point is given by

$$Z_{AF} = \frac{Z_{AMIN}}{\cos^2\left(90 \frac{x}{L}\right)} \qquad \text{(Eq 1)}$$

where

$Z_{AF}$ = antenna impedance at the feed point

$Z_{AMIN}$ = antenna impedance where the current is a maximum

This is exactly what one gets with a voltage step-up transformer with a turns ratio of N:1 or an impedance ratio of $N^2$:1, where

$$N = \frac{1}{\cos\left(90 \frac{x}{L}\right)}$$

The value of N may be adjusted from 1 to a large number by moving the feed point along the wire, that is, by changing x. The term "invisible" is used to describe the transformer because its function is realized by feed-point position, not by a physical transformer at the feed point. Fig 2 shows an equivalent circuit of the invisible transformer.

The impedances of Eq 1 are complex.

$$Z_{AMIN} = R_{AMIN} + jX_A$$

and

$$Z_{AF} = R_{AF} + jX_{AF}$$

At resonance, $X_A$ and $X_{AF}$ are zero, so from Eq 1 we have

$$\frac{R_{AF}}{R_{AMIN}} = \frac{1}{\cos^2\left(90 \frac{x}{L}\right)} = N^2 \qquad \text{(Eq 2)}$$

or

$$\frac{x}{L} = \frac{1}{90} \cos^{1}\sqrt{R_{AMIN} / R_{AF}} \qquad \text{(Eq 3)}$$

An estimate of $1/N^2$ is plotted in Fig 3. Also shown in that figure is the result of a MININEC simulation for a half-wave 160-m dipole in free space and 60 feet above a perfect ground. Notice the close agreement. Simulations for many other cases involving free space, real and perfect ground, and harmonic operation, using both MININEC and NEC, show similar agreement. The result summarized in Eqs 2 and 3 is the same as that given in Kraus.[5]

An example will illustrate the principle. Assume that a center-fed half-wave dipole has a feed-point resistance of 73 Ω at resonance. Suppose it is desired to have a perfect match to a 200-Ω transmission line at resonance. Then

$$\frac{R_{AF}}{R_{AMIN}} = \frac{200}{73} = 2.74$$

From Eq 3 or Fig 3,

$$\frac{x}{L} = \frac{1}{90} \cos^{-1}\sqrt{1/2.74} = 0.587$$

This means that an 80-m dipole with L = 65 feet would be fed off center by 65 × 0.587 = 38.2 feet to achieve the 200-Ω feed-point resistance.

For the transformer analogy to be accurate, it is not only necessary for the result summarized in Fig 3 to be true, but the off-resonance behavior must also be consistent with that of a transformer. A good demonstration of the validity of the invisible transformer model is shown in Fig 4, where it is seen through MININEC simulations that antenna Q (at resonance) does not change significantly as the feed point is moved off center. This property is essential for the invisible transformer analogy to be valid.

Since the antenna Q does not change as the feed point is moved off center, one should conclude that maximum match bandwidth is not increased through the use of the OCF antenna, a result that contradicts some claims. Actually, off-center feed does provide, through the invisible transformer principle, the freedom to select the right feed-point impedance to obtain maximum match bandwidth. However, no other fundamental match bandwidth broadening mechanism results from off-center feed.

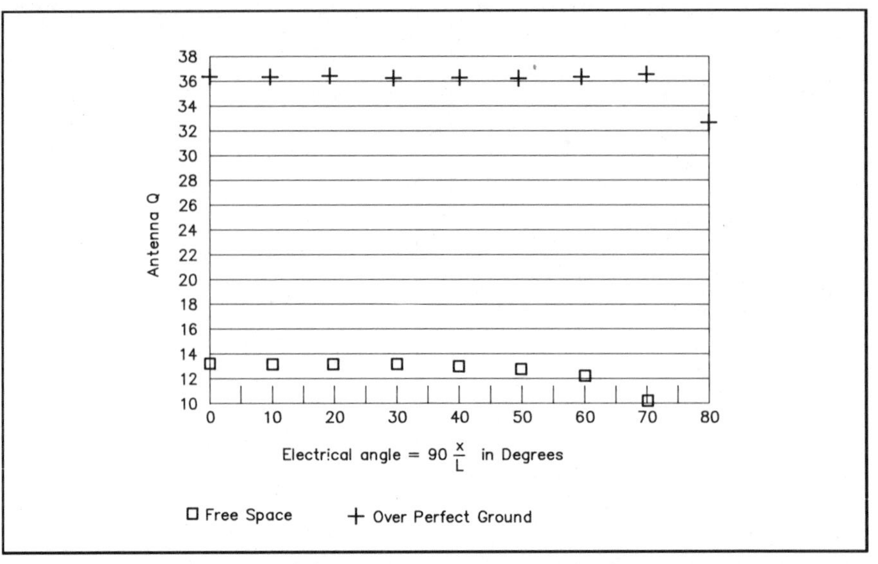


Fig 4—Antenna Q variation as the feed point is moved along the antenna. Note that the Q of a half-wave 160-m dipole hardly changes for both the free space case and the 60 feet above perfect ground case.

### Full-Wave and Multiple Half-Wave Antennas

The same invisible transformer principle applies for other resonant antennas whose overall length is a multiple of one-half wavelength. Fig 5 shows a 1-wavelength antenna fed at a current maximum point (Fig 5A) and fed a distance x away from the current maximum (Fig 5B). The current direction reverses at the center of the antenna. Now it is obvious why I used the terminology of the previous section. By referencing x and L to points where the standing current wave has maximum amplitude, Eqs 1, 2 and 3 apply to all cases. Notice that since there are two current maxima on a full-wave antenna, there are three other points where the feed-point resistance at resonance is the same as that of Fig 5B. These are shown by an X in that figure. This provides some added flexibility in the design of multiband antennas.

Through a judicious choice of the feed-point position and antenna dimensions, multiband resonant antenna operation is possible. Even more flexibility is achieved by adding one or more wires.[6]

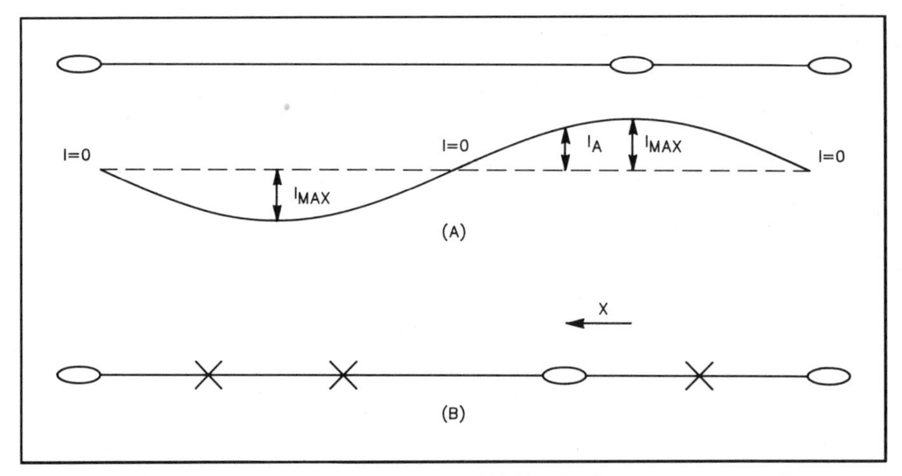


Fig 5—Current distribution on a full wave wire. (A) Feed point at a current maximum; (B) feed point at a distance x from a current maximum.

### Matching the Antenna to the Transmission Line

Since off-center feed usually results in a high feed-point impedance, either a high-impedance balanced transmission line or a coaxial feed line with an impedance step-up transformer is used. However, there is another serious issue to be dealt with—that of feed-line radiation. Feed-line radiation occurs because the common-mode RF voltage at the feed point is not zero. This leads to unequal currents in a balanced transmission line, and hence feed-line radiation. With coax feed, radiation occurs when current flows on the outside surface of the outer conductor. Such currents exist when coax is directly connected to the feed point of any kind of wire antenna, center-fed or otherwise. Good antenna design practice minimizes feed-line radiation. Here we will take steps to reduce it as much as possible by the use of coaxial cable feed, a choke balun/transformer at the antenna, and additional line isolation.

As I explain later, for maximum match bandwidth, the impedance ratio of the choke balun/transformer must be high enough to provide a generator impedance greater than the feed-point resistance at resonance. There is no advantage gained by using a ratio higher than 4:1 as long as the optimum feed-point location for that impedance ratio is chosen. Since a broadband 4:1 choke balun is easier to realize than one with a higher ratio, it is an unnecessary complication to use something other than 4:1.

Fig 6 shows a 4:1 choke balun made from two 1:1 choke baluns.[7] Notice the windings on the low-impedance side are connected in parallel and the windings on the high-impedance side are connected in series. This type of balun is one form of the Guanella Transformer. It has the highly desirable property that either port may be balanced or unbalanced. For that matter, either port may be neither balanced nor unbalanced. The OCF antenna presents a

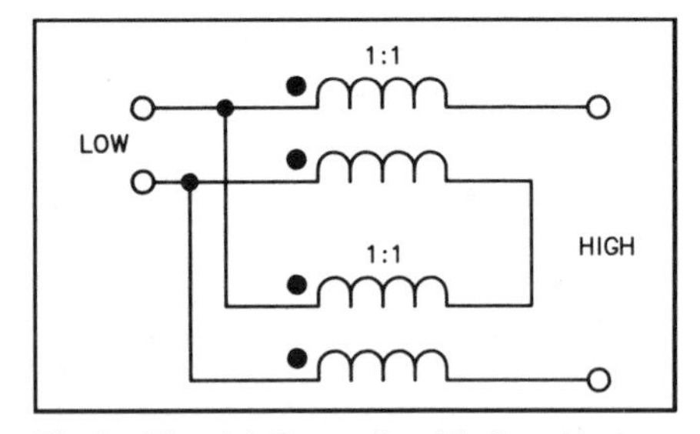


Fig 6—The 4:1 Guanella wide-band balun made from two 1:1 choke baluns.

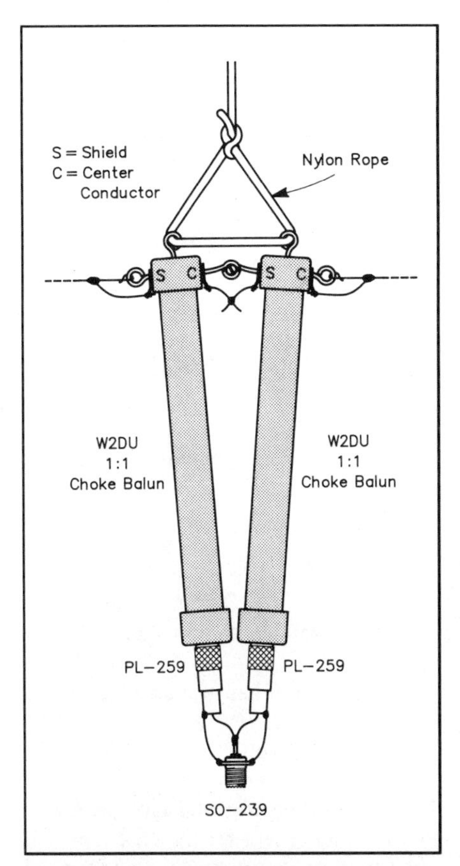


Fig 7—A 4:1 choke balun made from two W2DU HF 1:1 choke baluns. A no. 10 stainless steel screw, a lockwasher and a nut are used to fasten the two units together.

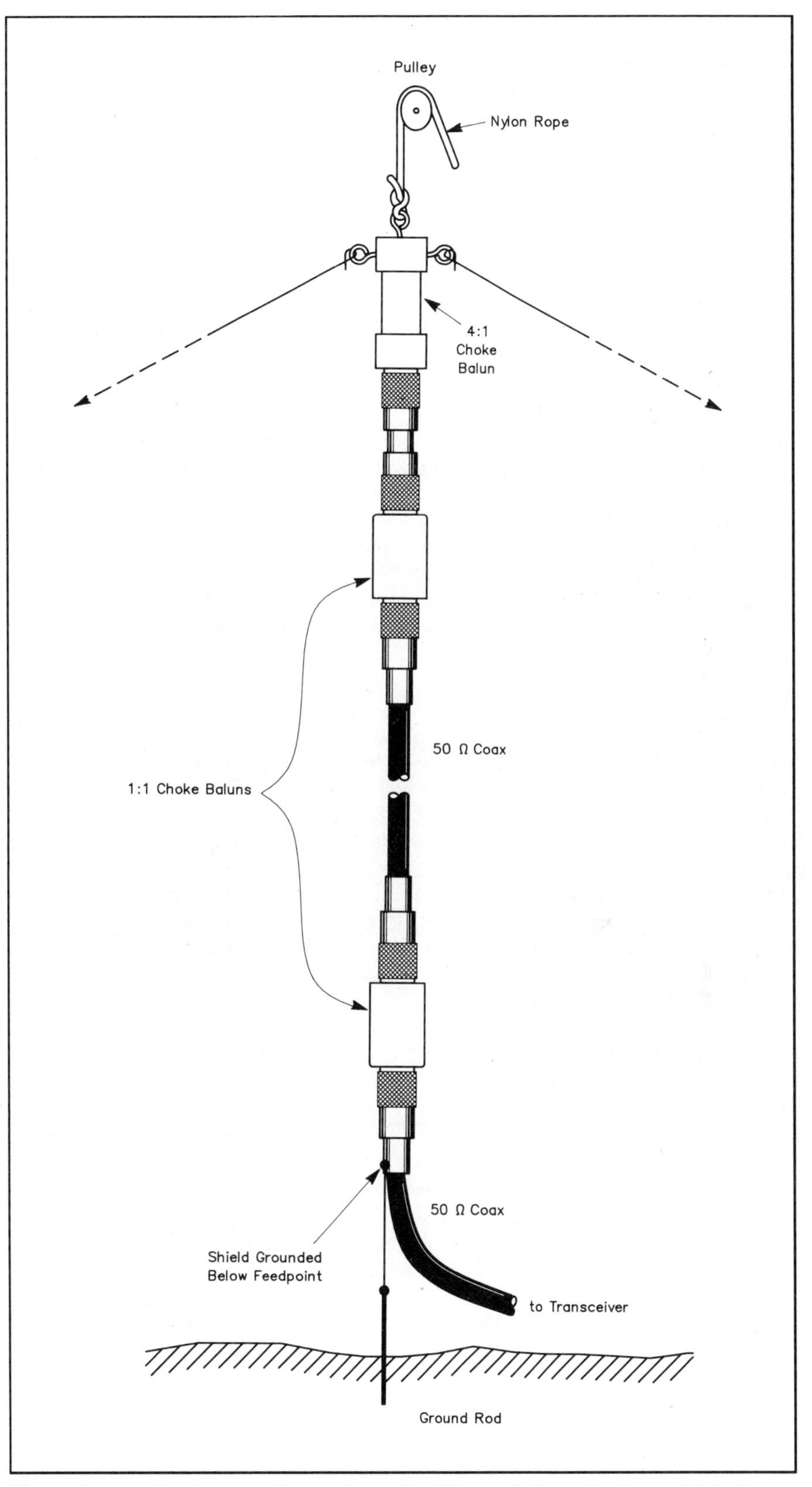


Fig 8—Antenna feed system. Convenience, feed-line radiation reduction and lightning protection are provided by this arrangement.

load on the transformer that is neither balanced nor unbalanced. Hence, the choke type of impedance transformer is ideally suited for OCF antenna applications.

One method of realizing a 4:1 choke balun is to use two W2DU HF 1:1 choke baluns (manufactured by Unadilla)[8] with a parallel connection at the 50-Ω side and a series connection at the 200-Ω side.[9] A continuity check must be made to identify the shield and center connector terminals so they may be connected as shown in Fig 7. The W2DU HF 1:1 baluns are made from a short piece of 50-Ω coaxial cable threaded through many small ferrite toroids.[10] Ideally, the characteristic impedance of the cable used in these baluns should be 100 Ω for this application, but the 50-Ω units perform quite well from 10 to 160 m.

I have observed that a 4:1 choke balun alone does not provide enough isolation, especially on 160 m. I routinely use two 1:1 choke baluns in addition to the 4:1 choke balun described above. W2DU HF choke baluns are available from the manufacturer (Unadilla) with SO-239 connectors at each end through special order. This arrangement is well suited for in-line 1:1 use. I have also used a Radio Works model B4-2KX 4:1 choke balun and their model C1-2K 1:1 in-line choke balun (with PL-259 connectors at each end) with similar results.[11]

A recommended feed arrangement is shown in Fig 8. Notice in the figure that the shield of the coaxial cable is connected to a rod driven in the ground directly below the feed point. Line isolators (1:1 choke baluns) are placed at the antenna end of the transmission line and next to the shield ground connection as shown. This technique, when used with choke baluns as shown, reduces feed-line radiation and keeps RF out of the shack. The technique also affords a degree of lightning protection since all antenna system elements have a low impedance path to ground at a point remote from the operating location. Note that if an antenna tuner is used with this system, it can be of the unbalanced variety, and no tuner balun is required. In many tuners, the balun is frequently the weakest link.

Although I have not yet tried it, I believe that there is another effective, and perhaps superior, way to prevent feed-line radiation. By threading the vertical portion of the feed line through several ferrite-type line isolation beads, spaced about 10 feet apart, feed-line radiation should be virtually eliminated. A good candidate for cable similar to RG-8 in diameter is the Chomerics type A637.

### Maximizing Match Bandwidth

An objective of the design procedure outlined here is to obtain the maximum match bandwidth on a particular band. Although there are several definitions of match bandwidth, the one I will use here is the bandwidth over which the SWR is less than 2:1. The antenna system includes not only the antenna, but also the matching balun/transformer and the transmission line. The match bandwidth we want to maximize is that measured at the transmitter end of the transmission line. The match bandwidth is not maximum when the antenna is matched to the transmission line plus balun/transformer at resonance.[12,13] More bandwidth is achieved when the antenna is mismatched at resonance and sees a generator impedance that is higher than $R_{AF}$. Further, the optimum amount of mismatch to achieve maximum match bandwidth depends on the loss in the transmission system.

Fig 9 shows the form of the SWR versus frequency curve for this antenna type. It is the familiar "bowl shape" characteristic of center-fed half-wave dipoles and many other antenna types. The term $S_L$ is used to describe the minimum SWR, which occurs at the antenna resonant frequency, $F_0$. The same shape occurs at the antenna end and at the transmitter end of the transmission line. I will use $S_{LANT}$ and $S_{LTRANS}$ to define the minimum SWR at the antenna and transmitter, respectively. In an article published in *QST* I explain the amount of mismatch in the presence of feed system losses that gives maximum match bandwidth.[14] The optimum mismatch is independent of antenna Q. These results are summarized here, since they are needed to optimize the match bandwidth of the antenna on a particular band.

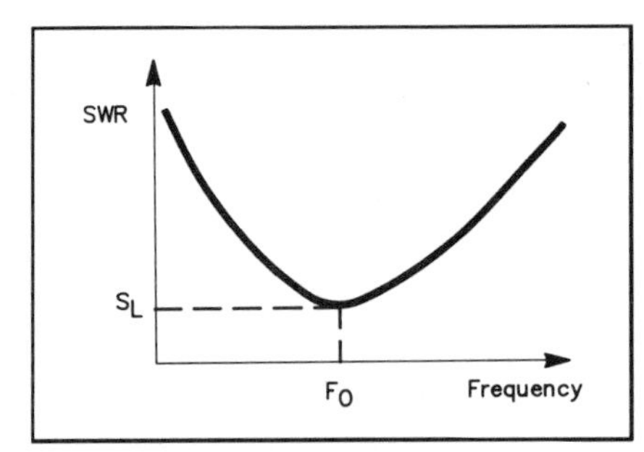


Fig 9—SWR versus frequency for resonant OCF antennas.

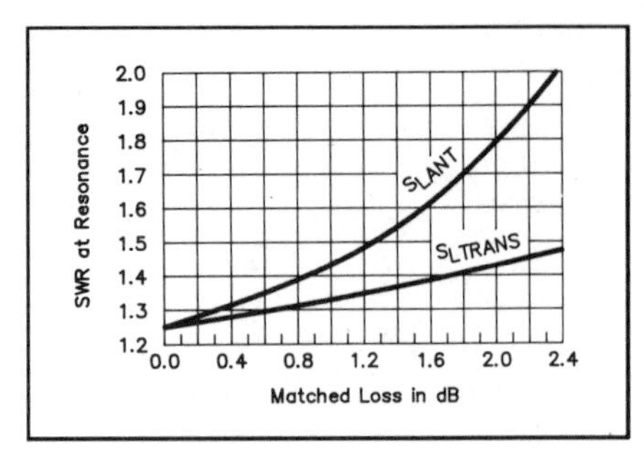


Fig 10—Optimum $S_{LANT}$ and $S_{LTRANS}$ as a function of matched transmission system loss.

The key result is shown in Fig 10, where the optimum values of $S_{LANT}$ and $S_{LTRANS}$ are plotted versus the matched feed-line plus balun/transformer loss. Matched loss refers to the loss when the feed system is properly terminated, which in general does not occur when an antenna is the termination. If one knows the approximate matched feed system loss, the optimum amount of mismatch may be found with the aid of Fig 10. For example, if the matched loss is 1 dB, $S_{LANT}$ = 1.43 and $S_{LTRANS}$ = 1.33. The matched loss in any particular application is easy to obtain by using published graphs in handbooks for the transmission-line loss and by using the manufacturer's specification for the loss of the balun/transformer. A good design will usually result in negligible balun/transformer losses. If those are not small, energy dissipation will be concentrated at the transformer, and, in high-power applications, the balun/transformer may be damaged.

Match bandwidth is not strongly dependent on $S_L$. For example, if the matched loss is very small, the optimum value of $S_{LANT}$ and $S_{LTRANS}$ is 1.25. See Fig 10. As long as the generator resistance is greater than $R_{AF}$ at resonance, $S_L$ can vary from 1.0 to 1.5 with a negligible effect on match bandwidth. Similar insensitivity to $S_L$ exists for higher matched loss values.

### Antenna Dimensions

Although operation on several bands is possible, it will be possible to achieve maximum match bandwidth on only a single band. I will refer to this band as the "design band."

Because of the influence of surrounding objects, such as the earth, trees, towers, guy wires and other antennas, it is often necessary to fine tune a resonant antenna. Hence, the physical design should allow for some wire length adjustment. A good starting point for the overall length may be found in *The ARRL Antenna Book*[15]

$$L_T = \frac{492\,(n - 0.05)}{F_0} \qquad \text{(Eq 4A)}$$

where

$L_T$ = overall length of the antenna in feet

n = number of half wavelengths at resonance

$F_0$ = resonant frequency in MHz

I found close experimental agreement with this formula except for the case n = 1. A good fit to my experimental data for n = 1 is

$$L_T = \frac{481}{F_0} \qquad \text{(Eq 4B)}$$

L, the length of a quarter of the current standing wave in feet, is given by

$$L = \frac{245.9}{F_0} \qquad \text{(Eq 5)}$$

This is the same as the free-space quarter-wavelength.

The overall length should be changed to bring the design band resonant frequency to the desired value. The length adjustment should be accomplished in a way that moves $S_L$ toward the intended value for $S_{LTRANS}$. In this way maximum match bandwidth will be achieved.

### Other Resonant Frequencies

After the antenna length, $L_T$, has been established, it is useful to know the other resonant frequencies of the antenna. These frequencies are determined by solving Eq 4 for $F_0$.

$$F_0 = \frac{481}{L_T} \quad \text{for } n = 1 \text{ and}$$

$$F_0 = \frac{492\,(n - 0.05)}{L_T} \quad \text{for } n \neq 1 \qquad \text{(Eq 6)}$$

Although it is not likely that the SWR will be low at all of the resonant frequencies, there may be some additional bands for which an antenna tuner will not be needed. Also, it will become clear how one can change $L_T$ to move a resonance to be within an amateur band. Small changes in $L_T$ will move higher frequency resonances without causing large movement in the resonant frequencies within lower frequency bands.

### Estimating $R_{AMIN}$

The radiation resistance at a current maximum, $R_{AMIN}$, must be estimated. A good start is found in *The ARRL Antenna Book*, which discusses the harmonic wire antenna in free space.[16] The radiation resistance increases for increasing n as shown in Table 1. The values given are those for antennas in free space. The actual amount depends on antenna height and ground conditions. For the design band, use the values of Table 1 as a starting point.

### Design Technique

We are now equipped with the information needed to design an OCF antenna. A step-by-step procedure follows. To illustrate the process, I will use an example of a dual-band 80-160 m OCF antenna. The design will be optimized for 80 m, but consideration will be given to providing satisfactory performance on 160 m. The antenna is fed with a 4:1 choke balun and a 50-Ω coaxial transmission line. At my installation, for 80 m, the feed-line plus balun loss is 1 dB, 0.1 dB (assumed) for the balun and 0.9 dB for the transmission line, which is 230 feet of RG-213 coax cable.

**Table 1**
**Radiation Resistance at Current Maximum for Increasing N**

| *n* | *$R_{AMIN}$* (Ω) |
|---|---|
| 1 | 73 |
| 2 | 94 |
| 3 | 106 |
| 4 | 115 |
| 5 | 121 |
| 6 | 127 |
| 7 | 131 |
| 8 | 135 |
| 9 | 138 |
| 10 | 141 |
| 11 | 144 |
| 12 | 147 |
| 13 | 149 |
| 14 | 151 |
| 15 | 153 |
| 16 | 155 |

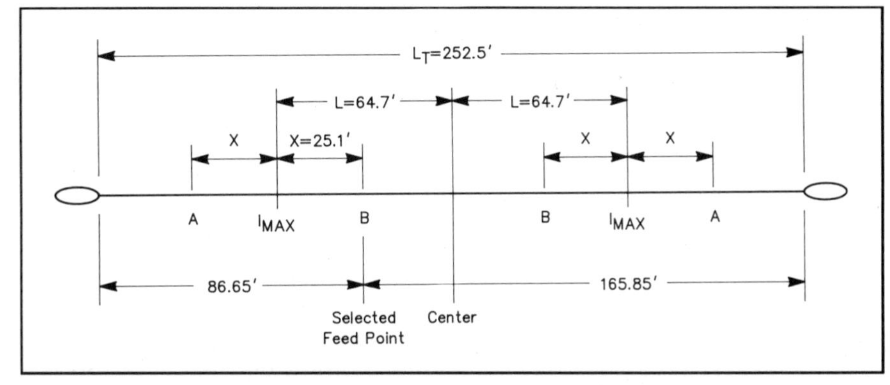


Fig 11—Calculated dimensions of the 160 plus 80-m OCF antenna. Two feed-point possibilities that lead to equal behavior on 80 m are marked A and B.

*Step 1: Select n and choose the design frequency.*

Since resonant operation is desired on both 160 and 80 m, I chose n = 2 on 80 m, the design band. I selected the design frequency to be 3.80 MHz since I like to operate around that frequency.

*Step 2: Calculate the overall antenna length, $L_T$, and the length of a quarter wavelength of the current standing wave, L.*

Using Eqs 4 and 5,

$$L_T = \frac{492\,(2 - 0.05)}{3.80} = 252.2 \text{ feet}$$

$$L = \frac{245.9}{3.80} = 64.7 \text{ feet}$$

*Step 3: Determine the location of the feed at the antenna point.*

First, the optimum amount of mismatch is determined by finding $S_{LANT}$ for a matched loss of 1 dB from Fig 10. Thus $S_{LANT} = 1.43$. Since the effective generator impedance at the antenna is 200 Ω (a 50-Ω transmission line and a 4:1 balun),

$$R_{AF} = \frac{200}{1.43} = 139.9 \text{ ohms}$$

From Table 1, $R_{AMIN} = 94$ Ω. Therefore

$$\frac{R_{AMIN}}{R_{AF}} = \frac{94}{139.9} = 0.672$$

From Eq 3,

$$\frac{x}{L} = \frac{1}{90}\cos^{-1}\sqrt{0.672} = 0.388$$

which leads to x = 0.388 × 64.7 = 25.1 feet. Fig 11 shows the two possible connections (marked A and B) that provide optimum match bandwidth on 80 m. The connection chosen will depend on which one provides the better match on 160 m. If from Table 1 $R_{AMIN}$ is assumed to be 73 Ω (n = 1), then the better choice would be the connection at A. However, since the antenna may be close to ground, the true value of $R_{AMIN}$ could be different from that shown in Table 1. At AI1H, the connection at B (shorter wire = 86.65 feet, longer wire = 165.85 feet) provided better performance on 160 m. This means that the value of $R_{AMIN}$ for 160 m is somewhat greater than 73 Ω. This will be discussed later. Both connections of Fig 11 (feed point at A or B) yielded essentially the same performance on 80 m.

Notice in Fig 11 that the feed-point location is not referenced to the wire ends, but is referenced to the center of the antenna. This procedure avoids errors caused by so-called end effects.

*Step 4: Calculate all resonant frequencies.*

Using Eq 6 and the final antenna length of Step 2, tabulate the expected resonances of the antenna. This is done in Table 2 for $L_T$ = 252.5 feet.

*Step 5: Construct the antenna system and measure the SWR.*

The antenna design in the previous section was constructed at AI1H and is shown in Fig 12. It was assembled in a way that the feed point could be lowered to the ground and that the wire lengths could be easily changed. The wire gauge is no. 14 AWG, although this is not critical. The feed system is the same as the one shown in Fig 8.

The degree of match achieved using the calculated wire lengths (86.65 and 165.85 feet) is shown in Fig 13. The measurements were made with a Daiwa model CN520 cross-needle SWR meter. The measured SWR on 160 and 80 m is close enough to warrant no changes. Also shown in Fig 13 is the SWR on 15, 12 and 10 m.

*Step 6: Tune the antenna.*

This step is performed on the design band, in our case 80 m, and is optional if

**Table 2**
**Expected Antenna Resonances**
**($L_T$ = 252.5 ft)**

| n | $F_0$ (MHz) |
|---|---|
| 1 | 1.905 |
| 2 | 3.800 |
| 3 | 5.748 |
| 4 | 7.697 |
| 5 | 9.645 |
| 6 | 11.594 |
| 7 | 13.542 |
| 8 | 15.491 |
| 9 | 17.439 |
| 10 | 19.388 |
| 11 | 21.336 |
| 12 | 23.285 |
| 13 | 25.233 |
| 14 | 27.182 |
| 15 | 29.130 |
| 16 | 31.079 |

Note that resonances fall inside four of the nine HF bands:

| | |
|---|---|
| n = 1 | 160 m |
| n = 2 | 80 m |
| n = 11 | 15 m |
| n = 15 | 10 m |

Also, one of the resonances is near the 12-m band (n = 13).

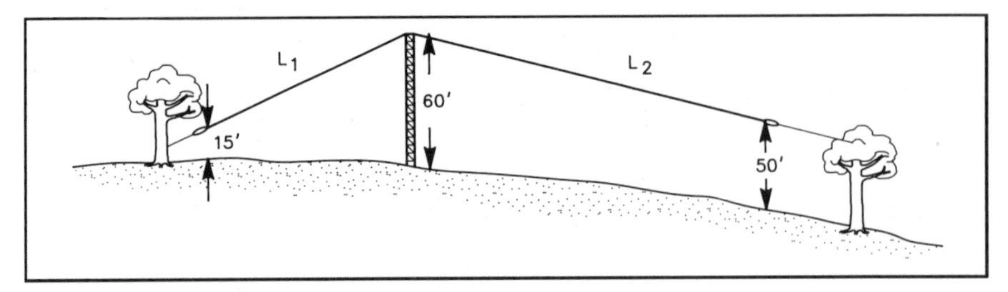


Fig 12—Multiband OCF antenna at AI1H.

one is satisfied with the match obtained in Step 5. First, SLTRANS is found from Fig 10. In our example the matched loss is 1 dB, so SLTRANS = 1.33. This value will yield the maximum match bandwidth as long as the mismatch at the antenna is caused by having the generator impedance (200 Ω) greater than the antenna impedance at resonance.

When the antenna is constructed, it is unlikely that the resonant frequency will exactly match the design value, 3.80 MHz. Further, it is unlikely that SLTRANS will equal the optimum value of 1.33. Changing one or both leg lengths could bring $F_0$ and SLTRANS to the design values. In all cases, $F_0$ will increase if the antenna is shortened and will decrease if it is lengthened. The adjustment of SLTRANS to the design value may be accomplished after the proper resonant frequency is obtained. The resonant frequency is kept constant by not changing the overall length of the antenna. SLTRANS is optimized by shortening one leg and by lengthening the other leg by the same amount. In some cases, however, like this one, initial performance may be close enough to the design values so no further adjustment to optimize the match on the design band is warranted.

*Step 7: Touch up design to match specific requirements.*

Using the principles outlined in this paper, one can modify the lengths to achieve a good match on additional bands. Do not expect low SWR readings on all of the "resonant" bands. Remember, not only must resonance occur, but also the feed-point resistance at resonance must be between 100 and 400 Ω for the SWR at the antenna to be less than 2:1. I recommend that SWR measurements be made to first establish the bands where a good match might be achieved, and that this be followed by some intelligent tweaking. In the design example, a low SWR was initially measured on only 160, 80 and 10 m.

By lengthening the longer wire by 3.1 feet, I moved the n = 13 resonance into the 12-m band while keeping resonances within the 15- and 10-m bands. This had little impact on the 160- and 80-m characteristics. While keeping the overall length constant, the feed point may be moved to achieve an optimum match on 15, 12 or 10 m. This technique makes use of the invisible transformer principle to achieve a variable impedance match at the antenna.

I varied the feed-point position in order to achieve a good match on 160, 80 and 12 m. The wire lengths are 87.8 and 167.8 feet. The SWR characteristics are shown in Fig 14. Also shown in that figure (the dashed curves) is the impact of removing the two 1:1 choke baluns and the ground connection below the feed point. The fact that the match characteristic changed indicates that some feed-line radiation is

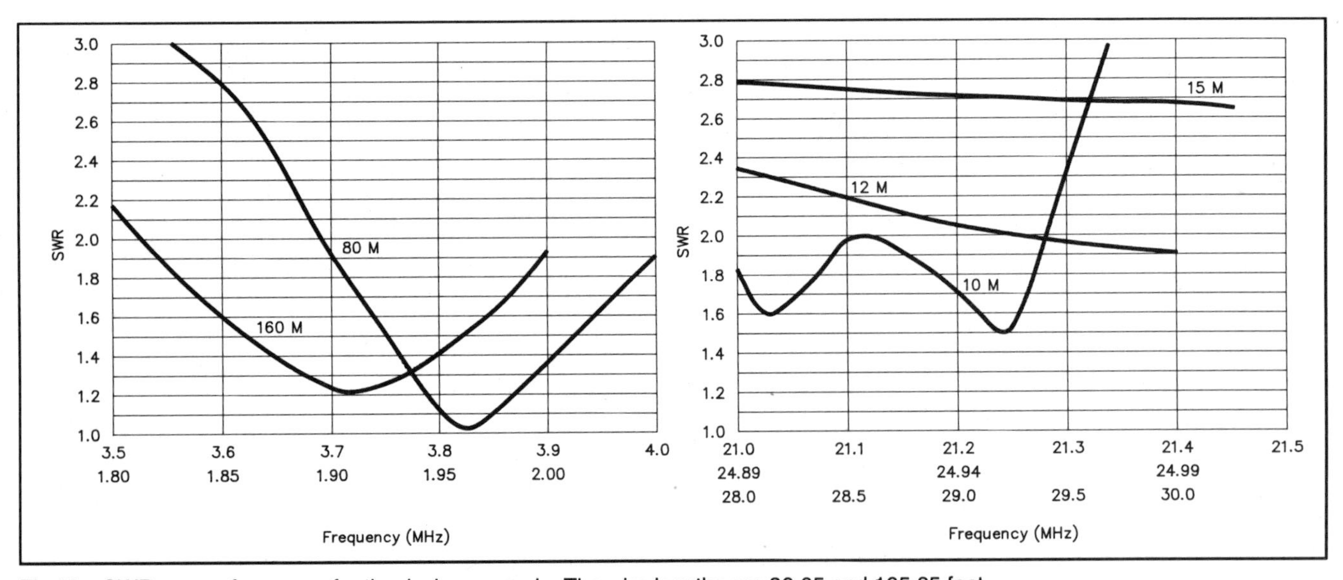


Fig 13—SWR versus frequency for the design example. The wire lengths are 86.65 and 165.85 feet.

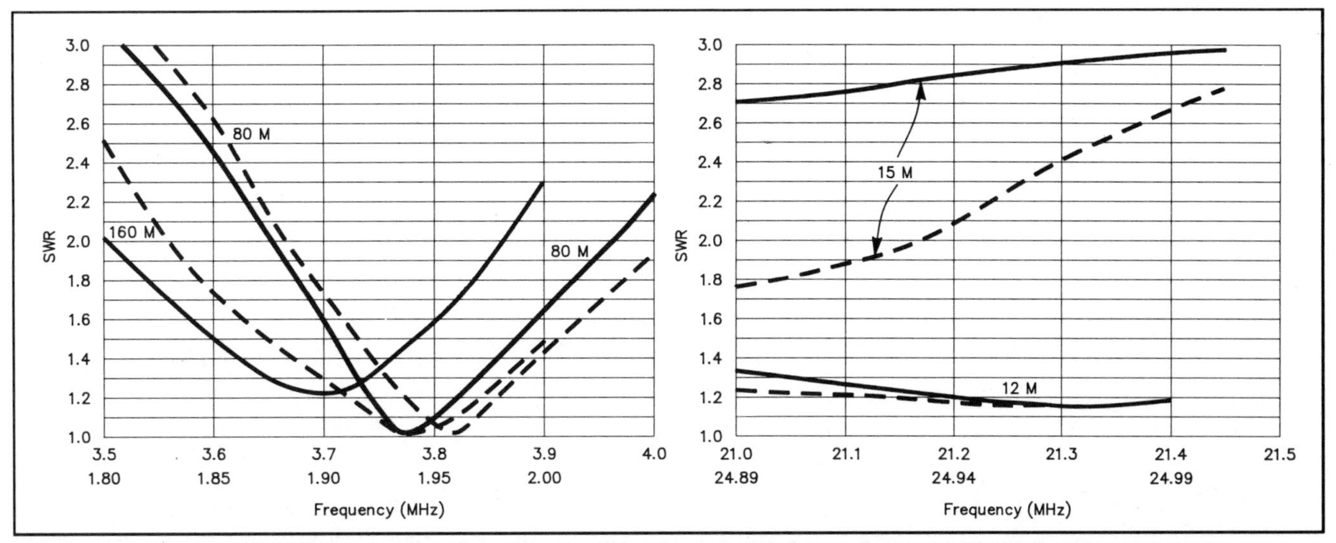


Fig 14—SWR versus frequency when the wire dimensions were changed to achieve a good SWR on three HF bands. The wire lengths are 87.8 and 167.8 feet. The dashed curves show the SWR when the two 1:1 choke baluns and the ground connection are removed.

present unless the additional line isolation is used.

I show in the Appendix how one can estimate the SWR at each resonant frequency. With this information, it is possible to modify the design to match a particular objective before physical changes are made to the antenna.

This is a good point to mention that one should not get carried away with fine tuning the multiband antenna design. One strives to achieve a reasonable match on several bands so that feed-line losses are low, an antenna tuner is not necessary, and band change and transmitter tune-up are simplified.

However, the type of antenna described here is also a good radiator on bands where the SWR is not less than 2:1. The transmission line and balun should be selected so the added stress on these components due to high SWR is tolerable. Hence, this kind of antenna plus an unbalanced antenna tuner will be a good performer on those bands for which a low SWR is not realized, as long as the transmission system losses are acceptable. The actual loss in the transmission system may be easily calculated from SWR measurements made in the shack and a knowledge of the matched loss of the transmission system.[17] I therefore recommend that when you use this type of antenna system with an antenna tuner, that two SWR meters be installed, one on each side of the tuner. In available tuners, the built-in SWR meter is on the transmitter side of the tuner.

### Antenna System Losses

We should have some idea of losses in our antenna systems. A major potential component is the loss in the transmission line. For this kind of antenna system, it is easy to calculate that loss element, since the SWR at the transmitter end of the feed line and the matched loss of the feed system are known. Use *The ARRL Antenna Book*, as mentioned earlier, or the accompanying formula (Fig 15) may be applied.

$$A = 10 \log \frac{4SWR_{TRANS}}{(SWR_{TRANS} + 1)^2 10^{-A_0/10} - (SWR_{TRANS} - 1)^2 10^{A_0/10}}$$

where

$A$ = attenuation of the feed line in dB

$A_0$ = attenuation of the feed line when it is matched at the antenna end in dB

$SWR_{TRANS}$ = SWR at the transmitter end of the feed line.

Fig 15—Formula for calculating loss in transmission lines.

$A_0$ may be determined by measuring the cable length and looking up the loss per 100 feet at the frequency of interest.

At HF, the matched loss of commonly used transmission lines is proportional to the square root of frequency. If one finds $A_0$ at one frequency, $F_1$, the matched loss at another frequency, $F_2$, may be determined by using the following formula.

$$A_{0F2} = A_{0F1} \sqrt{F_2/F_1} \qquad \text{(Eq 7)}$$

where $A_{0F1}$ and $A_{0F2}$ are the matched loss at frequencies $F_1$ and $F_2$, respectively.

Table 3 shows some interesting data for the antenna of Figs 12 and 14—that is, after the dimensions were tweaked to obtain a good match on 160, 80 and 12 m. On these bands the worst-case line loss is quite acceptable (0.9, 1.8 and 2.5 dB, respectively). Note from the table that except on 17 and 10 m the worst-case transmission-line loss is less than one S unit (6 dB) for all HF bands. The feed line used is 230 feet of RG-213. ($A_0$ = 0.9 dB at 3.80 MHz.) These losses are somewhat larger than would be obtained with a ladder line feeding a balanced antenna of similar length. However, such an antenna would not usually show a good match over three HF bands, as this OCF antenna does. It is clear that the coax-fed OCF antenna as an all-HF band antenna is a compromise design, especially when a long feed line is required.

Table 4 shows the percent of the band for which $SWR_{TRANS}$ is less than 2:1. The match bandwidth (also called the 2:1 SWR bandwidth) is enhanced by the loss in the transmission line. For comparison, I show the effect of having a very short feed line. If there is less feed-line loss, there will be less match bandwidth. I do not advocate using feed-line loss to achieve broadband operation; the effect is always present to some extent, however. I believe that many impressive claims we see for broadband operation are due to either feed-line loss,

**Table 3**
**OCF Antenna Data**

| *Band* | *Worst Case Frequency (MHz)* | *SWR at Transmitter* | *Matched Loss (dB)* | *Line Loss (dB)* |
|---|---|---|---|---|
| 160 m | 2.0 | 2.2:1 | 0.7 | 0.9 |
| 80 m | 3.5 | 3.3:1 | 0.9 | 1.8 |
| 40 m | 7.0 | 4.3:1 | 1.2 | 4.0 |
| 30 m | 10.1 | 2.7:1 | 1.5 | 2.8 |
| 20 m | 14.0 | 2.9:1 | 1.7 | 3.8 |
| 17 m | 18.068 | 3.5:1 | 2.0 | 6.6 |
| 15 m | 21.45 | 3.0:1 | 2.1 | 5.7 |
| 12 m | 24.89 | 1.3:1 | 2.3 | 2.5 |
| 10 m | 29.7 | 3.1:1 | 2.5 | 9.0 |

**Table 4**
**Percent of Band with SWR Less Than 2:1**

| | *Percent of band SWR <2:1* | |
|---|---|---|
| *Band* | *230 ft RG-213* | *Zero cable length* |
| 160 m | 92% | 78% |
| 80 m | 63% | 48% |
| 12 m | 100% | 100% |
| 10 m | 59% | 0% |

inaccurate (and often generous) SWR meters or unmentioned loss producers, such as baluns.

In a *QST* article,[18] I discuss and quantify the trade-off between feed system loss and match bandwidth.

**The 160-m Feed-Point Impedance and Match Bandwidth**

The measured SWR on 160 m (Fig 14) may be used to calculate $R_{AMIN}$, the feed-point resistance at resonance at the current maximum point, which is the center of the antenna. This calculation yields $R_{AMIN}$ = 124 Ω, which is much higher than the free-space value for a half-wave dipole (73 Ω). For low-height half-wave dipoles, it is well known that as the height is decreased, the feed-point resistance will be lower than the free-space value.[19] Why then the high value of 124 Ω?

Another observation is the relatively large match bandwidth of the antenna. Since match bandwidth and antenna Q are inversely related, it is sufficient to look at the antenna Q. The Q of the antenna calculated from the data of Fig 14 using the procedure given in the reference of Note 20 is about 9.3, which is lower than the free space value of about 13. For low-height dipoles, Q should increase as the height of the antenna decreases, again the opposite effect of what I observed.

Incidentally, a simple way of resolving the calculation ambiguity mentioned in that reference is to insert a 10-Ω noninductive power resistor in series with the antenna feed-line center conductor at the antenna. Observing whether the SWR at resonance goes up or down will resolve the ambiguity.

Simulations of OCF antennas I have done using the MN antenna analysis program have revealed that losses in the 4:1 balun and 1:1 isolators, especially at frequencies where the choke impedance is too low to guarantee low currents on the outside of the shield of the coax, will lead to high feed-point resistance and low antenna Q. These losses can be substantial. I conjecture that the high feed-point resistance and wide match bandwidth I observed on 160 meters are due to losses in the ferrite material in the 4:1 balun and 1:1 isolators. It's analogous to adding a resistor in series with the feed line at the point of connection to the antenna. Further work is needed in this area. It will help to explain why there is often a difference between the match bandwidth of an OCF antenna and the match bandwidth of a center-fed antenna of the same length and height above ground.

**On the Air Performance**

The design example has turned out to be a very good HF antenna. I have received excellent reports on all the HF bands. In a sense, it is a compromise design, but the material I have presented allows one to know the effect of the compromises. There has been no sign of RF in the shack. On the higher frequency bands, the patterns have many lobes and nulls, so performance depends on the direction of the station being contacted. The number of lobes equals twice the number of half waves of the current standing wave ($2 \times n$). MININEC simulations, properly done, will give one a good feel for the directional patterns of the antenna. The antenna is mostly omnidirectional on both 80 and 160 m. On 160 m the performance should be identical to that of a center-fed half-wave dipole at the same height.

**Summary**

This article has covered a lot of antenna topics, all of which were focused on the design of an OCF multiband HF wire antenna. First, the invisible transformer idea was explained. Feed-line radiation, ways to avoid it, and 4:1 choke baluns were addressed. I presented a detailed design procedure with a specific example. The very important, but often ignored, topic of antenna system losses was reviewed in some detail.

An important conclusion to be reached is that the OCF antenna conforms to existing theory. When feed-line radiation is eliminated, which is good design practice, the measured behavior is explained by simple concepts that have been around for some time. Except on 160 meters, I found no experimental evidence that off-center feed leads to match bandwidth enhancement. I believe that losses in the balun and isolators are significant on 160 meters for the antenna described here.The material presented here will enable radio amateurs to fashion their own OCF designs to match their particular needs. Properly executed, the OCF antenna will satisfy many requirements and is well worth considering for an upcoming project.

This project has benefited from the support of several people: first and foremost, Barbara, my wife, N1DIS. Discussions with Ted Provenza, W3OWN, Jack Belrose, VE2CV, Wally Luke, WA2NVG, and Al Simpson, WA1VHD, led me to a better understanding of the OCF antenna. NEC simulations by Dudley Chapman, WA1X, provided further confirmation of the invisible transformer principle and insight into the effect of a nearby lossy ground on feed-point impedance. Chris Kirk, NV1E, brought the earlier work on broadband choke baluns by W. A. Lewis of Collins Radio to my attention.

**Notes**

[1] *The ARRL Antenna Book*, 16th Edition (New-

ington, CT: The American Radio Relay League, 1991), pp 2-7 to 2-10.

[2]John Belrose and Peter Bouliane, "The Off-Center-Fed Dipole Revisited: A Broadband, Multiband Antenna," *QST*, August 1990, pp 28-34.

[3]Jerry Hall, "Garant Enterprises GD-8 'Windom' Antenna," *QST*, September 1990, pp 30-32.

[4]William B. Wrigley, "Impedance Characteristics of Harmonic Antennas," *QST*, February 1954, pp 10-14, 100.

[5]John D. Kraus, *Antennas*, Second Edition (McGraw Hill, 1988), pp 227-228. This result also appears in the reference of Note 4, p 11.

[6]See Note 2, p 30.

[7]Jerry Sevick, *Transmission Line Transformers*, Second Edition (The American Radio Relay League, 1990), pp 1-5 to 1-7.

[8]W2DU HF 1:1 choke baluns and line isolators are available from Unadilla, a Division of Antennas Etc, PO Box 4215 BV, Andover, MA 01810.

[9]Note 2, p 32. Also see J. Belrose, "Transforming the Balun," *QST*, June 1991, pp 30-33. I had independently conceived and used two 1:1 choke baluns to make a 4:1 choke balun (as well as three 1:1 choke baluns to make a 9:1 choke balun) prior to learning from Jack Belrose, VE2CV, that he had a similar idea. I have since learned that 4:1, 9:1 and 16:1 impedance transformers made from 1:1 transmission line choke baluns are described in a book by W. E. Sabin and E. O. Schoenike, *Single-Sideband Systems and Circuits* (McGraw Hill, 1987), pp 409-411. These authors credit W. A. Lewis of Collins Radio with first conceiving the idea in 1965, but they say the material was not published outside the company.

[10]M. W. Maxwell, *Reflections* (The American Radio Relay League, 1990), Chapter 21. This excellent book explains antenna matching and is recommended reading.

[11]The 4B-2KX 4:1 choke baluns and C1-2K line isolators are available from The Radio Works, Box 6159, Portsmouth, VA 23703.

[12]Frank Witt, "Broadband Dipoles—Some New Insights," *QST*, October 1986, p 30.

[13]F. J. Witt, "Optimum Lossy Broadband Matching Networks for Resonant Antennas," *RF Design*, April 1990, pp 44-46.

[14]Frank Witt, "Match Bandwidth of Resonant Antenna Systems," *QST*, Oct 1991, pp 21-25.

[15]Note 1, p 2-8. This result also appears in the reference of Note 4, p 12.

[16]Note 1, p 2-9. This result also appears in the reference of Note 4, p 10.

[17]Note 1, p 24-13 and 24-18.

[18]Note 14.

[19]Note 1, pp 3-10 and 3-11.

[20]Frank Witt, "The Coaxial Resonator Match," *The ARRL Antenna Compendium, Volume 2*, p 117.

## APPENDIX
## SWR at Other Resonant Frequencies

It is possible to estimate the SWR at the other resonant frequencies. This is very useful information since one may wish to modify the paper design before actually building the antenna in order to increase the likelihood that the SWR will be low on a particular band.

This calculation requires knowledge of:

n, the number of half-waves on the antenna, end to end

$L_1$, the length of the shorter wire, in feet

$L_2$, the length of the longer wire, in feet

$R_{AMIN}$, an estimate of the feed-point resistance, in ohms, at a current maximum at resonance

$A_{0F1}$, the matched loss at some reference frequency, $F_1$, in dB

First, using Eq 6, and that $L_T = L_1 + L_2$, $F_0$ is calculated. The off-center distance of the feed point $L_{OC}$ is calculated from

$$L_{OC} = \frac{L_2 - L_1}{2}$$

Then L, a quarter wavelength, is calculated using Eq 5. Now m, the exact number of quarter wavelengths between the antenna center and the feed point, is calculated from

$$m = \frac{L_{OC}}{L}$$

The integer value of m, which we will call $m_I$, is the number to the left of the decimal point.

At this point, x/L is calculated. The calculation is a bit tricky since the steps taken in the process vary depending on whether certain numbers are even or odd. (Zero is an even number.) There are two cases:

Case 1: n odd, $m_I$ odd or n even, $m_I$ even:

$$\frac{x}{L} = m_I + 1 - m$$

Case 2: n odd, $m_I$ even or n even, $m_I$ odd:

$$\frac{x}{L} = m - m_I$$

With x/L and $R_{AMIN}$ from Table 1, $R_{AF}$, the feed-point resistance at resonance, is found from

$$R_{AF} = \frac{R_{AMIN}}{\cos^2\left(90\frac{x}{L}\right)}$$

Now $S_{LANT}$, the SWR at the antenna at resonance, is given by

$$S_{LANT} = \frac{R_{AF}}{200} \quad \text{or} \quad S_{LANT} = \frac{200}{R_{AF}}$$

Choose the formula that yields $S_{LANT}$ greater than 1.

By using Eq 7, $A_0$, the matched loss in dB at $F_0$ is calculated. Using the following formula, $S_{LTRANS}$, the estimated SWR at resonance at the transmitter, is found:

$$S_{LTRANS} = \frac{(S_{LANT} + 1)10^{A_0/10} + S_{LANT} - 1}{(S_{LANT} + 1)10^{A_0/10} - S_{LANT} + 1}$$

The above sequence of calculations was applied to predict the SWR at resonance for the antenna of Figs 12 and 13. Table 5 shows the results of these calculations as well as the measured data. The close agreement over the entire HF region is further evidence that existing theory is adequate to analyze and design OCF resonant antennas.

**Table 5**
**Comparison of Calculated and Measured Data**

| | *Resonant Frequency (MHz)* | | *SWR at Transmitter* | |
|---|---|---|---|---|
| *n* | *Calculated* | *Measured* | *Calculated* | *Measured* |
| 1 | 1.905 | 1.915 | 1.90 | 1.20 |
| 2 | 3.800 | 3.822 | 1.33 | 1.00 |
| 3 | 5.748 | — | 5.95 | — |
| 4 | 7.697 | 7.682 | 1.35 | 1.18 |
| 5 | 9.645 | 9.582 | 1.04 | 1.47 |
| 6 | 11.594 | — | 3.71 | — |
| 7 | 13.542 | 13.522 | 1.25 | 1.23 |
| 8 | 15.491 | 15.463 | 1.24 | 1.40 |
| 9 | 17.439 | — | 2.69 | — |
| 10 | 19.388 | 19.503 | 1.21 | 1.10 |
| 11 | 21.336 | — | 1.44 | — |
| 12 | 23.285 | 23.048 | 2.12 | 1.90 |
| 13 | 25.233 | 25.247 | 1.17 | 1.00 |
| 14 | 27.182 | — | 1.64 | — |
| 15 | 29.130 | 29.180 | 1.76 | 1.45 |

# The K4VX 10-Meter Elephant Gun Yagi

**By Lew Gordon, K4VX**
**PO Box 105**
**Hannibal, MO 63401**

In the early 1900s the ivory hunters in Africa demanded a weapon that could kill a charging adult elephant with a single shot. There were high-powered rifles capable of inflicting serious wounds, but quick, clean kills were another matter. Out of this necessity a .600 caliber double-barreled rifle was created. It was the first true "Elephant Gun."

In the fall of 1986, I acquired a used KLM 6-element "Big Stick" for 20 meters. I already owned two very effective 20-meter Yagis, so I had planned to scavenge the KLM and use its aluminum for other projects. After looking at its 57.7 foot, 3 inch diameter boom, however, I suddenly envisioned a long-boom Yagi for 28 MHz. In '86 we were on the upswing of the sunspot cycle, and 10 meters looked like the ideal band for a "one-call" killer antenna.

It took only a few days in December to design and build the antenna. Between Christmas and New Years it was erected with the help of my daughter and her future husband. My first QSO with the new antenna was with Dave, ZS6DN. The new Yagi boomed over the top of a gigantic SSB pileup. Dave said I was the strongest signal he had heard during the new sunspot cycle. When I described what I was using, he dubbed it the "Elephant Gun" of antennas!

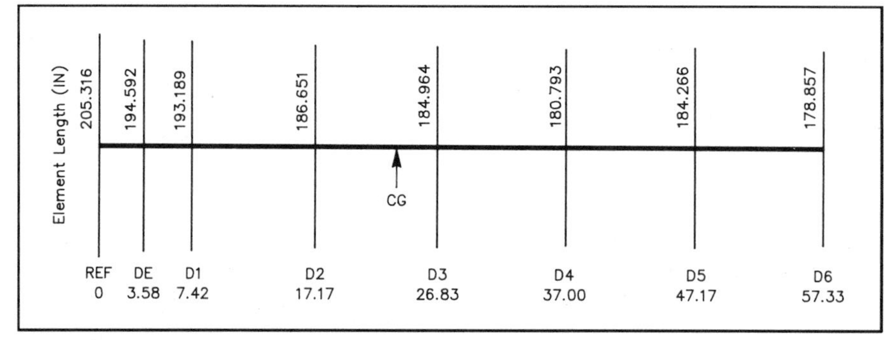


Fig 1—Element spacing (in feet from the reflector) and element length in inches. The center of gravity (CG) = 24.57 feet.

**Table 1**
**Cylindrical Element Construction**

| *Element* | *Length (in.)* | *Spacing from Ref. (in.)* | *Diameter (in.)* |
|---|---|---|---|
| Reflector | 205.3160 | 0 | 0.625 |
| Driven Element | 194.5920 | 43.0 | 0.625 |
| Director 1 | 193.1891 | 89.0 | 0.625 |
| Director 2 | 186.6513 | 206.0 | 0.625 |
| Director 3 | 184.9637 | 322.0 | 0.625 |
| Director 4 | 180.7936 | 444.0 | 0.625 |
| Director 5 | 184.2663 | 566.0 | 0.625 |
| Director 6 | 178.8570 | 688.0 | 0.625 |

## Design

My initial motivation for the design was based upon a 4-element Yagi created by the late Jim Lawson. It was known as the "PV4" and appeared in an issue of the *National Contest Journal.*[1] Jim's design placed the reflector quite close to the driven element (about 0.125 λ). This approach challenged all of the previous concepts of "good" Yagi design.

My design goals for the long-boom Elephant Gun Yagi included a minimum of 12.5 dBi forward gain, a 20 dB front to back ratio and an SWR not exceeding 2.5 to 1 from 28.000 to 28.750 MHz. Starting with these goals, I spent the next several days performing many iterations with my Yagi antenna design program, YAGINEC—a user friendly version of MININEC for Yagis. I eventually arrived at an 8-element design with fairly close spacing for the reflector and first director. Although I was never satisfied with the bandwidth, the gain and front to back ratio equaled or exceeded my goals.

I later developed a new program called YAGIMAX. It is a much faster Yagi design system than the old YAGINEC program and it allows the user to maximize performance on an element-by-element basis. Naturally, I used YAGIMAX to redesign the Elephant Gun in an attempt to improve the bandwidth. I finally arrived at the equivalent 0.625 diameter cylindrical tube dimensions given in Table 1 and Fig 1.

### Construction

Construction of a large antenna requires careful attention to the strength of the boom since it presents the largest cross section to wind resistance (approximately 10 square feet). In its original application, the KLM used three boom guys to offset the use of relatively thin tubing. To be on the safe side, I did the same. The tapered element construction is provided in Table 2. These dimensions were provided by my TAPER program which uses the algorithms developed by W2PV to correct for the effect of tapered elements upon the equivalent cylindrical length.[2]

The elements were attached to the boom by a small section of aluminum channel stock and held in place by two 3-inch cadmium-plated U bolts. (As a substitute, a 4-inch by 6-inch aluminum plate ¼ inch thick will work just as well.) Each element was then secured by two cadmium-plated 1.25-inch U bolts. To reduce the effect of the element mounting hardware on the effective length of each element, I decided to insulate all of the parasitic elements from the U bolts and channel stock. I accomplished this by wrapping a piece of recovered PVC from some old ¾-inch Hardline around each element. This "cheap and dirty" approach actually works.

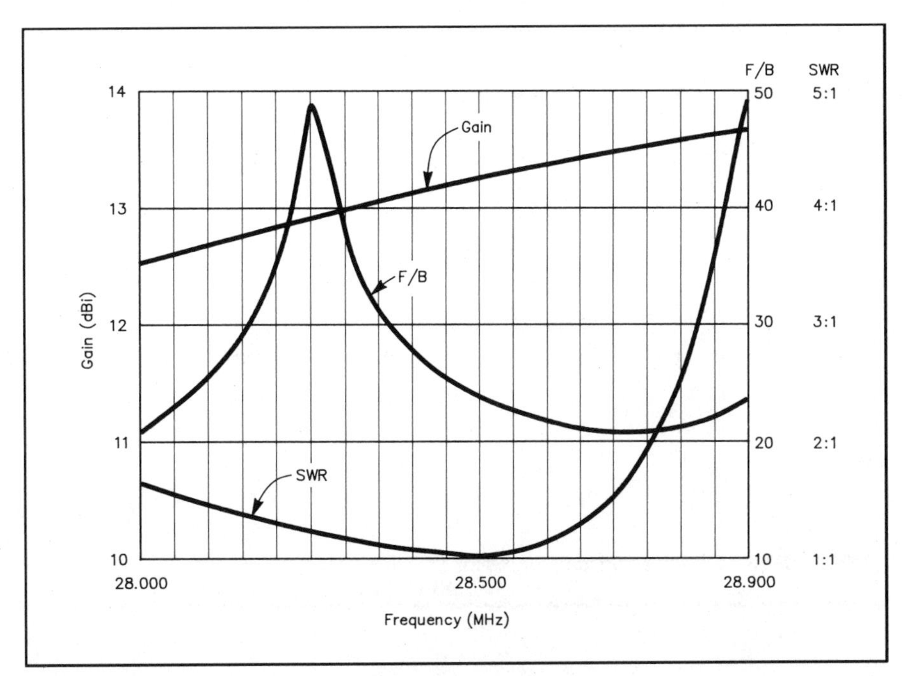


Fig 2—Gain, front to back ratio and SWR v frequency.

**Table 2**
**Half Element Construction**

| | *Section Diameter (inches)* | | | *Spacing from* |
|---|---|---|---|---|
| *Element* | *0.750* | *0.625* | *0.500* | *Ref. (in.)* |
| Reflector | 18.0 | 33.0 | 53.500 | 0 |
| Driven Element | 18.0 | 33.0 | 49.000 | 43.0 |
| Director 1 | 18.0 | 33.0 | 47.500 | 89.0 |
| Director 2 | 18.0 | 33.0 | 44.125 | 206.0 |
| Director 3 | 18.0 | 33.0 | 43.250 | 322.0 |
| Director 4 | 18.0 | 33.0 | 41.125 | 444.0 |
| Director 5 | 18.0 | 33.0 | 43.000 | 566.0 |
| Director 6 | 18.0 | 33.0 | 40.000 | 688.0 |

### Assembly and Mounting

As with all of my rotaries, I used the PVRC offset technique conceived by W3AU and W3GRF to mount the Yagi to the rotary mast. (This approach has been covered in detail in recent *ARRL Handbooks*.) Using the PVRC method allows one person to construct or repair a large antenna on the tower. This is a definite requirement for retirees like me!

The boom, which I previously marked for the exact location of each element, was hauled up vertically with the three boom guys securely taped at the center and U-bolted to the offset pipe. (The reflector position should be on top and the forward-most director at the bottom.) After attaching the last three directors—with the aid of my future son-in-law on the ground to assure element alignment—a small hand winch was used to turn the boom 180°. Next, the reflector, driven element and the first three directors were aligned and attached.

While the boom was vertical with the coaxial feed line securely taped to it, an SWR bridge was inserted at the gamma feed point. Unless you have access to a crane or cherry picker, this is the position at which the gamma must be tuned. For contesting requirements I chose 28.5 MHz as the point to obtain a 1:1 SWR. This was a compromise between the phone and CW portions of the band. After the antenna was matched to a 1:1 SWR, the U bolts on the boom-to-mast plate were loosened to allow the boom to be twisted (using the third director as a lever arm) to align the elements parallel to the ground. All that remained was to attach the boom guys, secure the mounting hardware, and make sure that the elements and boom were horizontal.

### Rotator Considerations

Since all of my rotary antennas are large in comparison to most installations, I use surplus propeller pitch motors exclusively for my rotators. Anyone attempting to construct an antenna of this size must be concerned with the torque that will be exerted upon the tower, mast and rotator. Nothing smaller than Rohn 45 should be used for the tower. If Rohn 45 is used, it should be braced with torque bars at the rotator. In addition, the mast should be secured to the tower with a thrust bearing that is capable of slipping under high torque conditions. (Slipping is always preferable to the alternative: a twisted tower section or a broken boom!) My position indicators (selsyns) are driven by the mast itself, so I do not need to climb the tower to realign the mast after slippage has occurred. The same is not true with most commercial rotators. In extreme wind con-

**Table 3**
**Elephant Gun Performance v Frequency**

| *Freq (MHz)* | *Gain (dBi)* | *F/B (dB)* | *Impedance (ohms)* | *SWR* |
|---|---|---|---|---|
| 28.000 | 12.50 | 20.79 | 19.40 – *j* 35.36 | 1.66 |
| 28.100 | 12.67 | 25.38 | 19.67 – *j* 32.82 | 1.46 |
| 28.200 | 12.83 | 34.36 | 19.86 – *j* 30.38 | 1.30 |
| 28.300 | 12.98 | 37.65 | 19.99 – *j* 28.17 | 1.16 |
| 28.400 | 13.12 | 27.65 | 19.89 – *j* 26.41 | 1.07 |
| 28.500 | 13.25 | 23.70 | 19.12 – *j* 25.23 | 1.00 |
| 28.600 | 13.36 | 21.64 | 16.98 – *j* 24.24 | 1.14 |
| 28.700 | 13.47 | 20.80 | 13.24 – *j* 22.32 | 1.52 |
| 28.800 | 13.57 | 21.21 | 8.88 – *j* 18.41 | 2.52 |

Note: Impedance normalized to 52 ohms for SWR.

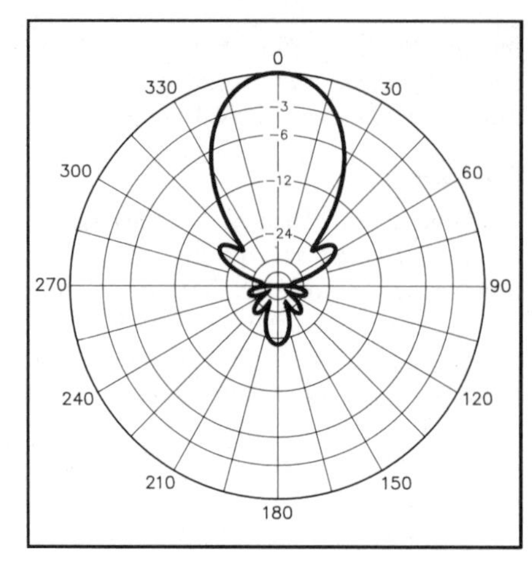

Fig 3—Free-space E-plane plot of the Elephant Gun Yagi at 28.5 MHz. The 0-dB reference is 13.2 dBi.

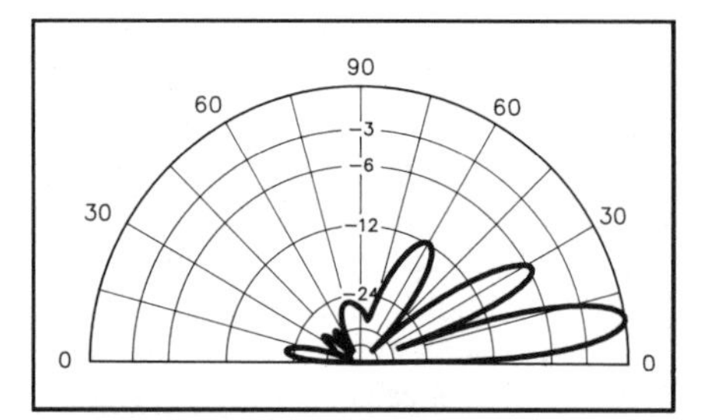

Fig 4—H-plane plot of the Elephant Gun Yagi at 28.5 MHz for a height of 50 feet over perfect earth. The 0-dB reference is 18.7 dBi.

ditions I often use the indicators to watch my antennas as they rotate by themselves!

### Performance

The performance of the Elephant Gun as predicted by YAGIMAX is profiled in Table 3 and illustrated in Fig 2. The free-space E-plane pattern at 28.5 MHz is shown in Fig 3. (The 3-dB beamwidth is approximately 40 degrees.) For my installation, I chose a height of 50 feet for this antenna. The H-plane pattern at this height over perfect ground is given in Fig 4. Notice that the main lower lobe covers 5 to 15 degrees at the 3-dB points. This is very effective for most of the propagation angles encountered on 10 meters.

How well does the Elephant Gun Yagi work? In the 1989 ARRL 10 Meter Contest my chief operator, WO0G, clinched the highest CW-only score. In the 1990 contest he racked up 2.2 million points in the mixed CW/SSB category. From the standpoint of my QTH in the Midwest, the Elephant Gun Yagi appears to be the perfect antenna for the Far East and Africa. Because my antenna arsenal also includes a monster quad with seven elements on a 40-foot boom, there are always comparisons being made between the two antennas. Into Africa or Japan, the Elephant Gun is the clear winner! I have fixed stacked arrays for Europe and South America, so I rarely need to point the Elephant Gun in those directions.

Without a doubt, this is a serious antenna for serious contesters and DXers. I probably should have kept it a secret!

### Notes

[1]B. Myers, "The 4-element PV Yagi," *NCJ*, Jan/Feb 1985, p 9.

[2]A similar TAPER program by B. Myers appears in Chapter 2 of *The ARRL Antenna Book*, 15th (1988) or 16th (1991) editions.

# Rear-Mount Yagi Arrays for 432-MHz EME: Solving the EME Polarization Problem

By Steve Powlishen, K1FO
816 Summer Hill Rd
Madison, CT 06443

Many EME operators observe periods of time when they are unable to make EME QSOs. They may start to think the predicted lunar path loss is an optimistic figure most of the time. However, an investigation into the EME signal-strength problem confirms that the predicted path loss (except for some short-term fading phenomena) is quite accurate. The real problem is the linear polarization in use at frequencies below 1296 MHz.

Fig 1, the theoretical polarization misalignment loss, shows how devastating significant polarization mismatch can be. There exist two causes of polarization misalignment. The first is called spatial polarization. This is a relatively straightforward three-dimensional geometry problem. The spatial polarization alignment problem is created because most EME stations use azimuth and elevation motion to track the moon with their arrays. This az-el motion is referenced to the array's local location on the earth's surface. When conducting space communications such as EME or satellite operation this earthbound reference is a hindrance. Reduced to its simplest explanation: Two stations using az-el array mounts may not look at the moon with the same relative polarization sense as seen from the moon.

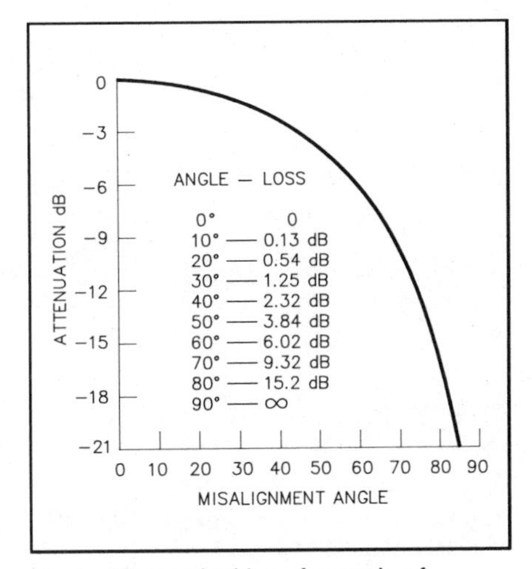


Fig 1—Theoretical loss for angle of polarization misalignment. The curve shows how fast the loss due to polarization misalignment increases after 45°. The curve repeats through 360°, giving no loss at 0° and 180° and maximum loss at 90° and 270°.

This can be pictured through the following example. An East Coast US station is trying to work a station in Europe. If they schedule during a high moon declination day and at moon rise for the East Coast station, both stations will have their arrays pointed somewhat to the East to see the moon. This will result in spatial polarity alignment. If they schedule a few hours later, the US station will still be looking to the East but the European station is now pointing near South at a high elevation angle. When the radio waves from the European station reflect back to the earth, they will appear to be near vertical in polarization to a US station with a conventional az-el array mount.

Intuitively, one should note that his own echoes will always have spatial alignment. This explains in part why many stations report hearing their own echoes as loud as anyone on the band. The other part of the explanation is probability. Since you are likely to listen for your own echoes more often than for any single other station, chances are you will hear more unusual signal strength peaks on your own echoes.

Spatial polarization effects generally work out such that given similar latitudes, the farther away in longitude the other station is, the less time spatial polarization will align. When crossing the equator, the effect can be different. In general, an East Coast US station has spatial alignment only to Europe for an hour or so at his moon rise and another short alignment period at European moon set. The East Coast US stations never have spatial alignment into Japan but have good spatial alignment for most of their window into Australia.

Spatial polarization can be easily calculated. The ability to do so is now incorporated into many computer programs, such as the VK3UM EME Planner and W9IP RealTrak programs. Spatial polarization changes continuously with the moon's motion relative to the earth's rotation. The rate of change is normally faster the farther apart stations are.

The second polarization effect is much less predictable. Faraday rotation is caused by the ionization in the earth's atmosphere. This electrical charge literally turns the radio waves in a corkscrew fashion as they pass in and out of the ionosphere. Faraday rotation is not yet precisely predictable for us amateurs. In general, at 432 MHz and above there is very little Faraday rotation during the solar activity minimum years. During the peak years of the sunspot cycle it is accepted that at 432 MHz Faraday rotation is often in the range of one full rotation (360°) for the two-way trip to the moon and back. During the very peak periods of solar activity and the resulting ionospheric disturbances over two full rotations may be possible on the two-way trip, although these effects are fairly short lived. In the years of minimum solar activity it may be unusual to see as much as 90° of Faraday rotation at 432 MHz.

The amount of Faraday rotation also changes with time. At the peak of the solar cycle Faraday rotation typically seems to move at around 30 to 40° per hour, given no unusual solar or ionospheric disturbances. Under disturbed conditions, especially in combination with sunrise or sunset, I have observed rates of change over 180° per hour. Under quiet conditions at the sunspot minimum years during nighttime operation, the amount of Faraday rotation may change only a few degrees over several hours. Since the amount Faraday rotation changes is inversely proportional to the square of the frequency, these figures are valid for 432 MHz only.

Faraday rotation will not be easy to predict in the near future, either. Complicating the matter is that Faraday rotation is anisotropic. That is, it doesn't matter if the wave is coming into or going out of the ionosphere. In either case, the wave front will turn in the same rotation direction. If Faraday rotation did not behave in this manner, it would not be a problem, as the rotation would simply cancel out.

This anisotropicity complicates the QSO process. If one has to move polarization to one side of spatial alignment on receive in order to hear the other station, one will then have to move it the opposite way for the other station to hear him on transmit. The combination of spatial polarity misalignment and Faraday rotation causes the classic EME problem in that station A can hear station B but station B cannot hear station A, or vise versa. Faraday rotation can also create the situation when neither station A nor station B can hear each other. Faraday rotation sometimes can be beneficial in that it will allow two az-el mount fixed-polarization stations with significant spatial polarity misalignment to hear each other and make a contact. Advanced mathematics students familiar with probability theory (along with the sometimes less capable students of Murphy) understand that this condition is the exception.

If we wanted to determine the actual amount of Faraday rotation at any given moment, we would have to build two EME arrays, one fixed in polarization and the other rotatable. The fixed array would be used to transmit, and the echoes would then be peaked on the rotatable array. Because of the way polarity peaks repeat at 180° intervals, however, it may not be possible to determine polarity rotation any more accurately than to the nearest 180°. This means if 10° of rotation was measured you really would not know if it was 10, 190 or 370°.

The K1FO 16 × 14-element 3.6-λ 432-MHz EME array shown at 2° elevation and vertical polarization.

A very ambitious operator could solve the problem by setting up another EME station with two arrays at a frequency that would never have more than 180° of Faraday rotation, such as 1296 or 2304 MHz. In the example, if around 40° of rotation was measured on 1296 MHz one would then know that the 10° measurement at 432 MHz really corresponded to 370°. To further complicate the Faraday measurement problem, this operator would have determined the amount of rotation only at his array's specific location. Local ionization effects can vary greatly across the earth at the same time, especially if one end of the EME path is in daylight and the other end is in darkness.

Thinking about the situation, a given station could have 45° of Faraday rotation on his signal when it leaves the atmosphere. Another 45° of rotation would occur when the echo returned, making the total perceived rotation 90°. At the other end of the path, however, if it were a significant distance away, Faraday rotation could be 90° each way. Therefore, if the first station was able to do this sort of ionospheric sounding, he would be able to know Faraday rotation at his location is a total of 90° (two-way path). But he would have no way of knowing that 5000 miles away the effective two-way Faraday rotation is 135° (on that particular path only!), unless he heard a station 5000 miles away transmitting. He then would have to find the polarity at which signals peaked, calculate the relative spatial polarization and finally deduce how much Faraday rotation there was, given our previous 180° accuracy problem. All this assumes he knew what polarity that DX station was transmitting on!

### Solving the Problem

There are several methods available to solve this polarization dilemma. Spatial polarization alone could be remedied if everyone were to use polar mounts. The use of polar mounts would not help to work around Faraday rotation, however. In addition, polar mounts can really be used effectively only on rear-mounted arrays, that is, parabolic dishes, collinear arrays or rear-mounted short Yagis.

The second way to solve the problem would be to use circular polarity. Circular polarity sense is reversed when the signals reflect off the moon. This means you must transmit in one polarization sense (for example, right hand) and receive in the other (for example, left hand). To accomplish this, one would need two separate helix arrays, crossed Yagis set up for circular polarity with the ability to switch polarization sense or a parabolic dish with a two-way circular feed.

Unfortunately, in practice at 432 MHz only the dish solution is truly practical. A pair of helix arrays is unmanageable for most. Crossed Yagis with their associated phasing lines have not yet proved to be able to work effectively on EME receive. An additional problem with circular polarization is that it would be a discouragement in creating new EME activity, since well-equipped tropo stations would find it much harder to make their first EME QSOs with circular polarized EME stations versus linear polarized stations.

If linear polarization is to be the standard at 432 MHz, then it is necessary to be able to rotate polarity at will or at least be able to switch between horizontal and vertical polarization (or simply two polarity arrangements aligned 90° apart). Dish stations have a distinct advantage, as it is simple to feed a dish with two sets of dipole feeds positioned mechanically 90° apart. Most dish stations prefer to rotate their feeds, which is easy to do considering their small size. This allows one to be able to track all possible polarization angles.

Parabolic dish antennas therefore seem like the way to go at 432 MHz. There is only one slight problem with using a dish at 432 MHz. To have a good signal, a dish

Fig 2—Horizon view at K1FO from 20 ft elevation is typical of the eastern part of the US. It is not very conducive to EME operation at 432 MHz unless the moon is at a high elevation.

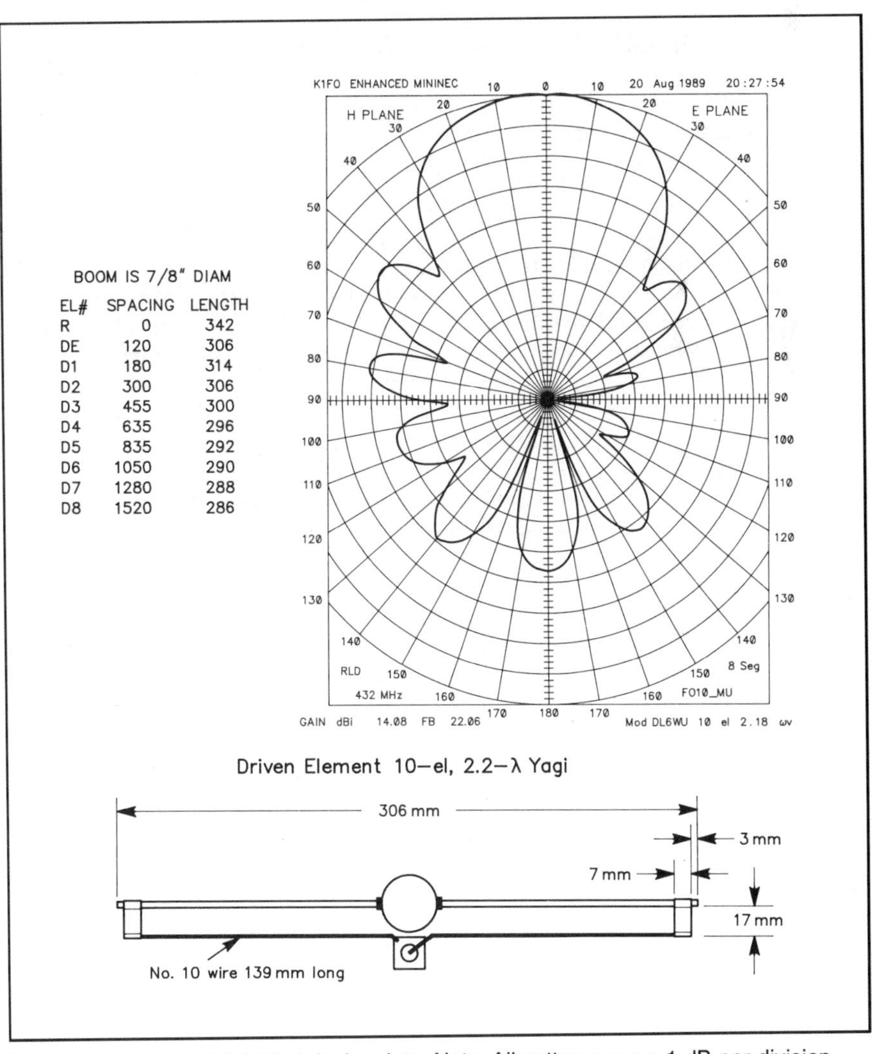


BOOM IS 7/8" DIAM

| EL# | SPACING | LENGTH |
|---|---|---|
| R | 0 | 342 |
| DE | 120 | 306 |
| D1 | 180 | 314 |
| D2 | 300 | 306 |
| D3 | 455 | 300 |
| D4 | 635 | 296 |
| D5 | 835 | 292 |
| D6 | 1050 | 290 |
| D7 | 1280 | 288 |
| D8 | 1520 | 286 |

Fig 3—10-element, 2.2-λ Yagi design data. Note: All patterns use a 1-dB-per-division linear scale instead of the standard ARRL pattern in order to show lobe detail more clearly

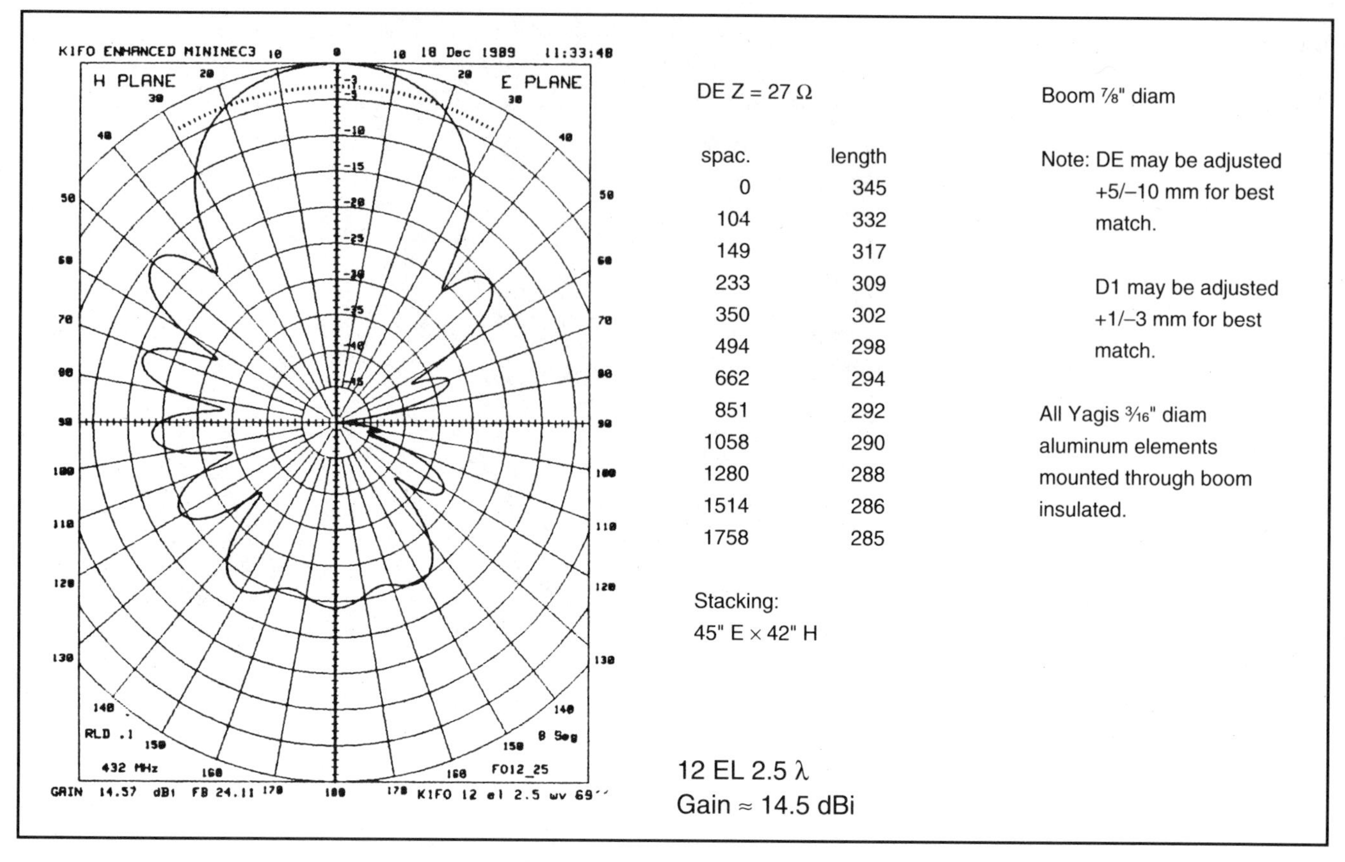


Fig 4—12-element, 2.5-λ Yagi design data.

in the 20-ft (6 m) diameter range is needed (>26 dBi gain). To be a big gun we are looking at 24 feet (7.3 m) diameter (~28 dBi) or larger dishes. This is all possible of course if you put your dish low in height, as a 28-ft (8.5 m) diameter dish (~29 dBi) probably has around 200 square feet (19 $m^2$) of wind load area.

For most of us located outside of the Great Plains, Fig 2 demonstrates this problem. This would be the horizon view at my QTH if I were to put up a parabolic dish centered at 20 ft (6 m) above the ground. My moon window would be effectively limited to elevations above 40°. I suppose if I had a quarter for every time I listened to distant EME stations trying to schedule around window restrictions I could afford to go purchase a brand new tower capable of supporting such a dish 100 ft (30 m) high!

Yagi arrays have the advantage of being much lower in wind load and weight for a given amount of gain. A 24-ft (7.3 m) diameter dish has about 28 dBi gain and 160 square feet (15 $m^2$) of wind load. A Yagi array that could be mechanically turned in polarization could be built to have 28 dBi gain with less than 25 square feet (2.3 $m^2$) of wind load area. This allows the average operator to locate that Yagi array above the tree level. Moreover, such an array is easily assembled without a team of helpers or mechanical assistance from equipment such as cranes. All of my Yagi arrays, with fixed and adjustable polarization, were assembled by two people with only the assistance of a gin pole. I could have assembled them alone if longer assembly time was allowed.

## A Practical Yagi Solution

Having been educated to be an engineer and having been born an analytic, I spent a long time analyzing the possible approaches to the design of a Yagi array with adjustable polarization. I decided my criterion was to have an array that was at least equal in performance to my old 12 × 22-element 6.1-λ Yagi array (27.8 dBi) or in the ballpark of a 24-ft (7.3 m) diameter parabolic dish. Although less gain would do a respectable job, I felt this was the minimum gain required to work single-Yagi stations repeatedly. In addition, I felt I was already having enough trouble merely holding my frequency during the ARRL EME contest. Anything less would reduce my contest capability from also-ran status to the pitiful category. I then spent far too much time analyzing what my options were.

## The "Make it Bigger" Approach

This is not a very scientific solution to the EME signal strength problem. The thought process uses a rationale of "if the signals are not as strong as they should be, I'll simply make it bigger until I get the results I'm looking for." When good long-Yagi designs for 432 MHz started to appear, many stations rushed to this plan without gaining an understanding of the problem they were trying to solve. Another look at the polarization alignment loss curve (Fig 1) shows that increasing your array's gain will initially have very positive results. You will now be able to work stations at significantly greater misalignment angles, but soon there reaches a point when almost nothing will help.

For example, if you are just able to make EME QSOs with small stations at perfect polarity alignment (for example, you have a 23 dBi array), increasing your array gain by 3 dB will allow you to make QSOs at ±45° polarity alignments, which is quite an improvement. If you then increase the array gain by 3 dB, again your polarity window is increased by an additional 15°

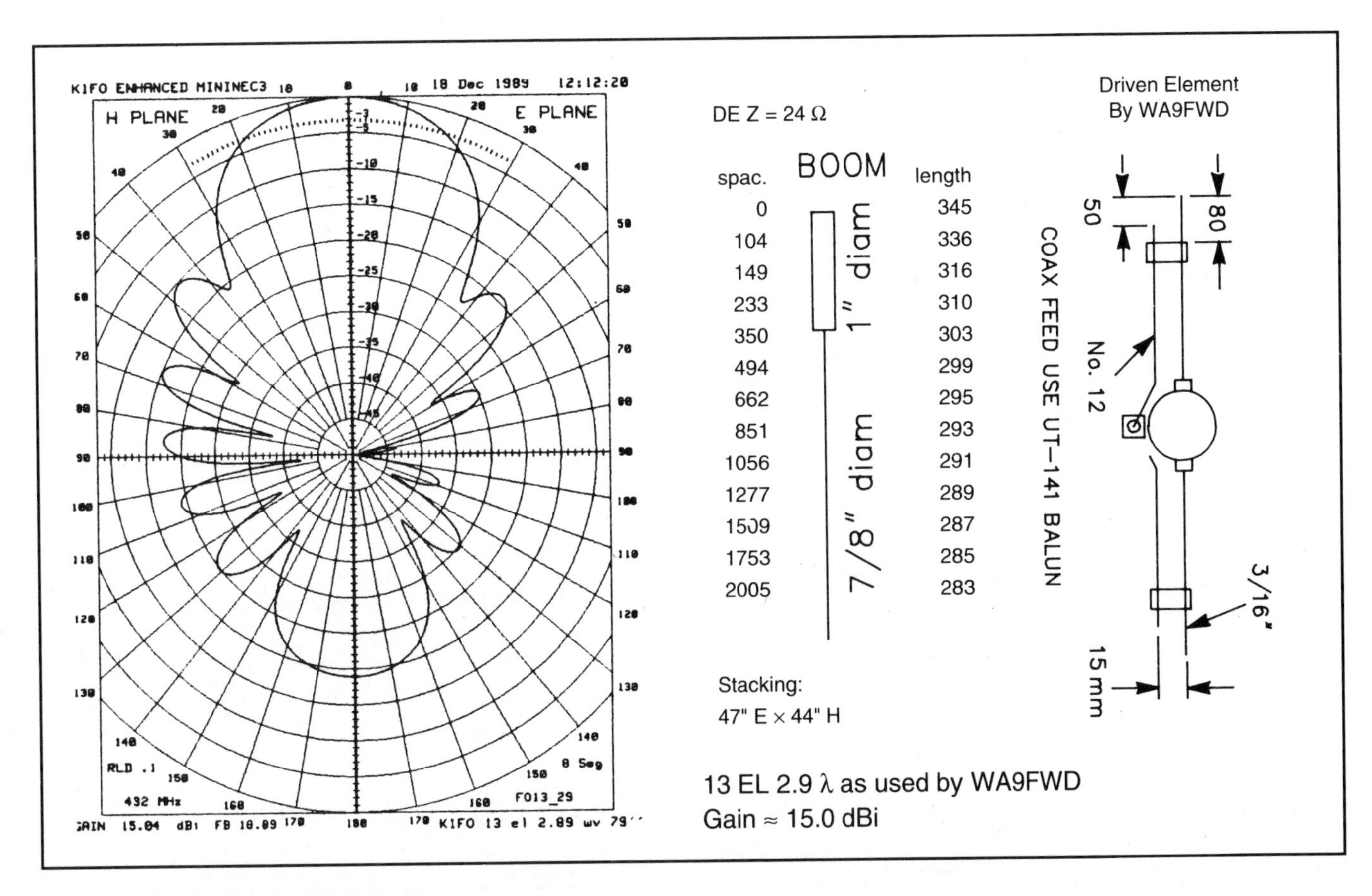


Fig 5—13-element, 2.9-λ Yagi design data.

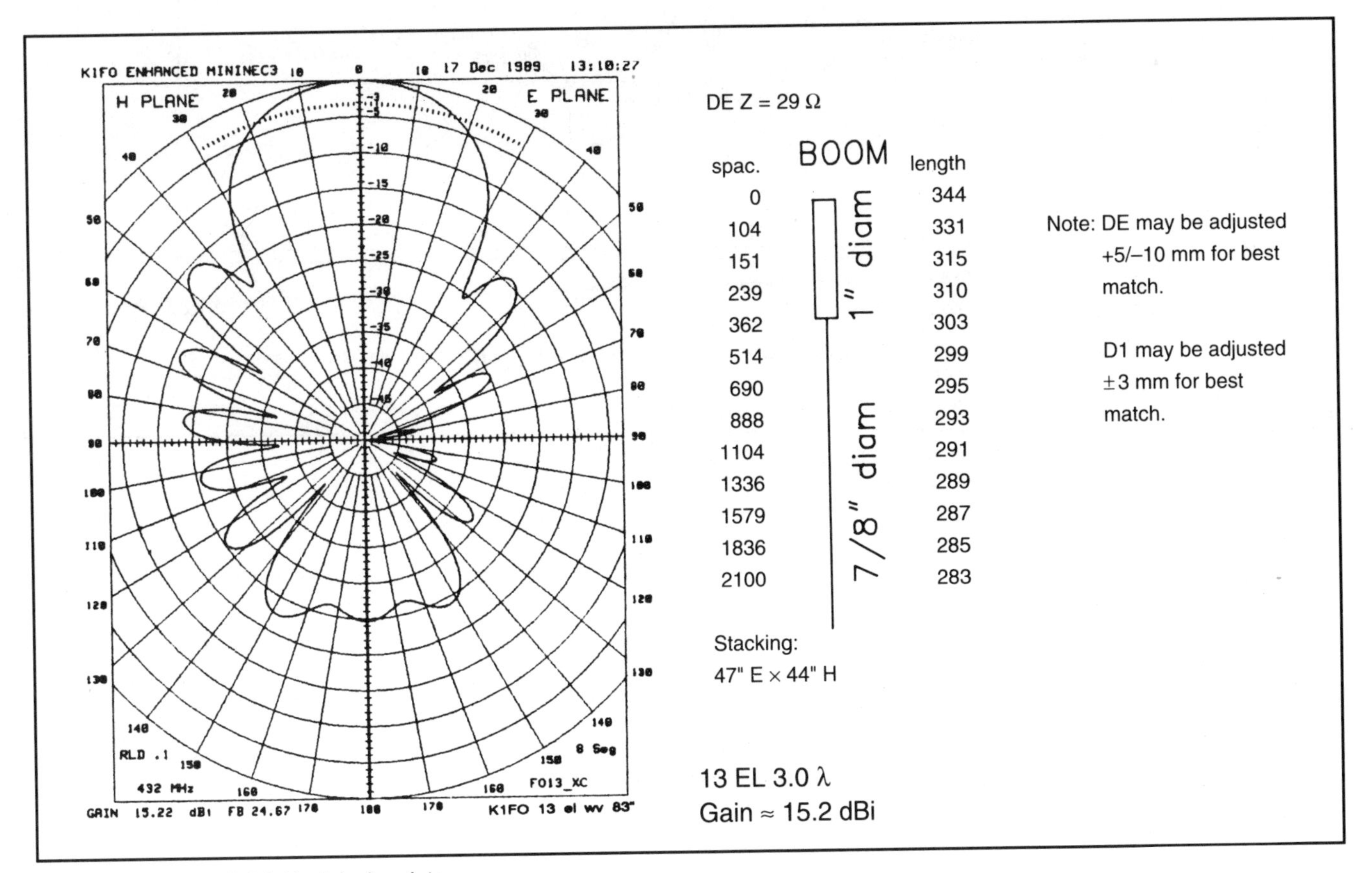


Fig 6—13-element, 3.0-λ Yagi design data.

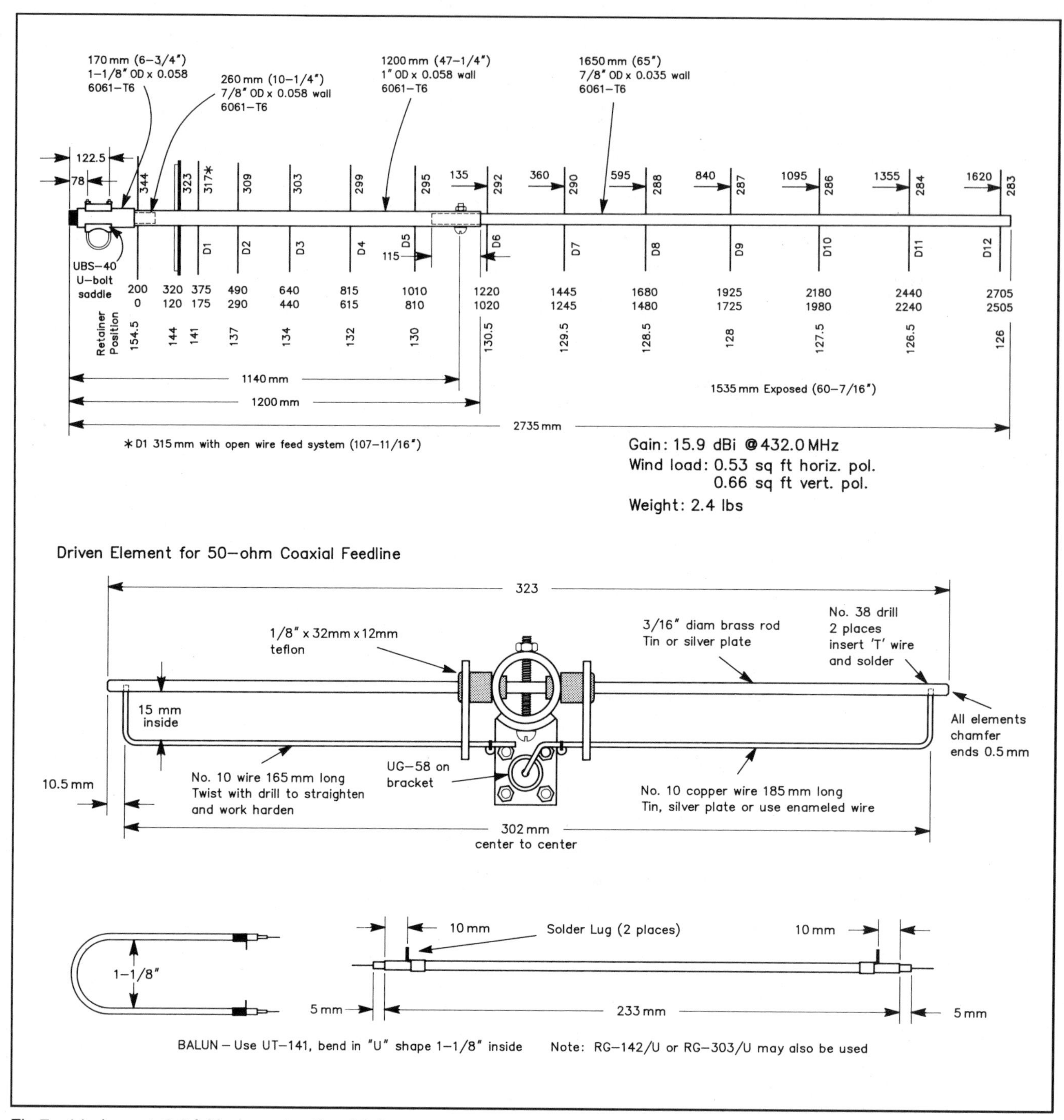


Fig 7—14-element, 3.6-λ Yagi construction details.

each way to ±60° or only ⅓ the amount the first 3 dB increase of array gain gave. However, you now have a 29-dBi-gain array, which is fairly substantial in size. If you again increase the array size by another 3 dB to 32 dBi, your polarity window has only increased by another ±9° to ±69°. Going all the way to 35 dBi, which is now an immense array, the polarity window increases by only ±4° more to a ±75° total polarity window.

In other words, even with a 35-dBi array you could wind up locked out from very small stations. Several stations (WBØTEM who had a 24 × RIW-19 Yagi array, DL9EBL with 16 × 10-λ Yagis, NC1I with 16 × 10.5-λ Yagis and DL9KR with 16 × ever-growing Yagis) have typified this "make it larger" approach. It is interesting to note that both WBØTEM and DL9EBL have gone to parabolic dishes while NC1I has started work on an adjustable-polarization array.

DL9KR has been the most successful of the fixed-polarization stations. I attribute much of his success to his tireless search for perfection in the Yagi and feed system losses, the ultimate preamplifier, excellent operating skills plus complete focus of his amateur VHF operation on 432-MHz EME. Some of his success must also be credited to his European location which assures Jan good spatial alignment to the majority of active 432-MHz EME stations. In spite of all this, sometimes even DL9KR needs several tries to work a new station.

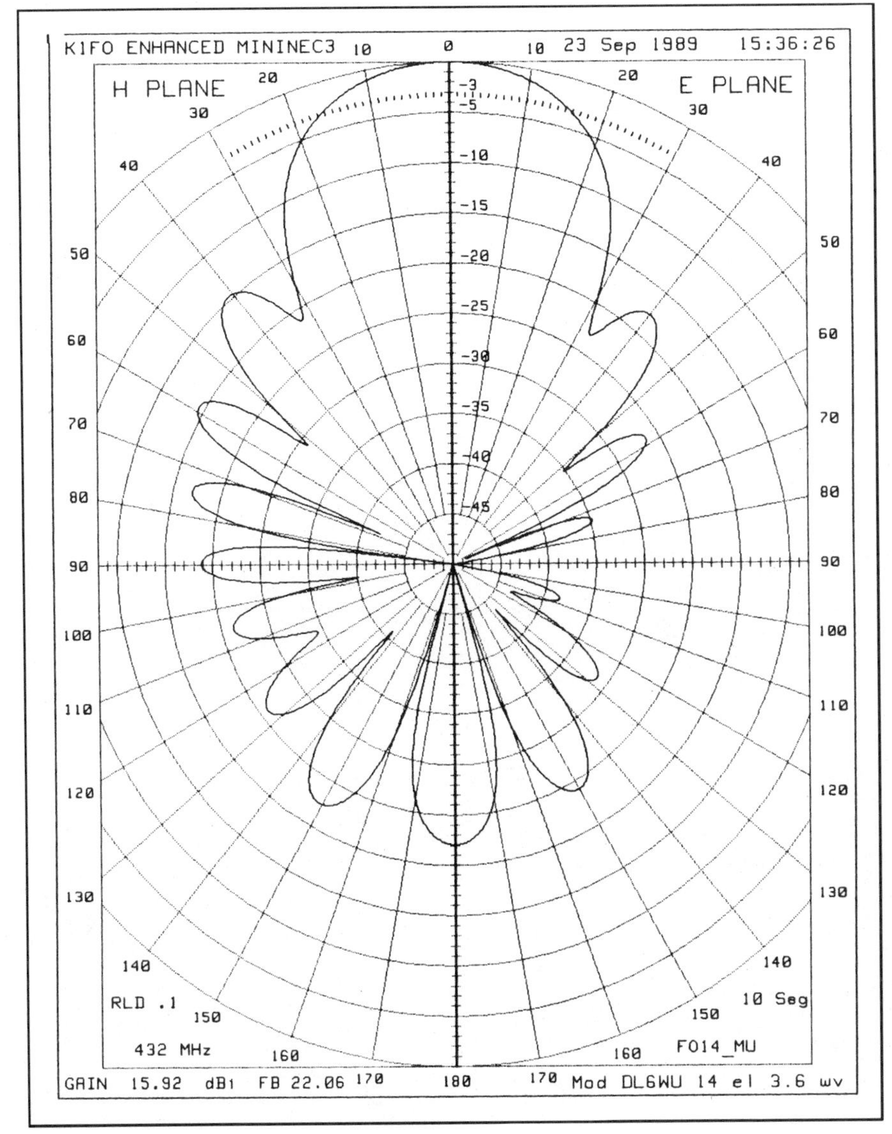


Fig 8—Calculated E- and H-plane patterns for 14-element Yagi.

Thoughts of going to 16, 24 or even 32 of my 22-element Yagis were considered. The lingering anxiety of winding up perfectly misaligned at the wrong moment weighed heavily against this approach. The "make it bigger" approach is also not the best plan for someone who desires to mount the array 80 ft (24 m) above ground level in order to have a reasonable moon window. If the magnitude of the support tower does not provide a significant enough discouragement, the thought of attempting to assemble or repair such a large array on top of a high tower should.

### The Crossed-Yagi Approach

The next idea was to use crossed elements on a common boom, allowing polarization to be switched 90°. I really liked the idea of the capability to switch polarization instantaneously. A minor consideration was the 3 dB loss when signals were polarized 45° out. My enthusiasm for this type of an array was tempered by its associated mechanical and electrical problems. Earlier work on Yagi design demonstrated that 432-MHz Yagis simply cannot tolerate significant amounts of conductive material within the Yagi aperture, which is parallel to the Yagi elements. Although it takes quite a bit of offending material to lower the forward gain of the Yagi significantly, the pattern rapidly deteriorates even when only phasing lines are run near the elements. This expected receive performance loss simply could not be tolerated.

The only possible way to minimize the problem would be to use fiberglass stacking frames and run the phasing lines out at a 45° angle. Fiberglass is a poor substitute for 6061-T6 aluminum in terms of strength, weight and cost. In addition, fiberglass is very flexible, which would compound the problem of keeping all the individual Yagis aligned at all elevation angles. The next problem was that my 22-element Yagi spacings did not lend themselves well to interlacing the vertical and horizontal elements. To accomplish this, the booms would need to be made longer to allow for an offset difference between the two sets of elements.

The small physical size of 432-MHz Yagis makes for another problem in having adequate space to connect the two sets of driven elements to the two sets of phasing lines without interacting with the nearby directors. Commercial OSCAR antennas made from crossed Yagis have traditionally fared very poorly on the antenna range, demonstrating this problem.

The final problem is the complexity and cost. Due to the above-described performance limitations, I felt that 16 Yagis would be a minimum for this type of array. The cost of Heliax-type phasing lines and power dividers would be prohibitive (remember, there are two sets of 16-Yagi feeds on this array). If smaller, higher loss phasing lines were used, receive performance would be compromised. To maximize performance, one would also want separate TR relays and preamplifiers for the vertical and horizontal Yagis to avoid the loss of the array selecting relay and jumpers. Having climbed my tower countless times in the middle of the night to remove water from my 12-Yagi array phasing lines, just the thought of increasing the connector count by nearly 3 times (84 connectors!) led to several sleepless nights. Finally, since I subscribe to the DL9KR "stupid simplicity" school of thought, any plans for a crossed-Yagi array were dropped.

### Rotating the Individual Yagis

I don't know if he originated the idea, but the ability to rotate the individual Yagis in an array was suggested many years ago by K2UYH in the *432-MHz EME Newsletter*. I never gave the idea much thought until KDØGT called me a few years ago to discuss the approach. I did my best to discourage Marty, but in spite of me he proved the method is workable. The primary disadvantage is the complexity of the mechanism needed to rotate the Yagis. Marty opted to use the "windshield wiper" lever arm approach to gang two Yagis on a single motor. With a little engineering innovation it should be relatively easy to extend the lever arm principle to turn four Yagis off a single motor. For arrays using more than four Yagis, keeping the multiple polariza-

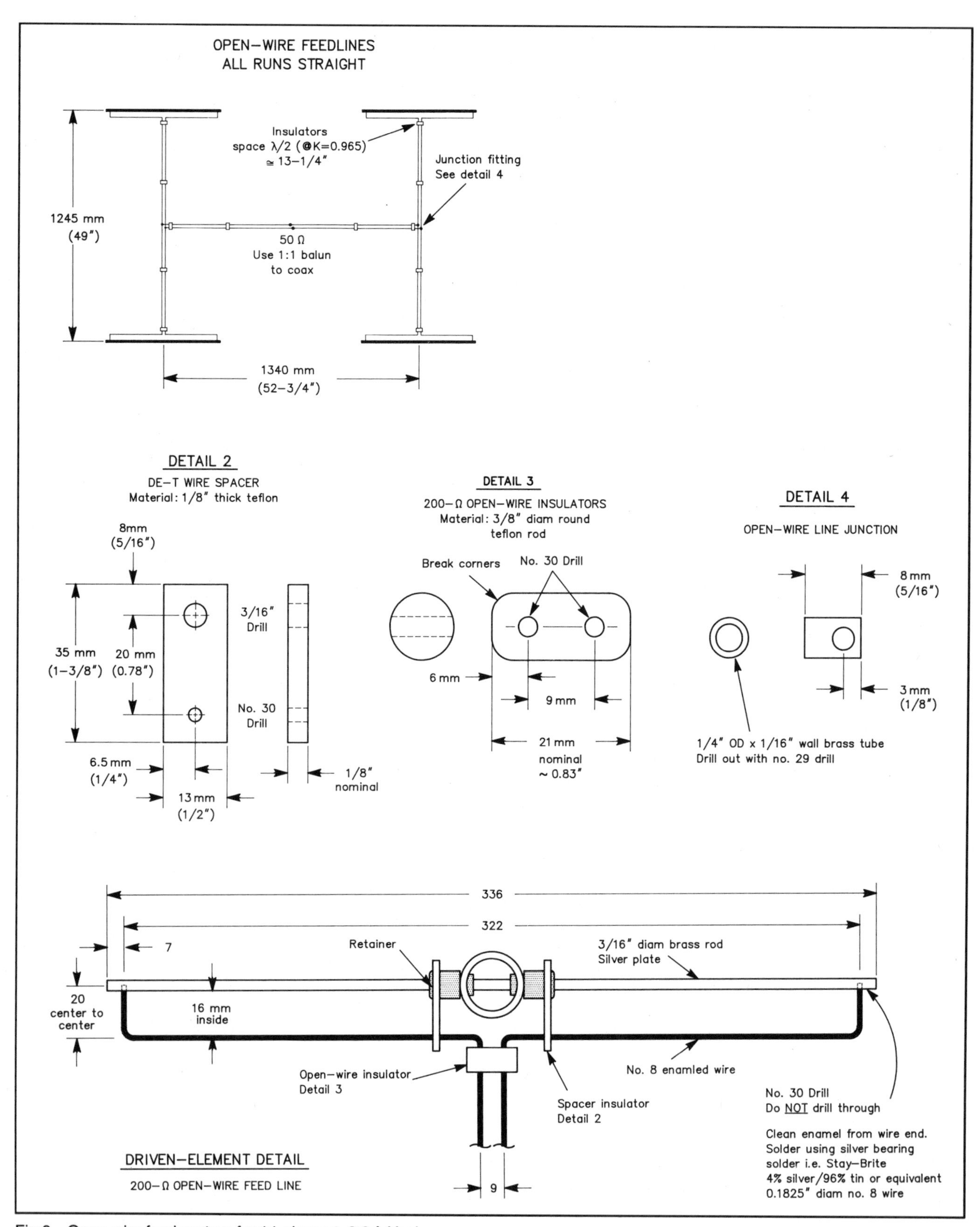


Fig 9—Open-wire feed system for 14-element, 3.6-λ Yagi.

tion drive motors synchronized is a problem. The builders of these type arrays so far have chosen to use 0 to 90° of rotation with a mechanical stop at 0 to force synchronization.

My antenna range tests and KDØGT's array experience agree that if metal masts are used performance is good until the Yagi elements get within 30° of parallel to the masts. At that point, the pattern (especially the front to back) rapidly deteriorates and about 1 dB of gain is lost on a typical long

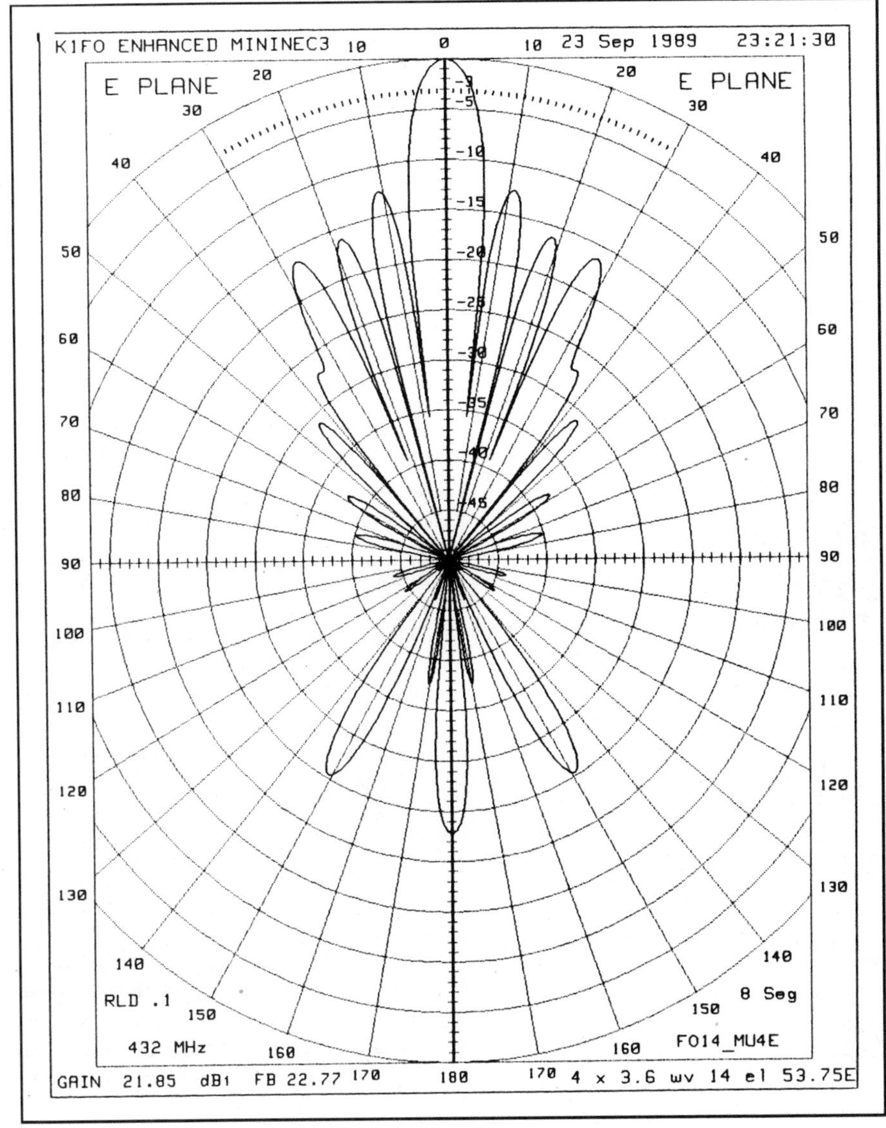


Fig 10—Calculated E-plane pattern for 4 × 14-element Yagis stacked 52.75 in. apart in the E plane.

Yagi. Marty also reports a significant SWR change making the usable polarity range 0 to 60°. KB3PD has attempted to solve the problem by using fiberglass masts. The fiberglass approach has been only partially successful as it has been difficult keeping the Yagis aligned due to the flexibility of even heavy-wall fiberglass poles.

If someone were to use this approach I would suggest a solution to the motor synchronization problem be found as a first priority and the array be set up for ±45° rotation or ±60° if possible. This polarity range will be more of an advantage as the full mechanical rotation range would now be useful. Attention must also be paid to routing the phasing lines such that they don't wind up parallel to the elements in close proximity to them. This points up another limitation in the "rotate the individual Yagis" approach. The phasing lines must be made long enough and flexible enough (read that "small in diameter") to allow for the rotation. This naturally implies their loss will be higher than fixed-mount Yagis, hence limiting receive performance. Those using four Yagis would do well to consider this approach, as mechanical complexity would be manageable. A half dB loss on receive would be more than made up for by even ±45° polarity rotation.

**Rotating the Entire Array**

Mechanically rotating the entire array around its boresight axis to change polarity is not a new approach, either. W1JR used the method with his extended expanded collinear array as far back as 1972. The Mount Airy VHF Club's HK1TL expedition adapted polarity rotation to Yagis in 1976. They used 16 × 3.5-λ Yagis that were semi rear mounted, allowing the array to be rotated with elevation above 20°. In the late 1970s, when good long-Yagi designs appeared (>5 λ), this approach was quickly forgotten. WBØYSG later had some success with a mechanically impressive but somewhat impractical array of 16 K2RIW 19-element Yagis. It incorporated a rather stout frame that held the array from behind to allow for polarity rotation. If only he knew that the second reflector on the RIW Yagi did nothing, history might have been different.

I'm sure many of us thought about using smaller rear-mount Yagis, and rotating the array in a similar manner. I can claim credit for contemplating this approach in 1985. Unfortunately, at that time my mind set was still focused in the "make it bigger" approach. I quickly dropped any thoughts I had for this type of array as at that time I thought there was no way to obtain enough gain for reliable EME communications with a little rear-mount array.

It's an old cliche that necessity is the mother of invention. That being the case, the burden of proving that a small rear-mount Yagi array could work was left to Tim Pettis, KL7WE. Since Tim is located in Anchorage, Alaska, he has the misfortune of being spatially aligned to no one for any length of time save himself. It was lucky for Tim that the high-performance DL6WU log-taper Yagi and its computer-assisted derivatives happened to be available to solve his dilemma. While many Yagi designs of suitable boom length had existed for years, even those with near maximum gain per their boom length lacked the clean pattern and "stackability" required for successful 432-MHz EME operation. Just think of how many 16 short-Yagi-array builders were either totally unsuccessful or barely EME-capable with those old designs.

The rear-mount Yagi array approach had the advantages of "stupid simplicity." There is only one moving point in the array. This lets all the Yagis be permanently aligned in their proper spot. It also allows for the lowest-loss phasing lines, be they coaxial or open wire, since the driven elements are at the rear of the array. Any possible offending metal in the Yagis' aperture is behind the array, permitting maximum receive performance to be obtained.

There are only two disadvantages of this type of array. The first is that some level of metal fabrication is required for the elevation/polarity rotation mount. The second is

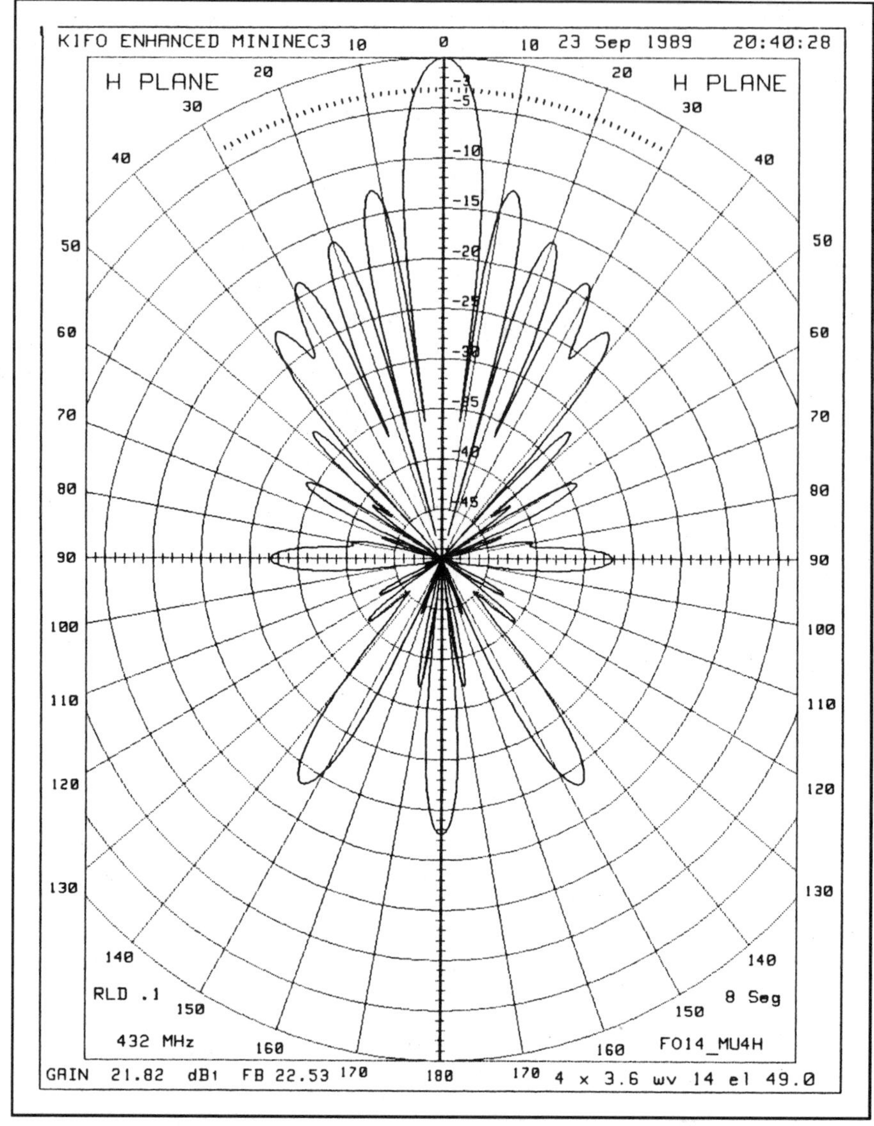


Fig 11—Calculated H-plane pattern for 4 × 14-element Yagis stacked 49 in. apart in the H plane.

that it will be difficult to become king of the band with a rear-mount array. A gain of 26 dBi can be had in a very modest array size. A 28-dBi-size array will be at the limits of most EME hopefuls in terms of size and complexity. Higher gains than that can be obtained, but only by using arrays with mechanical complexities approaching that of a substantial-size dish installation or by someone with considerably more mechanical ingenuity than myself.

### The Split-Array Solution

Barry, VE4MA, had been suffering through the polarization blues almost as bad as KL7WE. Located in Winnipeg, Manitoba he faces long periods of spatial isolation. His approach was to split his array in half, making the 16-Yagi array into separate arrays of 8 vertical and 8 horizontal Yagis. Since then others like N4GJV have emulated the approach by using separate vertical and horizontal arrays. Although a workable solution, the split- or two-array solution has the disadvantage of reducing the array gain by 3 dB plus having the potential of losing another 3 dB for 45-degree polarity alignments. The split array also has the same double preamplifier or array-switching complexity problems as the crossed-Yagi array. The two-array approach adds the further complexity of a second tower and complete EME system cost. For those ambitious enough to build a second array (assuming they have the real estate), the second array adds a nice bit of redundancy should anything happen to one of the arrays.

### The Chosen Plan

The first year I was operational on 432 MHz EME (1984) was near the sunspot minimum. I quickly learned when to run schedules and intuitively knew when spatial polarity was aligned before I got around to calculating it scientifically.

EME is a competitive business and you quickly start comparing yourself to similar stations. For a good, highly active Yagi station with a similar moon window, N4GJV became the comparison. It's interesting to note that in 1984 Ron worked one initial that I didn't, and in 1985 he didn't work anyone new that I didn't. With 1986 the situation started to change. That year Ron worked 11 initials that I didn't catch. In 1987 the number was down to 2 but back up to 8 in 1988 and 7 through the first 9 months of 1989.

Something had happened due to the sun and its rising flux. Suddenly, spatial polarization wasn't working as Faraday rotation became greater and greater. Since I often accuse N4GJV of operating under a golden Faraday window, the decision to go rotatable polarity was made in August of 1988. The event that forced the decision was WA9FWD's expedition to Nevada to give me my 50th state. It took me over two hours to work John. Although I was speaker copy in Nevada for over an hour I couldn't hear a trace of John out here. We luckily caught a polarization rollover when I worked John as I faded out in Nevada and he came up to solid copy in Connecticut. It took N4GJV about 15 minutes to work John and to reinforce the Faraday window thought. NC1I was not able to work WA9FWD/7 even with 3 dB more array gain, as I had stolen the brief polarization window.

Examining why N4GJV had been easily able to work WA9FWD/7 came down to spatial polarity. Since Ron is 500 miles closer to Las Vegas than I am, he had less spatial polarization misalignment. It just happened that the combined Faraday rotation and spatial alignment caused a 90° polarity lockout for me. Ron's location probably experienced a similar amount of Faraday rotation. However, since Ron had about 20° less spatial misalignment, even when the Faraday rotation was added in, Ron didn't see the full 90° cross polarization I had and therefore he was able to work Nevada. I had worked my 50th state but if I had only a standard 30-minute schedule slot the story would have been different. With 50 states worked it was time to start counting countries. Having missed at least two of them during my 432 days so far, I didn't want to take any more chances.

It should also be obvious why these efforts to overcome the polarity problems have been driven from North America. The

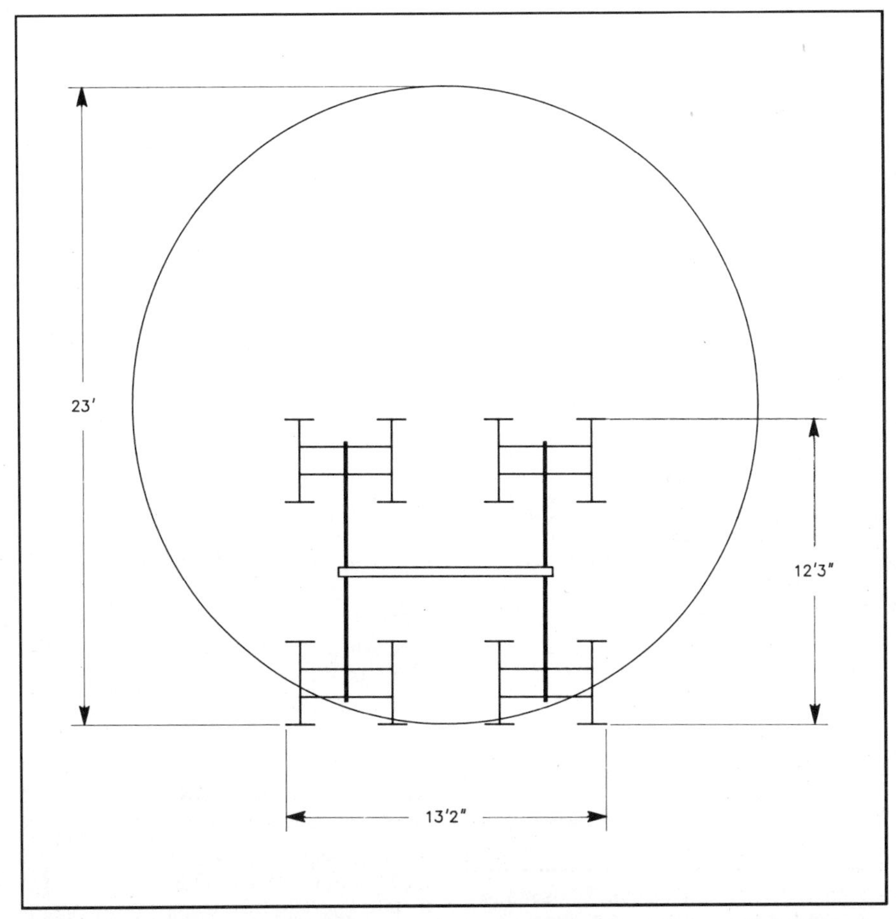


Fig 12—A comparison of a 23-ft diam parabolic dish to the 16 × 14-element, 3.6-λ Yagi array shows how much smaller the Yagi array is. Both the dish and the Yagi array have approx 27.5 dBi gain at 432 MHz and similar array temperatures.

greatest amount of EME activity on 432 MHz is in Europe, where most operators have over 100 active stations located within a relatively good spatial alignment window. The best situation a North American station can have is to be on the East Coast of the US. Even in that case he would have long-term spatial alignment only to around 25 stations. This creates more of a need for control over one's polarity in this part of the world.

When I finally came to this conclusion I would have liked to have polarity movement immediately available to me. I came very close to splitting my array as VE4MA had done but I decided a more elegant approach was desired. After some long consideration I chose the rear-mount array approach. This type of array allowed me to get the maximum receive performance given the size of the array. This was consistent with my thought process of getting the most performance for the minimum amount of wind loading, thus satisfying my requirement to locate the array above the trees. I also felt that further development of the rear-mount array would make for the most significant contribution to 432-MHz EME technology development by giving other operators who may have limited space an array with the highest performance-to-size ratio.

### The Yagi Design

The KL7WE array provided a starting-point role model. It used 16 × 11-element 2.5-λ DL6WU Yagis for a calculated array gain of 25.7 dBi, not up to the gain levels I was looking for (28 dBi). Even 24 of them would wind up short of my target. The next step was to determine how long a boom could be supported from the rear. My initial decision was to go with 16 × 4.4-λ (121 in./3 m) long, 17-element Yagis based upon one of the W1EJ spacing patterns. With an individual Yagi gain of over 16.6 dBi, the complete array of 16 would make my 28-dBi target. A test Yagi was built and tested. The Yagi proved to be manageable, but as I started to mock up the array the torque loading on the frame by the 10-ft (3 m) long Yagi booms quickly became unmanageable.

I decided to back down to a 16-element 4.0-λ Yagi. At the same time John, WA9FWD, was bugging me to come up with a roughly 7 ft (2 m) long Yagi for him to use in a rear-mount array. I became involved in a rather lengthy distraction of looking at Yagis from 2 λ to 5 λ for use in rear-mount arrays. It was a bigger problem to change boom length with these shorter Yagis than with the long Yagis I was more familiar with. The front-to-back ratio of a short Yagi would oscillate quite dramatically as elements were added or removed. This effect made sense since adding or deleting an element to a 3-wavelength Yagi will change the boom length by a greater percentage of the total boom length than will the same process on a 10-λ Yagi.

I then took a look at the DL6WU design and found that at certain boom lengths a tweaked DL6WU design gave a better overall pattern, while at other specific boom lengths my modified W1EJ design was better. The result of this work was a 13-element 2.9-λ Yagi that WA9FWD has quite successfully used in his rear-mount array. Also designed, but not yet used in arrays, are 12-element 2.5-λ, 10-element 2.2-λ and 13-element 3.0-λ Yagis.

The 3.0-λ 13-element Yagi is an improved version of the Yagi WA9FWD is using. It is a result of my discovery that a given Yagi design can have its pattern dramatically changed by simply narrowing or widening all spacings by a small percentage. Thus, there is no overall perfect absolute spacing pattern for all Yagi boom lengths, but there are preferred exact spacing patterns for any given boom length. All of the Yagis documented in this article are good EME performers; it's just a matter of how large an array you want to build. Fig 3 has the design data and calculated pattern for the 10-element, 2.2-λ Yagi. The 12-element, 2.5-λ model is in Fig 4. The 13-element, 2.9-λ information is in Fig 5 and the 3.0-λ Yagi is covered in Fig 6.

With WA9FWD on track with his array it was now time to return to finishing the design of my own array. I had decided that I wanted to use open-wire phasing lines for the groups of four Yagis. This not only would give significantly lower phasing line loss but would be lower in cost and eliminate 32 coaxial connectors. The 4.0-λ Yagi was a little too high in gain to use 2.0-λ spacings (at the open-wire velocity factor of 0.965). If I understacked the 4.0-λ Yagis in order to use the open-wire line the array would only have about 0.1 dB more gain than the 3.6-λ Yagis due to the lost stacking gain.

I was also still concerned about the weight and torque loading of even these

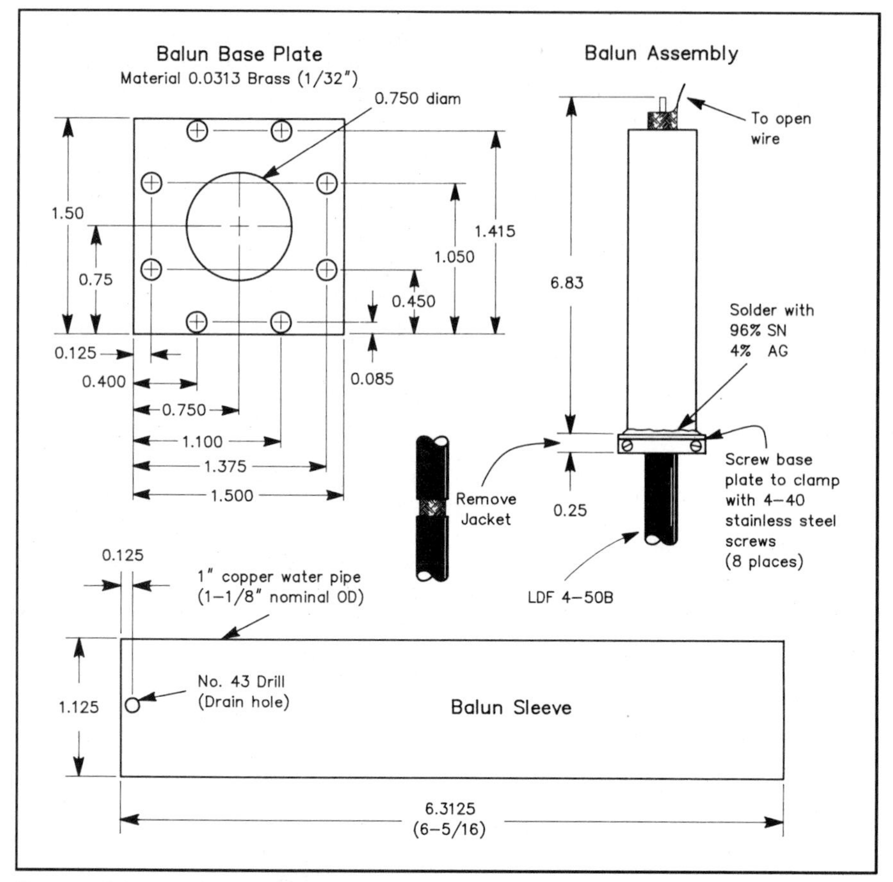


Fig 13—Construction details for the balun baseplate, assembly and sleeve. Tests made by the author showed losses in both sun noise and earth noise without the sleeve baluns.

110-in. (2.8 m) long Yagis. A decision was then made to back down still another director to a 3.6-λ Yagi. At 15.9 dBi per Yagi the 16-Yagi array was expected to have 27.5 dBi gain. Although this was 0.3 dB lower than my target, I felt that the open-wire phasing lines would lower the system temperature enough to more than make up for that difference on receive. The missing transmit gain could be had by reducing my 2.3 dB of transmit feed-line loss.

The 3.6-λ boom length was a unique length in that with a very small amount of element spacing squeezing and stretching from the original designs both the W1EJ and DL6WU derived designs worked well at that length. I built test antennas of both designs, and as expected, measured results were hard to separate. The DL6WU-based design was finally chosen because it used one less director, which would reduce construction time and weight. The DL6WU-based Yagi also had a slightly higher natural driven-element impedance, which I felt would be easier to adapt to the open-wire phasing lines. Array temperature analysis by DJ9BV of both designs showed only a 0.2 K difference in predicted array temperature (in favor of the W1EJ design) out of predicted array temperatures of approximately 28 K for 16 Yagis. Fig 7 covers complete construction details for the 14-element, 3.6-λ Yagi including a driven element for coaxial feed. The calculated E- and H-plane patterns for the 14-element Yagi is in Fig 8. In Fig 9 the open-wire feed system is explained. Note that with open-wire phasing lines the lengths of the driven element and first director are different than for coaxial feed.

### Array Construction

The array of the 16 × 14-element, 3.6-λ Yagis is amazingly small, considering its performance. Its overall size is only 13 ft 2 in. wide by 12 ft 3 in. high by 8 ft 9 in. deep (4.0 × 3.7 × 2.7 m). I feel that the array gain and performance are equivalent to a 23 ft (7 m) diameter parabolic dish. In terms of physical attributes, the Yagi array covers just over 1/3 the area and has a maximum wind load (excluding the mount) of 16 square feet. A 23 ft (7 m) diameter dish with even a relatively open mesh reflector surface may run from 50 to 150 square feet (4.6 to 14 m$^2$) of wind load area. Even with its polarity mount the Yagi array is under 25 square feet (2.3 m$^2$) of area.

Calculated array patterns for stacks of four Yagis in the E plane at 52.75 in. spacing and four Yagis in the H plane at 49 in. spacing are given in Figs 10 and 11. The net 16-Yagi array pattern (4 × 4 stacking) should be similar to these four-stack simulations. The computer analysis implies that the calculated gain for the 16-Yagi array would be over 27.7 dBi, making the 27.5 dBi estimated real world appear to be a realistic figure. Fig 12 is a comparison drawing of the relative size of the Yagi array and dish.

Even with these relatively minuscule array dimensions (at least by EME standards), some careful engineering is required to control the array weight since it is to be rear mounted. Any amount of weight can be counterbalanced. When the mass is distributed over a large area, however, its inertia is multiplied into angular momentum, which can get out of hand when the array is moved. The polarity rotation creates further demands since the array has to do more than simply track the very slow-moving moon. During QSOs the array often will have to be moved back and forth by 90° at each sequence. Even more demanding is the process of searching for schedule stations and replies to one's CQs. In this case the array may be in nearly continuous motion through the entire 2½-minute receive period.

The Yagis were built out of ⅞ in. diameter × 0.035 in. wall front boom sections and 1 in. diameter × 0.058-in. wall rear boom sections of round 6061-T6 aluminum tubing. I'm sure 0.049 in. wall rear sections would have also been strong enough and saved a couple of pounds of array weight. I prefer the tighter boom joints made with the 0.058 in. wall tube, however. The parasitic elements were made from 3⁄16 in. aluminum rod. I considered using tubing for the elements to save weight but decided to use the rod, as I have had bad experiences from insects building nests in open tubing. The driven elements are made from brass rod so they could be soldered directly to the open-wire line. The brass has been covered with a clear lacquer coating. Do not use unprotected brass or copper in either the elements or phasing lines, as once the oxide coating develops receive performance will be significantly degraded due to the increased resistive losses.

The open-wire line was made from no. 8 double enameled wire. This size provides the best loss to weight and wind load compromise. At 200-Ω impedance, the loss of the line is about the same as ⅞ in. diameter 50-Ω Heliax, but at a fraction

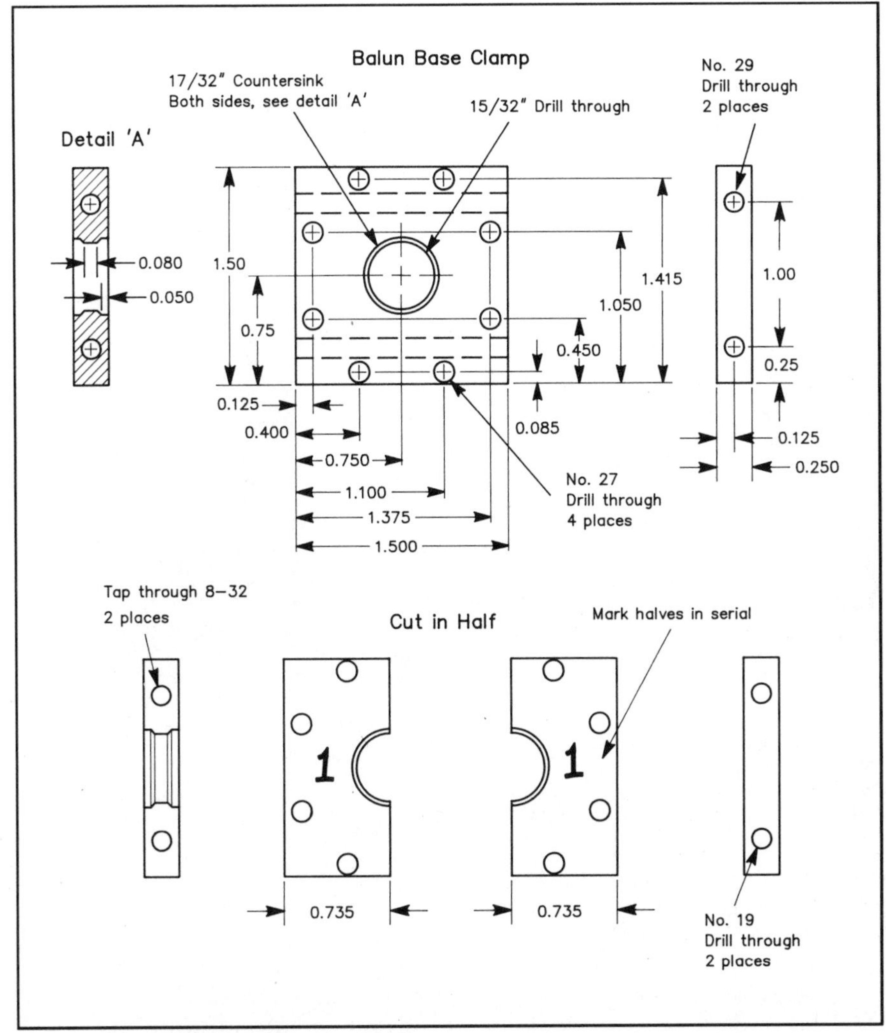


Fig 14—Construction details for the balun base clamp. Some builders have soldered the baluns directly to the shield.

of the cost, weight and wind load. Fig 9 includes details of the Teflon insulators, which are spaced at roughly ½-λ intervals. Dimensions on the brass line junctions are also given. The inner four phasing lines are made from Andrew LDF4-50A ½-in. low-density foam Heliax, which is soldered directly to the open-wire line junctions. The 2-λ spacing of each bay of four Yagis converts their nominal 200-Ω folded dipoles to 50-Ω center impedance. This allows direct connection to the Heliax inner phasing lines. Eliminating connectors, along with the minimal driven-element junctions that are made near the voltage maxima points on the driven elements, contributes to the outstanding receive performance of the array.

Quarter-wave sleeve baluns (that is, ¼-λ chokes) are attached directly to the Heliax shield. The lines are cut to 66 in., which is electrically 11 quarter wavelengths at 432 MHz. The length of the phasing lines (1¼ λ) was chosen because theoretically, on odd quarter-wavelength multiple of feed line should act as a choke to prevent current from flowing on the outside of the shield. This current could degrade array performance, especially on receive. At UHF, it can be hard to establish where the real RF ground points are, and the effect of odd quarter-wavelength transmission lines.

To establish the effect of the baluns and phasing line length, sun noise and earth noise measurements were made with and without the sleeve baluns on the array. The removal of the balun sleeves causes a loss of 0.5 dB in sun noise and 0.2 dB in earth noise, thus validating the effect of the sleeve baluns. Details of how the baluns were made to clamp directly on the LDF-50A phasing lines are given in Figs 13 and 14. Some builders of the array have made baluns that were simply soldered to the LDF-50A shield. I recommend that silver-bearing solder with at least 2% silver and little or no lead be used at all junctions. This will greatly reduce corrosion problems. Don't forget to cover all junctions with clear lacquer or enamel.

The center power divider is a 3⁄2-λ 4-way unit, otherwise known as a half-wave 2-meter divider. This long power divider was chosen both to take advantage of the low loss of the power divider's 1¼-in.-diameter air line and to move the phasing line connections out of the clutter of the center of the array (preamplifier box frame mounting hardware, etc).

The open-wire line required a design compromise in the stacking frame. The mechanically simplest, lowest cost, lowest weight and lowest wind-load stacking frame would use one large cross boom and 4 long vertical masts. This type of frame does not lend itself well to the open-wire line as the line sections would have to be assembled on the frame up on the tower. I wanted to be able to assemble and test the open-wire line and bays of four Yagis on the ground. To accomplish this, the four small H and one large H frame arrangement was used. It can be seen in Fig 15. This method is very convenient, since in the air assembly is reduced to a minimum. The penalty is about 15 pounds of additional aluminum tubing and U bolts.

The small H frames are made of 1½-in.-diameter × 0.065-in.-wall aluminum tubing. I chose the 1½-in. diameter because it was very strong for the 54-in. and 56-in. lengths used in the frames. In addition, the larger diameter would resist twisting at its connection points better than a smaller diameter. Fig 16 gives a better view of how these small H frames are assembled. The 0.065-in. wall was selected simply because I had some pieces of it on hand. KA1ZE was nice enough to give me the rest of the tubing I needed, left over from one of his projects. I'm sure that 0.058-in.-wall tubing and even 1⅜-in.-diameter tubing would work fine for these frames. The frames are held together with Radio Shack Archer U bolts with saddles (see Table 1, Array Parts List). The Yagi booms are held to the small frames using UBS-40 U-bolt saddles which are available from Rutland Arrays. The booms were reinforced with a short section of 1⅛ in. × 0.058-in.-wall tubing at the U-bolt mounting points.

These saddles have been the source of the only problem with the array so far. During a very severe fall wind storm, all of the Yagis ended up leaning over at about a 15-degree angle. The problem is that the saddles tend to spread over time. After tightening the U bolts a couple of times things seems to be staying in place, but we have fortunately not again experienced the

Fig 15—This view of the 16 × 14-element 3.6-λ rear-mount 432-MHz Yagi array shows the small-H/large-H stacking frame construction. The combination open wire and Heliax phasing lines can also be seen.

Fig 16—The small H-frame construction can be seen in more detail here. The open-wire phasing lines including their connection to both the driven elements and coaxial phasing lines are also visible.

Fig 17—The prop-pitch azimuth rotator installation can be seen in this view. The rotator plate is made from 1/4" steel plate and 2" angle iron.

Fig 18—The mast mounting and hinge plate are seen in this photograph, as well as the connection of the main array frame cross piece to the polarity rotation shaft. The large hole in the mount U-channel is needed to clear the azimuth mast to allow the array to be pointed below the horizon in order to make earth-noise measurements.

Fig 19—More details of the polarity mount are apparent in this view. The TVRO actuator mount, $T^2X$ rotator location and the counterweights are visible. The two U bolts and V-blocks on the polarity shaft are safety catches to keep the array in place should the thrust bearing and rotator fail when the array is pointed on or below the horizon.

**Table 1**
**Parts List for 16 × 14-Element Rear-Mount Array**

Array frame, phasing lines and preamp box
(Yagis not included)

16—1½ in. U bolts with saddle (RS Archer 15-826).
16—1½ in. long U bolts (RS Archer 15-820, 8 pkgs of 2).
8—4½ in. × ½ in.-13 grade 8 bolts.
20—2 in. × 5/16 in. U bolts.
4—2 in. U-bolt saddle.
16—½ in. flat washer.
8—½-13 nut.
8—½ in. lock washer.
16—⅜ × 1¼ in. bolts.
32—⅜ in. flat washer.
⅜ in. lock washer.
⅜-14 nuts.
8—2½ in. pipe hold-down clamp, B-Line B2400-2½.
2—2 in. diam × 0.125 wall × 10 ft, 6061-T651 tube.
1—3 in. diam × 0.125 wall × 10 ft, 6061-T651 tube.
8—1½ in. diam × 0.058 wall × 56 in., 6061-T651 tube.
8—1½ in. diam × 0.058 wall × 54 in., 6061-T651 tube.
6 ft 1⅜ in. diam × 0.058 wall 6061-T651 tube.
2—6 in. × 9 in. × ⅜ in. 6061-T651 or 2024-T3 plate.
8—4½ in. × 4½ in. × 3/16 in. 6061-T651 plate.
150 ft no. 8 double enameled copper wire.
4 ft ⅜ in. diam. Teflon rod.
1—12 in. × 2 in. × ⅛ in. Teflon sheet.
1 ft ¼ in. OD ⅛ in. ID brass tube.
3 ft 1 in. copper water pipe.
2—3 in. U bolts (for holding preamplifier box mount).
4—1½ in. × 1½ in. × 0.032 brass sheet.
6 in. 1½ in. × ¼ in. brass bar stock.
1—3/2-λ 4-port power divider.
2—1 in. × 0.125 in. × 12 in. long 6061-T651 angle stock.
2—1 in. × 0.125 in. × 14 in. long 6061-T651 bar stock.
2—1½ in. U bolts.
4—¼-20 × 1 in. hex bolt.
8—¼ flat washer.
4—¼ split lock washer.
4—¼-20 hex nut.
1—8 in. × 8 in. × 4 in. PVC weatherproof junction box.
4—Andrew L44-N connectors.
25 ft Andrew LDF4-50A cable.
8—8-32 × 1½ in. bolts.
8—no. 8 lock washer.
16—4-40 × ½ in. screw.
16—no. 4 lock washer.
16—4-40 hex nut.
¼ pound 2% silver solder.

sustained 65 mi/h winds since that storm.

The large frame is made out of two 10-ft-long pieces of 2-in. OD × 0.125-in.-wall 6061-T651 tubing for the sections that hold the two bays of four Yagis. The cross piece is made of 3-in.-diameter × 0.125-in.-wall 6061-T651 tubing. The frame is very strong, which is necessary given aluminum's elasticity. Keep in mind that on an array that will be turned in polarization, any physical sag in the frame that lets the Yagis flop around in alignment can lead to pattern skewing. On a conventional array one can easily compensate array azimuth and elevation indicators for this skewing. When you are rotating the array in real time, this pattern skewing can move your main lobe off the moon as the array is rotated in polarity. Since the strength of tubing increases at almost the fourth power of its diameter, I suggest using larger diameter tubing.

Those wishing to save weight and cost might consider using 0.065-in.-wall (thin wall) irrigation tubing along with Phillystran or similar stress rigging. This irrigation tubing is readily available from farm equipment supply outlets. The large frame is held together and also held to the polarity rotation shaft using galvanized pipe hangers. The size for 2½-in. pipe (2⅞-in. nominal OD) is a nice tight fit on the 3-in. tube and gives much better load distribution than U-bolts would at these high stress points.

The total array weight ended up just over 150 pounds, which is a fair amount to hold from behind. During the design and construction process I felt like an aerospace engineer as every little piece used in the array was evaluated for strength and weight. When the mast, polarity mount and other required pieces are totaled up the array with mount is in the 300-pound range. The azimuth rotator is a prop pitch motor (either small or medium size depending who you are talking to). The prop pitch motor with its nice slow start and coast-down turning characteristics, along with its immense torque and virtually nil backlash and slop, is the best EME array rotator I have used. Its only drawbacks show up on tropo operation in the form of slow (90-second) rotation speed and noisy operation. A look at the prop pitch azimuth drive installation is given in Fig 17. I was fortunate enough to have the main drive gear from the prop pitch motor. A ½-in. steel plate 9 in. diameter was welded to the gear. Mast clamps for the Hy-Gain HDR-300 were used to clamp the mast to the plate.

### Polarity Mount

What originally appeared to be the toughest part of the job, designing a mount for the polarity rotation was solved in a straightforward manner thanks to the work of WA9FWD. This type of mount simply uses a steel U-channel with a hinge plate welded to it along with a mounting plate for a standard ham type rotator. Henceforth, this mount will be referred to as the WA9FWD-type polarity mount. Since my array was substantially larger than John's,

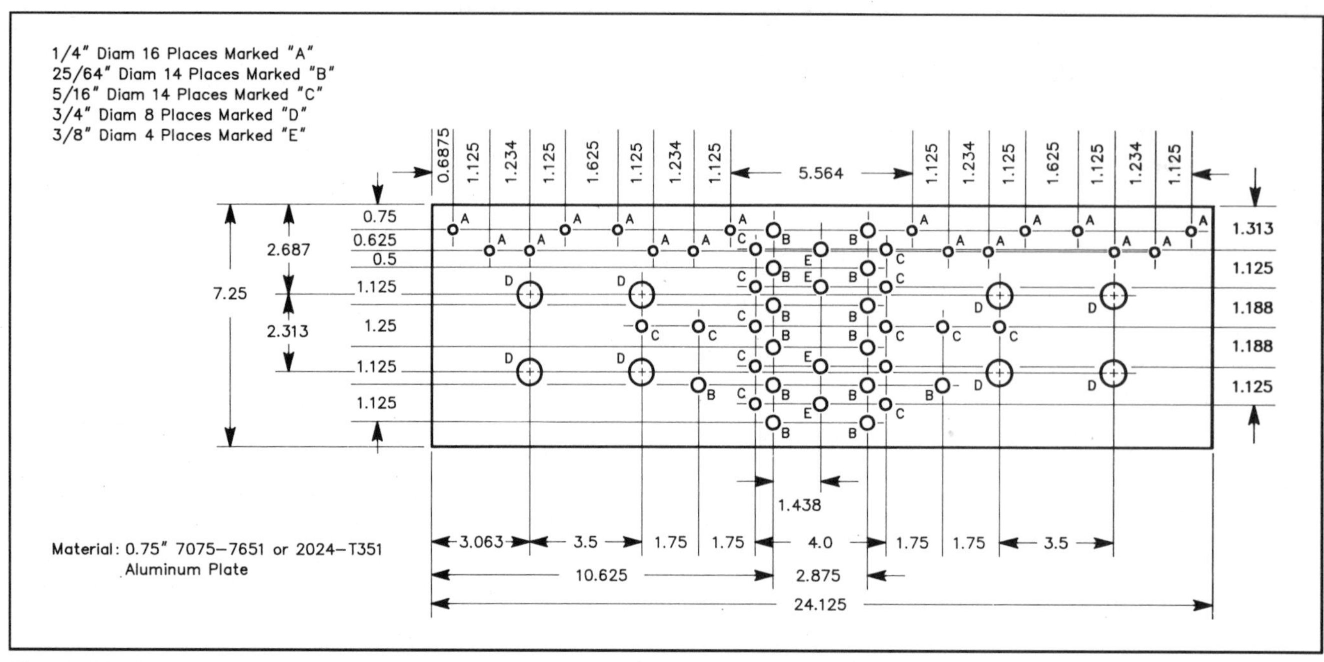


Fig 20—Machining drawing of mast mounting plate.

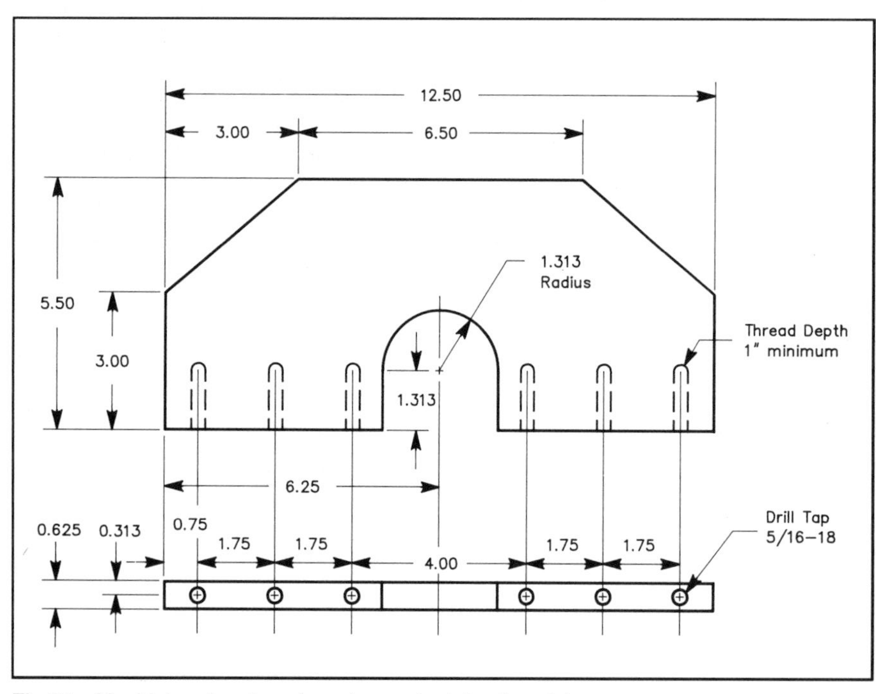


Fig 21—Machining drawing of mast mount reinforcing plate.

I scaled up his mount in size and changed the mast mounting method. John originally used a hinge plate welded to the top of the azimuth mast. I decided to bolt my hinge plate to the azimuth mast differently than John did both to allow mast changing and the ability to point the array into the ground for earth noise measurements. The general construction of the mount and mast connection is shown in Figs 18 and 19. The steel material was obtained from a local steel fabricating shop, which also cut it to size. The aluminum pieces were obtained at a scrap sale. Suitable steel is standard structural steel like 1010, 1015 or 1020. High strength alloys like 4130 are also good to use. The aluminum can be 2024-T3, 6061-T6 or 7075-T5. All of these materials can be purchased in small quantities from the Dillsburg Aeroplane Works in Harrisburg, Pennsylvania, tel 717-432-4589. The assemblies were welded together by the same local fabrication shop.

Machining details of the mast plate are covered in Fig 20. Fig 21 is a bracing piece to keep the mast hinge plate from bending when the U-bolts are tightened. This bracing is extremely important to keep the mast hinge plate from distorting and binding the hinges. The hinges are heavy-duty full mortise stainless-steel commercial door hinges. There are several extra holes in the mast mounting plate. These were put in to give clearance to the additional tropo antenna mounts I plan to add to the polarity mount.

Polarity motion is obtained from a Hy-Gain $T^2X$ rotator. Although the rotator is adequate I am not overwhelmed with these type rotators. I had to disassemble the $T^2X$ rotator to disable the friction brake. Unfortunately I discovered the effectiveness of the friction brake, which would bring polarity motion to a near destructive screeching halt, only after the array was first put together on top of the tower. The removal of the friction brake leaves full braking responsibilities to the wedge brake. Although it has not yet failed I have that nagging feeling every time I see the array rock around in the wind, that sooner or later the brake teeth will eventually be reduced to aluminum filings. Keep in mind that these type rotators were not designed for this type of polarity rotation, especially when the array is down on the horizon. I don't know of a better ham type commercial rotator for this purpose available at a reasonable cost.

The rotator is held in place by the rotator mounting plate (Fig 22). This plate is

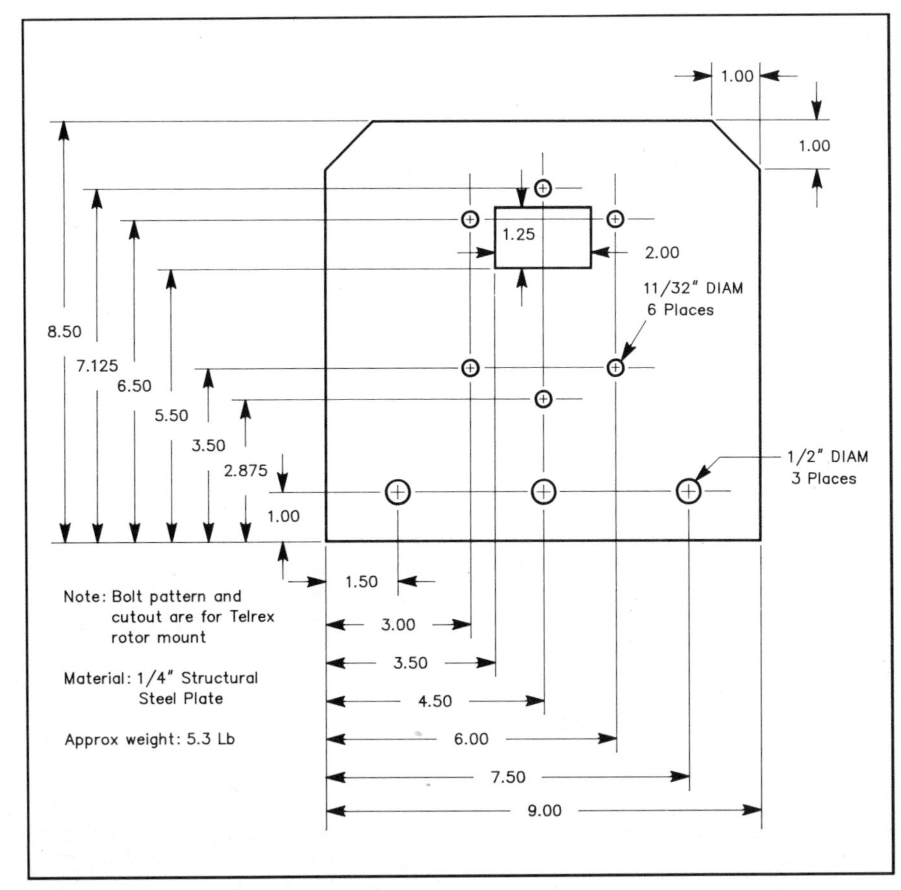


Fig 22—Machining drawing of polarity mount rear plate.

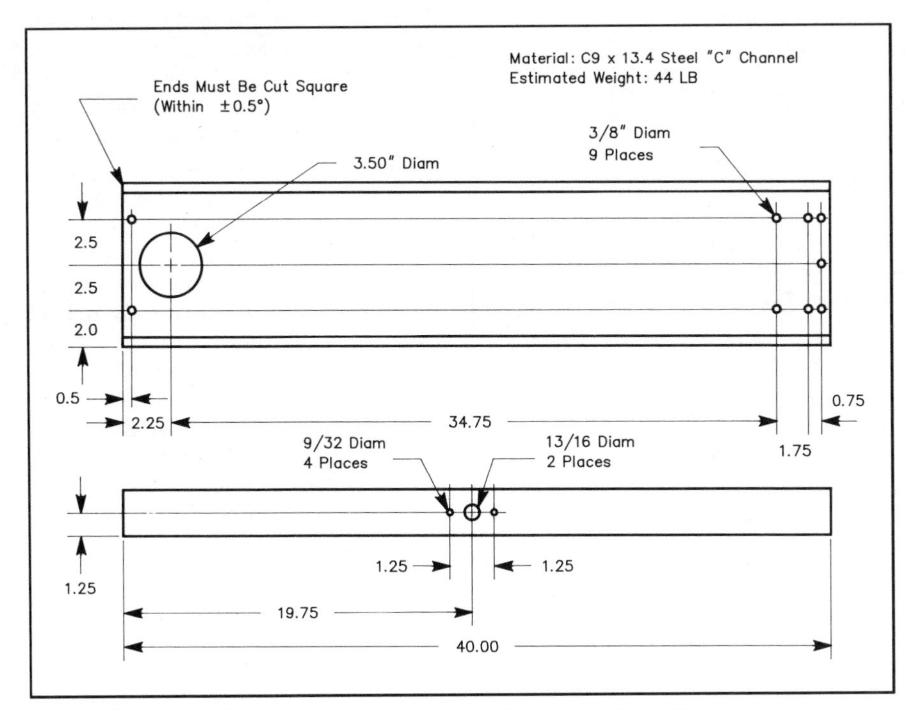


Fig 23—Machining drawing of polarity mount main support channel.

**Table 2**
**Parts List for 16 × 14-Element Rear-Mount Array**

**Array Mount**

4—heavy-duty full mortise commercial door hinges, Hager 1191.
32—¼ × 20 × ¾ in. flat-head bolts for hinges.
32—hex-head bolts for hinges.
68—¼ × 20 hex nut.
68—¼ × 20 split lock washer.
112—¼ × 20 flat washer.
4 ft 2 in. × 0.156 4130 "N" tube.
2—2 in. × 5⁄16 in. U bolts.
1—¾ in. × ¾ in. × 3 in. V-block for 2 in. U bolts.
8—2½– ⅜ in. U bolts.
4—3½ in. × ⅜ in. grade 8 bolts.
8—⅜ in. flat washer.
4—⅜ in. split lock washer.
2—V-block ½ in. × ½ in. for 2½ in. U bolt.
2—25-pound barbell weights.
2—5⁄16-18 × 2½ in. hex bolt.
2 ft 1¼ in. diam steel bar stock for barbell weights.
14 in. 1 in. diam steel rod for actuator mount.
4—1 in. flat washer.
4—¼-20 × 1 in. hex bolt.
7—5⁄32 in. hitch pins.
6—5⁄16 × 1½ in. grade 8 bolts.
8—5⁄16 split lock washer.
8—5⁄16-18 hex nut.
16—5⁄16 in. flat washer.
1—Rohn TB-3 thrust bearing.
1—Hy-Gain $T^2X$ rotator.
1—4 in. × 4 in. × 0.020 aluminum shim stock.
1—24 in. × 8 in. × ¾ in. 7075-T3 (or 2024-T3) plate.
1—14 in. × 6 in. × ⅝ in. 2024-T3 (or 6061-T6) plate.

welded to the main 9-in. support channel (Fig 23). At the other end the thrust bearing plate (Fig 24) holds the polarity shaft in place. The thrust bearing is a standard Rohn TB-3 that so far has worked perfectly. A mating hinge plate to that on the azimuth mast (Fig 25) is welded to the 9 in. U-channel as explained in the mount assembly drawing (Fig 26). The array is strapped to the polarity shaft plate (Fig 27) by the previously mentioned pipe hangers. This plate in turn is welded to the polarity shaft with some substantial bracing (Fig 28). The plate must be welded squarely to the polarity shaft as must the entire polarity mount be true. Consider that the array has an E plane –3 dB beamwidth of 6.5° and an H plane –3 dB beamwidth of 7°. If the plate was only 3.25° out of square signals could drop by 3 dB as polarity was turned—assuming the array was properly peaked on the moon!

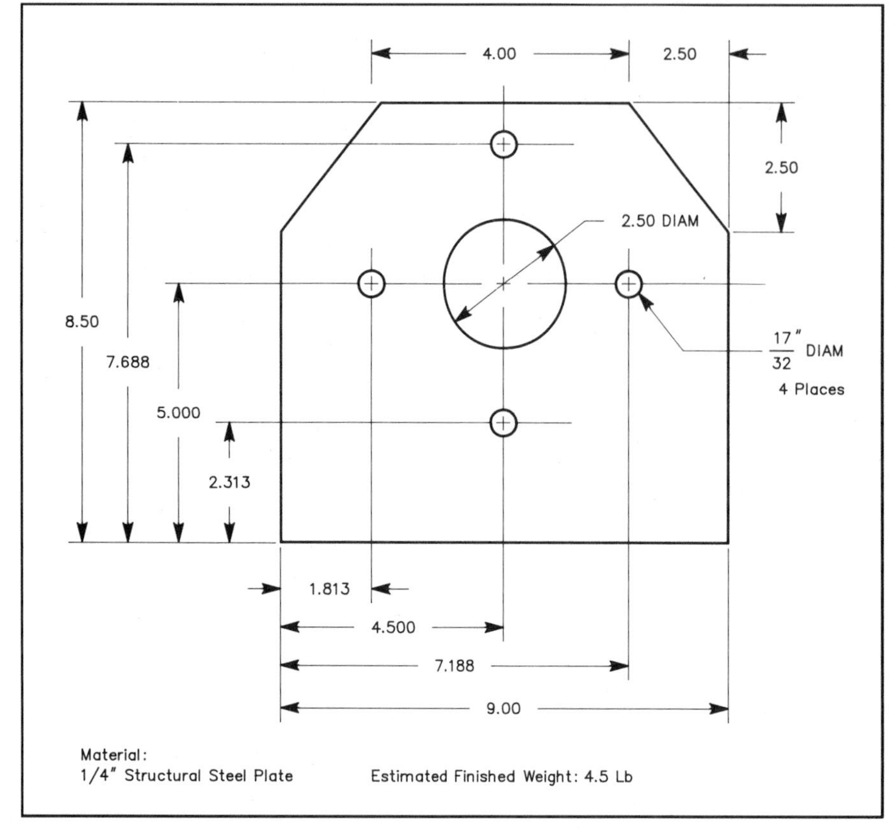


Fig 24—Machining drawing of polarity mount front plate.

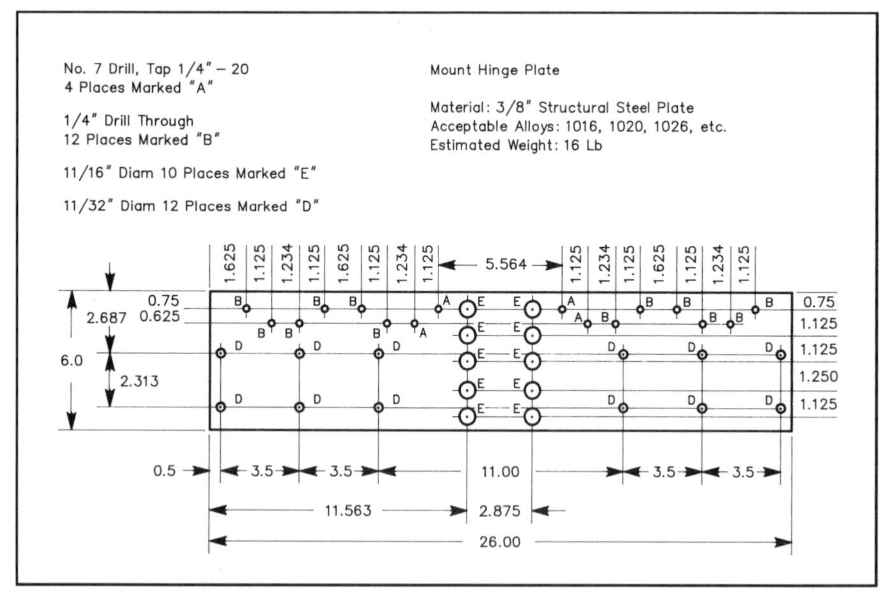


Fig 25—Machining drawing of polarity mount hinge plate.

Although the weight of the mount is substantial it is not a problem, other than getting the mount on top of the tower in the first place. To counterbalance the array, an additional 50 pounds of iron in the form of barbell weights were attached to the rear of the mount via a 1¼-in.-diameter steel rod. The linear actuator is always loaded in slight compression. If you use a TVRO linear actuator the ball screw type is preferred and it is desirable to keep it loaded. Even with the loading there is some rocking of the array: The elevation is run down from vertical due to the starting inertial and weight distribution. Storage position for the array is straight up in order to evenly distribute wind loading on the tower. This will avoid undue twist loading on the tower which could lead to failure in high winds.

### Other Rear-Mount Array Options

The design goal of the 16 × 14-element Yagi array was to come as close as possible to the performance of a 24-ft (7.3 m) dish and to at least meet the performance of my old 12 × 22-element Yagi array. Those not quite so adamant about their EME capability level should consider smaller rear-mount arrays. You might be surprised how small and light a rear-mount array can be while exceeding the performance of four long Yagis.

Sixteen of the 10-element, 2.2-λ Yagis will be as good or better than 4 × 24-ft long Yagis with less wind load and the benefits of polarity rotation. The gain of the 16-Yagi array will be about 25.7 dBi, which is enough to make many easy EME QSOs. This gain figure is equal to an array of 4 × 10.5-λ (24 ft) long optimum design Yagis. The 2.2-λ Yagis are also the right size to adapt to an open-wire phasing line system which further reduces cost and construction complexity. Twenty-four of these Yagis will have performance close to my 16 × 14-element Yagi array at 27.3 dBi, but with substantially less torque loading due to their very short booms.

Twelve of the 14-element, 3.6-λ Yagis should be noticeably better than 4 long Yagis with a total array gain of 26.2 dBi. The mechanical complexities will be substantially less than my 16-Yagi array, although 12 Yagis will not lend themselves to open-wire phasing lines.

The 12- and 13-element Yagis will also work well in 16-Yagi arrays while their much shorter booms (than the 14-element Yagis) make for a much more manageable array. Gain of 16-Yagi arrays will be around 26.2 dBi for the 12-element Yagis and 26.8 dBi for an array of the 13-element, 3.0-λ models. WA9FWD went from 4 × 12-λ long Yagis to 16 × 13-element, 2.9-λ Yagis. Although the net array gain is only 0.5 dB higher with the rear-mount array, John feels the new rotatable array is vastly better than the 4 long Yagis.

Longer Yagis could also be adapted to polarity rotation if one didn't need to have the polarity rotation available down to 0° elevation. These Yagis could be supported a few feet from the rear of the boom. The elevation mount could also be extended on the array side of the hinge to accommodate this type of array.

One final benefit of the rear-mount array approach is that stacking arrangements using odd numbers of Yagis in either plane of the array can be accommodated. This is because when the array is rear

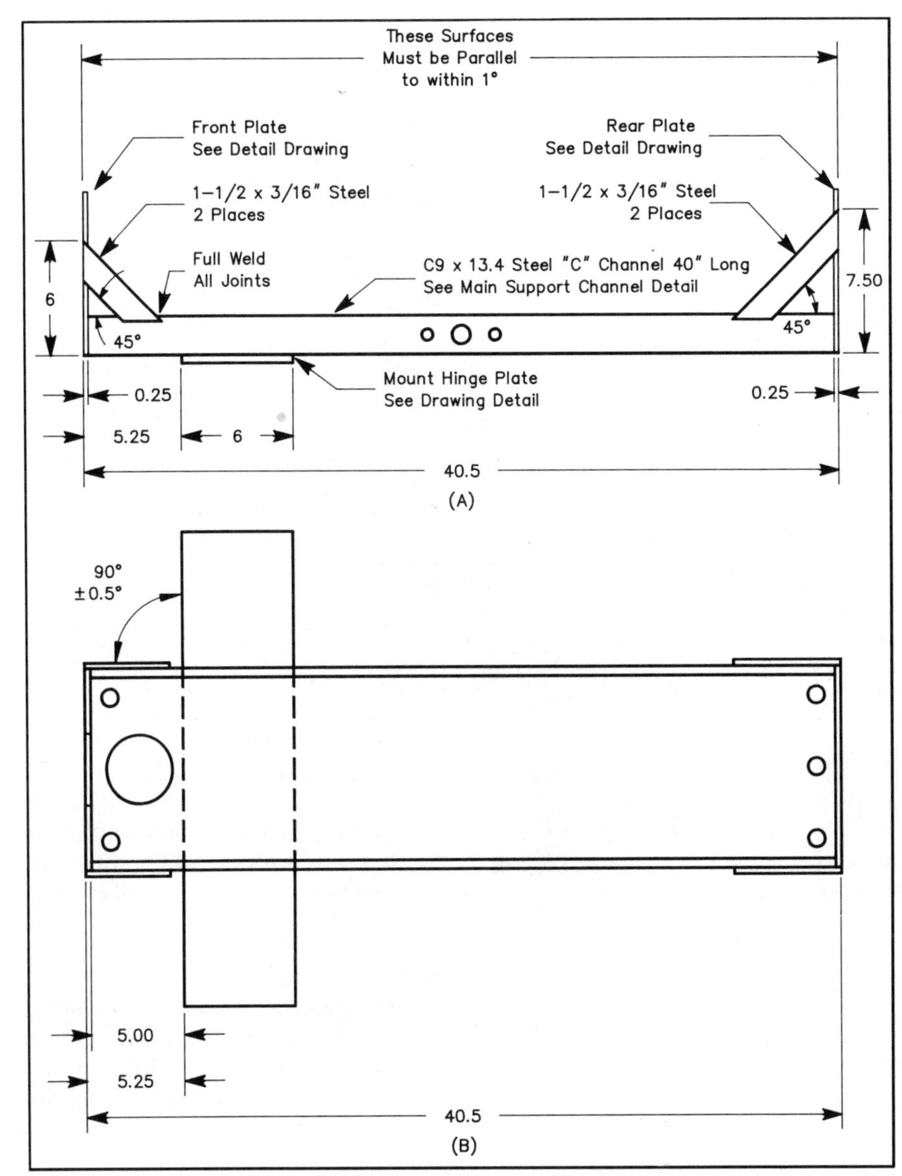


Fig 26—Assembly drawing of adjustable-polarization mount. Side view is at A; top view is at B.

**Table 3**
**Noise Measurements**

| *Source* | *16 × 14R* | *12 × 22* |
|---|---|---|
| Sun | 18.5 dB | 18.0 dB at flux = 180 |
| Earth | 5.5 dB | 5.0 dB at –14° elevation |
| Sagittarius | 5.6 dB | 5.3 dB |
| Cygnus A | 3.2 dB | 3.0 dB |
| Cassiopeia | 2.9 dB | 2.9 dB |
| Taurus | 1.3 dB | 1.2 dB |

mounted there is not a problem in having the center Yagi clear the mast. Array arrangements of 3 × 3 (9 Yagis), 3 × 4 (12 Yagis), 5 × 4 (20 Yagis) or 5 × 5 (25 Yagis) can all be used. The only limitation is one's ability to make an efficient power distribution system. Similarly open-wire phasing lines do not restrict array configurations due to mast clearance problems. Virtually any grouping of bays of four Yagis can be built using a total of 16, 24, 32 or 36 Yagis.

The basic WA9FWD polarity mount can be easily scaled up or down in size to accommodate the size array one intends to build. John's mount uses 8 in. channel and a Hy-Gain HAM-4 rotator. With a little extra thinking about the polarity mount most builders should easily be able to adapt the mount to the materials available locally.

### The View From Up Here

After completing a project of this magnitude one has to justify his efforts. Consider that absolutely no pieces from the old array were able to be reused except for the elevation actuator (not including mounting hardware) and array electronics such as the preamplifiers and relay. Everything else including the azimuth rotator, preamplifier box, mast stacking frame, phasing lines and Yagis were all new. It's easy to see why many stations just keep adding Yagis to their existing arrays even if they didn't pick the best design to begin with. So have the results justified the effort?

The rear-mount Yagi array has been an unqualified success at K1FO. Predicted net gain of the array was that it would be 0.3 dB below the old 12 × 22-element Yagi array at 27.5 dBi. Receive performance was expected to be equal to the old array due to lower phasing line losses. The best way to attempt to compare array performance is through noise-source measurements. In reality, all noise measurements on the new array are measurably higher than with the old array. See Table 3.

Obviously, there is a limitation to the accuracy of my measurements, but in any case it is consistent that the new array is better on receive than the old one. I also feel as if I am hearing better, but this may be due in part to having polarity rotation. Since my entire receive setup (relay, preamps, filters, etc) are the same ones from the old array, it is possible that something had deteriorated in the old array. Likewise, it is possible that I am overstating the array temperature, understating the loss of the old coaxial feed system or overstating the open-wire feed-line loss. Since most stations are reporting that my transmit signal is as good or better than with the old array, one would have to conclude that the new array is indeed working as well or better than expected.

From these noise measurements and the estimated array gain I believe that my total system temperature is 73K. This is composed of 33K array noise reception (at cold sky), 15K of phasing line power divider and relay losses and 25K total system noise

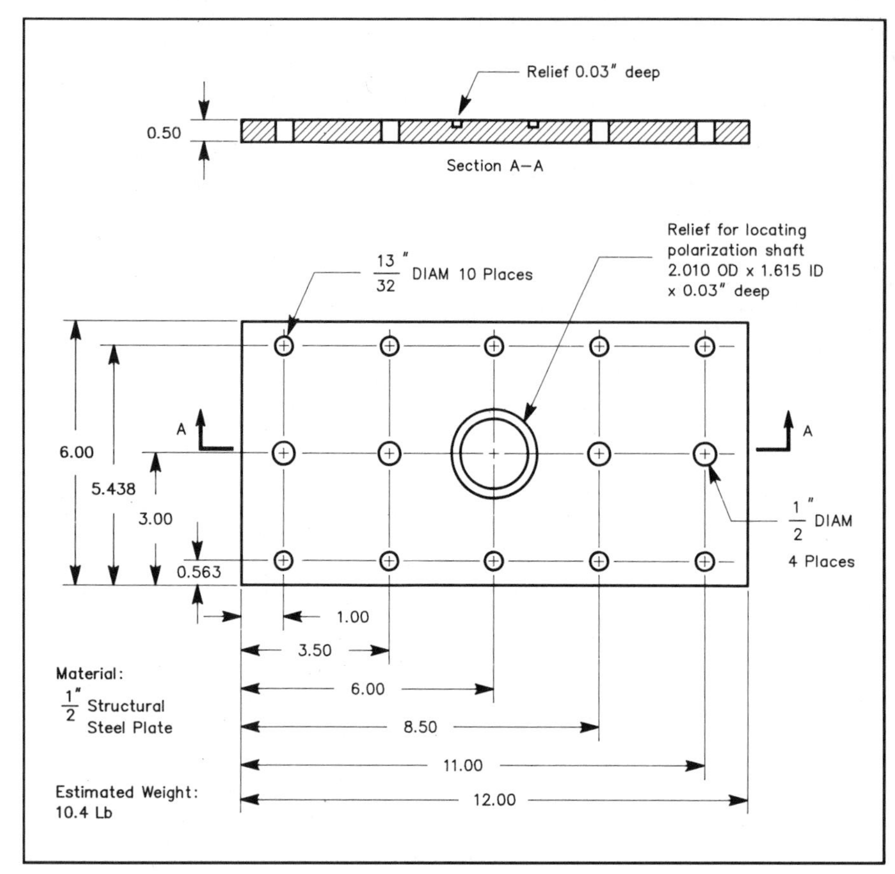


Fig 27—Machining drawing of the array mounting plate.

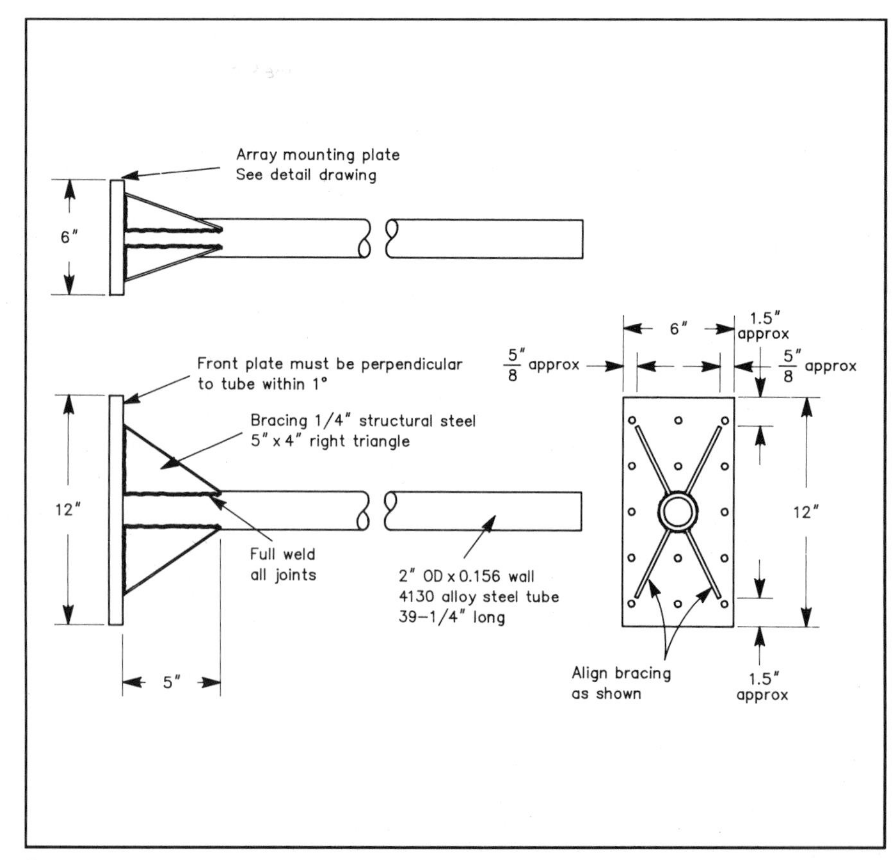


Fig 28—Assembly drawing of polarization shaft.

Fig 29—The view at K1FO from 70 ft above ground level, the height of the 16 × 14-element Yagi array, is quite a bit better than at 20 ft, where a large dish would be mounted. Not only does having the array at this height allow EME operation down to the horizon, but with tropospheric bending, EME signals are frequently heard 10 minutes before visual moon rise and for the same time after visual moon set.

temperature.

In the first 9 months of operation I have made 322 EME QSOs with 138 different stations, 47 of which were new initials. As everyone knows, the more initials you have, the harder it is to work new ones. In the last year with the old array I only worked 20 new stations. As further proof, consider that of those 47 stations 6 are single Yagis, 1 has 2 Yagis and 20 are using 4 Yagis. In addition 27 of the 47 were worked on random and compared to my old benchmark, N4GJV, I have been able to work every new station that Ron has worked, a significant change from the last few years. The only stations I haven't been able to work are running either a single Yagi and moderate power (<500 W) or four Yagis and low power (< 200 W). Of the 47 initials I believe that about 22 of them would not have occurred without polarity rotation.

Having polarity rotation is like cheating. Instead of spending hour upon hour straining to copy signals buried in the noise, it's simply peak and go. As a final note we refer to Fig 29 and observe the horizon view from my array center level of 70 ft (21 m). It's a much more pleasing view than the one down at dish-mount altitude.

# Loop Antennas: The Facts, Not the Fiction

By A. J. Henk, C Eng, FIEE, G4XVF
10 Aston Way
Epsom, Surrey KT18 5LZ
United Kingdom

Not another lot of words on loop antennas, surely? So much has been written and said on the subject—at least in the amateur journals—but how much of it is of any use to the serious radio amateur? This article is intended to clear away some, at least, of the intellectual fog surrounding these devices and to address areas which have been generally neglected, such as how efficient are they? How effective are they (not the same thing) and how can they be made better?

Notwithstanding the proliferation of references to the subject, it is not intended to provide an extensive bibliography of papers; others have already done this. The references which are included have been selected as having something useful to say which is appropriate to this article, and not as an attempt to impress.

### Debunking the Myths

Let's make one thing clear at the outset: There is no mystique of any sort associated with loop aerials. They obey the standard laws of physics like anything else and the principles are quite simple. Anyone who tries to make it sound otherwise and wraps it all up in arcane language or presents it as a "black art" should have his credibility suspected immediately. It is hoped that this article will explain the operating principles, without too many shortcuts, and in language which is easily understood by anyone with sufficient technical knowledge to have passed (honestly!) the licensing tests. There will be some math presented but, again, the simplest and most basic knowledge of ac theory will be quite adequate to reveal many of the loop's innermost secrets.

### What is an Aerial?

An antenna (or aerial if you prefer—both terms will be used synonymously here) is a device for coupling an electrical circuit to a medium through which radio waves propagate, in our case free space, or vice versa. It was the development of this component (by Marconi) which took wireless communication out of the realms of a laboratory curiosity and turned it into practical reality. Without the "elevated electrode" the signals don't go very far. A pair of aerials can form a simple four-terminal transmission network: four terminals because you feed the signal into one pair of terminals (the first antenna) and a smaller version of it emerges from the other pair. Such a network is reciprocal, that is to say it works both ways. Any two antennas, of any sort, will exhibit this reciprocity, a feature which has been of great value to the author during the study of loops.

Reciprocity works like this. First, the two antennas must be correctly matched to the transmitter and the receiver, respectively, and this matching must be preserved. If this is the case and you feed, say, a one-watt signal into one aerial and then find that you get one microwatt from the second (receiving) antenna, you can be sure that, if you now reverse things and feed your one watt (matched, of course) into the second antenna, you will have a microwatt out of the first. There are no exceptions to this rule.

Most common transmitting antennas are "electric" or "E-field" devices, and work basically by making as large an electrical disturbance as possible across a chunk of space. The bigger the chunk the better the aerial because the disturbance is more widespread. This is what you might expect; after all, particularly at low frequencies, big aerials work better than small ones if everything else is the same. Extracting big performances from little antennas is the aim of many amateurs, particularly in the UK, and smallness is one of the main attractions of the loop.

Where there is an oscillating electric field, there is also an oscillating magnetic field produced by it in space. A radio wave must contain both magnetic and electric components in order to exist. In the case of the loop, a strong magnetic field is generated by passing a heavy current through the radiating conductor and this magnetic field then generates a corresponding electric field in space, thus providing the two elements needed. For this reason the antenna is sometimes referred to as a "magnetic loop."

### What is a Loop Aerial?

Let's stop for a moment and have a look at loops from another perspective. First, what do we mean by a "loop aerial"? In this article the term is taken to refer to a loop of one or more turns of conductor whose dimensions are small compared with the wavelength of the signals being transmitted. It may be circular, square or any other shape. Anything larger than about one-eighth of a wavelength across is becoming a bit big for this definition; such devices as "delta-loops" are a different kettle of fish and this article does not apply to them.

Magnetic loop aerials are as old as wireless itself; indeed some of us remember very clearly the frame aerials used quite effectively by the early broadcast receivers before they became modernized by the availability of efficient magnetic materials and turned into ferrite rods (indeed the US term "loopstick" testifies to their origins). They have not gained widespread popularity for transmitting because of the great difficulty of achieving high efficiencies, but for receiving they have many desirable properties. It should be noted that, even though reciprocity applies (yes, also to ferrite rods) you cannot just interchange transmitting and receiving aerials and achieve the same communications quality or performance. However, that's another story. Loops which are used for transmitting do not generally use ferrite and therefore need to be larger to be effective. In addition, they have fewer turns, typically below five and quite often only one. Whereas you can't necessarily transmit on a receiving loop, you can normally receive very well with a transmitting unit. This, of course, applies equally to E-field aerials.

### The Loop as a Radiator—Radiation Resistance

We have spent quite enough time and space looking at aerials in general and seeing how the loop fits into the scene. Let's now examine it more carefully and see how it works. There is a simple but very useful concept which we can use to help us, and that is the idea of "radiation resistance." This can be used in connection with any antenna but here we are going to restrict ourselves to the loop. As we have seen, this works by pumping as much RF current as we can muster through a conductor, thus generating our oscillating magnetic field in true Faraday tradition. In order to do this we need RF *power*, and this normally comes from our transmitter—let's assume it's a 100-W rig in the interests of simplicity. Let's also assume that the transmitter, feeder and aerial are all properly matched so that all of our 100 watts goes into the aerial system.

With any luck, some of this power will be radiated off into space—the rest disappears into the system losses and is dissipated as heat. If we have been very lucky and manage to radiate ten out of our 100 watts, 90 must go in loss resistances of one sort or another. Note that all losses are resistive, as pure inductance or capacitance cannot dissipate power. Since power only dissipates in resistance, it is seductive to consider the *radiated* power also as being absorbed into a resistance. Our circulating current might be (say) 10 A. We know from basic theory (power = $I^2R$) that, because we are losing 90 watts in losses, our total loss resistance (from all sources) must be 0.9 ohm, as $10^2 \times 0.9 = 90$ watts. Therefore, why not consider the other—radiated—10 watts as being in a resistance of 0.1 ohm? (Again $10^2 \times 0.1 = 10$ watts.) This is called the "radiation resistance" and every antenna has some.

It must be remembered that, because this is not a real resistive loss, it is not an embarrassment (not in an aerial, anyway). Indeed, our aim is to push as much power into it as we can. It is therefore in our interests to achieve as high a value for this radiation resistance as possible. Low values are bad news.

Once we know the radiation resistance of our antenna and the current flowing through it (ie, that flowing in the loop), we can calculate the power actually radiated and, since we know how much power we are feeding in, the efficiency follows very easily. More of this later—it is a key point in assessing the performance of a magnetic loop.

### Basic Operation

We can now start to look at the antenna in a little more depth and see what makes it tick. Fig 1A shows the physical arrangement normally employed, stripped of all the nonessentials. It is shown for a single-turn loop but the principles are the same irrespective of the number of turns. A conductor is bent round into a circle (or some other suitable shape) and the free ends are connected across a capacitor. The arrangements for connecting this to the transmitter or receiver are discussed later. This forms a tuned circuit, resonant at the operating frequency.

This can be considered, to a high degree of accuracy, as comprising three simple components, as shown in Fig 1B. The inductance, L, is that of the bent conductor and the capacitance, C, is that of the tuning capacitor, plus any strays, which resonates with L at the desired frequency, f. The third component is resistive and represents, as one component, all the resistive parts of the system, or $R_t$ (for *total* resistance). $R_t$ will mainly be losses of one sort or another, no circuit being perfect, but there is a further nonlossy part of this component which we mentioned earlier, the radiation resistance, $R_r$.

In connection with these resistive components it should be realized that they are not all physically in series as shown in the figure. The conductor losses are clearly series losses, but losses in the capacitor dielectric appear across the capacitor terminals as a shunt resistance. Even so, we can still use a series resistance to represent these parallel losses by giving it a value which will have the same effect on the circuit as the parallel losses themselves, and doing it this way brings great simplicity to the analysis. Also, in with this series resistance, we can include any losses which occur outside the physical confines of the loop such as ground losses (which will affect the loop performance and Q) and therefore bring all losses, from whatever source, together in a single component. This keeps life very straightforward.

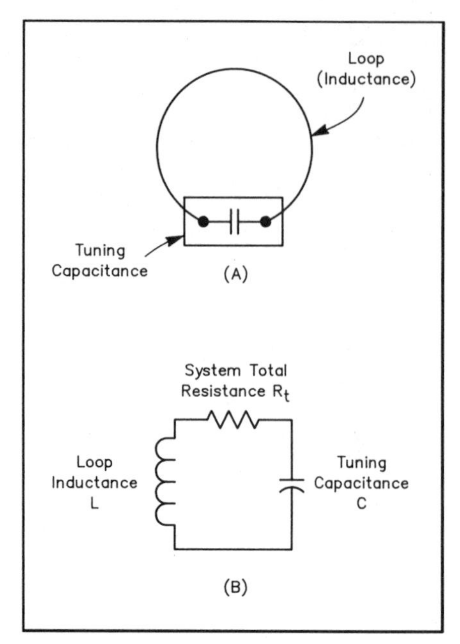


Fig 1—At A, the arrangement of a loop antenna. B shows the loop-antenna equivalent circuit.

It is our avowed intention to make as much current as possible flow in this circuit. This will, of course, occur at the resonant frequency of L and C, given by

$$f = \frac{1}{2\pi\sqrt{LC}} \qquad \text{(Eq 1)}$$

This current flows equally through all the circuit elements (their being in series) but actual *power* appears only in the resistive part as you cannot dissipate power in a pure reactance. Therefore the total power in the system is

$$P_t = I^2R_t \text{ watts}$$

If a transceiver, for example, is correctly matched to this system, then $P_t$ is equal to the power output of the transmitter, let's say 100 watts.

Now, $R_t$ is the total resistance of the circuit and is made up of many different parts. All but one of these parts will correspond to losses of various sorts and can be considered as forming a single aggregate loss resistance $R_l$. The one remaining non-lossy part is—yes, you've guessed it—our old friend $R_r$, the radiation resistance. Therefore we can say, because the total loss is the sum of all these separate parts,

$$R_t = R_l + R_r$$

### Efficiency

Power dissipated in $R_l$ is, of course, lost as expensively generated heat. However, power "dissipated" in $R_r$ is our precious radiated power and we cherish it tenderly as that is precisely what we are trying to achieve with our antenna. Suppose that we radiate 25 watts out of our 100 W, then our aerial is 25% efficient. The other 75 W (ie, three times as much) goes in the loss resistance. This means that $R_l$ must be three times as big as $R_r$ in this example since I and therefore $I^2$ is the same value for both $R_l$ and $R_r$. We can write a simple equation to define our efficiency if we know these resistance values:

$$\text{Efficiency}(\eta) = \frac{R_r}{R_r + R_l} \times 100\% \quad \text{(Eq 2)}$$

We can now define the term "efficiency" as the proportion, as a percentage, of the power fed to the system which is actually radiated. For transmitting, this efficiency is obviously of great interest because you will need to pump 1000 watts into an antenna which is 10% efficient in order to achieve the same results as 100 W into a 100% efficient radiator. There's money in there somewhere, as always. Notice that we have not needed to bother with L or C in arriving at this conclusion. That is not to say that they don't affect efficiency. Far from it. But they do it by altering the values of these resistances and the current flowing through them.[1]

### Coupling the Radio to the Loop

Let's take a break from the theory for a while and look at another practical aspect of our aerial. So far we have not considered how best to connect, or couple, the transmitter or receiver to the loop. While obviously very important, it is, in practice, not at all difficult. There are three basic methods, all of which have one thing in common: They form an impedance-matching function between the transmitter feeder and the resonant circuit. They are shown in Fig 2. It is assumed that 50-ohm coaxial feeder is used, although other types can be accommodated. It is possible to achieve a virtually perfect match at a spot operating frequency with any of these coupling systems.

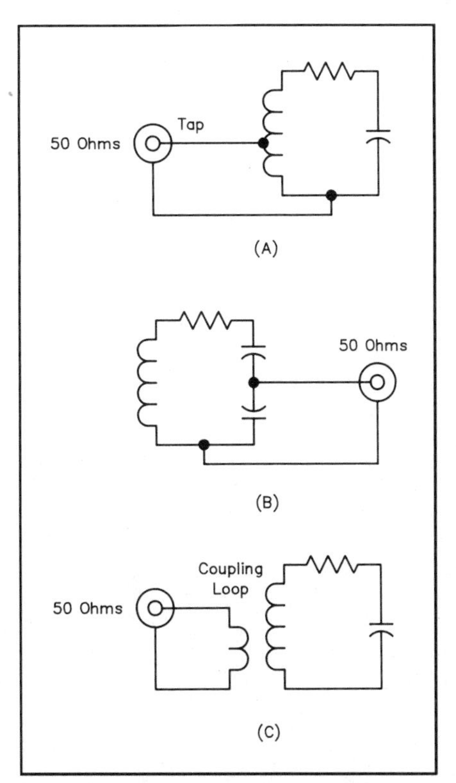


Fig 2—Methods of coupling to a loop antenna: A—inductive tap; B—capacitive tap; C—transformer.

The inductive tap method (Fig 2A) is quite convenient; simply connect the coax outer to the electrically neutral point on the loop (this should be the mechanical center if the construction is symmetrical) and move the inner along the loop conductor until the SWR at resonance becomes 1:1. If you are actually doing this in practice you will find it necessary to move, re-tune, move, re-tune, etc, until it reaches the right point. This is not difficult although it might be a bit tedious.

Capacitive tapping (Fig 2B) is similar in principle to inductive tapping but is far less easy to achieve. It is also less easy to synchronize the feeder connection point to the electrically neutral point. The author is not aware of any practical transmitting systems using this method. It is worth noting, in connection with the electrically neutral point, that the system will work wherever the connection is made but, if it is not at the neutral point, current will flow down the feeder and this will introduce performance differences which are not readily calculable and, more importantly, will introduce extra losses.

Transformer coupling (Fig 2C) is one of the most popular methods and involves connecting the radio equipment to a small loop which is placed near the main conductor, the arrangement acting as a tuned transformer. The small loop can be moved in and out to affect the matching, and a position will be found at which the correctly matched situation results in a precise 1:1 SWR at the selected frequency.

It is a happy coincidence that, with any of these coupling methods, the matching condition stays much the same over a very wide bandwidth. That is to say, if you change frequency or frequency band you will have, of course, to re-tune the loop but the coupling will remain correct. For those interested in the reason for this, Ramstrom (Ref 1) gives a useful qualitative treatment.

### Radiation Resistance—How Much?

Much reference has been made to radiation resistance, but this concept is only useful if we know what the value is in real engineering units, like ohms. Fortunately, it is a very simple matter to calculate it for a loop of any shape. It depends only on the area of the loop ($\pi r^2$ for a circular loop of radius r), the number of turns N, and the frequency. If the frequency is expressed as a wavelength, no units need be specified providing the same are used for wavelength and area. The simple formula is

$$R_r = 320\pi^4 \left(\frac{A}{\lambda^2}\right) \times N^2 \text{ ohms} \quad \text{(Eq 3)}$$

where A is the area and $\lambda$ is the wavelength, both in the same units, and N is the number of turns.

This is true *for any shape of loop*.

For example, take a single-turn loop, 1 meter in diameter, at 10 MHz. Here the wavelength is

$$\lambda = \frac{3 \times 10^8}{10 \times 10^6} = 30 \text{ meters}$$

and the area is

$$A = 0.5^2\,\pi = 0.785 \text{ meters}^2$$

Since N = 1 for the single turn, the radiation resistance is

$$R_r = 320\,\pi^4 \left(\frac{0.785}{30^2}\right)^2$$
$$= 0.0237 \text{ ohms (23.7 milohms)}$$

Note that 23.7 milohms is not a large value and the problem of making enough current (65 amperes) flow in this to radiate 100 W of RF can be appreciated.

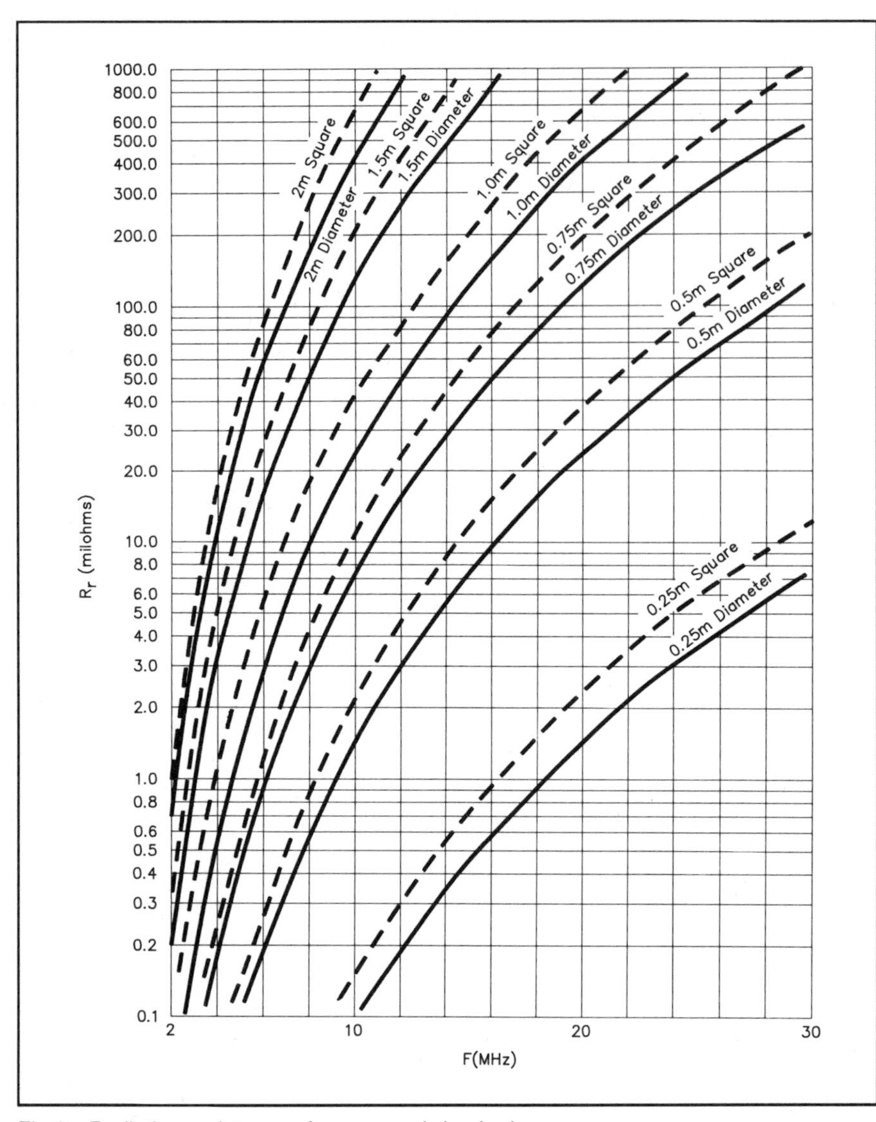


Fig 3—Radiation resistance of square and circular loops.

A graph showing $R_r$ against size for most practical loop sizes is given in Fig 3.

### The Missing Link

We have had a brief (qualitative) look at coupling the transmitter into the loop and we have seen how we can determine the radiation resistance. We still need to know more, though, before we can claim to have a full picture of how the system works. What happens between the RF feeder and the radiation resistance? Here we need to consider the loop inductance, the tuning capacitance and our loss resistance. It is interesting to note that the physical loop of conductor serves two different electrical purposes. One is to provide the radiating action ($R_r$, the aerial function) and the other is to provide inductance (L) for the matching function we are now going to examine. This is where everything finally comes together.

Returning to our diagram of Fig 1B, we have a standard tuned circuit. The loop provides the inductance L, the value of which can be deduced in a number of ways. The loop dimensions are, particularly at low frequencies, generally chosen to maximize $R_r$, and the resulting inductance accepted whatever its value is. This is not always the case, but happens far more often than not. The inductance value can be measured on a bridge or other inductance meter, it can be calculated (see accompanying sidebar) or it can be derived when it is resonated at a known frequency by a known capacitor using the relationship

$$L = \frac{1}{4\pi^2 f^2 C} \qquad \text{(Eq 4)}$$

where L is in henries if C is in farads and f is in Hz.

Because the circuit is resonant (at our operating frequency), we know L, C and f. The relationship of Eq 1 applies, of course. What we don't know is the resistive component. We know part of it, though, because we can calculate $R_r$ from Eq 3, which only leaves the loss resistance, $R_l$.

### Efficiency Revisited—The Significance of Q

The Q of a tuned circuit can be defined in several ways, but the most helpful here is to use the ratio of inductive reactance to resistance, assuming all the resistance to be in the inductor. (In practice this is not the case, but the assumption does not introduce errors if we are only looking at the total losses and don't start trying to allocate different bits to the capacitor, etc. Later we will have a brief qualitative look at where losses arise; wherever they do, though, they can be included in the inductor for calculation purposes.) In our case the resistance is the total $R_t$ and therefore we can say

$$Q = \frac{\omega L}{R_t}$$

We also know that $R_t = R_l + R_r$, so we can deduce that

$$R_l + R_r = \frac{\omega L}{Q}$$

It is now but a tiny step to the final denouement, the efficiency of the antenna overall. If we substitute for $(R_l + R_r)$ in Eq 2 above, we obtain the relationship

$$\text{Efficiency}(\eta) = \frac{QR_r}{\omega L} \times 100\% \qquad \text{(Eq 5)}$$

This result is interesting because it is so simple. It does, however, beg the question, "how do we know the Q?"

### Losses—Where Do They Come From?

Before going into that any further, let's take a slightly closer look at the losses we are likely to find in a practical system. The most obvious one is the RF resistance of the conductor, and this is not difficult to calculate. In practice, however, it is probably not worth the effort because it is but one of several sources, most of which are not calculable and may well exceed the conductor resistance contribution. These sources include:

- capacitor losses (several components here)
- insulation losses
- other constructional losses (asymmetry, etc)
- losses due to the environment.

Capacitor losses are, of course, determined by the quality of the capacitor being used. When you are looking for extremely

## The Inductance of a Loop

An important feature of a loop, whether used as an antenna or not, is its inductance. This can be calculated accurately if the dimensions are known and the right formulas are used. Most formulas are traceable to an early document published by the USA National Bureau of Standards (Ref 3) back in the days when the triode valve was very much a novelty. It says much for this document that it has not been surpassed for these kinds of data. For circular single-turn loops it gives

$$L = 0.01257 \times a \left[ 2.303 \log_{10} \frac{16a}{d} - 2 \right] \text{ microhenries}$$

where

a = mean radius of loop in cm
d = diameter of conductor in cm

and for square single-turn loops it gives

$$L = 0.008 \times a \left[ 2.303 \log_{10} \frac{2a}{d} + \frac{d}{2a} - 0.774 \right] \text{microhenries}$$

where

a = length of one side in cm
d = diameter of conductor in cm

Inductance values for single-turn circular and square loops with dimensions between 0.1 and 10 meters are given in Figs A and B, respectively. There are plots for three different conductor sizes covering the range of most practical designs. Values for intermediate sizes can be interpolated by eye.

If you want to be very precise it will be necessary to appreciate that, particularly at low frequencies, inductance is frequency dependent. This would mean that the conductor wall thickness (of the tubing) would need to be taken into account. However, these differences are very small and the formulas given above are the high-frequency ones which assume that the current density is zero (due to skin effect) on the inside of the tubular conductor. They can, of course, be used with both solid and tubular conductors.

Ref 3 also gives formulas for multiturn loops and coils although these are not reproduced here. The characteristics of such coils require one or more tables of correction factors whose values depend upon the dimensions of the winding. Unfortunately there is no simple accurate alternative and the reader is referred to the NBS document for further details. An approximate formula for multiturn loops, derived from NBS, is given in Ref 2.

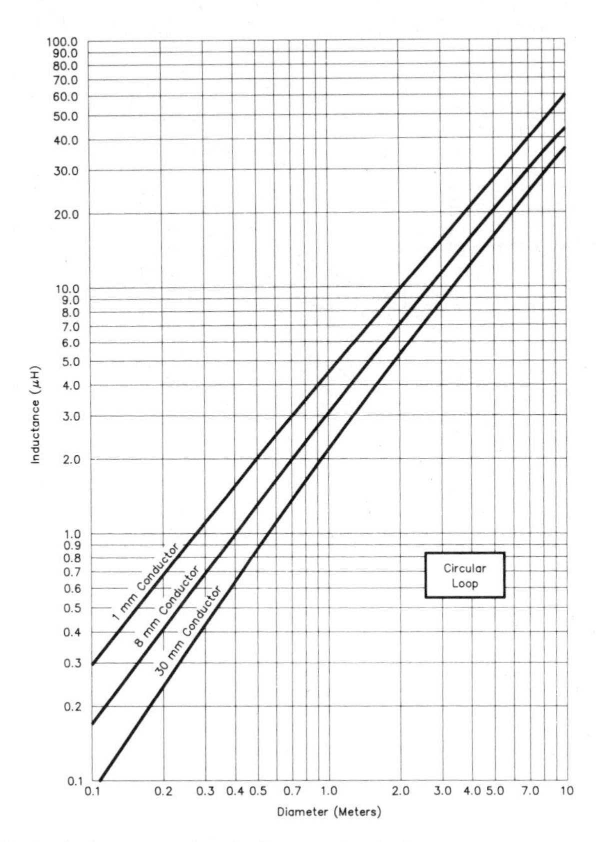


Fig A—Inductance of a single-turn circular loop.

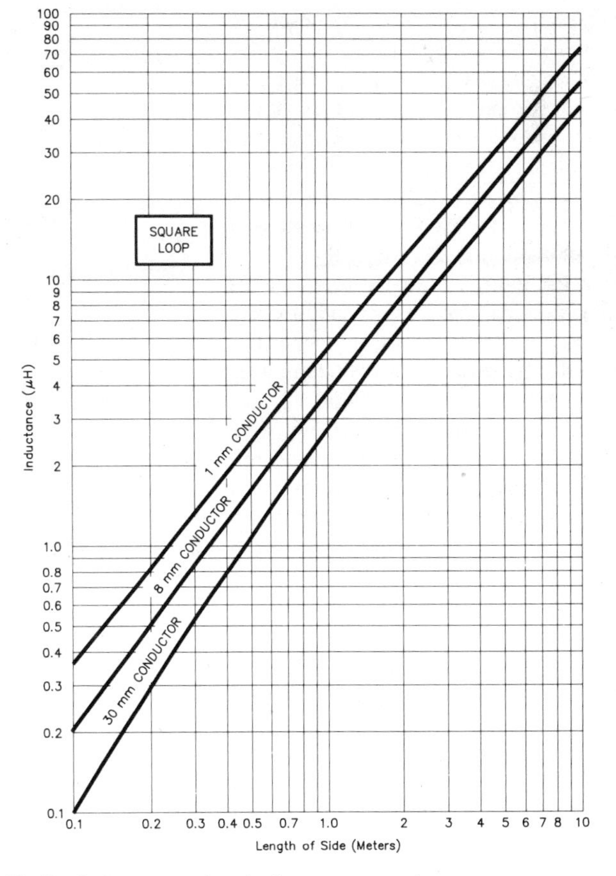


Fig B—Inductance of a single-turn square loop.

high Q values coupled with the high working voltages which go with these on loop transmitting antennas, the need for a good quality capacitor for this voltage and Q makes the construction of the antenna a very expensive privilege. However, without the costly component you will not approach the Q as calculated solely on the RF resistance of the conductor. You probably won't anyway.

Insulation losses in other parts of the assembly cannot be ignored as the loop has to be supported and terminated somehow. The best material the author has found which is both available and easy to work is polystyrene. This material has a dissipation factor of 0.00007 at 1 MHz and 0.0001 at 100 MHz. This is only equaled by strontium titanate ceramic, other ceramics being 2-10 times as lossy. Polycarbonate scores an unimpressive 0.01 at 1 MHz. Teflon is almost as good as polystyrene but very expensive (Ref 2).

A departure from perfect (electrical) symmetry will result in currents flowing in parts of the system where no currents should flow. Such parts of the system can be expected to be more lossy than the conductor and primary insulation, adding to the total system loss. It is to maintain this electrical symmetry that the need to identify the electrically neutral point arises when tapping into a loop, as described above.

Losses due to the environment are perhaps the most difficult to handle. To isolate a loop from its electrical environment (most notably the ground and supporting steelwork) by, say, half a wavelength is difficult to say the least at 14 MHz and well nigh impossible at 3.5 MHz. By its very nature the loop is designed to couple as well as possible into its surroundings, and such spacings as these will be needed to ensure that the presence of other items does not reflect losses into the system.

More confusion? Just when we were reaching what appeared to be a final conclusion? Not really. After all, what we need to know is $R_l$ as a total of all these effects without being too interested in how much comes from each source (unless we are going into the design business, which is an entirely different matter). We can determine this by direct measurement of Q and calculation (knowing $R_r$) or, better still, if we insert the values into Eq 5 above we will have the efficiency directly. The procedure for the Q measurement will not be described here; for the moment we will assume that we know its value.

### Voltages and Currents in the Loop

The big disadvantage of the magnetic loop antenna for transmission is the low radiation resistance, typically a few milohms. Large currents are required to achieve useful radiated powers and these are very difficult to generate. Because the tuned elements are in series, these currents also flow through the inductor (the loop itself forms the inductor) and the tuning capacitor. The reactances of these two components are equal (but opposite in sign) when the loop is correctly tuned, so the voltages across them are also equal (and opposite). The tuning capacitor therefore needs to be rated at this voltage, as well as needing extremely low losses—not an easy combination to achieve.

If a 100-W CW signal is correctly matched into the loop, the voltage across the capacitor can be calculated from the expression

$$V = 10\sqrt{Q\omega L}\ \text{V RMS}$$

and the current flowing around the loop from

$$I = 10\sqrt{Q / \omega L}\ \text{A RMS or}$$

$$I = V / \omega L\ \text{A RMS}$$

where $\omega = 2\pi f$; f is in MHz and L in microhenries.

Corrections can be made for other power levels. For example, a 400-W PEP SSB signal will double the values.

A set of curves giving these values for circular loops of three different diameters is given in Fig C. The curves are plotted for a frequency of 3.5-MHz: All the results are inversely proportional to frequency, so halve the 3.5-MHz voltage value for 7 MHz, and so on. The current is affected in the same way. Again, the figures should be doubled for 400 W PEP SSB.

It is worth noting that, for a high-efficiency (100%) 0.5-meter diameter loop operating on 80 m with 100 W CW, there can be up to 89.8 MEGAvolt-amperes circulating in the system, but, of course, never more than 100 watts. Food for thought.

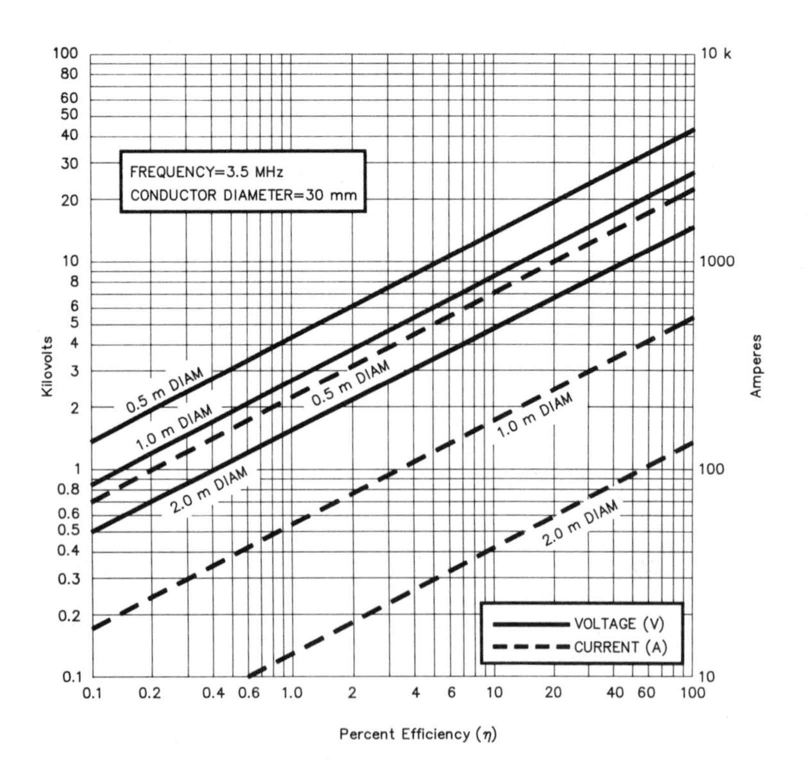


Fig C—Voltages and current in single-turn circular loop with 100-W CW input (double the figures for 400 W).

## All Losses Revealed Through Q

The Q is influenced by all the effects listed above, including, significantly, the operating environment of the loop. Therefore, if the measurement is taken with the antenna in its normal working position, all the effects will be present and the $R_l$ and therefore efficiency will be that which is obtained in normal operation. It will be found to be very different from that calculated from conductor losses alone, unless the conductor is very thin and/or the loop is very small. To avoid errors due to feeder and transmitter loading it is necessary for the feeder to be disconnected from the aerial for the measurements so that nothing is connected to the feeder socket on the antenna. Thus it is the unloaded Q which is important here.

It should be noted that this method is perfectly valid for all magnetic loops, of whatever construction, shape, number of turns and coupling method. Any losses, however introduced, will show up in the Q measurement and will therefore be taken into account in the efficiency.

## The Easy Way

To save a lot of computation, a set of graphs has been prepared which enables efficiency to be read off against measured Q for the 40-m and 80-m amateur bands for reasonably sized loops, both square and circular. These are presented as Figs 4, 5 and 6.

Although the formulas used, and given above, are capable of high accuracies, the need for such precision in practice is questionable. After all, an inaccuracy of 10% in the efficiency corresponds to a change in radiated power of about 0.4 dB. Results read off the graphs will be substantially better than 10% accurate provided all the parameters correspond to those stated on the curves. In practice, within a band, things don't change much if the loop size and Q are constant, so there's not much to be gained by a lot of long-haired toil with correction factors for conductor diameter, etc.

An exception to this general statement is the frequency sensitivity, and it is worth correcting for the frequency within the band at which the efficiency is needed, particularly at 3800 kHz. The curves in Figs 4 and 5 are plotted for the low-end frequencies of 3500 kHz and 7000 kHz respectively, and the correction necessary for other frequencies can be found from Fig 6. The correction factor from the figure is used as a multiplier for the efficiency. For example, an efficiency of 5% at 3500 kHz becomes 5.5% when the correction factor is 1.1. The correction for the 7-MHz band is much smaller, although rather more dependent on the Q-value. A single value of 1.04 at 7100 kHz should be used, and is accurate in the mid-Q ranges where practical designs lie. Interpolation is again straightforward and is linear.

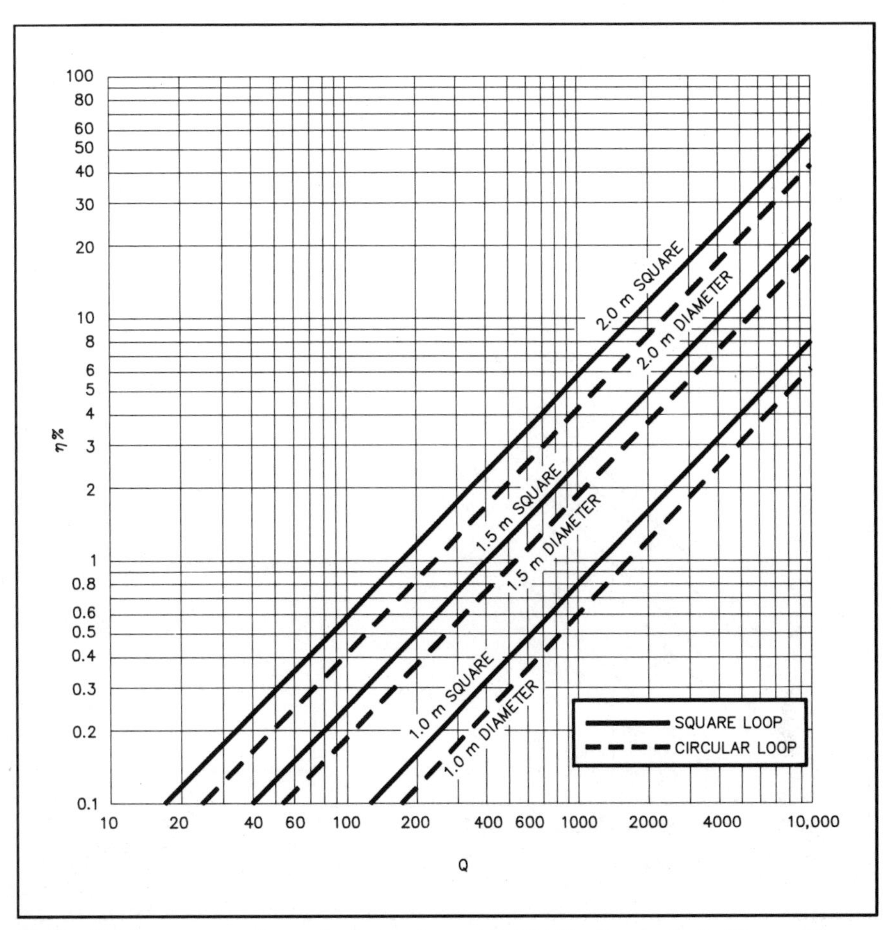


Fig 4—Efficiency curves for a single-turn loop at 3500 kHz. Conductor is 15-mm diam copper.

There will not be much correction required over the bottom ends of the bands but, particularly in the case of 80 m where the width of the band allocation is a significant percentage of the low-end frequency, some correction may be wise up in the upper-frequency region of the band. There will be small efficiency differences if the conductor diameter is not the same as for the plots but, in the main, these will be far less than the difference in Q performance (which will be taken care of by the measurements) caused by the same diameter changes and will be secondary effects.

## Performance and Effectiveness

Loop aerials can be very effective indeed given the right circumstances. Their strength lies in their receiving performance which does not require high efficiencies. On the MF and HF bands the achievable signal-to-noise ratios (and hence readability) which can ultimately be achieved with a practical antenna system depends on how much stronger the wanted signal is than the background noise on the band. This noise, a combination of general noise and identifiable interference, is normally strong enough for the receiver-generated noise to be completely submerged in it. Indeed, from a highly efficient aerial it can be several tens of dB higher than the receiver background.

This means that the efficiency of the receiving antenna can be greatly reduced without affecting the signal-to-noise ratio, as both the signal and the noise are reduced together and the "missing" strength is returned noiselessly by the receiver AGC. Thus, our loop can provide good service here, as was found in the days of early portable wirelesses. It is compact, unobtrusive and can be very portable.

This feature means that we no longer have to strain every muscle to achieve astronomical Q values, nor to construct large and ungainly loops to achieve large values of radiation resistance. (Note, incidentally, that receiving aerials need radiation resistance—that's reciprocity for you.) Also the bandwidth is limited, not by a steeply

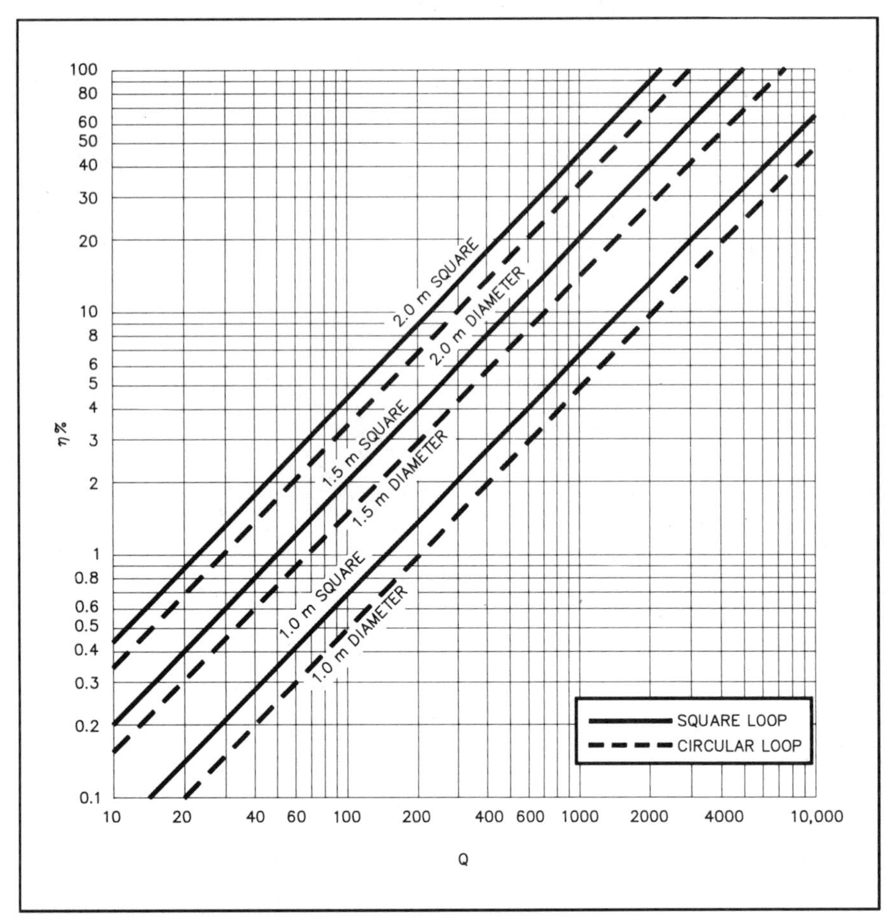


Fig 5—Efficiency curves for a single-turn loop at 7000 kHz. Conductor is 15-mm diam copper.

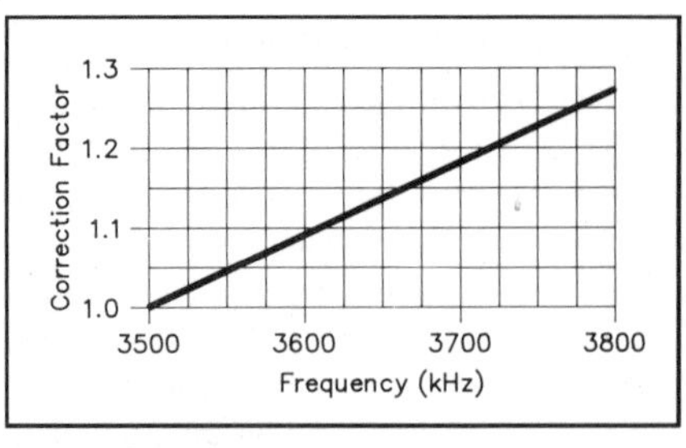


Fig 6—Corrections to be applied to Figs 4 and 5 for frequency. For 7100 kHz use 1.043 with linear interpolation.

rising SWR which can pose a risk to a transmitter, but by how much sensitivity you can afford to lose before receiver noise becomes a problem. This makes the loop a far more tame animal and promising prospect than when seen as a transmitting antenna, where efficiency is all.

Its advantages don't end here, though: There are two further features which are very valuable at the receiving end. The first is the directional property of the loop. It will not respond to signals arriving at right angles to the plane of the loop (ie, "sideways-on"). Thus there is, as it were, a line of zero sensitivity running through the center of the loop at 90° to its plane and passing through the center. This can sometimes be arranged to point directly at a source of local interference. With the plane of the loop vertical which is the normal case, this line is horizontal and parallel to the ground. Since local interference generally originates at ground level, rotating the loop can often cause a useful reduction of its effects. Signals from the same general direction but arriving via the ionosphere (normal radio signals) will enter the loop at a higher angle of elevation and therefore not be attenuated to the same extent, if at all. The signal-to-interference ratio is thus improved.

The second is more subtle in origin but can be very dramatic in effect. Since the loop is primarily sensitive to the *magnetic* component of the received wave and the majority of local interference comprises *electric* fields (they are generally induction rather than fully radiated fields) the loop signals are surprisingly clean and QRM-free.

Don't expect miraculous performances when transmitting on your 1.5-meter diameter loop on 40 m or 80 m. The author was surprised at the low efficiencies he was apparently achieving (based on Q measurements) and even had little enough faith to suspect the formulas (or, to be precise, how he had applied them). It was not until carefully controlled field trials of two identical loops in the far-field region had been conducted over a line-of-sight test range (on Epsom Downs) using the reciprocity principle, and with appropriate correction for ground proximity, that he became fully satisfied with the mathematical model. Off-air signal reports on normal QSOs were obtained, suggesting quite a good performance; it was only later that the truth emerged. The two factors contributing to the apparent performance were the "QRP reporting effect" (and reports became far less optimistic when the other station was not told it was a new loop) and the poor performance of the average amateur wire aerial with which one inevitably compares reports.

Returning to the concept of improving the signal-to-noise ratio at the receiver, the only way the transmitting station can contribute to this (within a fixed bandwidth) is by increasing the average transmitted power. Given, in the case of Amateur Radio, that there is a limit to the permissible RF power of the transmitter, the performance of the aerial is obviously paramount. Thus, we need high efficiency as our primary requirement, and we have seen how difficult it is to achieve this using a magnetic loop.

The aerial designer who strives to obtain, say, 10% efficiency out of his 1.5-m diameter loop on 80 meters will need to achieve a Q value (including ground losses) of over 5000. This designer needs to be hot stuff on the key because, with this high Q and correspondingly narrow bandwidth, his SSB rig is not going to last long with the SWR hitting over 2.6:1 only 350 Hz either side of the loop's center frequency, even supposing the other station can decipher the sharply filtered speech. He will also have great fun tuning the thing—standing within 5 feet of it will throw it off resonance by its full bandwidth—and he will also have to save up for that very expensive tuning capacitor because there will be over 6 KV across it with 100 watts going in, and nearly 13 KV (!) at 400-W PEP on SSB.

Ramstrom makes a telling observation in Ref 1 where he is describing field tests on a three-turn loop: "On 4 MHz the loop gain was 20 dB below a full size dipole and ... on 6 MHz the loop was [also] 20 dB below."

This was a loop constructed professionally using the highest grade materials and suggests an achieved efficiency of about 1%. The potential user of a transmitting loop, particularly at the lower frequencies, is very strongly recommended to perform his own measurement of Q and check his loop's efficiency for himself rather than to take optimistic claims at their face values. There may be surprises in store!

To finish on a purely personal note, the author has designed a mobile antenna making the best use of the features of magnetic and electric antennas. One of his problems in achieving a satisfactory mobile performance was, as is often the case, the interference from the car's electrics and the likes. A 750-mm circular single-turn loop made out of 3-mm aluminum rod and tuned with a small trimmer was coupled into the rig with a smaller coupling turn. This can be oriented to reduce the car-generated noise almost to zero. It is, of course, self defeating now to try to use this for transmitting and so a second antenna was mounted alongside; a 2-meter-long base-loaded whip. The combination of the high-efficiency whip for sending and the quiet receiving loop makes for one of the most effective mobile antenna systems the author has ever encountered.

In another example the author's small battery-operated 80-m receiver, which has a small (400 mm) loop on the top and a low-noise front end, can frequently copy signals indoors which are inaudible on a 30-ft external vertical whose measured efficiency exceeds 50%. The efficiency of the 400 mm loop is about 0.001%.

Small loops can be great fun.

## Note

[1]As a real-life example, careful measurements were made on a commercially manufactured loop antenna 0.8 meter in diameter. It was in new condition as delivered. The frequency of measurement was 14.1375 MHz and the measured Q was 592.6. The calculated results are therefore:

Loop inductance: 1.643 μH
Tuning capacitance: 77.117 pF
3-dB bandwidth (SWR = 2.6 at band edges): 23.857 kHz
Radiation resistance: 38.841 milohms
Effective series loss resistance: 207.5 milohms
Power actually radiated (with 100 W input): 15.767 W
Power lost (with 100 W input): 84.233 W
Current flowing through loop (with 100 W input): 20.148 A
Voltage across loop terminals (with 100 W input): 2941.2 V

This makes the efficiency of the radiating loop just under 16%; the manufacturer's claim for this figure is 62%! This all goes to show that you can't be too careful: Don't take too much on trust where loops are concerned.

## References


S. Ramstrom, "HF Loop for Transmitting and Receiving," *IERE Conference Proceedings*, No. 50, 1981. Radio Receivers and Associated Systems, pp 455-487.

*Reference Data for Radio Engineers* (Indianapolis, IN: Howard W. Sams & Co).

US Department of Commerce, National Bureau of Standards, Washington, DC, Technical Note C74—Jan 1937, *Radio Instruments and Measurements*.

D. Foster, "Loop Antennas with Uniform Current," *Proc IRE*, Oct 1944, pp 603-607.

# Fun With Small Loop Antennas On 80 Meters

by James E. Taylor, W2OZH
1257 Wildflower Drive
Webster, NY 14580

Small loop antennas, having maximum dimensions less than one-tenth wavelength, have been used successfully for a variety of lower frequency applications. They are especially useful where space is limited. Patterson pointed out some of the limitations of such antennas.[1] The author extended these principles to mobile application for the 80-meter band.[2] That installation was used with pleasing results during some 40,000 miles of mobiling. (A number of users of the MobiLoop have remarked about the surprisingly good results. A possible explanation is given later in this article.)

This article points out certain less-recognized characteristics of such antennas and reviews some relevant experiments in the 80-meter band.

## First, Some Theory

*Induction and Radiation Fields*

It is known from Maxwell's Equations[3] that

$$dB = k \times I(1/R^2 + 2\pi/R\,\lambda) \qquad \text{(Eq 1)}$$

where

- $I$ = RF current of wavelength $\lambda$ in a conductor (an antenna)
- $dB$ = resulting incremental magnetic induction at a distance $R$

The first term is the normal electromagnetic induction field, as in an electromagnet. The second term is the RF radiation field.

Here we see that, for large distances compared with the wavelength, the induction field is negligible—we have only radiation. Conversely, at short distances the induction field is much greater than the radiation field. If the distance from the antenna is equal to $\lambda/2\pi$, the induction field is just equal to the radiation field. This "radian distance" has been used to develop what is called the "radian sphere" for small antennas.[4]

The radian sphere is a theoretical spherical volume having a radius of $\lambda/2\pi$ with its center coincident with the center of either a capacitive or an inductive antenna. This sphere is the space occupied mainly by the stored energy of the electric or magnetic field—it defines the boundary within which the reactive power density exceeds the radiative power density.

For a wavelength of 249 feet (3.953 MHz) the radian distance is about 40 feet. This means that any object (including the earth) placed closer than 40 feet from any 80-meter antenna, including the small loop type, will have a field induced in it stronger than the radiation field. The power loss in the object increases rapidly as it is brought closer to the antenna. If it is a lossy dielectric, the E-field losses will prevail and if it is a lossy magnetic material, the H-field losses can present problems. Typical earth displays both types of loss.

Knowing this gives us a good reason to place the entire 80-meter antenna at least 40 feet from the ground or from any other objects which may produce losses! Since a small loop antenna is essentially a one-turn coil, it displays a strong induction field perpendicular to the plane of the coil. Such an antenna should be placed as far as practical from lossy magnetic materials, especially along the axis of the coil.

*Difference between Loop Antennas and Dipoles*

It can be further derived from Maxwell's Equations[5] that when a small loop antenna is placed in an RF field, the induced current is normally much less than that induced in a short dipole wire of the same maximum dimension. A similar reduction of efficiency applies if the antennas are used for transmitting. The radiation resistances of the two antennas vary as the squares of the respective induced currents and are related by the following equation.

$$R_d/R_\ell = (\ell \times \lambda/2\pi \times A)^2 \qquad \text{(Eq 2)}$$

where

- $R_d$ = radiation resistances of the short dipole wire
- $R_\ell$ = radiation resistances of the small loop
- $\ell$ = length of the wire
- $A$ = area of the small loop

For a wavelength of 249 feet, if the loop is a circle of diameter 12 feet, its radiation resistance compared with that of a 12-foot short dipole will be, from Eq 2, 18 times less than that of the dipole! Thus, assuming perfect matching, the short dipole is a much more efficient radiator than the small loop of the same size. In terms of power, in this case, the loop can be expected to be down 12 decibels compared with the dipole of equal size.

Let us calculate what this means for a practical example to be used later. The radiation resistance for a small, single-turn, loop antenna is given by[6]

$$R_r = 31200 \times (A/\lambda^2)^2 \qquad \text{(Eq 3)}$$

Assuming a circular coil of diameter 12 feet and a wavelength of 249 feet as before, the calculated radiation resistance turns out to be approximately 0.1 Ω. From Eq 2, a twelve-foot long dipole has a radiation resistance 18 times this or 1.8 Ω. If it were not for inefficiencies of coupling or impedance matching the small dipole would be

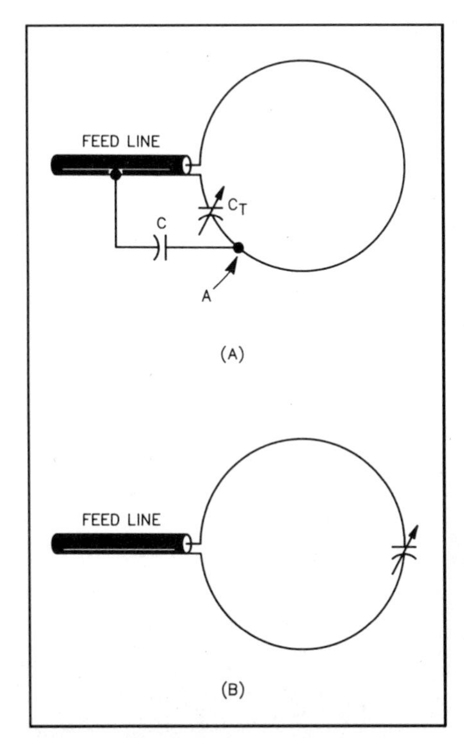


Fig 1—At A, an unbalanced loop antenna. $C_T$ is the resonating capacitor, while C is stray capacitance from the loop to the feed-line shield. Moving $C_T$ to the opposite side of the loop, as at B, places half the loop reactance in series with each side of the capacitor, effectively isolating it from the feed line.

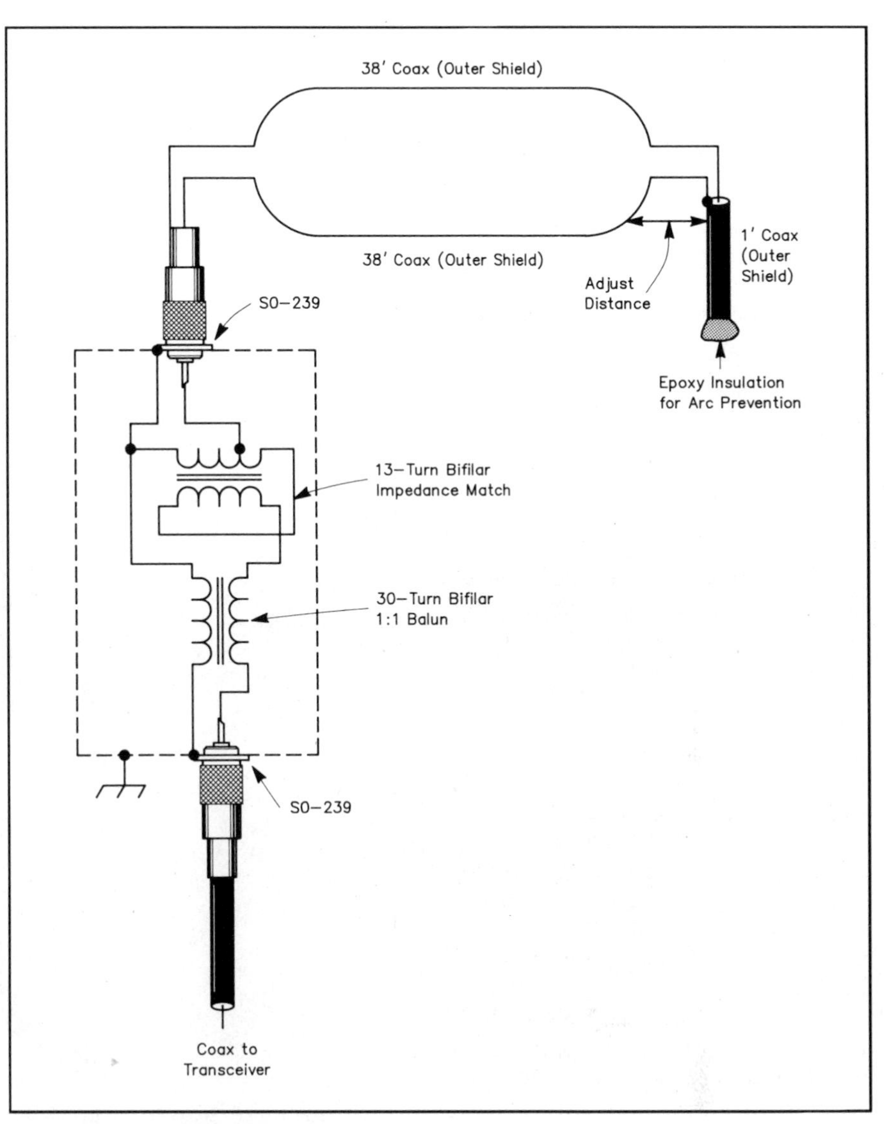


Fig 2—A practical balanced loop antenna. Because the loop elements are flexible, this antenna can be easily relocated.

the more efficient antenna by 12 decibels.

It is informative to reflect upon what this means, physically. Why is a symmetrical loop less efficient than a wire dipole of the same size? The key is the word "symmetrical." We realize that because the loop is small the RF current is essentially of constant magnitude throughout its circumference.

Since the elements of current on diametrically opposed sides of the loop are equal in magnitude and opposite in direction, their fields tend to cancel; on the axis of the loop the cancellation is total. When the loop dimensions are a fraction of the wavelength, the incremental fields, even for off-axis positions, also tend to cancel. They cancel because they are almost out of phase. Therefore, the effectiveness of the symmetrical small loop as a radiator is much lower than that of a single dipole wire which does not have this symmetrical cancellation.

The closed symmetry of the loop is less efficient than the open symmetry of the dipole. Therefore, we should try to design toward open symmetry rather than closed symmetry. Later on, this insight will be applied in a practical situation.

*The Q Value of Loops*

The Q of an inductor is the ratio of the inductive reactance to the effective resistance in the circuit. The Q is given by the equation

$$Q = 2\pi \times f_r \times L/R \qquad \text{(Eq 4)}$$

where

$f_r$ = resonant frequency
R = effective resistance

The universal resonance curves can be used to evaluate Q for a practical loop antenna installation by observing the deviation from resonance on an SWR bridge. The applicable equation is

$$Q = 0.5 \times f_r/df(3\ dB) \qquad \text{(Eq 5)}$$

where df(3 dB) is the deviation from the resonant frequency for the voltage to be down by $\frac{1}{2}^{1/2}$, that is, for the power to be down by a factor of 2, that is, 3 decibels. Equating the two relationships of Eqs 4 and 5 for Q yields

$$df(3\ dB) = R/4\pi L \qquad \text{(Eq 6)}$$

We now have a direct method of calibrating the SWR bridge to measure the Q of a loop antenna. A standard coil is connected in a series circuit with a known resistance. The resistance is chosen to approximately match the calculated and measured constants for the typical loop antenna.

These values turned out to be 8.5 µH and about 2.7 Ω, and the corresponding measured frequency increment was df(3 dB) = ±25 kHz, yielding a Q value for this standard coil of $Q_s$ = 79. Once the Q was known, the SWR bridge calibration figure was measured to be 9% of full scale. The $Q_s$ for the loop antenna becomes

$$Q(loop) = 1980/df(9\%) \quad \text{(Eq 7)}$$

where df(9%) is the measured frequency half-bandwidth for the SWR bridge to read 9% of full scale. These values are summarized in Table 1 for typical loop configurations.

### Practical Considerations

*Balanced versus Unbalanced Loops*

Small loop antennas are usually fed with coaxial cable. A balanced radiating system is preferable to an unbalanced radiator. The reason is depicted in Fig 1.

Fig 1A shows a small loop antenna having its variable resonating capacitor ($C_T$) connected directly to the feed-line shield. At resonance there will be a high RF voltage across the resonating capacitor. Point A is at a high potential with respect to the shield of the feed line. Therefore, there will be strong coupling through the stray capacitance C to the shield.

The resulting unbalanced common-mode current in the shield causes the feed line to radiate, which distorts the antenna pattern. The problem can be reduced by using the balanced configuration of Fig 1B. Here, the loop presents inductive reactance to either side of the feed line, in effect isolating the resonating capacitor, $C_T$, from the shield and reducing common-mode current.

It is interesting to note that the unbalanced arrangement of Fig 1A displays a hybrid characteristic—it wants to resonate as a simple tuned inductor but it also wants to establish an image, due to the capacitive coupling back to the feed-line shield. This is somewhat analogous to the two terms of Eq 1, where the simple tuned inductor tends to produce the induction term, whereas, the "image" in the feed line tends to emphasize the radiation term. Thus, even though we try to set up a pure unbalanced loop, it tries to become a balanced radiator as in Fig 1B.

This tendency has prompted the comment that there is no such thing as a monopole radiator—even a whip mounted on a car has an image on the car body! (The image-seeking tendency exists, even in a balanced dipole antenna.

### A Flexible Balanced Loop

Small loop antennas ideally combine low RF-loss resistance with mechanical flexibility. A flexible antenna is easier to move. Coaxial-cable outer conductor is a suitable material for the radiator. Fig 2 shows a practical arrangement. The radiating loop is formed from two 38-foot lengths of RG-8 type coaxial cable. A 1:1 balun provides feed-line isolation. The balun consists of 30 bifilar turns on a 2-inch-diameter ferrite core (T-200-2).

An additional 15-turn bifilar autotransformer matches the input impedance. Typically, the tap on this transformer is 10 turns up from the ground connection, for an input impedance of 50 Ω as measured on a noise bridge. Since the impedance varies as the square of the turns ratio, the input impedance of the loop is $(10/30)^2 \times 50$ or about 5.5 Ω. This figure, which is much larger than those usually quoted for small loop antennas, includes losses due to ground penetration, which are substantial.

The transformers were mounted in a covered plastic box approximately 6 × 3 × 2 in. (Radio Shack 270-223), with SO-239 connectors mounted on either end. A simple method of tuning the antenna to resonance is a one-foot length of RG-8 coaxial cable (capacitance = 29.5 pF/ft) mounted so that the spacing between the shield and the opposite side of the antenna can be varied for trimming adjustments. After initial adjustment of the length of this coaxial capacitor, a spherical dome of clear epoxy is formed at the open end to eliminate the tendency to arc at high power levels. See Fig 2.

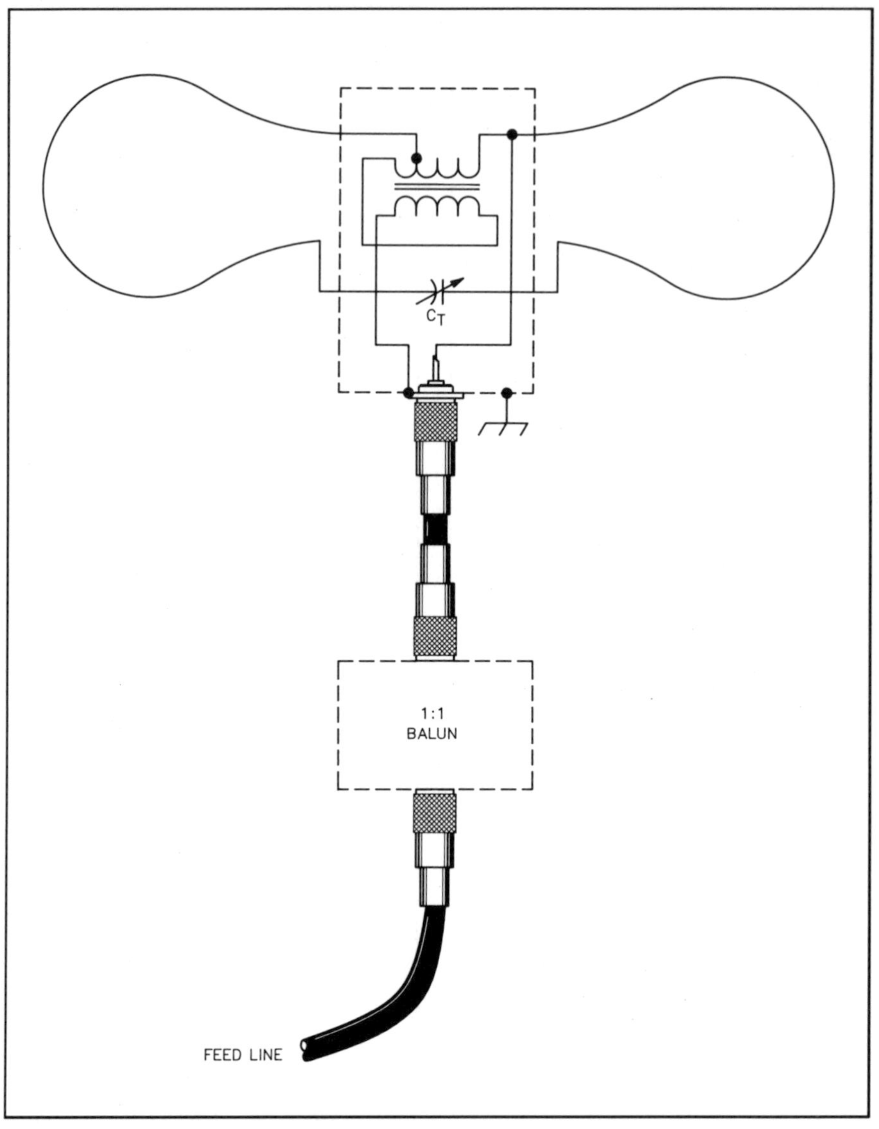


Fig 3—The DiLoop places impedance-matching and isolating components at one location. Loop lengths are not critical.

### The DiLoop

It is sometimes convenient to have the feed-line connector, the tuning capacitor, and the impedance-matching components in a single housing. Fig 3 shows such a

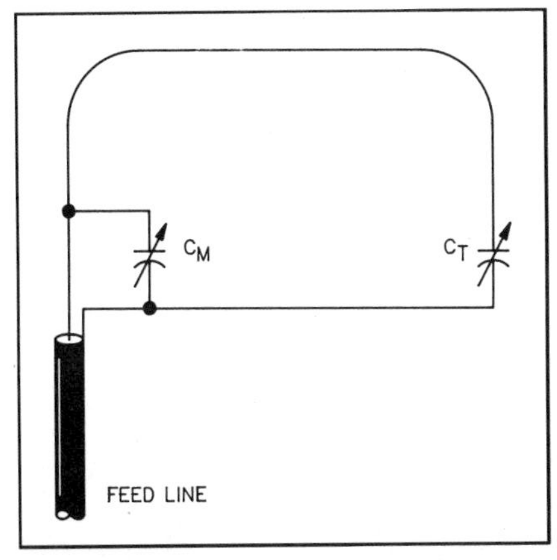

Fig 4—The ArchLoop is self supporting, which makes it suitable for roof-top installation. A metal roof may be used as the return conductor.

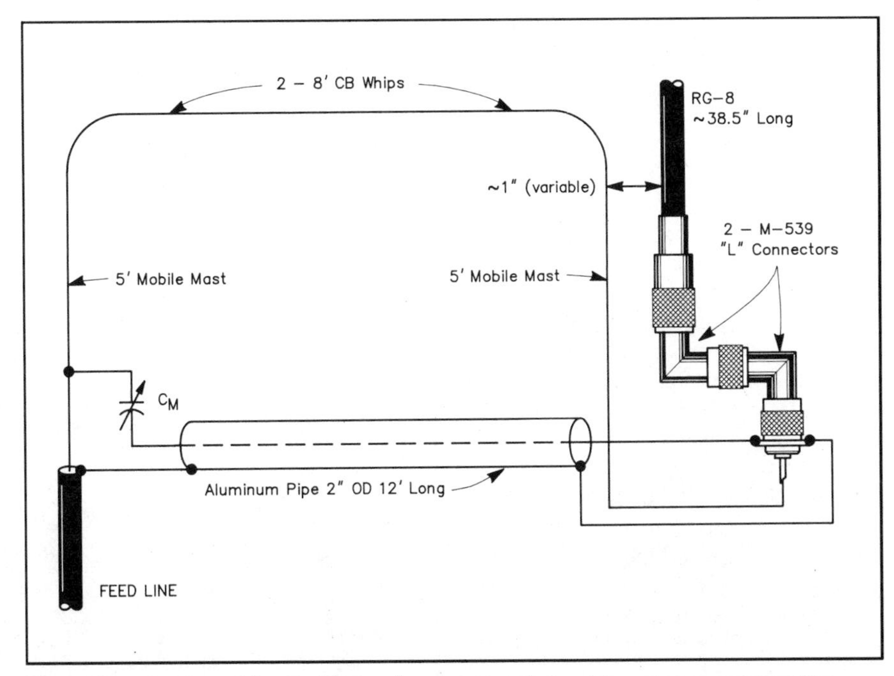

Fig 5—Construction of the ArchLoop. An assortment of mobile antenna components was used in this version, but any of the construction methods used for the other loops are suitable.

butterfly-shaped or DiLoop arrangement. Matching can be accomplished by a transformer, as in the flexible balanced loop. A wide-spaced air variable capacitor, $C_T$, may be used for resonance adjustment.

See Fig 3. The impedance-matching transformer is an autotransformer with 15 bifilar turns, as described earlier. The tuning capacitor is a transmitting type air variable (Fair Radio Sales M7463006-1, 22-118 pF). These components are mounted in a covered plastic box approximately 7¾ × 4¼ × 2¼ (Radio Shack 270-232). Five SO-239 coaxial connectors mounted on the sides and end of the box provide reliable connection to the shields of the two radiating loops.

The lengths of the loops are not critical and they need not be equal—I used lengths of approximately 30 feet and 40 feet. With these lengths, the value of the tuning capacitance at resonance is approximately 30 pF. With the impedance-matching tap at 8 turns up from ground, the SWR was approximately 1.2:1 at resonance.

### The ArchLoop

For some installations where a self-supporting loop is desirable, for example on the roof of a building, the ArchLoop configuration (Fig 4) may be convenient. $C_T$ is the resonating capacitor and $C_M$ is an impedance-matching capacitor. If the roof is metal, it may be used as the horizontal return conductor. In this case, the similarity to its ancestor, the MobiLoop,[7] is apparent.

You can see that this antenna is a version of the balanced loop. The arch and the horizontal return lines provide the respective inductive impedances required to balance the RF current paths. There is an obvious asymmetry however, because the roof is a conductive plane, not just a wire. Thus, we should expect such a loop antenna to be more efficient than a loop having totally closed symmetry. This is a plausible explanation for the high performance of the MobiLoop.

An ArchLoop was constructed as shown in Fig 5. The arch is supported by two 5-foot mobile masts. Mounted on each is an eight-foot-long stainless-steel whip. These whips are curved to form the arch. The overlapping ends are bound together with wire. The entire arch is then covered with shielding braid or aluminum foil to reduce the RF resistance.

At the test installation the mobile masts were mounted on mobile bumper-mounts, which were supported on sawhorses about twelve feet apart. The masts also support the horizontal aluminum pipe, the input coax connector and the coaxial tuning capacitor. The capacitor is described in the next paragraph. The assembled ArchLoop is stabilized in the desired vertical plane by four lengths of nylon cord tied from the tops of the mobile masts to the ends of the sawhorses.

See Fig 5. Note that impedance matching is achieved simply by means of a capacitance connected across the coaxial feed line. This capacitance is initially determined by use of an air variable capacitor (Fair Radio Sales 4G-535, 4-section, 20 to 535 pF/section, all sections connected in parallel), and then replaced with a fixed 500-volt mica or ceramic capacitor. The value turned out to be 1800 pF. The value of the resonating capacitance was determined by first using an air variable capacitance, 120 pF which was later replaced with RG-8 cable, 38.5 inches long. At the bottom of this fixed capacitor I placed a PL-259 connector and two M-539 angle connectors.

The coax was placed inside a capped length of half-inch PVC pipe for stiffening. Trimming adjustment was accomplished by pivoting the angle connectors about a vertical axis, thereby changing the capacitance between the coax shield and the vertical mast. One-kilowatt input was used without arcing. It is important to keep the capacitor connections dry, as at this power level there is a peak potential of several thousand volts across this capacitor.

### Asymmetrical Loops

As already noted, the efficiency of a symmetrical small loop antenna is very low, in part because of its closed symmetry. If the loop is intentionally made less symmetrical its efficiency of radiation tends to improve, since the field cancellation is reduced. Referring to Fig 5, the pattern of the ArchLoop is somewhat asymmetrical due to its shape.

If we place this loop close to the ground it becomes more asymmetrical. If the antenna is mounted directly on a metal surface, such as a metal roof or a car body, the RF currents are rendered still more asymmetrical; that is, we are approaching open symmetry. As discussed earlier, the result is increased radiation efficiency, though the resistive losses may be greater.

### The TriLoop

A different loop with more open symmetry is shown in Fig 6. For illustration, see the lower horizontal section of Fig 5. To shield the current flowing in this conductor, we could place a wire inside the pipe, coaxially. Since such an arrangement effectively becomes a coaxial capacitor however, this capacitance can provide the principal resonance tuning of the loop. Now, a still greater degree of open symmetry can be achieved.

The main radiation from the antenna of Fig 5 tends to be confined to the upper, curved portion, while the horizontal pipe serves as a capacitive shield. This concept was tested by changing from the ArchLoop arrangement of Fig 5 to the TriLoop shown in Fig 6.

We wish to achieve a balanced current distribution either side of the feed point, as shown in Fig 1B, but with a nonsymmetrical field, due to the extended nature of the pipe capacitor. The aluminum pipe was capped at either end with PVC pipe caps, which provide good insulation. Brass ¼-20 bolts were mounted through the centers of the caps. The bolts support a taut length of 12-gauge stranded copper-clad wire along the axis of the pipe. The calculated capacitance is 63 pF. The matching capacitor, $C_M$, and the resonance trimmer, $C_T'$, are air variables. A good match was attained at resonance.

The relative RF currents in various parts of the radiator were measured with an MFJ-206 antenna current probe. The relative current on the outside of the pipe at the center was 47% (down 7 dB, compared with that on either side of the feed point) and the common-mode current induced in the feed-line shield was 27% (down 11 dB). The TriLoop was tuned to resonance with $C_T' = 28$ pF. For a measured input impedance of 50 Ω the value of $C_M$ was 2200 pF. A coaxial tuning capacitor similar to the one used for the balanced antenna was fabricated. Using this arrangement, an input power of a kilowatt was used on single sideband without arcing.

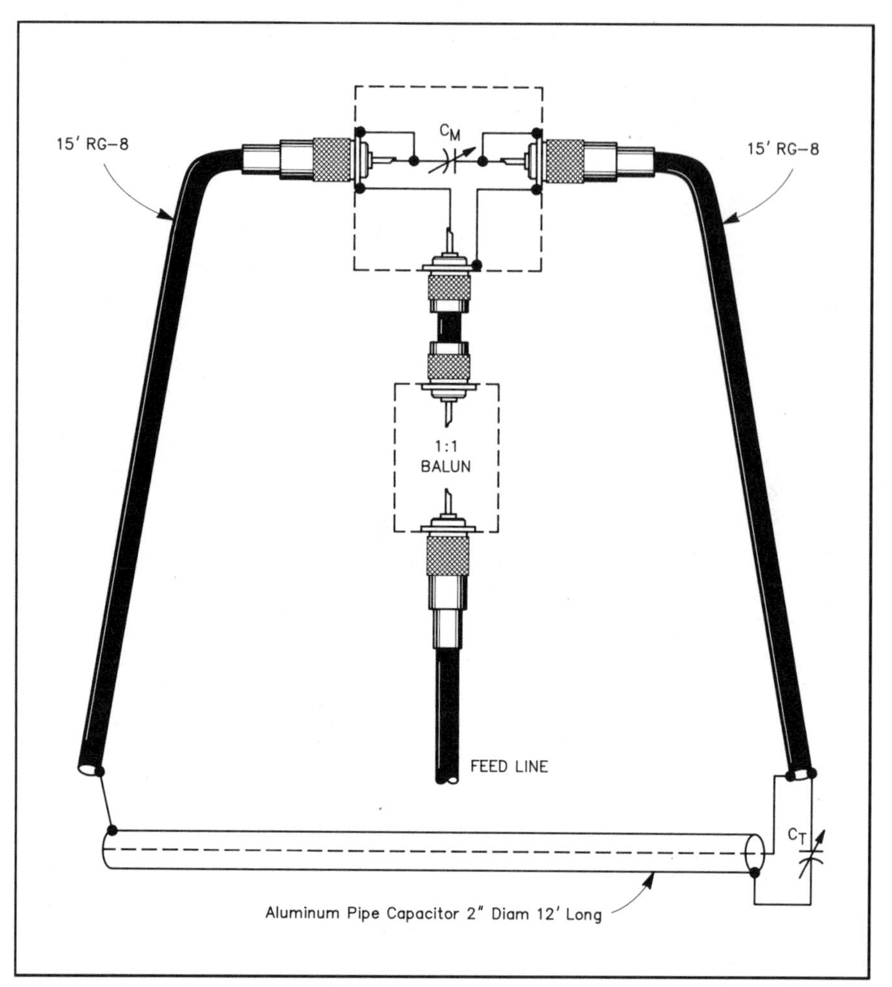


Fig 6—The TriLoop performed well in tests conducted by the author, perhaps because radiation is concentrated in the upper portion of the antenna. Stations over 200 miles away were only about 4 dB weaker on this antenna than on a high dipole.

**Table 1**
**Performance Comparison of Loop Types**

| | | | Distance | | |
|---|---|---|---|---|---|
| *Configuration* | *df(9%)* | *Q(loop)* | *>200 mi* | *10-200 mi* | *<10 mi* |
| DiLoop | 40 kHz | 50 | –8 dB | –12 dB | — |
| BalancedLoop | 32 kHz | 62 | –10 dB | –10 dB | — |
| ArchLoop | 40 kHz | 50 | –6 dB | –10 dB | — |
| TriLoop | 40 kHz | 50 | –4 dB | –10 dB | –2 dB |

### Results

Extensive comparisons of the various loops were made for both received and transmitted signal strengths. An accurate RF attenuator was used for the receiving tests. Reciprocity between received and transmitted signals was repeatedly confirmed. These results are summarized in Table 1. The measurements were made mainly during daylight hours in the summer time. Static and other interference are normally less severe than at night and more consistent results are attainable. It was observed that the more highly scattered daytime radiation for the more-distant stations produces some anomalous effects. These are mentioned later.

From Table 1 you can see that average performance for all of the small loops was about 6 dB down compared with a high dipole, for more-distant stations. For the limited data taken, the TriLoop appeared better than the average, but the DiLoop gave somewhat inferior results. Out to 200 miles, all loops are down about 10 dB. This is probably due as much to the superior high-angle radiation characteristics of the quarter-wavelength-high reference antenna as to the deficiencies of the loops.

There were certain noteworthy performance anomalies. For example, a station about ten miles away was noted to be down only 2 dB using the TriLoop. Also, there were a few times near midday when stations in the Midwest were surprisingly strong. This was noted first for the DiLoop suspended in a vertical plane under the overhang of the house, when a net of stations in the tenth call area were actually stronger on the DiLoop than on the dipole! Another time, I heard Michigan stations with equal signal strengths for the TriLoop, the ArchLoop and the dipole. I confirmed this by calling WT8J (using my phased array to get his attention). He reported "slightly readable" on all three of the test antennas.

## Conclusion

Four basic configurations of small loop antennas were constructed: The DiLoop, the Balanced Loop, the ArchLoop and the TriLoop. The difference in radiation efficiencies of small dipoles and small loops is obvious. The TriLoop arrangement approaches the greater efficiency of the dipole radiation. Efficient coupling can be had without an external tuner, by using either toroidal transformers or a simple capacitive match. Tuning to resonance can be accomplished by either an air variable capacitor or by the use of a coaxial-cable stub.

Practical operation of each loop is possible up to one kilowatt. Extensive receiving and transmitting tests showed some anomalies in radiation/propagation characteristics.

## References


[1] K. H. Patterson, "Down To Earth Army Antenna," *Electronics*, August 21, 1967.

[2] J. E. Taylor, "The Mobiloop," *QST*, Nov 1968, pp 18-19.

[3] G. P. Harnwell, *Principles of Electricity & Magnetism* (New York: McGraw-Hill), pp 552ff.

[4] H. A. Wheeler, *IEEE Trans on Antennas and Propagation*, Vol AP-23, No 4 (Jul 1975), pp 462-469.

[5] See p 557 of ref 3.

[6] F. E. Terman, *Radio Engineers' Handbook*, 1st ed (New York: McGraw-Hill), pp 814.

[7] See ref 2.

# 12-Meter Quad

**By Howard G. Hawkins, WB8IGU**
**915 Sanford Ln**
**Au Gres, MI 48703**

My continuing interest in loop antennas suspended between trees has led to the evolution of a 12-meter quad constructed from ½-inch aluminum Hardline tail ends. Bamboo spreaders contribute to the economy of its construction.

This design was inspired by Frank Lester, W4AMJ, who has used a 16-element, 2-meter circular quad consisting of four bays of four elements with considerable success. Many DX stations, intrigued by the simplicity of the 12-meter design, have expressed interest in its construction.

The circular design, in the author's experience, seems to be superior to the diamond, square or delta configuration, possibly being a better match to the RG-59 quarter-wave matching section. I am fortunate to have several trees available at both my summer and winter QTHs. The simplicity of design is enhanced by having the antenna suspended from a "skyline" between two trees. Rotation is obtained by use of an "armstrong" rotator, consisting of attached cords and small stakes.

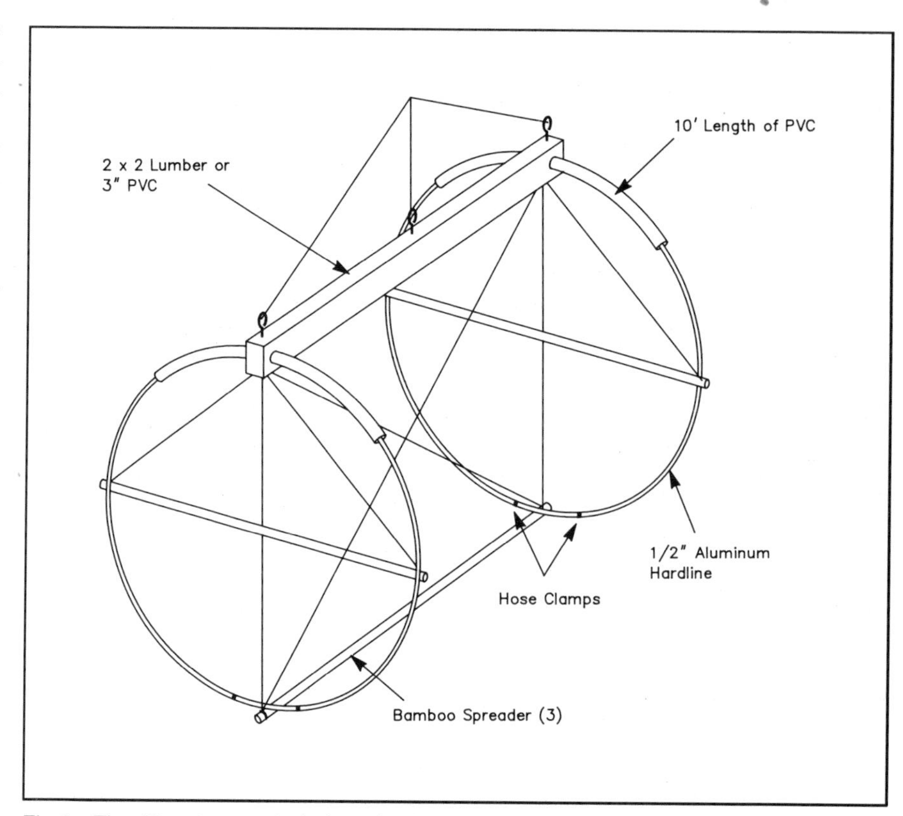


Fig 1—The 12-meter quad, designed to be economical to build and enjoyable to experiment with. The author has found the circular configuration to be a good performer.

Based on the formula $\ell = 1005/f_{MHz}$ for the driven element and $\ell = 1005 \times 1.05/f_{MHz}$ for the reflector, and a center frequency of 24.960 MHz, element lengths work out to 40 feet 4 inches for the driven element and 42 feet 4 inches for the reflector. An additional 2 feet on each element allowed for clamping and adjustment. I used stainless-steel aircraft-type hose clamps. The ¼-λ matching section of RG-59 (solid dielectric) used between the antenna and the RG-58 feed line is 6 feet 6 inches. Spacing between elements, based on 0.2-λ spacing, works out to 7 feet 10 inches. A 2 × 2 piece of wood or 3-inch PVC drain pipe is used as the top spreader and support member. Ten-foot lengths of PVC through the wood and enclosing the top of the Hardline loops distribute the load and help form the loop configuration. Dacron strings are used extensively as bracing, both in the vertical and horizontal planes, to help hold the loops parallel. Your local CATV company may provide the Hardline at little or no cost.

I worked several European stations from my Florida location during erection and tuning with the antenna about 5 feet above the ground. Have some fun; try some circles.

# The Log Periodic Loop Array (LPLA) Antenna

By Duane Allen, N6JPO
302 Massachusetts Ave
Riverside, CA 92507

I would like to introduce for experimentation and design refinement the log periodic loop array (LPLA) antenna of Fig 1. The antenna was conceived during the ARRL Antenna Design Contest (January 1984). One category of contest entries was antennas covering all amateur bands between 20 and 10 meters. The example used here to demonstrate design techniques covers 17 to 10 meters. The antenna is similar in construction to the log periodic dipole array (LPDA) antenna, except the radiating elements are loops and the feeder line between elements is placed on alternate sides of the loops. The LPLA should provide a 3-dB gain improvement over an equivalent LPDA.

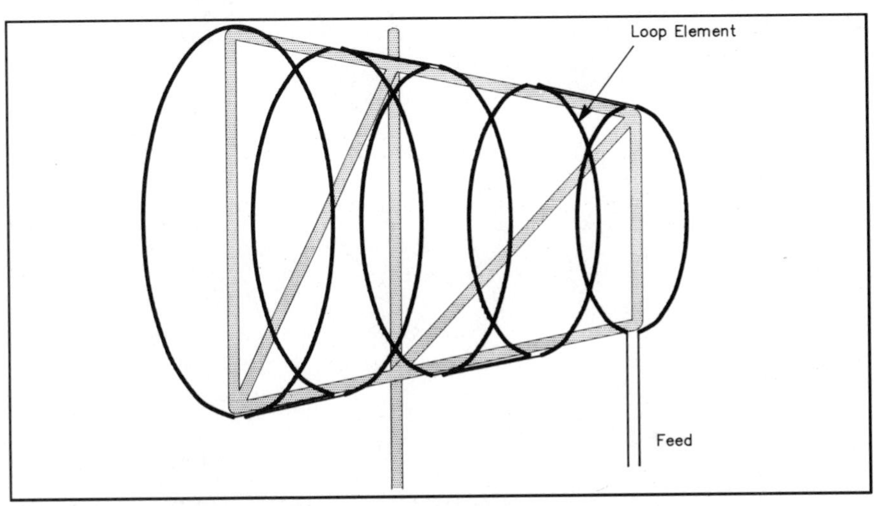


Fig 1—Log periodic loop antenna (LPLA). The shaded sections comprise a nonconductive frame support.

## Estimating the Number of Loop Elements

The first design step is to determine how many loop elements are needed to best fit the antenna to the amateur frequency bands. A derivation of element resonance for frequency is given in the Appendix at the end of this article. By minimizing the number of elements, the length of the antenna can be kept reasonably short. But a minimum number of elements requires that the elements resonate as well as practically possible on the frequencies of interest. Table 1 shows the results of a spreadsheet analysis of where the upper amateur HF bands would resonate on arrays with various numbers of elements. (The spreadsheet file, in Quattro and Lotus 1-2-3 WK1 format, is available on diskette; see **Diskette Availability** on a page at the front of this book.)

Inspection of the table shows that the five-element array best fits the four upper amateur HF bands. This would be a reflector element and the four nearly resonant elements. Arrays with other numbers of elements result in the 15- and 12-meter bands not resonating near integer loop values.

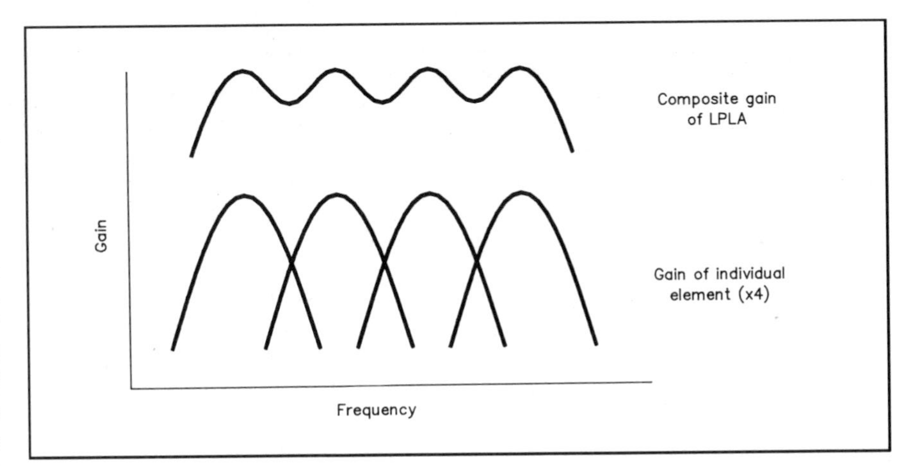


Fig 2—Gain versus frequency of a log periodic array.

## Fitting the Loop Elements

The resonant frequency $f_i$ for the loop i is characterized by the formula

$$f_i = f_0 \times \tau^i$$

where
$f_0$ = a base frequency
$\tau$ is a design constant, the ratio of element size and spacing

## Table 1
## Spreadsheet for Element Resonance for Upper HF Bands

Lower, mean and upper frequencies are in MHz. Tabular values indicate the element number for resonance at the associated frequency. (This spreadsheet is available as LPLA.WK1 on a companion diskette; see **Diskette Availability** on an early page of this book.)

| | A | B | C | D | E | F | G | H | I | J | K | L | M | N | O | P | Q |
|---|---|---|---|---|---|---|---|---|---|---|---|---|---|---|---|---|---|
| 1 | Ham band: | | | 17 M | | : | | 15 M | | : | | 12 M | | : | | 10 M | |
| 2 | | | | | | | | | | | | | | | | | |
| 3 | Total | : | Lower | Mean | Upper | : | Lower | Mean | Upper | : | Lower | Mean | Upper | : | Lower | Mean | Upper |
| 4 | Elements | : | 18.580 | 18.630 | 18.680 | : | 21.000 | 21.224 | 21.450 | : | 24.890 | 24.940 | 24.990 | : | 28.000 | 28.837 | 29.700 |
| 5 | | | | | | | | | | | | | | | | | |
| 6 | 3 | : | 0.994 | 1.000 | 1.006 | : | 1.274 | 1.298 | 1.323 | : | 1.663 | 1.668 | 1.672 | : | 1.933 | 2.000 | 2.067 |
| 7 | 4 | : | 0.988 | 1.000 | 1.012 | : | 1.548 | 1.597 | 1.645 | : | 2.326 | 2.335 | 2.344 | : | 2.865 | 3.000 | 3.135 |
| 8 | 5 | : | 0.982 | 1.000 | 1.018 | : | 1.822 | 1.895 | 1.968 | : | 2.989 | 3.003 | 3.017 | : | 3.798 | 4.000 | 4.202 |
| 9 | 6 | : | 0.975 | 1.000 | 1.025 | : | 2.096 | 2.193 | 2.290 | : | 3.652 | 3.671 | 3.689 | : | 4.730 | 5.000 | 5.270 |
| 10 | 7 | : | 0.969 | 1.000 | 1.031 | : | 2.370 | 2.492 | 2.613 | : | 4.315 | 4.338 | 4.361 | : | 5.663 | 6.000 | 6.337 |
| 11 | 8 | : | 0.963 | 1.000 | 1.037 | : | 2.645 | 2.790 | 2.936 | : | 4.978 | 5.006 | 5.033 | : | 6.595 | 7.000 | 7.405 |
| 12 | 9 | : | 0.957 | 1.000 | 1.043 | : | 2.919 | 3.088 | 3.258 | : | 5.641 | 5.674 | 5.706 | : | 7.528 | 8.000 | 8.472 |
| 13 | 10 | : | 0.951 | 1.000 | 1.049 | : | 3.193 | 3.387 | 3.581 | : | 6.305 | 6.341 | 6.378 | : | 8.460 | 9.000 | 9.540 |
| 14 | 11 | : | 0.945 | 1.000 | 1.055 | : | 3.467 | 3.685 | 3.904 | : | 6.968 | 7.009 | 7.050 | : | 9.393 | 10.000 | 10.607 |
| 15 | 12 | : | 0.939 | 1.000 | 1.061 | : | 3.741 | 3.984 | 4.226 | : | 7.631 | 7.677 | 7.722 | : | 10.325 | 11.000 | 11.675 |
| 16 | 13 | : | 0.932 | 1.000 | 1.068 | : | 4.015 | 4.282 | 4.549 | : | 8.294 | 8.344 | 8.395 | : | 11.258 | 12.000 | 12.742 |
| 17 | 14 | : | 0.926 | 1.000 | 1.074 | : | 4.289 | 4.580 | 4.871 | : | 8.957 | 9.012 | 9.067 | : | 12.191 | 13.000 | 13.809 |
| 18 | 15 | : | 0.920 | 1.000 | 1.080 | : | 4.563 | 4.879 | 5.194 | : | 9.620 | 9.679 | 9.739 | : | 13.123 | 14.000 | 14.877 |
| 19 | 16 | : | 0.914 | 1.000 | 1.086 | : | 4.837 | 5.177 | 5.517 | : | 10.283 | 10.347 | 10.411 | : | 14.056 | 15.000 | 15.944 |
| 20 | 17 | : | 0.908 | 1.000 | 1.092 | : | 5.111 | 5.475 | 5.839 | : | 10.946 | 11.015 | 11.084 | : | 14.988 | 16.000 | 17.012 |

Selected cell entries for the spreadsheet (as accomplished in Quattro and Lotus 1-2-3):

```
.
.
C4: 18.58
D4: @SQRT(C4*E4)
E4: 18.68
.
.
O4: 28
P4: @SQRT(O4*Q4)
Q4: 29.7
.
.
A6: (F0) [W8] 3
B6: [W1] ':
C6: ($A6-2)*@LOG(C$4/$D$4)/@LOG($P$4/$D$4)+1
D6: ($A6-2)*@LOG(D$4/$D$4)/@LOG($P$4/$D$4)+1
E6: ($A6-2)*@LOG(E$4/$D$4)/@LOG($P$4/$D$4)+1
.
.
A7: (F0) [W8] 1+A6
.
.
O20: ($A20-2)*@LOG(O$4/$D$4)/@LOG($P$4/$D$4)+1
P20: ($A20-2)*@LOG(P$4/$D$4)/@LOG($P$4/$D$4)+1
Q20: ($A20-2)*@LOG(Q$4/$D$4)/@LOG($P$4/$D$4)+1
```

**Notes:**

1. The elements are numbered 0 for the reflector to (N – 1) for the highest frequency band.

2. The mean frequency is the geometric mean of the band end frequencies:

$$f\,mean = \sqrt{f\,upper \times f\,lower}$$

3. Element resonance for

$$f = 1 + (N-2)\frac{\log\,[f\,/\,(\text{lowest band mean freq})]}{\log\,[(\text{highest band mean freq})\,/\,(\text{lowest band mean freq})]}$$

By taking the logarithm of the relationship, we gain a formula which can be solved using least-squares regression analysis. For the formula

$$\ln(f_i) = \ln(f_0) + i \times \ln(\tau)$$

the least-squares method yields estimates of $\ln(f_0)$ and $\ln(\tau)$. These estimates can be used to calculate resonant frequencies for the loops elements. The resonant frequency for element no. i is

$$f_i = e^{\ln(f0)+i \times \ln(\tau)}$$

Table 2 shows the calculation of best fit frequencies for our example, as well as physical dimensions of the resulting LPLA. These results are obtained from the spreadsheet. The spacing between elements is related to $\sigma$, the relative spacing constant.

### Physical Construction

A truss supporting a pair of nonconductive booms, one at the top of the array and one at the bottom, would provide stability for the elements loops and interelement feed line, which could be made of aluminum or copper tubing.

While the example here is designed for multiple band capability, the LPLA could be designed for single-band operation. The gain should show some ripple over the design frequency of the antenna, but will be flatter than a single driven element antenna. See Fig 2. For sufficiently high enough frequencies, an LPLA could be "printed" on flexible plastic and rolled into shape. See Fig 3.

### Conclusion

The LPLA antenna should provide the RF designer with another option in designing antennas. Some design characteristics, such as feed impedance and interelement feed-line spacing, still need to be optimized. Hopefully this introduction to the LPLA will encourage other antenna design articles.

## Table 2
## Spreadsheet Results of Regression Analysis to Fit Elements

```
      S          T          U          V          W          X         Y
1  Elmt No.   Freq.    Ln(Freq)   Calc f    Dia. ft    Sigma    Space ft
2        0                        15.982    19.601      0.05     6.158
3        1   18.630     2.925     18.517    16.918               5.315
4        2   21.224     3.055     21.454    14.602       Tau     4.587
5        3   24.940     3.216     24.856    12.603     0.863     3.959
6        4   28.837     3.362     28.798    10.878
7
8           Regression Output:             Array length ft    20.020
9  Constant                         2.771  Total tubing ft  274.411
10 Std Err of Y Est                 0.009
11 R Squared                        0.998
12 No. of Observations              4.000
13 Degrees of Freedom               2.000
14
15 X Coefficient(s)    0.147
16 Std Err of Coef.    0.004
```

Cell entries for the spreadsheet:

S1: 'Elmt No.
T1: ^Freq.
U1: 'Ln(Freq)
V1: "Calc f
W1: 'Dia. ft
X1: "Sigma
Y1: 'Space ft
S2: (F0) 0
V2: @EXP(V$9+S2*U$15)
W2: 984.18/V2/@PI
X2: (G) 0.05
Y2: 0.5*@PI*(W2-W3)*4*X$2/(1-X$5)
S3: (F0) 1
T3: +D4
U3: @LN(T3)
V3: @EXP(V$9+S3*U$15)
W3: 984.18/V3/@PI
Y3: 0.5*@PI*(W3-W4)*4*X$2/(1-X$5)
S4: (F0) 2
T4: +H4
U4: @LN(T4)
V4: @EXP(V$9+S4*U$15)
W4: 984.18/V4/@PI
X4: "Tau
Y4: 0.5*@PI*(W4-W5)*4*X$2/(1-X$5)
S5: (F0) 3
T5: +L4
U5: @LN(T5)
V5: @EXP(V$9+S5*U$15)
W5: 984.18/V5/@PI
X5: +V3/V4
Y5: 0.5*@PI*(W5-W6)*4*X$2/(1-X$5)
S6: (F0) 4
T6: +P4
U6: @LN(T6)
V6: @EXP(V$9+S6*U$15)
W6: 984.18/V6/@PI
W8: 'Array length ft
Y8: @SUM(Y2..Y5)
W9: 'Total tubing ft
Y9: @PI*@SUM(W2..W6)+2*Y8
S8 through U16: Regression output. In Quattro select /AR, or in 1-2-3 select /DR. Set S3..S6 as the independent variable, and set U3..U6 as the dependent variable.

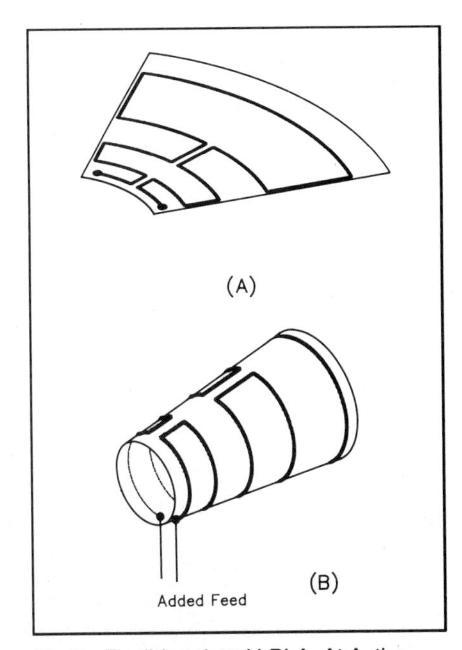


Fig 3—Flexible printed LPLA. At A, the printed conductors, and at B, the antenna rolled into its final form.

## Appendix
## Derivation of Element Resonance for Frequency f

Given N elements, one of which is a reflector (labeled element zero), lowest and highest frequency bands, and a frequency f, then the number of the element where f resonates are related as follows.

$$\left[\frac{f}{[\text{lowest band mean freq}]}\right] = \left[\frac{[\text{highest band mean freq}]}{[\text{lowest band mean freq}]}\right]^{\frac{(\text{Element number}-1)}{(N-2)}}$$

$$\left[\frac{[\text{highest band mean freq}]}{[\text{lowest band mean freq}]}\right]^{(\text{Element number}-1)} = \left[\frac{f}{[\text{lowest band mean freq}]}\right]^{(N-2)}$$

$$(\text{Element number}-1)*\log\left[\frac{[\text{highest band mean freq}]}{[\text{lowest band mean freq}]}\right] = (N-2)*\log\left[\frac{f}{[\text{lowest band mean freq}]}\right]$$

$$\text{Element number} = 1 + (N-2) * \frac{\log\left[\frac{f}{[\text{lowest band mean freq}]}\right]}{\log\left[\frac{[\text{highest band mean freq}]}{[\text{lowest band mean freq}]}\right]}$$

# The K4EWG Log Periodic Array

**By Peter D. Rhodes, K4EWG**
**3270 River Rd**
**Decatur, GA 30034**

With all of the HF bands available to us, the log periodic dipole array (LPDA) is an idea worth serious consideration—especially among amateurs who desire a single, high performance, multiband antenna.

Since the publication of my original paper in *QST* in 1973[1] (also republished in *The ARRL Antenna Book*[2]), I have done extensive HF work with these arrays using tubing elements.[3,4] This article represents practical solutions to certain problems which result from the electrical and physical constraints of HF log periodic dipole arrays.

The goals of this article are as follows:

1) Minimize the number of elements without sacrificing gain and/or SWR for the desired bandwidth.

2) Apply W2PV taper correction to the element lengths.[5]

3) Minimize boom length.

4) Simplify construction by removing the cross-wire feeder system and lumped balun.

5) Simplify the assembly by using "plumbers delight" techniques.

6) Reduce the weight of the 14-30 MHz array.

I will not repeat the design for the LPDA, as it can be found in the latest *ARRL Antenna Book.* However, it is sufficient to say that the LPDA performs as a frequency independent (broadband) array. When designed properly, it will provide the equivalent gain of a 3-element Yagi with a 15-dB front to back ratio on all bands within the array passband. Fig 1 depicts a typical LPDA with the specified design parameters. If the design constant $\tau$ is near 1.0, more elements are required. If the element-spacing constant $\sigma$ is greater than 0.1, a longer boom is required. A balance must be struck if goals 1 and 3 are to be satisfied.

I define this process as the *Logarithm Periodicity Method.* Basically this method determines a specific design constant $\tau$ which will permit element lengths to be chosen so their discrete resonances occur at or near a desired band (for example 14, 18, 21, 25, 28 and 29 MHz).

Fig 2 relates the H-plane (vertical) and E-plane (horizontal) half-power (3 dB) total beamwidths in degrees.[6] Notice that the E-plane beamwidth is almost constant

K4EWG Log Periodic Array atop the 120-foot tower (40-meter beam visible below). *(photos by the author)*

at 60° to 55° for $0.8 < \tau < 0.92$ and $0.05 < \sigma < 0.13$. Also, notice that the H-plane beamwidth varies from 85° to 155° under the same ranges of $\tau$ and $\sigma$.

The directivity in dBi can be computed using

$$G_{dB} = 10 \log \frac{41{,}253}{E^{\circ}\, H^{\circ}} \qquad \text{(Eq 4)}^{7}$$

It is not my intention to enter a gain controversy. However, Lawson states that the 3-dB total beamwidth is 100° in the H plane and 68° in the E plane for his 3-element Yagis on a 0.3-$\lambda$ boom.[8] Using Eq 4, this represents a directivity of 7.83 dBi. I have found that the gain of a 3-element monoband Yagi is quite impressive, particularly when one considers a single small antenna approximating such gain for five bands! Both $\tau$ and $\sigma$ (relative spacing constant) can be chosen from Fig 2, so that a directivity of 7 to 8 dBi can be realized. Since the E-plane beamwidth is almost

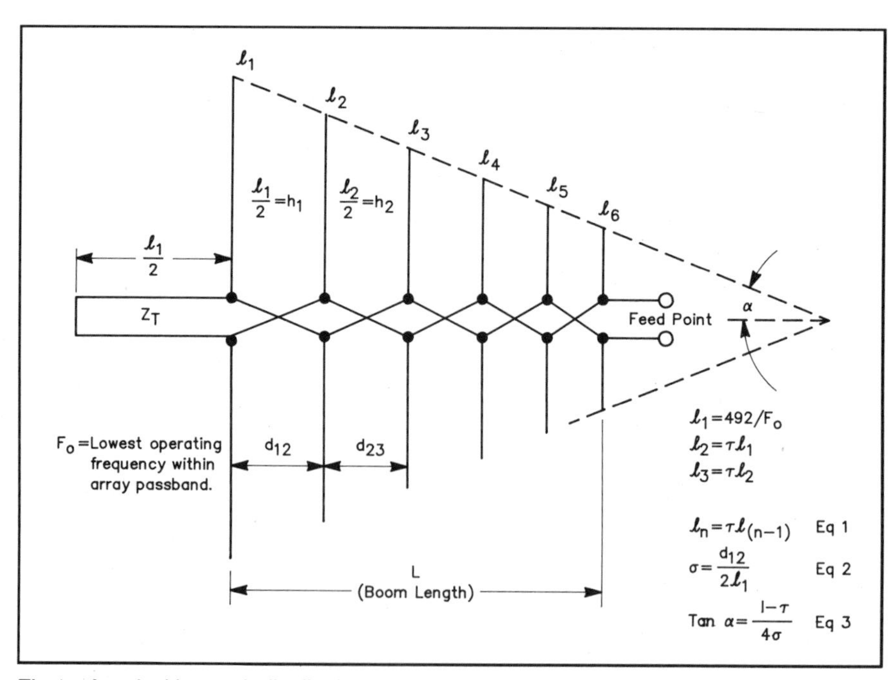


Fig 1—A typical log periodic dipole array.

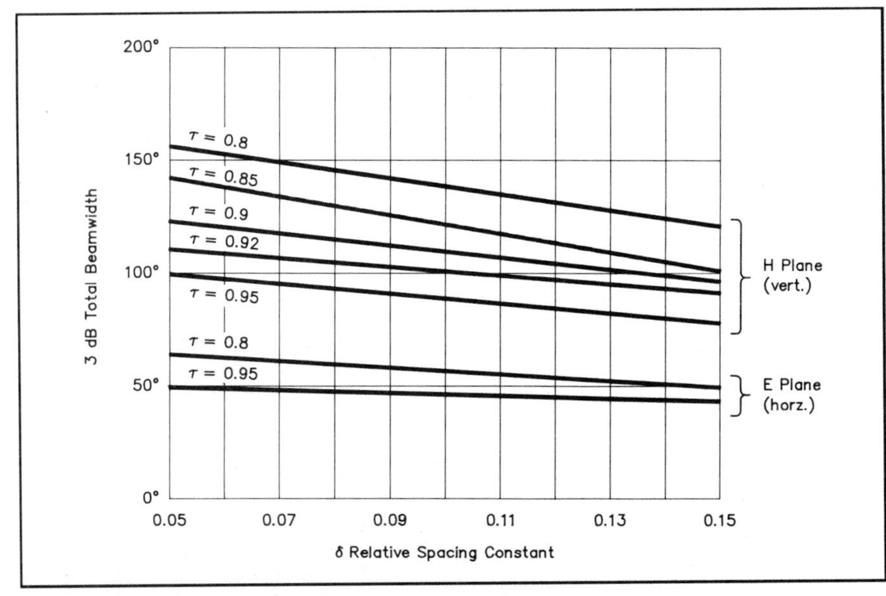


Fig 2—E- and H-plane versus half-power beamwidth.

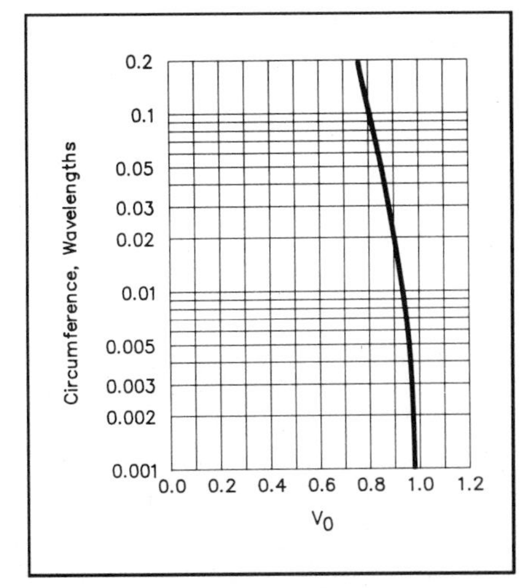


Fig 3—Element pipe circumference in wavelengths versus velocity factor $V_0$.

constant at 60°—55° and the gain desired is 7.83 dBi, the H-plane beamwidth is found to be 137°—109° using Eq 4. Also, Fig 2 reveals a number of combinations of $\tau$ and $\sigma$ which will yield the desired H-plane beamwidth.

The log periodic dipole array has the characteristic of repeating all electrical qualities such as directivity, impedance, front-to-back ratio, SWR and E- and H-plane 3-dB points at intervals of $\log(1/\tau)$. That is to say, there is a relationship between $F_1$ and $F_2$,

$$F_2 = [1 + \log(1/\tau)]F_1 \qquad \text{(Eq 5)}$$

$$F_3 = [1 + \log(1/\tau)]^2 F_1$$

where F = array frequencies of periodicity

Let $R = [1 + \log(1/\tau)]$

Then $F_2 = RF_1$

and $F_3 = R^2F_1 = RF_2$

$$F_n = R^{(n-1)}F_1 = RF_{(n-1)} \qquad \text{(Eq 6)}$$

Also,

$$1/\tau = \log^{-1}\frac{F_2}{F_1} - 1$$

$$1/\tau = \log^{-1}\frac{F_n}{F_{n-1}} - 1 \qquad \text{(Eq 7)}$$

Let's consider a 14-30 MHz LPDA with 7 to 8 dBi gain and determine the *least* number of elements required. From Fig 2, $\tau$ ranges from 0.850 ($\sigma$ = 0.05, H = 137°) to 0.920 ($\sigma$ = 0.07, H = 109°). Given this range of $\tau$, we can proceed to iterate and optimize the periodicity of the array. I will use 1-inch diameter pipe as a referenced normalized length for all elements. Referring to Fig 3[9] at 14.0 MHz, a 1 inch diameter pipe has 0.003-$\lambda$ circumference and the velocity factor $V_0$ = 0.94. The resonant frequency of $\ell_1$ is $f_1$, which is also the first periodic frequency of the array, $F_1$.

$$\therefore F_1 = f_1 = V_0F_0 = 0.94 \times 14.0 = 13.160 \qquad \text{(Eq 8)}$$

where $F_0$ = lowest desired operating frequency within the array passband in MHz

Also,

$$\lambda_1 = \frac{11{,}808}{f_1}$$

= free-space wavelength of $f_1$ and/or $F_1$ (Eq 9)

Combining Eqs 5 and 8, we can formulate Table 1. As you can see, trial no. 3 looks especially promising. In order to speed up the design process, the discrete dipole element resonant frequencies, $f_1$ to $f_n$ (specific ham band frequencies), are related to an average $\tau$. For example, if you choose $\tau$ = 0.85 (Table 1) with a 14-30 MHz array passband ($F_0$ = 14, $f_n$ = 30), Eq 10 will determine that seven elements are required (a 7-element array).

$$n = \frac{\log\dfrac{0.822\,F_0}{F_n} + 1}{\log \tau} \qquad \text{(Eq 10)}$$

where

n = number of elements

nth element = 0.411 $\lambda_n$

$\lambda_n = 984/F_n$

**Table 1**
**Array Periodic Frequencies for Different $\tau$ Values**

$F_1$ to $F_n$ = Array periodic frequencies within the array passband.

| *Trial No.* | *Trial $\tau$* | *1 + log (1/$\tau$)* | *$F_1$* | *$F_2$* | *$F_3$* | *$F_4$* | *$F_5$* | *$F_6$* | *$F_7$* | *$F_8 \rightarrow F_n$* |
|---|---|---|---|---|---|---|---|---|---|---|
| 1 | .82 | 1.08619 | 13.160 | 14.294 | 15.526 | 16.864 | 18.317 | 19.896 | 21.610 | etc → |
| 2 | .83 | 1.08092 | 13.160 | 14.225 | 15.376 | 16.620 | 17.965 | 19.419 | 20.990 | |
| 3 | .85 | 1.070581 | 13.160 | 14.088 | 15.083 | 16.148 | 17.287 | 18.508 | 19.814 | |

**Table 2**
**Element Half Lengths**

| | |
|---|---|
| $f_1 = 13.160$ | $\ell_{Eq\,1} = \frac{0.94 \times 5904}{13.16} = 210.857$ inches |
| $f_2 = 15.433$ | $\ell_{Eq\,2} = \tau[\ell_{Eq\,1}] = 179.801$ inches |
| $f_3 = 18.100$ | $\ell_{Eq\,3} = \tau[\ell_{Eq\,1}] = 153.318$ inches |
| $f_4 = 21.224$ | $\ell_{Eq\,4} = 130.737$ inches |
| $f_5 = 24.891$ | $\ell_{Eq\,5} = 111.481$ inches |
| $f_6 = 29.190$ | $\ell_{Eq\,6} = 95.061$ inches |
| $f_7 = 34.232$ | $\ell_{Eq\,7} = 81.060$ inches |
| | $\ell_{Eq\,n} = \tau\,[\ell_{Eq\,n-1}]$ (Eq 12) |

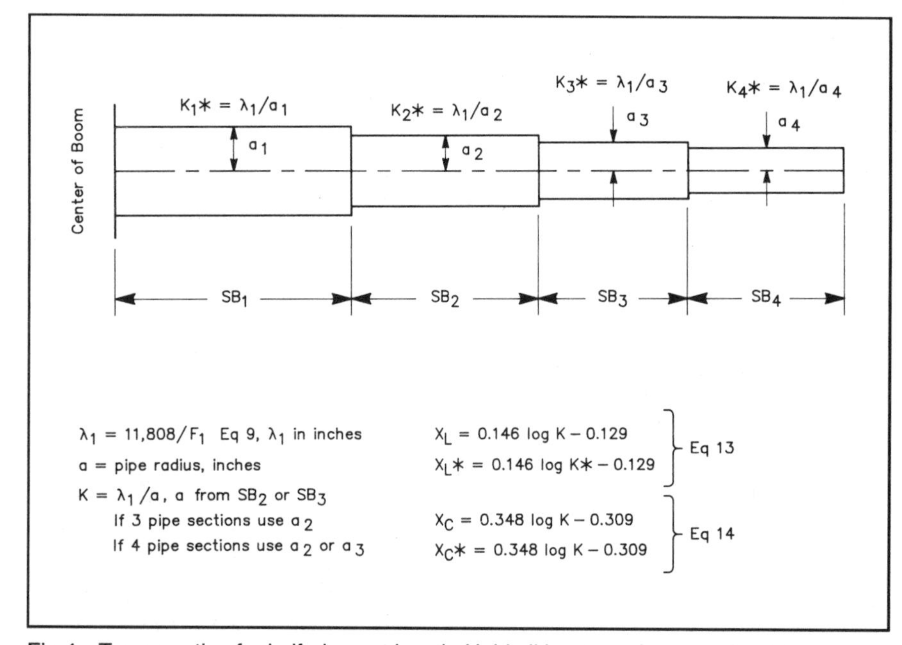


Fig 4—Taper section for half-element length. Hold all inner sections at a fixed length and vary the end section as needed.

K4EWG Log Periodic Array before being raised to the top of the tower, with XYL Tana assisting.

Since dipole $\ell_1$ is resonant at 13.160 MHz, desired (element) dipole frequencies may be assigned as follows.

$F_1 = f_1 = 13.160$ MHz
$f_2$ = unknown
$f_3 = 18.100$
$f_4 = 21.250$
$f_5 = 24.891$
$f_6$ = unknown
$f_7$ = unknown

Then $\tau_{3-4} = \frac{18.100}{21.250} = 0.85176$

$$\tau_{4-5} = \frac{21.250}{24.891} = 0.85372$$

The average value of $\tau$ is

$$\tau_{AVG} = \frac{\tau_{3-4} + \tau_{4-5}}{2} = 0.85274$$

and $f_n = \frac{f_{n-1}}{\tau}$ (Eq 11)

Then, using Eq 11 and the averaged $\tau$, we have

$$f_2 = \frac{f_1}{\tau_{AVG}} = 15.432$$

$$f_6 = \frac{f_5}{\tau_{AVG}} = 29.189$$

$$f_7 = \frac{f_6}{\tau_{AVG}} = 34.230$$

Of course, once a final $\tau$ has been found, Table 1 could be repeated for that specific $\tau$ to examine the $F_1$ to $F_n$ frequency periodicity over the entire array passband. Obviously, it is desirable that the five 14-30 MHz ham bands fall at or near these periodic frequencies. The final $\tau$ used in the K4EWG Array design was 0.85271318.

Now, using the W2PV approach to the taper corrections, I have chosen 1-inch outside diameter pipe for the electrically normalized lengths. I will call these *element half lengths*, $\ell_{Eq}$, per W2PV's designation. (See Table 2.)

Hence,

$\ell_{Eq} = \frac{V_0\,5904}{f_1}$, $V_0 = 0.94$ (Fig 3), $F_1 = f_1$,

and $f_n = F_n$.

Derivation of the taper correction equations can be simplified. However, the element taper should not exceed 0.125 inch in diameter for adjacent pipe to pipe telescoping sections within the element. Hence, a light or smooth taper eliminates the tedious m and $f(\theta)$ calculations.

The simplified method can be under-

**Table 3**
**Taper Equation for $\ell$ (K4EWG Array)**

SB2 is normalized section.

| Sect. No. | SB inches | 2a inches | K* | $\frac{X_{L*}}{X_{L2}}$ | $\frac{X_{C2}}{X_{C*}}$ | SA inches | $K_2 = \lambda_1 / a_2 = 1{,}794.13$ |
|---|---|---|---|---|---|---|---|
| 1 | 16 | 1.125 | 1595.14 | 0.9784 | — | 15.655 | $X_{C2} = 0.3461$ |
| 2 | 64 | 1.000 | 1794.53 | 0 | 0 | 64.000 | $X_{L2} = 0.8234$ |
| 3 | 60 | 0.875 | 2050.89 | — | 0.9761 | 58.565 | $f_1 = F_1 = 13.160$ MHz |
| $4_1$ | 70 | 0.750 | 2392.70 | — | 0.9468 | 66.489 | $\lambda_1 = 897.24$ in., $a_2 = 0.5$ in. |
| $4_2$ | 84 | 0.750 | 2392.70 | — | 0.9468 | 79.787 | $SA = (X_L^* / X_{L2})SB = (X_{C2} / X_C^*)SB$ |

$SB_{TOTAL} = 224$ in., trial no. 1 — $\ell_{Eq} = 218.006$, trial no. 1

$SB_{TOTAL} = 210$ in., trial no. 2 — $\ell_{Eq} = 204.709$, trial no. 2

stood using Fig 4 and the associated equations from W2PV's earlier paper.[10]

For example, to find taper equation $\ell_1$ in the K4EWG Array, choose $SB_2$ as normalized pipe section, $a_2 = 0.5$ inch, and $\lambda_1 = 897.264$ inches.

By looking carefully at Table 3, you'll see that $SB_4$ (the last pipe) was varied in length while all other pipes were fixed. Also, you will see that it would require 224 inches total length for the tapered element to have the same resonance as a 1-inch diameter pipe 218.006 inches in length. Trial no. 2 reveals 210 inches tapered to 204.709 inches for the 1-inch pipe.

**Table 4**
**Taper Equations for K4EWG Array**

| Element No. | Physical Half Lgth Inches | Taper Correction Eq for $\ell/2 = h$ | Pipe Section – Inches 1.125 | 1.00 | 0.875 | 0.750 |
|---|---|---|---|---|---|---|
| 1 | 216.473 | $1.0529\,\ell_{Eq1} - 5.5318$ | 16 | 64 | 60 | 76½ |
| 2 | 183.898 | $1.0344\,\ell_{Eq2} - 5.6767$ | 16 | 64 | 60 | 43⅞ |
| 3 | 155.606 | $1.0261\,\ell_{Eq3} - 1.7074$ | 16 | 64 | 60 | 15⅝ |
| 4 | 132.494 | $1.0269\,\ell_{Eq4} - 1.7611$ | 16 | 64 | 52½ | — |
| 5 | 113.065 | $1.0278\,\ell_{Eq5} - 1.5174$ | 16 | 64 | 33 1/16 | — |
| 6 | 95.911 | $1.0287\,\ell_{Eq6} - 1.8758$ | 16 | 64 | 15 15/16 | — |
| 7 | 82.476 | $1.0297\,\ell_{Eq7} - 0.9879$ | 16 | 32 | 34½ | — |

$\ell_{Eq\ 1\text{-}6}$ from Table 2. Use 8-inch minimum pipe overlay

## The Taper Correction Formula

The taper correction formula is

$$\eta = P\ell_{Eq} + q \quad \text{(Eq 13)}$$

where

$\eta$ = physical half length of the tapered element

$\ell_{Eq}$ = equivalent cylinder length, Table 3

$$P = \frac{\eta_1 - \eta_2}{\ell_{Eq\,1} - \ell_{Eq\,2}} \quad \text{(Eq 14)}$$

and

$$q = \frac{\eta_2 \ell_{Eq\,1} - \eta_1 \ell_{Eq\,2}}{\ell_{Eq\,1} - \ell_{Eq\,2}} \quad \text{(Eq 15)}$$

Note: Numbers 1 and 2 refer to trial selections 1 and 2 for the end pipe, not to be confused with element designations $\ell_1$ and $\ell_2$.

Since the elements are mounted above the boom using a single muffler clamp, there is very little, if any, need for boom-to-element mounting taper adjustment. Lateral element to boom stability is enhanced by the cross connecting hose clamps at the feed pipe for each element (Fig 5). If the elements are mounted by plates and U bolts as specified by Lawson, a plate taper must be developed for each element and included in the element taper schedule for section $SB_1$.[5]

The taper schedule for the K4EWG Array is given in Table 4. Boom length minimization is simplified by use of Fig 2. Actually, $\sigma$ can range from 0.05 to 0.08 with little change in array gain. Now,

$$\tan\alpha = \frac{1 - \tau}{4\sigma} = \frac{\ell_1 - \ell_n}{L}$$

and

$$\sigma = \frac{d_{12}}{2\ell_1} \quad \text{(Eq 2)}$$

where $d_{12}$ = spacing $\ell_1$ to $\ell_2$

Rearranging,

$$d_{12} = \frac{(1 - \text{tau})\,\ell_1 L}{\ell_1 - \ell_n}$$

where L = boom length, same dimensions as $\ell$.

Using the K4EWG Array as an example: L = 20 feet, $d_{12} = 4.78$ feet and $\sigma = 0.0674$.

A 20-foot boom length is quite a reduction from my earlier design of 26.5 feet.[1,2] Since $\tau = 0.8527$ and $\sigma = 0.0674$, the H-plane beamwidth is approximately 130° from Fig 2, giving a gain of 7.61 dBi from Eq 4.

The cross feeder is not necessary. It is, however, necessary to switch feeder connections electrically as shown in Fig 1 for end-fire (toward shortest element) directivity. If the switching were not done, broadside directivity (perpendicular to the plane of the array) would occur.[11] I have accomplished this switching as shown in

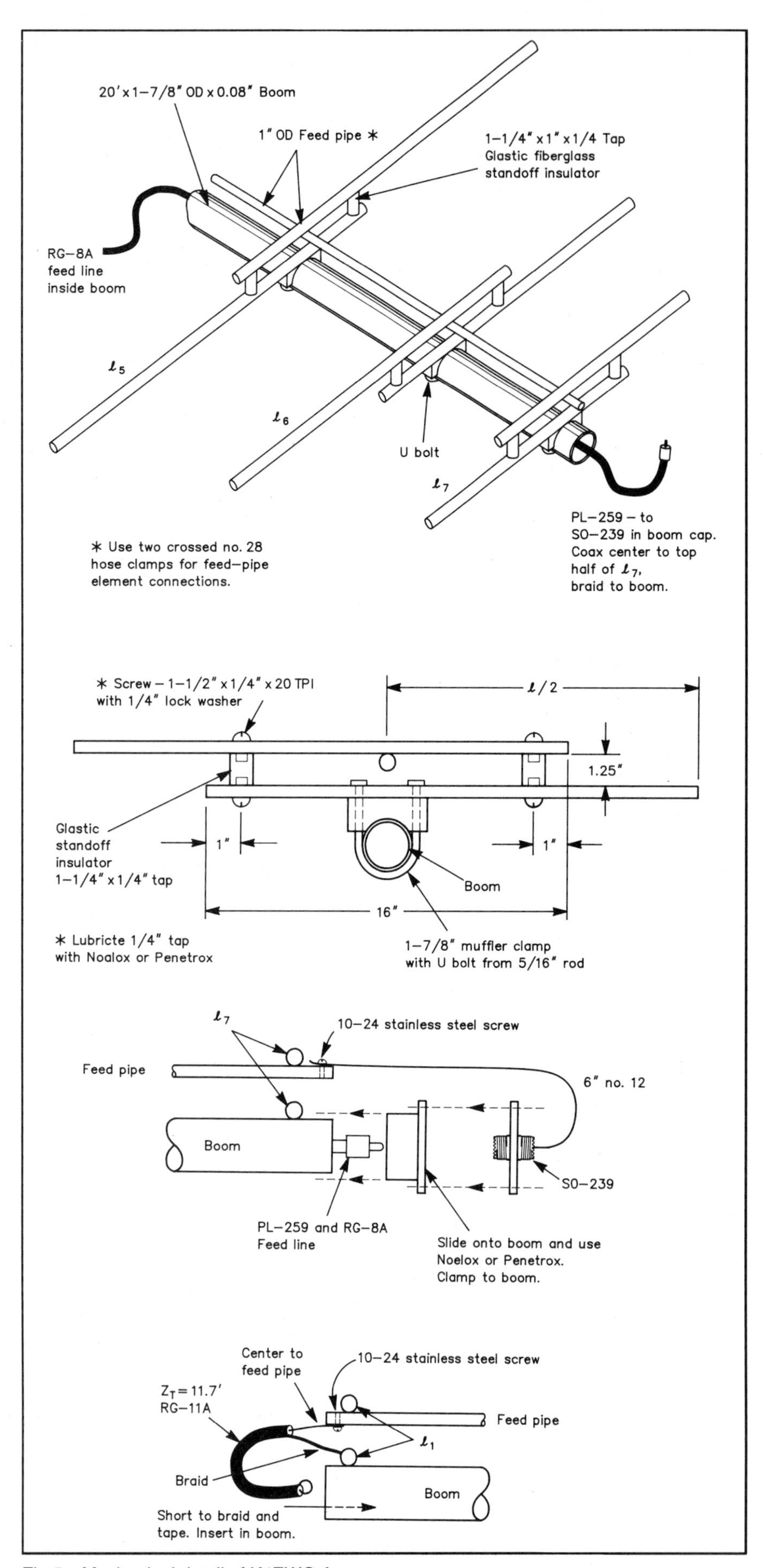


Fig 5—Mechanical detail of K4EWG Array.

Close-up detail of feed point.

Fig 5 (patent pending). I define this feeding system as *Eccentric Linear Feeding for Log Periodic Dipole Arrays* (patent and trademark pending).

Examination of Fig 5 and the accompanying photos reveals that goals 3, 4, 5 and 6 are satisfied (the total completed weight of the K4EWG Array is 56 lb). The lumped balun is not necessary since the feeder is unbalanced by design. Also the coax runs through the boom, which acts as a decoupling device for any unbalanced currents on the outside of the coax.

I will not develop the $Z_0$, or characteristic impedance, since it can be determined from the published data in the *ARRL Antenna Book*[1,2]. However, the SWR at the feed point of this array is listed for each of the five bands in Table 5. Needless to say, they are quite acceptable and are primarily due to the periodicity and feeder design approach. A summary of materials used in the K4EWG Log Array can be found in Table 6.

Astute readers will note that I have also listed SWR figures for the 10-MHz band.

**Table 5**
**SWR Versus Frequency (MHz) for K4EWG Array**

| *Freq MHz* | *SWR at Antenna* |
|---|---|
| 10.1 | 2.0 |
| 10.15 | 2.0 |
| 10.2 | 2.5 |
| 14.0 | 1.7 |
| 14.2 | 1.5 |
| 14.5 | 1.4 |
| 18.0 | 1.4 |
| 18.2 | 1.2 |
| 18.4 | 1.2 |
| 21.0 | 1.0 |
| 21.2 | 1.1 |
| 21.5 | 1.2 |
| 24.5 | 1.2 |
| 24.9 | 1.1 |
| 25.1 | 1.0 |
| 28.0 | 1.4 |
| 28.5 | 1.5 |
| 29.0 | 1.2 |
| 30.0 | 1.2 |

**Table 6**
**Materials List for K4EWG Log Array**

*Aluminum tubing—6061ST—0.047 in. wall thickness elements*
1.125 in. OD—192 in. (2 ea—8 ft pcs)
1.000 in. OD—1,224 in. (17 ea—6 ft pcs)
0.875 in. OD—816 in. (10 ea—6 ft pcs, 1 ea—8 ft pcs)
0.750 in. OD—312 in. (3 ea—8 ft pcs, 1 ea—4 ft pcs)
1.875 in. OD—240 in. (1 ea—20 ft pcs—0.08 in. wall)

*Clamps*
11 ea—1⅞ in. muffler clamps (U bolts not used for elements)
28 ea—no. 28 stainless steel hose clamps
7 ea—5⁄16 in. × 12 in. threaded rod (bend to form U bolts for elements)

*Miscellaneous*
1 ea ¼ in. × 8 in. × 12 in. aluminum boom to mast plate
44 ea—no. 6 stainless steel sheet metal screws
1 ea—boom cap, aluminum, homemade
1 ea—SO-239 coax connector
4 oz.—Noalox or Penetrox
28 ea—¼ × 1½ in. stainless steel screw, 20 TPI and ¼ in. SS lock washers
1 ea—$Z_\tau$, 11.7 ft RG-11A inserted inside boom, $Z_\tau = 0.33\ \ell_1$
14 ea—1¼ × ¼ in. tap Glastic standoff insulator*

*Glastic Manufacturing Co, Cleveland, Ohio, tel 216-486-0100. Distribution Headquarters: Electrical Insulation Supplies, 1255 Collier Rd NW, Atlanta, GA 30318, tel 404-355-1651 (Glastic part number 2165-1B).

(Now we have a 6th band on one antenna!) Operation on 10 MHz is possible due to the loading effect of $Z_\tau$ on $\ell_1$. $Z_\tau$ is a shorted quarter wave section for 14.0 MHz and, therefore, it acts like an open circuit at that frequency. At higher frequencies, most of the RF energy is radiated by the elements within the active region of dipole elements. Therefore, $Z_\tau$ receives very little energy from the feeder. At 10.1 MHz, however, $Z_\tau$ provides an inductive loading reactance for $\ell_1$, making it resonant at 10.1 MHz. Surprisingly, $\ell_2$ contributes as part of a 10-MHz active region and endfire directivity occurs with $\tau$ near 0.6, $\ell_1$ to $\ell_2$. Of course, with such a low $\tau$ the gain is not comparable to the design array passband (14-30 MHz), but it is interesting to see what extremes can be tolerated in such arrays. The log periodic array exhibits an interesting quality in that elements shorter than the nearest resonant element ($\tau > 0.75$) carry 80% of the radiated energy.[6] The array simply "wants to work"! $Z_\tau$ also provides a dc ground for lightning protection.

My previous papers have dealt with the technical and electrical aspects of such arrays. I'd like, however, to diverge from my normal closings and relate a brief note of nontechnical interest. After completing the array I left it sitting on the saw horses (approximately 3 feet above ground) and put it on the air. While due respect is given for propagation and my location, a single hour of operation resulted in numerous QSOs while running only 100 watts. Some of the stations worked included D68TW and 3B8CF (21 MHz CW), TA3F and VU2AU (21 MHz SSB), TY1OR (14 MHz CW), DL9JI (10 MHz, CW) and five Siberian stations on 21 MHz CW. I was surprised at how well this array performed so close to the ground! Subsequent to the "saw horse QSOs," I have been enjoying the antenna in its proper location on the tower at about 120 feet.

Ground view of tower assembly and antennas. Note the rotator mast pipe visible at left.

At the time of writing of this paper, I am building a second K4EWG Log Periodic Array which will stack 0.6 λ at 14 MHz (42 feet) under the top array (120 feet). It will be interesting to vary stacking distances with such arrays and plot the E- and H-plane patterns. As can be seen in the photos, the rotator is ground mounted with a 2-inch outside diameter pipe mast running along the entire length of the tower. Such a mast simplifies stacking arrays and I must credit my good friend, Bill Maxson, N4AR, for this neat idea. However, I'll leave the stacking details for another paper.

**Bibliography**


[1]P. D. Rhodes, "The Log Periodic Dipole Array," *QST*, Nov 1973.

[2]P. D. Rhodes, "The Log Periodic Dipole Array," *The ARRL Antenna Book*, 13th ed (1974) to the latest edition.

[3]P. D. Rhodes and J. R. Painter, "The Log-Yag Array," *QST*, Dec 1976.

[4]P. D. Rhodes, "The Log-Periodic V Array," *QST*, Oct 1979.

[5]J. L. Lawson, *Yagi-Antenna Design*, 1st ed. (Newington: ARRL 1986) pp 7-5, 7-11.

[6]R. Mittra, and J. D. Dyson, "Log Periodic Antennas," *Technical Report 76* (Urbana: Univ of IL Ant Lab, 1964). Also reprinted in *Electronics Industries*, May 1965.

[7]J. D. Kraus, *Antennas*, 2nd ed. (New York: McGraw-Hill, 1988).

[8]See Ref 5, p 8-5.

[9]Dept of the Army, *TM-11-666, Antennas and Radio Propagation* (1953), p 66.

[10]J. Lawson, *High Performance Antenna Systems*, May 11, 1978 (private distribution).

[11]King, Mack and Sandler, *Arrays of Cylindrical Dipoles* (London: Cambridge Univ Press, 1968).

# A Streamlined Mobile HF Antenna For Vehicles With Fiberglass Tops

**By Steve Cerwin, WA5FRF**
**10227 Mt Crosby Dr**
**San Antonio, TX 78251**

This article describes a sleek, base-loaded mobile HF antenna for use on vehicles with fiberglass tops. The antenna offers an aesthetically pleasing alternative to somewhat gaudy center-loaded designs, while giving outstanding performance. The fiberglass top permits the loading coil to be mounted inside the vehicle, which greatly enhances appearance with no sacrifice of efficiency.

Fig 1 shows the antenna installed on a Ford Bronco. A full-length (8-ft) stainless-steel CB whip is used as the radiator. In antennas, bigger is better, and an 8-ft whip provides an operational advantage over smaller elements. The whip is bumper mounted atop a ball and spring and a 12-inch-long fiberglass rod. It is secured to the fiberglass car top by an aluminum mounting bracket at eye level.

The mounting bracket mechanically braces the antenna and provides a feed path from the loading coil inside the car top. The portion of the whip above the bracket is the main radiator; the part below the bracket acts as capacitive loading. The capacitance between the car body and the bracket plus lower whip section allows fewer turns to be used in the lossy loading coil, thus improving efficiency.

Fig 2 shows the loading coil and capacitive match as seen from inside the vehicle. The loading coil is secured to the inside of the car top with a clamp made from two strips of fiberglass. The 3-inch-diameter coil was cut for 40-meter operation. It consists of 15½ turns of no. 12 wire wound at 6 turns per inch. The coil was made "long" by several turns to allow for tuning. An adjustable tap tunes the antenna to reso-

Fig 1—The streamlined mobile whip mounted on a Ford Bronco.

Fig 2—A view from inside the vehicle shows the 40-meter loading coil, shunt capacitors (below the coil), mounting bracket and antenna.

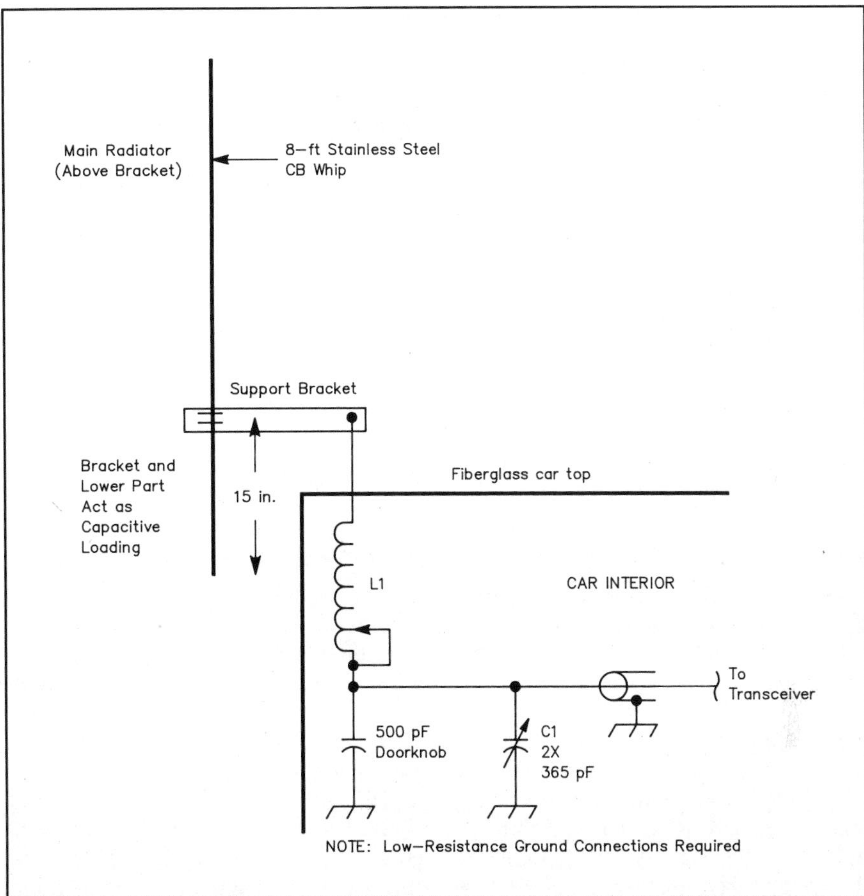


Fig 3—Schematic of the streamlined mobile whip and its matching network. C1 is a dual-section (365 pF each) receiving capacitor; the sections are parallel connected for 730 pF total capacitance. L1 is 20 turns of no. 12 copper wire wound 6 turns per inch with a 3-inch diameter (such as B & W no. 3033 or 2406T). The 40-meter version shown resonated with the tap 4½ turns from the grounded end.

nance at the desired operating frequency. The tap is permanently soldered to the coil once resonance is set. Operation on other bands (with the exception of 10 meters) is a simple matter of substituting another loading coil with the appropriate inductance.

A low-voltage air-dielectric variable capacitor and a parallel-connected 500-pF "doorknob" capacitor are mounted beneath the loading coil. This shunt capacitance transforms the feed-point impedance of the resonant antenna to approximately 50 Ω. The RG-58 cable leading from the transceiver cannot be seen in Fig 2. The cable runs underneath the plastic cover strip located below the rear window.

The ground connection for the capacitive match and the coax braid is made directly to the car body by means of a ground lug and sheet-metal screw. This ground connection must have very little impedance for efficient operation. As with all physically short antennas, the antenna radiation resistance is very low. Efficiency demands low-resistance conductors and connections throughout the entire antenna system.

Fig 3 is a schematic of the antenna. The design is a standard base-loaded whip with a capacitive shunt match. Tuning is accomplished by adjusting the coil tap for resonance, then adjusting the capacitor for best SWR. An excellent match results, and my solid-state transceiver has no trouble driving the antenna without a Transmatch. The acceptable SWR bandwidth is about 70 kHz on the 40-meter band.

I've used the antenna with a 140-W transceiver for over a year, and the results have been excellent. Signal reports have been consistently good, with many reports like, "Boy OM! You sure have a good signal for a mobile!" I used a large center-loaded antenna with a prominent capacitance hat before switching to this antenna, and this antenna seems to perform as well as that one. The antenna is mechanically well behaved, thanks to the top support bracket, and there have been no more of the "Man, that thing on your car sure is ugly!" comments from my XYL and nonham friends!

# A Simple Seeker Direction Finder

By Dave Geiser, WA2ANU
3710 Snowden Hill Rd
New Hartford, NY 13413-9584

Consider radio direction finding (DF) in general. Let us assume that any antenna is an elementary dipole that can receive radio signals. One dipole is needed to determine that a signal exists. Another dipole (or a second position of the first) is needed to determine a "line of bearing," a line through the receiving antenna and the signal source. A third dipole or position is needed to determine the direction of the signal source on the "line of bearing."

If the equivalent of two dipoles are used, the most common approach is to rotate the array, giving the equal of three or four antenna positions. Many simple commercial direction finders use four antennas, as the "software" for the display is simpler. All of the above assumes DF on a flat earth.

The above applies to both amplitude (signal strength) and phase (signal distance) methods of direction finding. This article concerns signal-distance (or phase) methods of direction finding.

It is important to remember two basic principles of radio direction finding: first that all of the significant signal should come through the searching antennas, and second, that remote structures may create reflections that confuse the search. Here, we consider only the first, and we will talk more of it later.

### The Receiver

Let us consider 2-meter FM DF. (This discussion also applies to other FM assignments.) Two dipoles at different distances from a transmitting source simultaneously receive different phases of the same signal. If we switch between dipoles that are less than ½ λ apart, there will be an apparent phase shift of somewhat less than 180° between the received signals. This phase difference, with an FM or PM detector, will give a tone at the switching rate. If the two antennas happen to be the same distance from the source (phase), there will be a null—no tone—at the rate the circuit switches between the two antennas. (Some second-harmonic energy from the tone may appear, but is no indicator.)

The indicated location of the transmitter is on a line that crosses (at 90°) the midpoint of the line joining the two switched antennas. It is not unusual to determine this null direction to an accuracy of one degree or better.

### Switching the Antennas

When we switch the antennas at a rapid rate (I prefer 800 to 1000 times per second) the switching of the transmitted signal can be heard as a tone in the FM receiver. The volume of the tone is greater when the phase difference between the antennas is greater (up to 180°). The volume drops to zero with zero phase difference (ideally).

Electronic switches are small, cheap and reliable. I like to use a timer (such as a '555) as a free-running multivibrator (see Fig 1). A CMOS version (for instance, a TLC555) uses less power than others and has enough output. (If the DF unit is battery powered, low power consumption is important.) I also like the switching pulses to be square waves, with the antennas alternately connected for equal times.

The actual switching of the antennas is done by semiconductor diodes. I use Philips ECG553 PIN diodes because they make decent low-capacitance switches and have a useful power-handling capability (should the antenna be used for transmitting). The switches need not be perfect, or even very good. Off-the-shelf 1N4148 silicon diodes work well in receive-only use. The important factor is the phase difference presented to the receiver.

A similar system was used in my earlier articles.[1,2] Parts of the Simple Seeker are interchangeable with, or may be added to, the Double Ducky. The antenna systems are interchangeable. The phase detector may be added to the corresponding point on the Double Ducky.

### Adding Sense to the Null

By finding a null, we find a line through the transmitting source. If we move off of the null, we find that the polarity of the receiver output changes (with respect to the switching waveform) depending on which antenna is nearer the source. Thus, we can compare the receiver-output phase to that of the switching waveform in order to find out which end of the "null" line really points towards the transmitter.

The common name for a circuit that makes this comparison is "phase detector." Any of several dozen circuits may be used; the one shown here is simple, noncritical, and uses widely available parts. Note that if there is either no receiver audio or no switching waveform, there will be no voltage difference in the rectified diode output, and therefore no indication on a zero-center microammeter. (I find a 100-0-100 μA meter satisfactory for both sensitivity and ruggedness. They are also fairly common as FM tuning indicators. A sensitive meter with zero at the scale end may be used by reversing the DPDT POLARITY switch to read right or left indications. If so, the user must remember which switch position represents left or right.)

A phase-detector BALANCE control (1-kΩ potentiometer) is provided, though it was not needed in my case. If there is a serious imbalance, it would probably be good to check the receiver tuning, the receiver audio or switching waveforms.

The phase-detector transformer

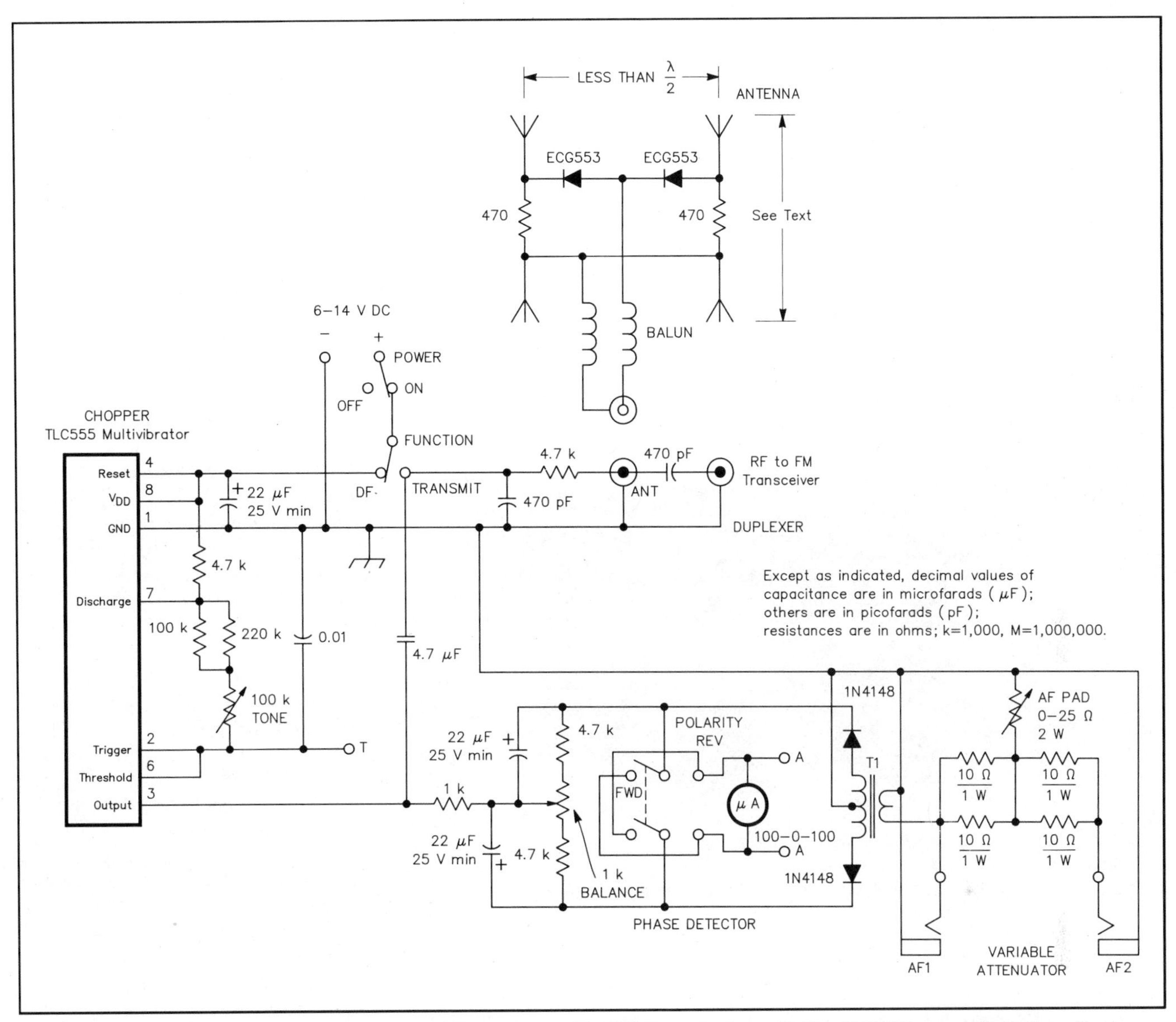


Fig 1—Schematic of the Simple Seeker. T1 is an AF transformer (8:1000 Ω, center tapped). The ECG553s are PIN diodes; silicon switching diodes may be substituted, so long as the unit is *not* used for transmitting. A capacitor may be added from point T to ground (or V+) to lower the tone frequency. Comparators may be driven by points A for other phase-detector displays (such as LEDs—see text). Either AF connector may be used for the receiver AF output or speaker/phones jack. AF1 is full strength; AF2 may be attenuated by adjusting the AF PAD control. All parts except the meter and PIN diodes should be available from Radio Shack. (The parallel 100-kΩ and 220-kΩ resistors may be replaced by one 68-kΩ ¼-W unit, if available.)

(1000 Ω—center tapped—to 8 Ω) has been common in Radio Shack inventory for many years, but it is not the only suitable part. Almost any AF transformer having around a 10:1 voltage step-up to a center-tapped secondary can provide the voltage step-up and push-pull audio needed for the phase detector.

### Other Phase-Detector Displays

The phase-detector output is a right-left indication, shown here on a meter. Often searchers in vehicles cannot both look at a meter and move safely. The following approaches have been used. They are mentioned to show what can be done.

The phase-detector output is a voltage: positive, zero or negative. Comparators, such as the common '339, can tell which voltage (of a few millivolts) is more positive or negative. Their output is capable of driving LEDs or switching circuits. Thus, nautical-scheme (red = left, green = right) LEDs can be used to indicate the corresponding direction, or a capacitor may be switched in (at point T, Fig 1) to give a *lower* tone for *left*, and higher for right.

Comparators can also provide a buffering action that allows the addition of time averaging to bearing information. A searcher, driving through a city or broken terrain, may experience several different indicated directions in a second or so. These diversions tend to center around the correct direction; thus an averaged reading is likely to be more correct. A comparator (or other buffering output) with low-pass filtering will result in a more effective and less frantic hunt. (Champion hunters often provide switching to display either time-averaged or instantaneous bearing information.)

### Antennas

Good antennas can be used with this system. This means that weak signals have a chance of being heard, and that transmissions may be made into a decent load. Both

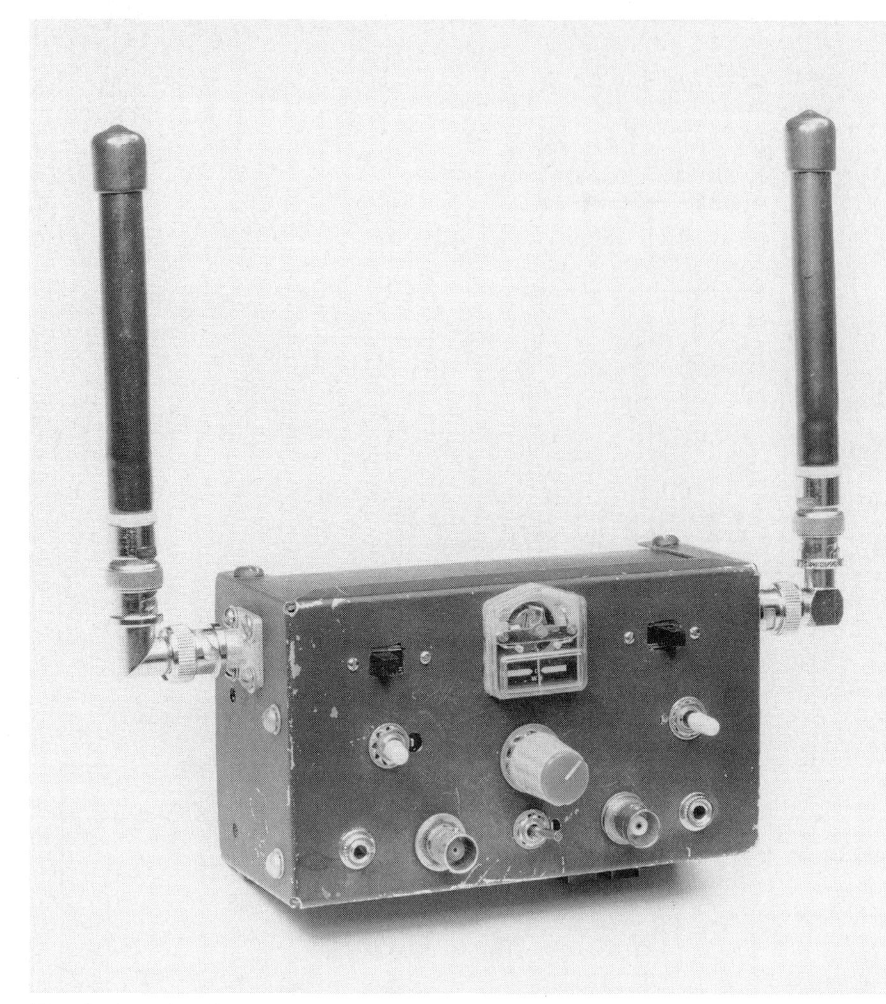

Fig 2A—Here is a field version of the Simple Seeker with attached antennas.

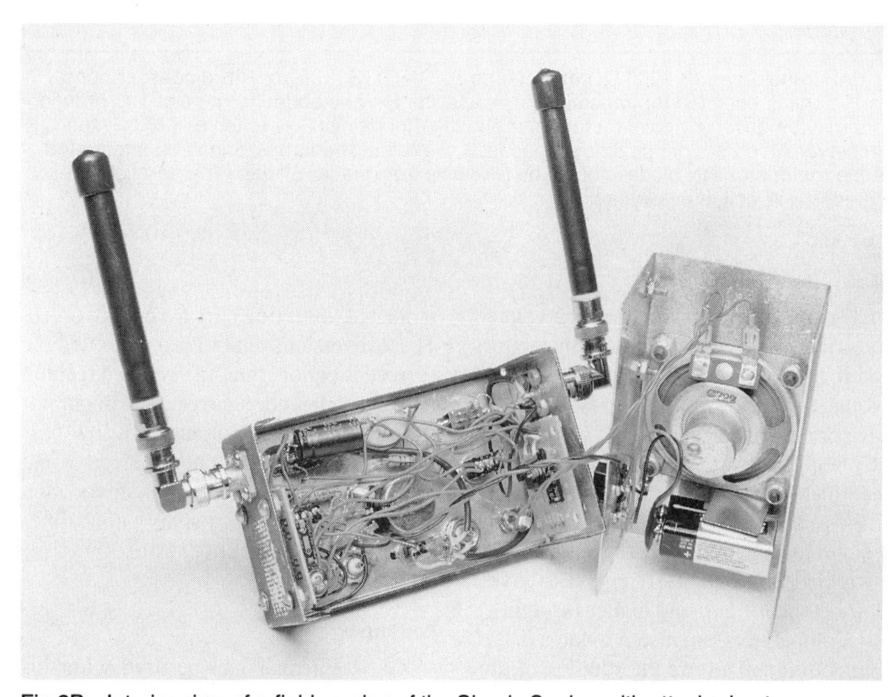

Fig 2B—Interior view of a field version of the Simple Seeker with attached antennas. The multivibrator and phase-detector circuits are mounted at the box ends. This version has a built-in speaker.

of these considerations are important if you are racing through the woods with an HT.

A ½-λ dipole is a decent antenna, and reasonably broadband. For receiving, antennas are effective over much more than the "half-wave plus five percent" sometimes called the "bandwidth." That limited bandwidth may be realistic for a transmitter that is critical of the load it feeds, but receivers do their best with whatever source is connected. My experience shows that a dipole cut for the middle of the 2-meter band covers the whole band, and rubber duckies do as well. The Coast Guard uses one dipole length from about 136 to 170 MHz. I have used 2-meter ½-λ (39 inches, 99 cm) dipoles, spaced 12 inches (30 cm) apart, for both 2-meter and 420-MHz DF. (On 420 MHz, a 1-m dipole is about 3⁄2 λ, which has a pronounced 90° lobe. It does not work very well on 222 MHz).

I am often asked, "What about effects from the inactive antenna?" If the diode switch were perfect and the parts ideal, that inactive antenna would see a 9:1 mismatch, and have little effect on the surroundings. Practically, that happens.

**Spurious Antennas**

Earlier we assumed that "all significant signal comes through the search antennas;" at least that is what we hope. That is necessary if we want the direction indication to be perfect.

It is theoretically impossible to attenuate a signal's strength to zero, though it may be practically unmeasurable. It is possible to equate two signal amplitudes, but for the result to disappear, the phases must also oppose. If only two sources (such as a pair of switched antennas) feed a detector that is sensitive only to phase differences, amplitude is unimportant. If some third path feeds the signal to the detector, however, the indication may not be predictable.

Thus, not only in amplitude-detecting DF systems, but also in this phase-detecting system, best accuracy demands that signals come only from the antenna system. Signals should not reach the receiver wiring either directly (through an unshielded case) or arrive on wiring other than the antenna cable. The phase-detecting system has the advantage that it is less amplitude sensitive than amplitude-detecting systems; but if the signal pickup is through small-aperture antennas such as "rubber duckies," a small signal leak may have a big effect.

I have found that a simple, tight wrap of aluminum foil will block unwanted signal pickup in some cases, but tighter shielding

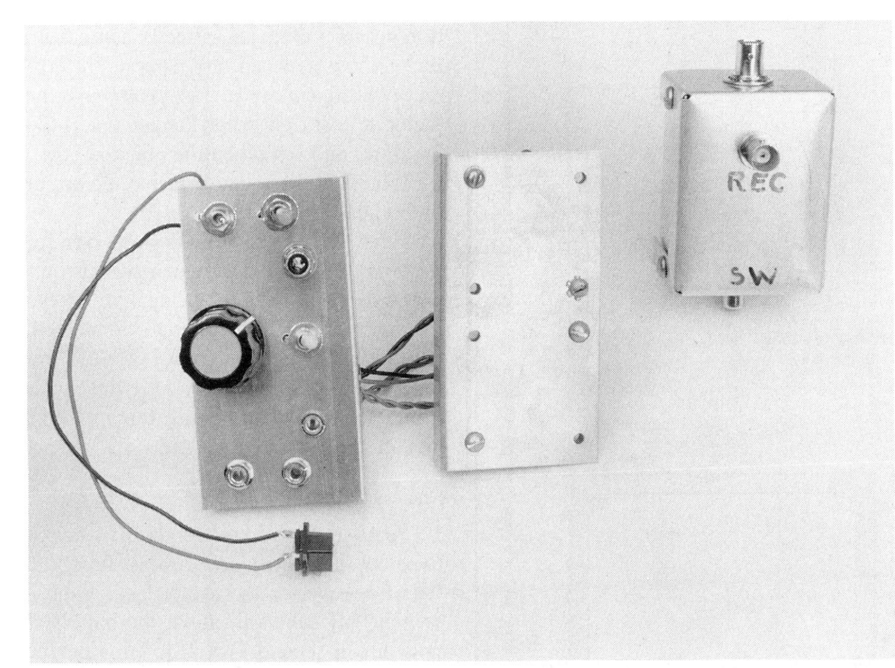


Fig 3A—Front-panel view of an "open" version used for presentations. Nearly all parts are readily visible.

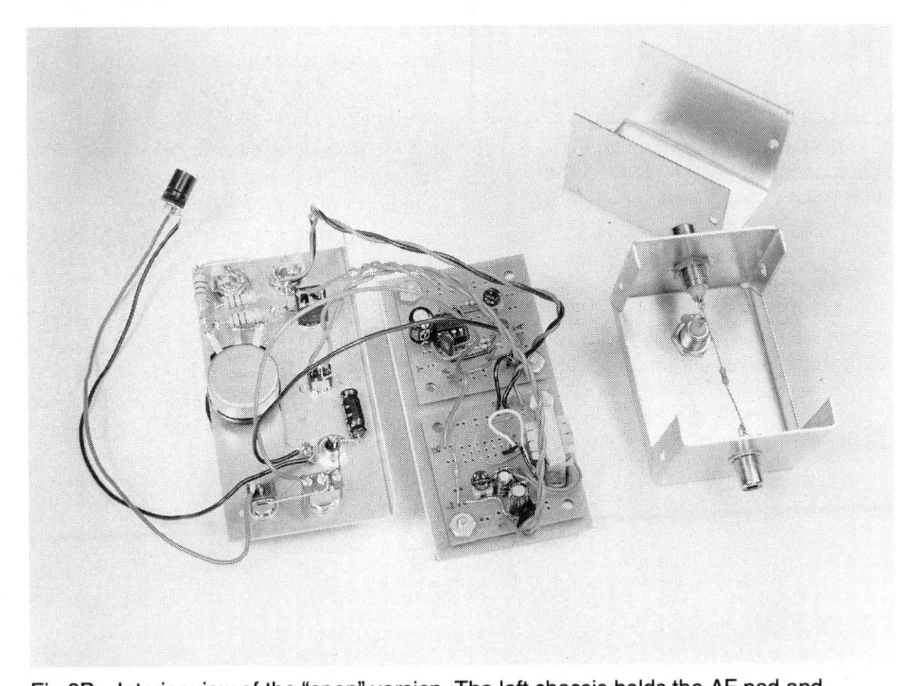

Fig 3B—Interior view of the "open" version. The left chassis holds the AF pad and connectors. The center chassis holds the multivibrator and phase-detector circuitry. The box holds the filter circuits, switching-lead and RF connectors. The shunt 470-pF capacitor is not visible; it is connected across the phono jack on the box.

may be needed. Sometimes a routine of holding all parts of the array and receiver in fixed relative position will maintain a known direction, but this is difficult if the antenna, receiver, feed line, chopper, phase detector and person must all keep the same position. Receiver shielding is probably the easiest approach.

Good shielding of a receiver allows the use of "duckie" antennas. I have had good results with five-inch duckies and eight-inch spacing, as in Fig 2.

### Direction Errors

It is not generally recognized that a constant error will cause a continuous searcher to move in a logarithmic spiral, which easily predicts the center (transmitting source). Similarly, a searcher who pursues left and right for right-angle indications will have a bearing to the transmitter after two right and two left indications.

### Construction

The electronic switch may be easily constructed on a 20-pin DIP pad, and the phase detector on another pad. (See Figs 2 and 3.) There is no need for either circuit to be close (in inches) to the antenna or receiver. There may be some advantage to placing the circuits near the various indicators, but these reasons do not seem demanding.

It is probably good to put the switching circuit in a shielded box, with the switching pulses fed through a low-pass filter (the series 4.7-kΩ resistor and shunt 470-pF capacitor) to the antenna and through a high-pass filter (470-pF capacitor) to the receiver or transceiver. (A capacitor of 470 pF has about 2 Ω of reactance at 2-meter frequencies.)

Batteries are always a problem. Paratroopers with similar gear place batteries outside of the case, for convenience. One writer places two batteries within the case and switches between them for charging or use. I prefer a storage battery (that may be either used or charged) within the case, with an external connector for another battery.

Three items have not been discussed: the antenna and its switching circuit, the antenna balun and the AF variable attenuator (or "pad").

The antenna is mounted on an "H" frame of ½-inch PVC tubing (see Fig 4) by the simple process of taping it to the frame with black plastic electrician's tape. The vertical arms are glued together and secured to the horizontal pieces with cotter pins, to permit easy disassembly for transport.

The antenna switching circuit consists of two diodes connected (in opposite polarity) to the coaxial cable from the switching circuit. Each of the diodes connects to a grounded 470-Ω film resistor, which provides a dc path to ground (in case the antenna configuration does not). Equal-length coaxial cables (in the system shown) lead to 39-inch dipoles. The spacing between dipoles is about 20 inches, but the spacing is not critical.

There is always a question about external currents flowing on the coax shield in coax-fed direction-finding antennas. I have not had problems when using a big antenna (such as in Fig 4), but I wrapped 3 turns (about 2-inch diameter) of the incoming coax to form a "choke balun." There were no problems.

Phase detectors may behave differently

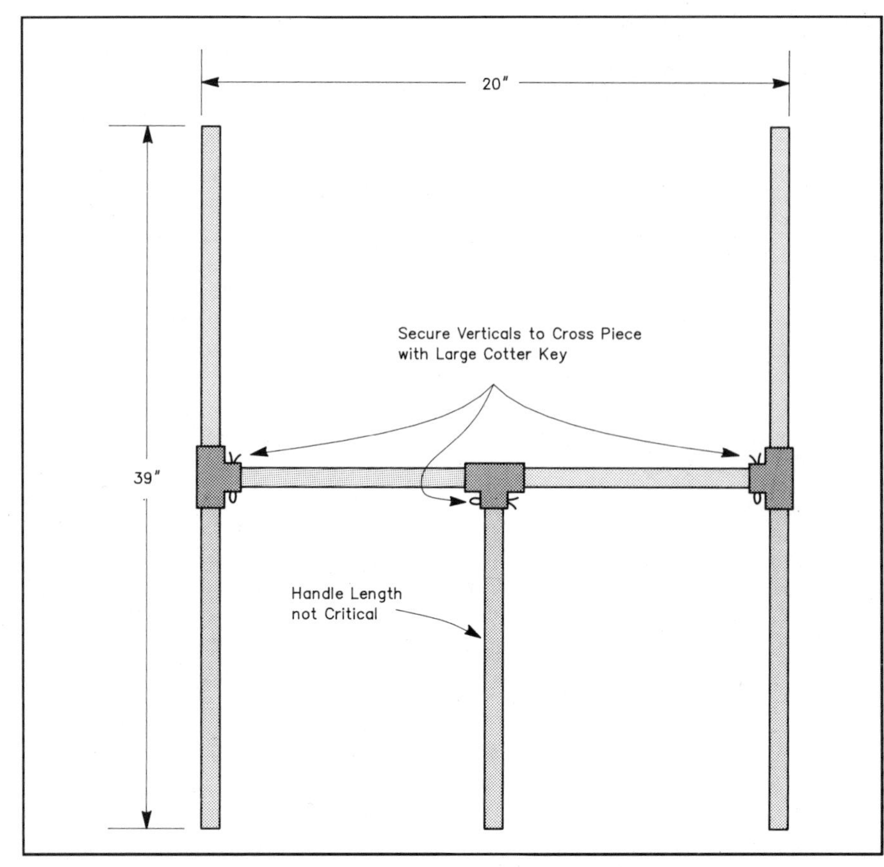


Fig 4—The antenna array described in the text. The "H" frame is constructed from ½-inch PVC tubing and tees. Glue the vertical components together. Connect the vertical tees and handle to the cross piece by drilling both parts and inserting large cotter pins. Tape the antenna elements to the tubes.

if one signal varies from weak to strong. Similarly, a listener may be deafened by one tone or strain to hear another. An AF attenuator allows either a full-strength signal or a lesser, adjustable, received signal to feed the phase detector. The receiver AF feeds into one jack, and the headphones or speaker are plugged into the other. I like to use phones and plug them into jack AF2, with the receiver audio connected to jack AF1.

## Meter Polarity

As discussed earlier, the line of direction points both to and directly away from the unknown transmitter location. The correct polarity (in my custom) is to have the meter deflect right when the antenna is too far right, and left when the antenna is too far left. The phase detector may do so, or the indication may be reversed.

Persons using a particular receiver for the first time do not know whether its output will be positive or negative at the instant one particular antenna is connected. Thus the direction-finder should be checked with each receiver. If the meter indication is backward, use the POLARITY switch to reverse the reading.

## Saving Batteries

Years ago I used a "center-off" switch to select the DF, OFF or TRANSMIT function. It is too easy for me to not leave such a switch "off" and run down the battery. I now use a second switch to provide the "on-off" function.

## Transmitting

The PIN diodes, when forward biased, show a low RF resistance and can pass up to perhaps 1 W of VHF/UHF power without damage. Thus the FUNCTION switch has a TRANSMIT position, in case transmission from an HT is desired. This applies a steady dc bias to one of the PIN diodes. This may be a needed function, but it should not be used casually.

## Notes

[1]D. Geiser, "Build a Double Ducky Direction Finder" (Jul 1981 *QST*, pp 11-14).

[2]D. Geiser, "Updating the Double-Ducky Direction Finder," (May 1982 QST, pp 15-17).

# The CCD Antenna—Improved, Ready-to-Use Construction Data

By Harry Mills, W4FD,
Box 409, Bullard Route
Dry Branch, GA 31020

and

Gene Brizendine, W4ATE
600 Hummingbird Drive SE
Huntsville, AL 35803-1610

Since we first described the controlled-current-distribution (CCD) antenna in 1978,[1] over a thousand copies of our data sheet have been requested by interested amateurs and professionals. Their questions and comments have given us an incentive to continue our experimentation with this unique antenna. The latest improvements in construction methods and performance are reflected in this update.

## Background

The CCD antenna concept of the first author has been undergoing continual refinement at his antenna test range since 1959. Antenna experimentation was a natural by-product of our careers as Federal Communications Commission engineers. Harry was Chief of Engineering, Renewal and Transfer Div of the Broadcast Bureau, serving at the Washington, DC, office for over 34 years. Needless to say, the unusual properties of the CCD design have held our interest since those early days.

At the heart of a full-wave CCD antenna is a string of series-resonant circuits which result from pairing each capacitor with an inductor (wire section) having equal reactance but opposite polarity at the operating frequency. The net result is a very low impedance to the flow of RF through the radiator. This idea is implicit in Harry A. Mills' US Patent 3,564,551.

## General Considerations

The recommended arrangement for a basic CCD system is shown in Fig 1. A 1:4 balun is connected to the transceiver through the shortest possible length of coax. Using a balun in this manner results in wider bandwidth and improved balance. (The impedance of the CCD antenna closely matches the characteristic impedance of space—376.7 Ω.[8])

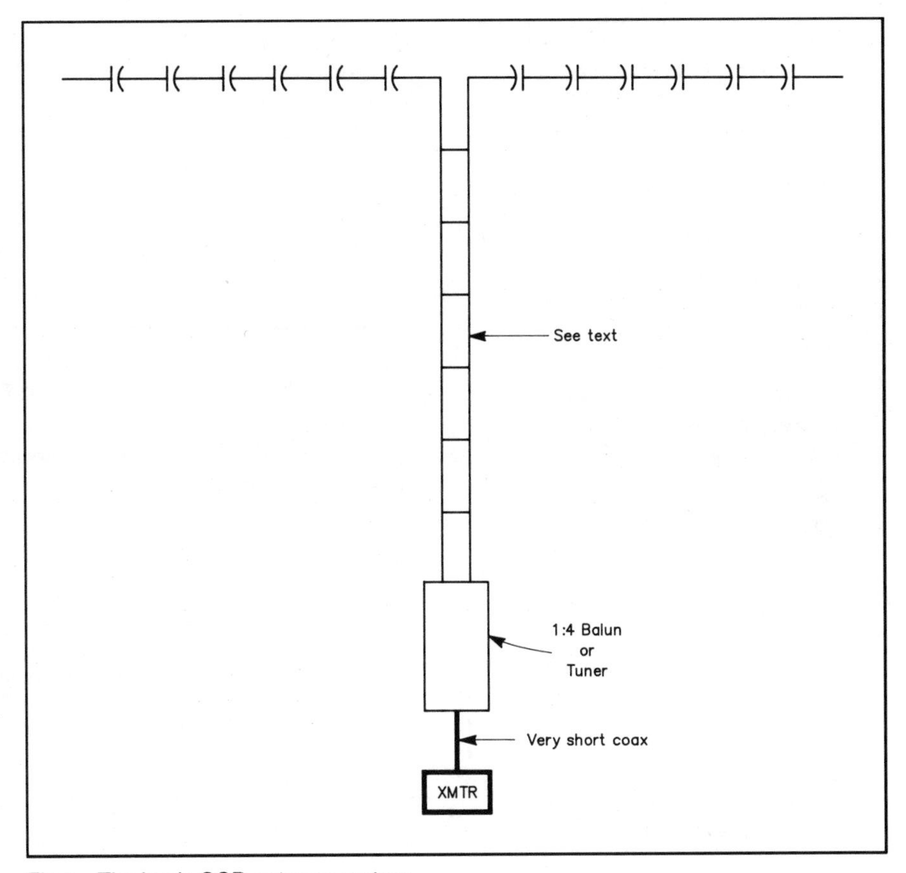


Fig 1—The basic CCD antenna system.

Ladder-type 450-Ω feed line is suitable for legal-limit power. For low-power operation (less than 300 watts), foam-type 300-Ω ribbon cable is adequate and provides an easier, neater installation. The line length is not critical in either case, but trimming it to any multiple of an electrical half-wavelength greatly improves the convenience of using a noise bridge *at the operating position.* This may reduce the need for repeated trips to the antenna.

Fig 2 compares the performance of two vertical CCDs against a conventional quarter-wave vertical standard. These field-intensity patterns were measured at 146 MHz on the W4FD range. Note the 12° vertical-plane main lobe from the CCDs—with *no* radials required!

## Component Considerations

There are two types of capacitors that have been found to offer the best performance in CCD applications. Silvered mica

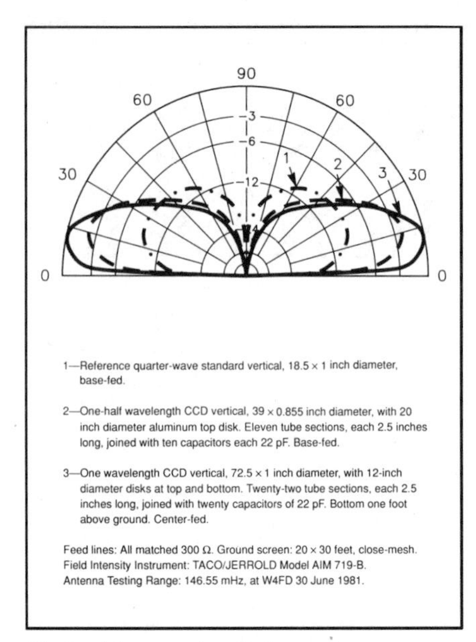


Fig 2—Measured radiation patterns of vertical 2-meter CCDs compared to a ¼-λ reference. Note the 12° main lobe of the full-wave CCD.

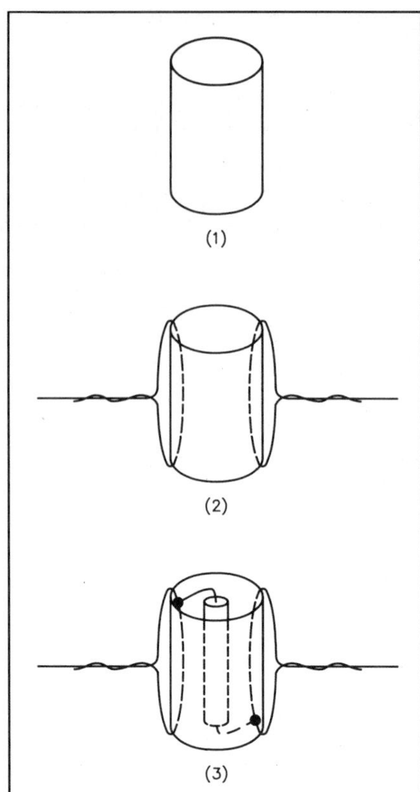


Fig 3—Capacitor/protector assembly in three easy steps. (1) Select a plastic water pipe diameter that is just large enough to accept the capacitor. Cut the pipe lengths ¼ inch longer than the capacitor to hold an adequate amount of silicone sealer. A pipe cutter or hacksaw is suggested. (2) Select the proper wire section length from Table 1, allowing enough for twists. Sand or scrape the wire ends in preparation for soldering. Loop the wire ends through pipes, making sure that the section lengths are exactly as shown in Table 1 for the desired band. The length of the wire section is measured between pipe centers as shown in Fig 4. Twist the ends securely. (3) Finally, solder the capacitors as shown. Do not apply the silicone sealer until after you've tested the CCD at a convenient height of 5 feet above ground.

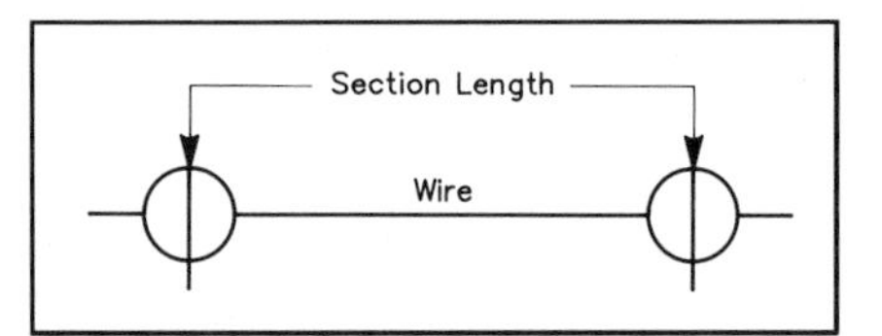


Fig 4—Measure the length of a wire section between the pipe centers.

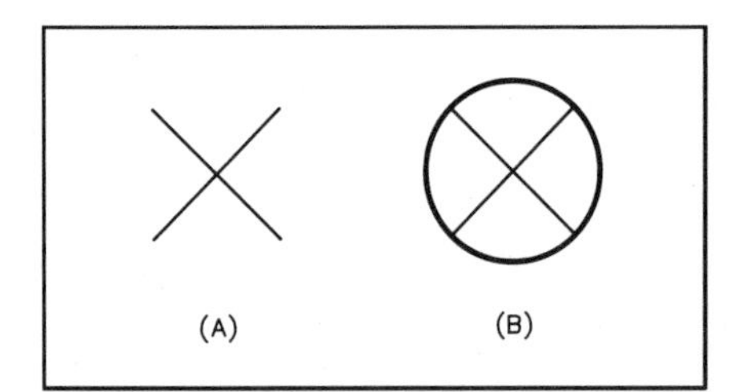


Fig 5—Two examples of simple end loads or hats that can be used if required. (A) Two stiff wires (such as brass welding rod) are soldered at their crossing points. Make two assemblies and solder one to each end of the CCD. (B) This design is similar to the crossed wires, but adds a rim for additional capacitive loading. The total end-load diameters can be found in Table 1.

**Table 1**
**Capacitor Sizes, Wire Sections and End Loads**

| *End-Load Diameter* | *Band (Meters)* | *Length (Feet)* | *Section (Inches)* | *Sections (Number)* | *Capacitor (pF)* | *Capacitors (Number)* |
|---|---|---|---|---|---|---|
| — | 160 | 560 | 140 | 48 | 1560 | 46 |
| 18° | 80 | 280 | 70 | 48 | 780 | 46 |
| 12° | 40 | 140 | 35 | 48 | 390 | 46 |
| — | 30 | 97 | 24.36 | 48 | 270 | 46 |
| 9° | 20 | 70 | 17.5 | 48 | 195 | 46 |
| — | 15 | 46.5 | 11.5 | 48 | 130 | 46 |
| — | 10 | 35 | 8.75 | 48 | 97 | 46 |

Note: Loads for other bands may be extrapolated from the values shown. (Capacitive load diameter increases current flow in end sections.)

capacitors have low loss, high Q, and are ruggedly constructed. The 300-V, 5% types are sold for as little as five cents each in quantities of 100.[7] Polystyrene capacitors are very stable, and offer high Q and very low loss. Regardless of which capacitor you choose, beware of those with fragile leads. Also be sure to check the tolerance of each capacitor (it must not exceed 5%). Try to locate an accurate instrument for your measurements. Club or school instruments are often available for testing in their labs.

Twelve-gauge wire was used to obtain the specifications in Table 1. Stranded wire with a tough plastic jacket will reduce precipitation static. Enameled solid copper is also a good choice. Avoid copper-clad wire since it often rusts through the coating and exposes the steel core, a poor conductor of RF.

### Setting Resonance

A correctly built CCD antenna for 80 meters, for example, will cover 3.5 to 4.0 MHz with a usable SWR and using *no* tuner. Because of the capacitor action, bandwidth is not symmetrical at each side of resonance. Resonance should be set at 25 kHz above the low end of any band by removing (or adding) equal numbers of capacitor sections at both antenna ends. If only partial band coverage is desired, set the resonance to 25 kHz above the lower limit of the band segment you intend to use.

### Recent CCD Successes

Excellent CCD performance on 160 meters has been reported by N4VL, W4FD and W9ALU. All have worked numerous VKs and ZLs on this challenging band. Among the many amateurs who have reported excellent performance on other bands are KK4X, KT3E, KD4CE, KK4EJ, N4VMB, NX4B and the husband and wife team of Delbert AB4TH and Dortha N4SHE.

### Acknowledgments

Without the support and encouragement of so many individuals, the CCD antenna in its present form would not have been possible. We wish to thank Dr Arthur Erdman, W8VWX, Professor of Electrical Engineering at Ohio State University; Dick Turrin, W2IMU, of Bell Telephone Laboratories for his MININEC3 modeling; and Harry A. Mills Jr, KK4X,[9] an engineer at Sprague Electric. Thanks are also in order for G4IAK, G8SBV, ZL2BRA and ZS5AP, who built and enjoyed the CCD. We appreciate the efforts of Tiziano Aspetti, I3ATE, whom we met at the Huntsville Hamfest. He translated the CCD story and published it in Italy. Lastly, the faster, more user-friendly antenna modeling ELNEC program of Roy Lewallen, W7EL,[10] was very helpful.

### References


1 Mills and Brizendine, "Antenna Design: Something New!," *73*, October 1978.

2 Mills and Brizendine, "The CCD Antenna-Another Look," *73*, July 1981.

3 Longerich, "The CCD Antenna Revisited," *73*, May 1982

4 Rennie, "Again, The CCD," *73*, September 1982.

5 Mills and Brizendine, "Antarctic CCD Antennas," *73*, July 1983.

6 Atkins, "The High Performance, Capacitively Loaded Dipole," *Ham Radio*, May 1984.

7 Fertik's Electronics, 5400 Ells St, Philadelphia, PA 19120.

8 *ITT Reference Data for Radio Engineers*, 4th ed.

9 Route 3, Box 654, West Jefferson, NC 28694

10 PO Box 6658, Beaverton, OR 97007.

# Controlled-Current-Distribution Antenna Performance: By Analysis

**By Bill Shanney, KJ6GR**
**19313 Tomlee Ave**
**Torrance, CA 90503**

I became interested in controlled-current-distribution (CCD) antennas after reading about them in *The ARRL Antenna Compendium, Volume 2*.[1,2] I later had a QSO with Jim Gray, W1XU, who was using a CCD antenna and putting out a good signal. Jim provided me with several references[3-6] which increased my curiosity. I found myself asking a lot of questions but getting inadequate answers concerning the theory of CCD antenna operation. The purpose of this article is to present the results of an extensive analysis performed using the MININEC antenna analysis program.[7]

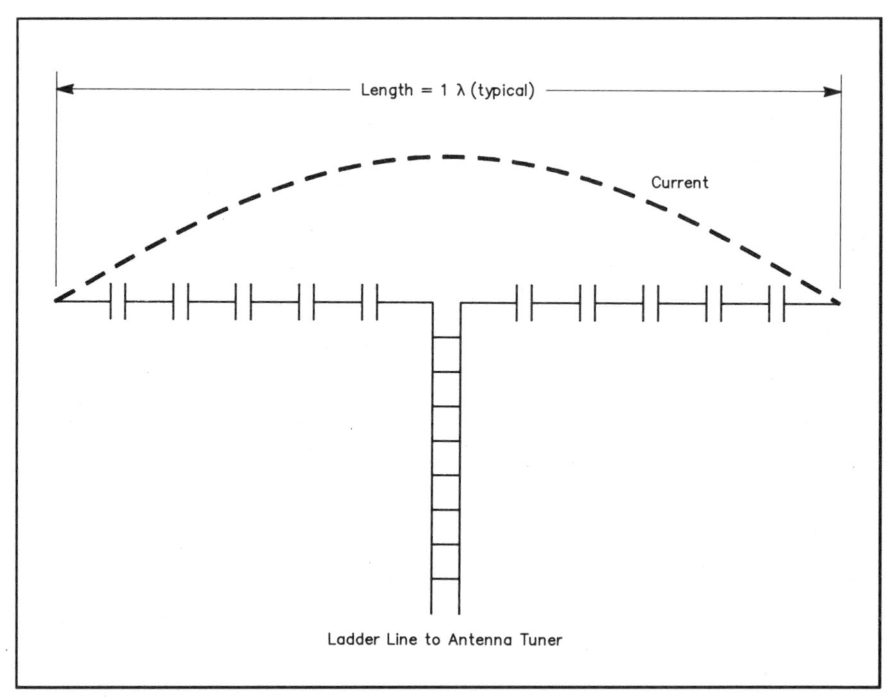


Fig 1—CCD dipole antenna configuration and current distribution. CCD dipoles are typically constructed with about 20 capacitors per half wavelength.

## Background

A CCD antenna consists of a length of wire broken into short sections separated by capacitors. As explained in the preceding article by the CCD inventors, the capacitor values are chosen to resonate or nearly resonate the inductance of the wire lengths at the design frequency, as shown in Fig 1 here. The capacitively loaded wire antenna may be configured as a dipole, vertical, loop or any other shape.

The main feature of a properly designed CCD antenna is the current distribution. For a CCD dipole of any length the current is at maximum at the feed point and decreases to zero at the open ends without the current reversals present in long-wire antennas. Since the phase reversals in a long-wire antenna are responsible for multiple lobes in the radiation pattern, the CCD antenna only contains the main lobes. One additional feature of a full-wave CCD is its low input impedance. A full-wave center-fed wire (two half waves in phase) has a center feed-point impedance of several thousand ohms due to the current minimum at its center, while a CCD dipole of the same length has a moderately low impedance.

The CCD dipole is simply a cascade of series-resonant circuits, each of which has no net phase shift through it. The phase shift through the inductive wire section is canceled by the opposite phase shift of the capacitor. It looks like a length of wire with current flowing through it but with no traveling-wave phase shift as observed in a plain length of wire. The open-circuit boundary condition at the dipole ends forces the current to zero at these points. The current distribution is *not* uniform in a CCD dipole, as some authors have suggested.

Above the design frequency the capacitive reactance decreases and the antenna behaves more like a plain wire. Below the design frequency the capacitive reactance gets larger. For the designs I have analyzed,

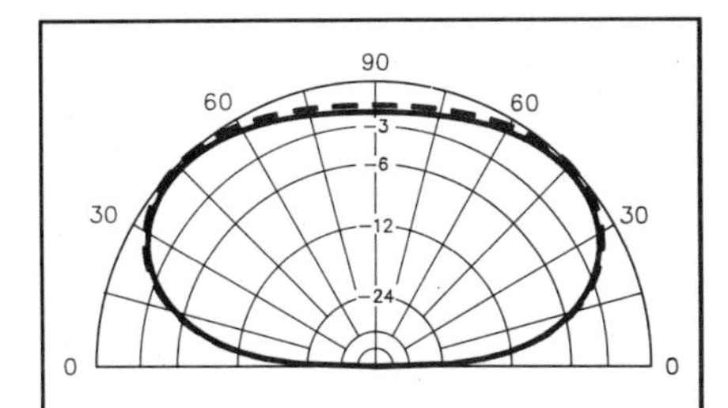


Fig 2—Vertical-plane radiation patterns of 40-meter center-fed antennas of 1 λ. Solid line: CCD dipole; broken line, center-fed wire (two half waves in phase). Note that the CCD dipole shows no performance advantage over the full-wave wire. CCD design: 48 equal-length wires, total length 140 feet, with 46 390-pF capacitors. Center-fed wire length = 130 feet. Maximum gain for both antennas is at a 51° elevation angle.

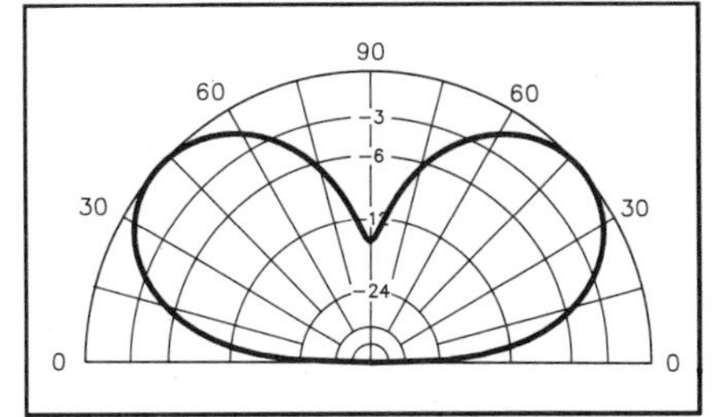


Fig 3—Vertical-plane pattern of an 80-meter horizontal full-wave CCD loop 40 feet above ground. The azimuth pattern is within 0.4 dB of being circular. This loop has its gain peak at 44° elevation. This is much better than the straight-up radiation characteristics of a wire loop at the same height. Design: 40 7-ft wires with 38 470-pF capacitors and a 235-pF capacitor opposite the feed point (to resonate two inductors). Calculated radiation resistance: 32 Ω.

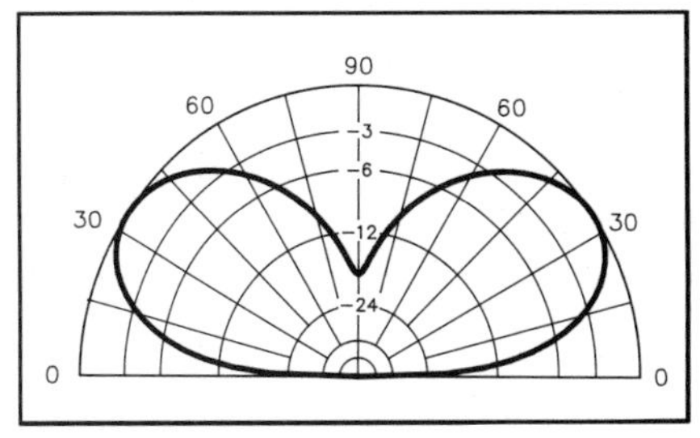


Fig 4—Vertical-plane pattern of a 40-meter horizontal full-wave CCD loop 40 feet above ground. Its azimuth pattern is within 0.8 dB of being circular. The gain peak of this loop is a very respectable 36°. Design: 40 3.5-ft wires with 38 270-pF capacitors and a 135-pF capacitor opposite the feed point. Calculated radiation resistance: 71 Ω.

however, radiation at half the design frequency does take place with performance comparable to a half-wave wire dipole.

**Table 1**
**Dipole Performance Comparison**

| *Antenna Type* | *Frequency (MHz)* | *Gain (dBi)* | *Impedance (ohms)* |
|---|---|---|---|
| Full-wave CCD | 7 | 6.4 | 306 |
| Full-wave wire | 7 | 6.8 | >4k |
| Half-wave CCD | 7 | 5.6 | 62* |
| Half-wave wire | 7 | 5.5 | 84 |
| Full-wave CCD at half frequency | 3.5 | 7.8 | 11.5* |

*Real part only (resistive component)

### Analysis

My interest in this unique antenna is a product of my desire for a better 40-meter signal. I use a beam for the higher frequency bands with good results, but it's hard to compete for DX on 40 meters with an inverted V up only 40 feet at its apex. Since I can't put my antennas any higher, I performed my analysis at 40 feet. This should be of interest to amateurs with limited space for antennas. I didn't compute free-space patterns since I was only interested in the performance over real ground. I used a ground conductivity of 4 mS/m and a dielectric constant of 13 for all of my calculations. There may be errors in the impedances calculated by MININEC since real-earth ground effects are not taken into account by this analysis program. The gains and antenna patterns *do* take real-ground reflections into account. [But the gain calculations in turn use the results of the impedance calculations, so cannot be taken as absolutely accurate.—*Ed.*]

### CCD Dipole Performance

A summary of the computed performance for CCD full- and half-wavelength dipoles and their wire counterparts is shown in Table 1. Calculated antenna radiation patterns for the full-wave versions are shown in Fig 2. There is no significant performance difference between the CCD dipole and a wire antenna of the same length. The only advantage of the CCD full-wave antenna is its low input impedance.

Table 1 does not show the large input reactance of a CCD dipole when operated at half frequency. The 3.5-MHz example shown has a 1500-Ω capacitive input reactance which may make matching difficult at high power levels since the input resistance is very low. The half-wave CCD dipole is also not resonant. The original authors pointed out this fact,[4] and the analysis provided verification.

### CCD Horizontal Loop Performance

A CCD loop differs from a dipole in a very significant way: it has no open-circuit ends to cause the current to be zero. A virtually uniform current distribution exists around the entire loop at resonance. The performance of a loop of any size with a constant current is described in Section 6-6 of *Antennas* by John Kraus.[8] The familiar donut-shaped pattern of a small loop is maintained for diameters up to ¾ λ.

The computed antenna patterns for 80- and 40-meter full-wave CCD loops are shown in Figs 3 and 4. The very respectable low-angle radiation performance of these loops makes them an attractive alternative for low-band enthusiasts. A full-wave wire loop at this height radiates straight up and exhibits a free-space impedance of about 100 Ω. (The MININEC computed impedances are shown in the captions for the figures.) The accuracy of the radiation resistance is questionable at 80 meters, however, since the antenna is less than ¼ λ high and since MININEC does not accurately account for the influence of real ground when calculating radiation resistance.

### CCD Horizontal Loop Analysis

I have analyzed a number of antennas with open-circuit ends, including a horizontal loop with a gap opposite the feed point. Their resonant frequencies all coincide with the resonance of the wire sections with the capacitors. You can imagine my surprise when the computed resonant frequency for a full-wave closed square loop was 10% low. I reduced all the capacitor values by 10% and the loop resonance moved up about 5%. Since the capacitors are ideal loads, this meant the inductance must be wrong. I went back to the original formula used to calculate the inductance.

$$L = 0.00508\,\ell[2.303\log_{10}(4\ell/d) - 0.75]\ \mu H$$

where
$\ell$ = wire length, inches
d = wire diameter, inches

I checked my calculations and found they were correct. Then it dawned on me that two inductors brought close together have mutual inductance. Terman's *Radio Engineers Handbook*[9] contains formulas for the mutual inductance of straight wire sections connected end to end and with a gap between them (designated as "D"). For equal-length wire sections connected end to end the formula reduces to

$$M = 0.00352\,\ell\ \mu H$$

For wire sections with a gap D,

$$M = 0.00585\,(2\ell + D)\log_{10}(2\ell + D) + D\log_{10}D - 2\,(\ell + D)\log_{10}(\ell + D)\ \mu H$$

The computer performs its analysis with no gaps in the wires, but practical CCD antenna construction techniques use insulating spacers to support the series capacitors. When the effects of mutual inductance were included, the calculated resonant frequencies for the CCD loop agreed within 1%. The inductance resonated by each capacitor is L + 2M. It is important to resonate the loop since it is narrowband with reactance increasing very rapidly off resonance.

I'm not sure why mutual inductance is not a factor in determining CCD dipole resonance, but it is important to note that the radiation mechanisms for the two antenna types are significantly different.[10] The CCD loop antenna radiates because the moving electric charge is accelerated as it moves in a curve. On the other hand, the dipole radiates due to the acceleration of the reflected charges at the open ends and the feed point. The charges oscillate back and forth and undergo periodic acceleration. Currents traveling in opposite directions may cancel whatever mutual inductance is present.

### Conclusions

CCD full-wave horizontal loops are an exciting option for the low-band enthusiast with limited space. Additional work needs to be done to measure or compute the impedance of loops close to ground. A CCD dipole, on the other hand, exhibits no real advantage over a wire dipole. Although not presented in this article, CCD verticals were analyzed and found to offer no advantage over conventional verticals, in my opinion.

During the course of this investigation, other CCD configurations such as rhombics, Vs, and various bent dipoles were analyzed. Some of these *did* produce superior performance compared to a wire antenna of the same shape. However, this was only true for particular antenna orientations and directions. If you are considering an irregularly shaped dipole installation, a MININEC analysis comparing the CCD to a wire antenna would be worthwhile.

### References


[1]S. Kaplan and E. Bauer, "The Controlled Current Distribution (CCD) Antenna," *The ARRL Antenna Compendium, Volume 2*, 1989, pp 132-136.

[2]S. Keen, "The End-Coupled Resonator (ECR) Loop," *The ARRL Antenna Compendium, Volume 2*, 1989, pp 137-139.

[3]H. A. Mills and G. Brizendine, "Antenna Design: Something New," *73*, Jul 1981, pp 282-289.

[4]H. A. Mills and G. Brizendine, "The CCD Antenna—Another Look," *73*, Jul 1981, pp 50-57.

[5]D. Atkins, "The High-Performance, Capacitively Loaded Dipole," *Ham Radio*, May 1984.

[6]G. Rennie, "Again, The CCD," *73*, Sep 1985, pp 12-14.

[7]ELNEC by Roy Lewallen, W7EL, was used for the dipole calculations. MNC by Brian Beezley, K6STI, was used to analyze the CCD loops.

[8]J. Kraus, *Antennas* (New York: McGraw Hill, 1988), Chapter 6.

[9]F. Terman, *Radio Engineers Handbook* (New York: McGraw Hill, 1943), pp 48 and 66.

[10]See pp 50-52 of Ref 8.

# The Cross Antenna

By R. P. Haviland, W4MB
1035 Green Acres Circle N
Daytona Beach, FL 32019

The cross antenna is a member of the quad family, and came to my attention through an IEEE paper.[1] As I had been working on quads, I studied the concept in some detail. This showed some interesting properties, suitable for amateur work in the frequencies above 144 MHz. The study is reported here in sufficient detail to allow application by scaling to the desired frequency band.

## General Concept

The basic cross antenna is composed of a number of identical U-shaped arms arranged radially about a central point, with the open end of the U toward the center and the ends of the adjacent arms connected together. For example, the 4-arm cross is shown in Fig 1. As in the loop antenna, there may be more than one turn around the cross. (The antenna is covered by French patent 85 10463 of July 9, 1985).

For each arm, the long sides are 180 electrical degrees in length, ½ λ. The short-side length depends on n, the number of arms, the length being 360/n electrical degrees. This is ¼ λ for the 4-arm type. As is common in large loop antennas, the electrical length is a few percent shorter than the physical length. The antenna is normally used with a reflecting screen. The cross antennas are of two types, of quite different characteristics. Both have interesting and useful properties.

The *traveling-wave cross* (TWC) produces a circularly polarized main lobe, and does this without any phase-shifting networks. At the best design point, the polarization ratio is essentially perfect, and the gain reasonably high. The polarization ratio shifts with frequency, so the antenna is relatively narrow band.

The *standing-wave cross* (SWC) produces a main lobe having both horizontal and vertical polarization, but does not have the phase relationship necessary for circular polarization. Gain is reasonably high at

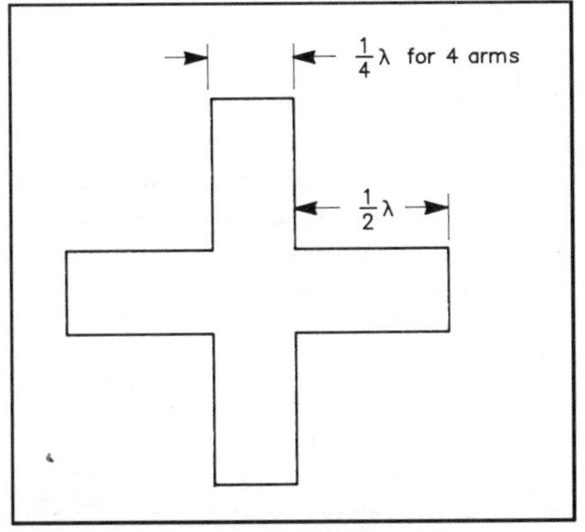


Fig 1—Four-arm cross antenna. The antenna is normally used with a reflecting screen. The conductor can be broken at any convenient place and the antenna fed with a balanced line, or broken at a short-arm conductor and one of the ends created fed against the ground plane. The other end is returned to the ground plane via a loading resistance.

the maximum gain point, but the interesting feature is that the gain exceeds that of a dipole over a frequency range of at least 6:1. Feed-point or drive resistance does vary, but is reasonably constant over frequency ranges in excess of three to one. Drive reactance is also variable, but can be low over a 2:1 frequency range.

## Construction

Both cross types are mounted above a ground-plane reflector. As with the more common bedspring antenna, the reflector should be at least ¼ λ larger in all dimensions than the cross itself; still larger is better. The cross can be mounted a convenient distance above the reflector, the range 0.1 to 0.2 λ being typical. If the reflector is not solid, mesh construction is necessary to give the same reflection for the horizontal and vertical components.

The cross can be of wire or tubing, but tubing is generally used for the higher frequencies to simplify element support. (The ratio of physical to electrical wavelength does depend on conductor diameter.) In the meter-wave bands, a combination of copper and plastic pipe and fittings can be used for a simple and rugged assembly.

The difference between the TWC and the SWC lies in the feed. For the standing-wave type, the cross conductor can be broken at any convenient place and fed with a balanced line. Delta or gamma matching is also possible. For a wide frequency range of operation, the feed impedance at any point is quite variable.

In the traveling-wave type, ideally, the standing wave must be completely suppressed. This is conveniently done by breaking a short arm conductor; one of the ends created is fed against the ground plane. The other is returned to the ground plane via a loading resistance. This resistance value is equal to the characteristic impedance of a single-wire unbalanced transmission line having the same conductor diameter d and spacing h from the ground plane, that is, $Z = 138 \log(4h/d)$. The dissipation rating of the resistor should be about 5% of the total input power for a single-turn 4-cross. The arrangement is related to the more common Beverage antenna.

## Performance

The following data is based on a MININEC analysis of a 4-cross with a long-arm length of 0.5 meter, short arm of 0.25 meter, and a conductor diameter of 0.012 meter. All data is for a spacing of 0.2 meter above a perfect infinite ground plane.

For the TWC configuration, the conductor of one short arm was interrupted at the center, with both ends turned 90° toward the ground plane. Feed is between one of the ends and the ground plane, and the load resistance is placed between the other end and the ground plane. The SWC

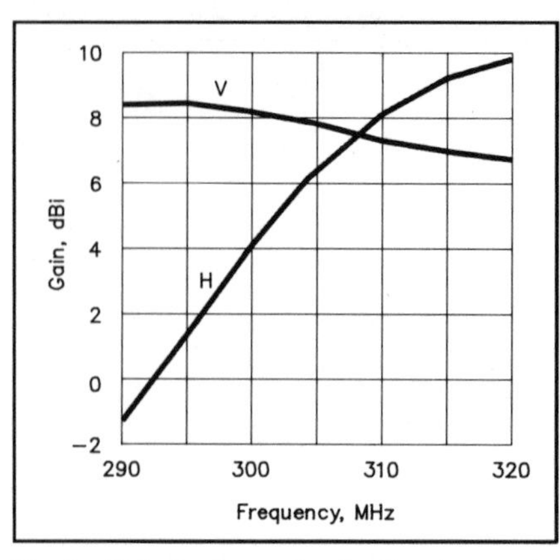

Fig 2—Polarization gains for the 4-cross TWC antenna.

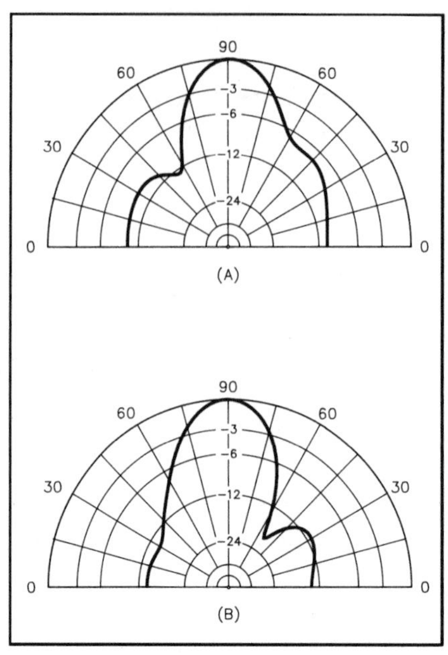

Fig 3—Calculated radiation patterns for the 4-cross TWC antenna. At A, for the vertical plane containing the open arm, and at B, for the plane at right angles to this.

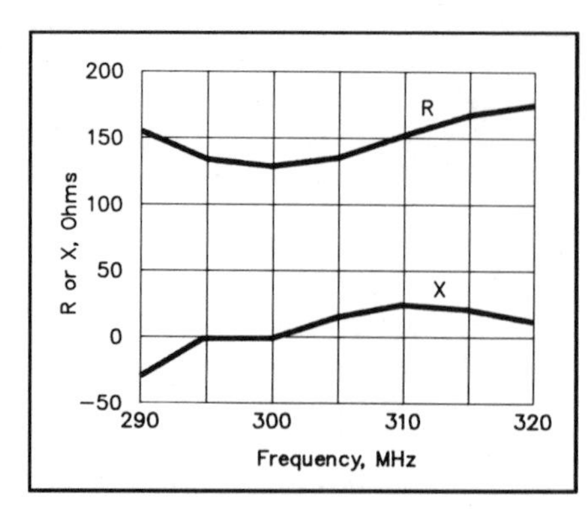

Fig 4—Feed-point resistance and reactance of the 4-cross traveling-wave antenna versus frequency.

loop is continuous, with feed at the end of a short arm.

Four analysis segments were used for the long arms, and two for the short. This is not a sufficient number for a truly accurate analysis, but experience with MININEC indicates that pattern and drive resistance accuracy should be reasonably good.

## TWC Type

The intensity of the main lobe normal to the ground plane is shown in Fig 2, for the horizontal and vertical components. Providing the phase is correct, the ratio of these gives the degree of polarization non-circularity. At the point of equal horizontal and vertical intensity, the calculated angle accuracy is within ±10 degrees of the ideal phasing, so the degree of circularity appears good. ("Appears" is used, since the number of calculating segments is too small for really good accuracy.)

Fig 3A shows the total component pattern for the plane containing the open arm, and Fig 3B for the one at right angles to this. The main lobe gain is 11.5 dB above isotropic; the main lobe beamwidth is about 32°.

The relatively large component along the ground plane, about −10 dB, appears to be from the feed and load segments acting as vertical radiators spaced ¼ λ apart. It would appear that this component could be reduced by restoring most of the open arm, bringing the feed and load segments close together, to act as a transmission line rather than a pair of radiators.

Fig 4 shows the drive resistance and reactance for frequencies near half-wave resonance of the long arm. Matching to a 50-Ω line can be done with an open stub to cancel reactance plus a 3:1 balun.

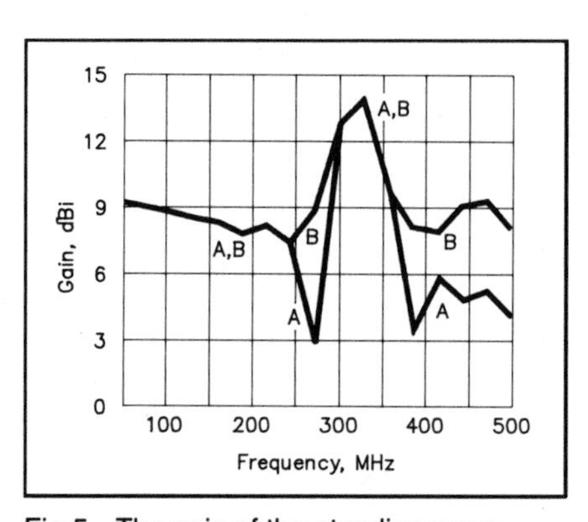

Fig 5—The gain of the standing-wave cross versus frequency. Curve A shows the gain normal to the ground plane, and curve B the gain for the largest lobe.

## SWC Type

The gain of the standing-wave cross is shown in Fig 5 for the range 75 to 475 MHz. Two values are shown, A being the gain normal to the ground plane, and B being the gain for the largest lobe. Maximum gain occurs when the long side is ½ λ, and is over 12.7 dBi, with a beamwidth of about 42°.

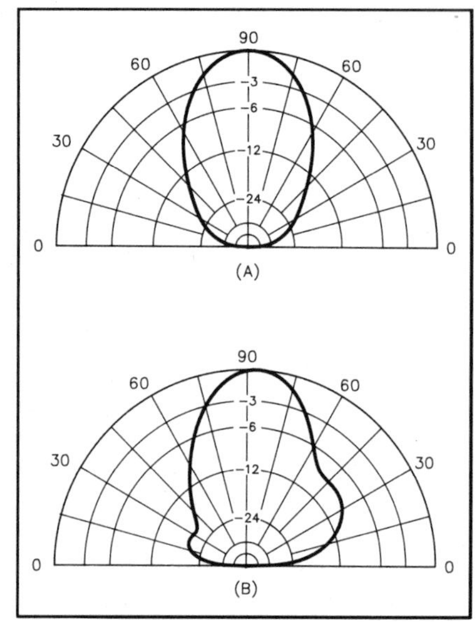

Fig 6—Radiation patterns for the 4-cross SWC antenna. A is the pattern in the vertical plane containing the open arm, and B for the plane at right angles to this.

At low frequencies the antenna is acting as a quad loop. As the frequency increases, the lobe normal to the loop starts to split; however, the length-shape relations cancel this splitting around half-wave long-side resonance. At still higher frequencies the lobe splitting again occurs.

The pattern at the maximum gain point is shown by Fig 6A for the plane through the open side, and 6B for the plane at right angles to this. The feed point for these patterns is at the junction of a long and short side. This nonsymmetry appears in the difference between the two patterns, and in the 90° pattern.

The drive impedance is shown in Fig 7. Below 150 MHz, the drive resistance is low, as in a quad well below resonance. Above about 275 MHz there is no true resonance, the reactance remaining low. This is also similar to a quad, which shows only the first resonance if the conductor diameter is large.

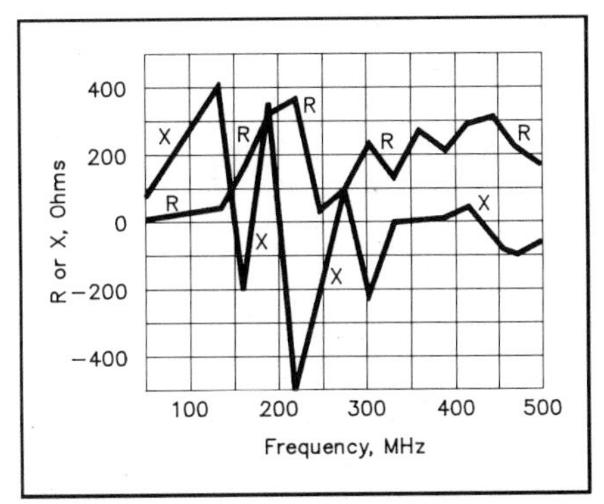


Fig 7—Feed-point resistance and reactance of the 4-cross standing-wave antenna versus frequency.

## Other Bands

If all dimensions are scaled by the ratio 300/new frequency, all of the performance data shown here applies to the new frequency. For example, the long side of a 900-MHz 4-cross design should be ⅓ meter, the short side ⅙ meter and so on. The frequency of maximum gain would be 924 MHz.

This scaling also applies to the conductor diameter. If it is not convenient to use the scaled size, another size may be used, but the frequency of maximum gain would shift somewhat. A MININEC analysis would give the shift.

## Further Study

Several of the unique characteristics of the cross antenna merit further study and experimentation. Circular and dual polarizations are used by satellite systems and broadcast stations to reduce fading. There are possibilities for meter-band DXing, for ATV and for OSCAR work.

Broadband antennas are always useful, and extending the already large frequency range of the cross would be worthwhile. Some of the possibilities are multiple-turn designs, perhaps using log periodic principles as suggested in the IEEE paper.

Optimum feed needs to be investigated. For example, would the addition of the fourth short side to the TWC type improve the pattern? And, since high gain is always useful, the best configuration for an array of 2 or 4 crosses should be worked out. A quick estimate indicates that an array gain approaching 20 dBi should be possible.

## Note

[1] A. G. Roederer, "The Cross Antenna: A New Low-Profile Circularly Polarized Radiator," *IEEE Trans on Antennas and Prop*, May 1990, pp 704-710.

# The Skeleton Discone

By D. Wilson Cooke, WA4RHT
PO Box 203
Tigerville, SC 29688

Do you need a coax-fed, broadband antenna that works on several amateur bands with little or no adjustment? This article describes experiments in which a discone antenna was stripped to a simple skeleton. These experiments suggest that a skeleton of a discone antenna can be constructed of wire, rod or tubing. It will work over several amateur bands without adjustment, and will maintain an SWR in a 50-Ω line of 2:1 or less. Sufficient data is given so that the element lengths can be calculated for any reasonable antenna frequency. Step-by-step construction details, however, are not provided.

The discone antenna was introduced in the mid '40s[1] and has been described infrequently in the amateur literature. Unlike dipoles, verticals, quads, Yagis and other popular antennas, the discone is not as well known or as widely used. The 15th (1988) edition of *The ARRL Antenna Book* contains a section on discone antennas for both HF and VHF use.[2] This material is recommended reading for anyone wishing further information on this unusual antenna.

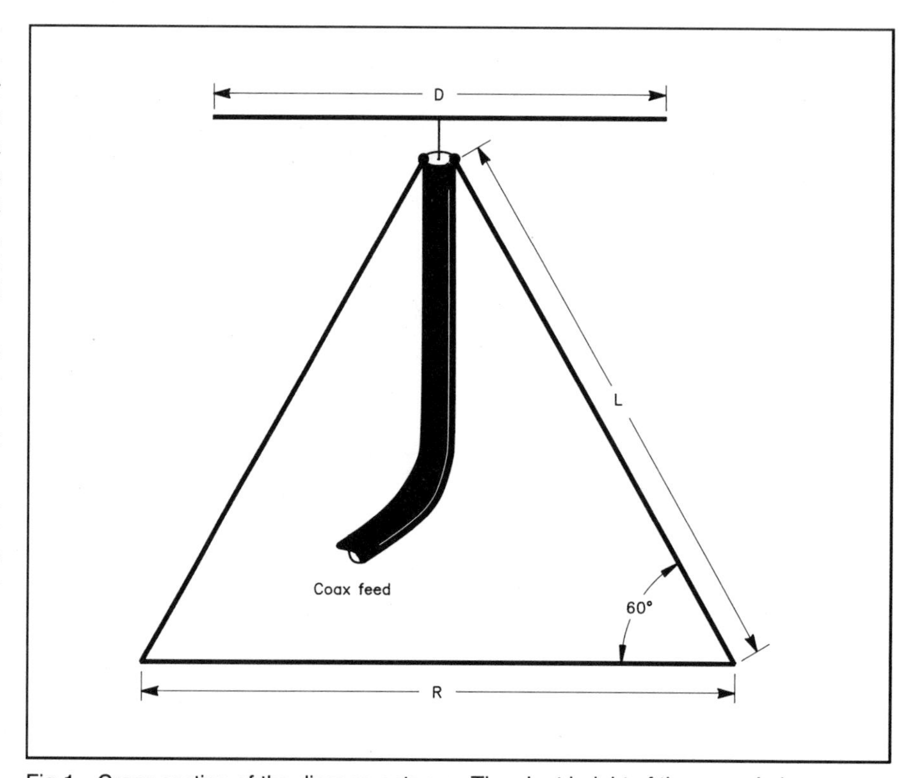


Fig 1—Cross section of the discone antenna. The slant height of the cone, L, is a quarter wavelength at the lowest operating frequency. The diameter of the disk, D, is 67-70% of L. The shield of the coax feed is attached to the top of the cone and the center conductor is attached to the center of the disk.

Fig 1 shows a cross section of the discone antenna. The bottom of the antenna consists of a metal cone whose slant height, L, is equal to the diameter of the base of the cone, R. Above the cone is a metal disk with a diameter (D) equivalent to about 67% to 70% of the slant height of the cone. The coax feed-line shield is attached to the top of the cone while the center conductor passes through a small hole in the top of the cone and is attached to the disk.

The slant height, L, is a free-space quarter wave ($246/f_{feet}$ or $75/f_{meters}$, where f is the frequency in MHz) at the lowest frequency for which the antenna is designed. Below this frequency the SWR increases rapidly. If the hole at the top of the cone is small with respect to the slant height, the antenna will provide a good match to 50-Ω coax over a frequency range of up to ten times the low-frequency cutoff.

As far as the feed line is concerned, the discone appears as a properly terminated high-pass filter. Radiation is vertically polarized at low angles over the lower frequencies. As the upper frequency limit of the antenna is approached, the radiation angle turns upward. Common vertical antennas like the groundplane and grounded vertical have the high current portion of the antenna at the bottom. The discone has the high current portion of the antenna at the top. The discone requires no ground or ground screen for its operation.

The size of a VHF discone is small enough to make construction and mounting of the antenna fairly easy. For example, an antenna with a low-frequency cutoff of 144 MHz might have a cone slant height of 20 inches and a disk diameter of 14 inches. In fact, a carefully constructed discone antenna might work through the 1.2-GHz band! It could be made of sheet aluminum or other metals.

A discone for HF is more difficult to construct. At 14 MHz the cone would require a slant height of at least 18 feet and a disk diameter of 12 feet. The construction and wind loading of such a discone would make the antenna impractical for most amateur applications.

Several skeletons of the discone have been made in which the disk and cone are simulated by wires or rods. Radio Shack and Procomm market similar VHF skeleton discones. The disk is replaced by eight horizontal spokes about one foot in length, giving a "disk" diameter of two feet. The cone is simulated by eight rods tilted down at a 60° angle similar to drooping radials. These are about three feet in length. The low-frequency cutoff of skeleton discone antennas is approximately 100 MHz. Above this frequency the SWR is usually below 1.5:1 well into the UHF region. Below 100 MHz, however, the SWR rises rapidly.

*The ARRL Antenna Book* describes a high-frequency skeleton discone with a low-frequency cutoff of 7 MHz. The "disk" consists of eight horizontal aluminum spokes and the "cone" is simulated by 24 Copperweld wires. (The wires also act as guys for the tower that supports the structure.) The tips of both the disk and cone spokes are connected by a wire loop. The *Radio Handbook* describes a similar antenna except that the cone is simulated by 48 wires.[3]

I became curious about the performance differences between various skeleton discones and full sheet-metal discones. I also wanted to know the minimum number of spokes in the skeleton that would yield acceptable performance.

Fig 2—One of the skeleton discones discussed in the text. Eight spokes simulate the cone and four simulate the disk.

### The Skeleton Experiments

It is relatively simple to model an antenna at VHF and then scale results to lower frequencies. I chose the middle VHF region as the principal frequency for investigation because of equipment availability.

Several skeleton discones were constructed using different numbers of spokes to simulate the disk and the cone. I selected a "cone" slant height (spoke length) of 38 inches and a "disk" diameter of 25.5 inches. These dimensions were selected because they were small enough to manage easily in testing, and yet large enough that cutting and measuring errors would account for a small percentage of the overall size of the antenna. I estimated that the low cutoff frequency point would be between the 6-meter and 2-meter amateur bands.

The foundation for the antenna was an aluminum funnel with a BNC connector mounted at the small end to serve as the junction between the disk and cone portions of the antenna. A machine screw was soldered to the center conductor of the BNC connector and it projected out of the top of the funnel. The spokes simulating the cone could then be quickly attached to the funnel and various disk-simulating spokes could also be attached to the machine screw. This arrangement would not be suitable for a permanent antenna, but it proved adequate for a temporary series of experiments. The experimental antennas were all connected to a 20-foot length of RG-58 coax leading to the measurement equipment.

Before discussing the actual measurements, a word should be said about measurement equipment. Like many amateurs, I do not have access to laboratory-grade equipment. Some of my measurement equipment is home constructed and calibrated, and some of it is low cost commercial equipment. I used a dip meter as a radio-frequency oscillator. It was coupled to a counter and SWR meter through a loop on a short length of coax. The output at the SWR meter was quite small. Consequently, the reflected power level was often in the range of a few millivolts. This means that the diodes in the SWR meter were probably operating in a nonlinear fashion. This would give rise to errors in SWR measurement.

Since all measurements were taken at approximately the same power levels while using the same equipment, the *relative* comparisons should be valid even though the absolute SWR measurements may somewhat inaccurate. All measurements were made in the yard with the antenna mounted on a wooden pole and the disk about seven feet above the ground. Several hundred measurements were made over a period of two weeks on a variety of different antenna configurations. I measured the SWR at 10-MHz intervals from 70 MHz to 320 MHz (the upper frequency limit of my dip meter). I tried configurations of eight, six, four and three spokes in various combinations. The results of these measurements are summarized in Table 1. Many other measurements were made that are not reported, but they were consistent with those in Table 1 and would contribute little additional information.

While a sheet-metal discone has a low-frequency cutoff when the slant height of the cone is approximately a quarter wave, the skeleton discone's low-frequency cutoff is significantly higher. The slant height of 38 inches used in these experiments should have given a cutoff at 77 MHz in a traditional discone, but the SWR was well above 3:1 at this frequency for most skeleton designs I attempted. It did not drop below 2:1 until the frequency reached 100 to 110 MHz (depending on the antenna spoke configurations). This represents a slant height close to *one-third* wavelength for the low-frequency cutoff.

The first configuration I tried was a copy of the commercial Radio Shack and Procomm discones. Eight spokes were used in the cone and the disk. The tips of neither the disk nor the cone were connected. The SWR dropped below 2:1 just above 100 MHz and remained low up to 320 MHz (column A of Table 1). At some points the SWR was below 1.1:1 for several MHz and was below 1.5:1 over most of the range. It did rise slightly at certain frequencies with a maximum SWR of 2.1:1 at 310 MHz. This leads me to conclude that this design would provide a perfectly acceptable antenna for scanner use, as well as for amateur transmissions on the 144, 222, and probably the 420-MHz bands. (No measurements were made on the latter band.)

I substituted a four-spoke disk at the top of the antenna while leaving the eight-spoke cone unmodified. The SWR measurements are reported in column B of Table 1. The SWR profile is slightly different, being higher on some frequencies and lower on others. Even so, this configuration would be as acceptable as the eight-spoke disk.

Column C reports the SWR profile of an antenna with a four-spoke disk and a six-spoke cone. The SWR is generally higher, though still below 2:1 throughout most of the range above 110 MHz. Three peaks are reported in the table where the SWR rose above 2:1. None of these are within amateur bands.

**Table 1**
**SWR Profiles of Several Skeleton Discones Discussed in the Text**

| *Frequency (MHz)* | *Antenna SWR* A | B | C | D | E |
|---|---|---|---|---|---|
| 70 | 4.9 | 4.8 | 5.8 | 4.8 | 5.7 |
| 80 | 2.7 | 3.2 | 4.5 | 4.1 | 4.2 |
| 90 | 2.0 | 3.4 | 4.4 | 4.4 | 3.7 |
| 100 | 1.7 | 2.7 | 6.6 | 4.1 | 4.2 |
| 110 | 1.7 | 1.7 | 1.9 | 1.7 | 1.9 |
| 120 | 1.0 | 1.3 | 2.6 | 1.2 | 1.5 |
| 130 | 1.1 | 1.3 | 1.3 | 1.4 | 1.5 |
| 140 | 1.0 | 1.1 | 1.1 | 1.8 | 1.2 |
| 145 | 1.2 | 1.3 | 1.5 | 2.1 | 1.5 |
| 150 | 1.4 | 1.3 | 1.5 | 2.0 | 1.5 |
| 160 | 1.2 | 1.3 | 1.4 | 2.2 | 1.5 |
| 170 | 1.1 | 1.2 | 1.3 | 1.7 | 1.3 |
| 180 | 1.1 | 1.2 | 1.8 | 3.2 | 2.0 |
| 190 | 1.2 | 1.4 | 1.4 | 2.1 | 1.3 |
| 200 | 1.2 | 1.2 | 1.5 | 2.8 | 1.6 |
| 210 | 1.6 | 1.6 | 2.0 | 3.0 | 2.0 |
| 220 | 1.1 | 1.0 | 1.1 | 1.6 | 2.1 |
| 225 | 1.5 | 1.6 | 1.5 | 3.7 | 1.6 |
| 230 | 1.1 | 1.1 | 1.2 | 2.7 | 1.1 |
| 240 | 1.4 | 1.5 | 1.2 | 2.5 | 1.3 |
| 250 | 1.2 | 1.9 | 1.1 | 1.2 | 1.8 |
| 260 | 1.5 | 1.4 | 1.2 | 1.2 | 1.4 |
| 270 | 1.2 | 1.4 | 1.2 | 1.2 | 1.4 |
| 280 | 2.0 | 2.2 | 2.5 | 2.1 | 2.6 |
| 290 | 1.6 | 1.8 | 1.8 | 2.1 | 1.8 |
| 300 | 1.3 | 1.5 | 1.4 | 1.4 | 1.7 |
| 310 | 2.1 | 1.9 | 2.4 | 2.2 | 2.4 |
| 320 | 1.4 | 1.6 | 1.6 | 1.7 | 1.3 |

A—Skeleton with eight spokes in the cone and the disk.

B—Skeleton with eight spokes in the cone and four in the disk.

C—Skeleton with six spokes in the cone and four in the disk.

D—Skeleton with four spokes in the cone and the disk.

E—Skeleton with six spokes in the cone and three in the disk.

Column D reports the SWR profile of an antenna with four spokes in the disk and in the cone. The SWR peaks are in the wrong places for the 144- and 222-MHz bands, though they are acceptable for many applications. When using my 10-watt, 144-MHz transmitter, the SWR was high enough to activate the SWR shutdown feature. (The transmitter output dropped about 15%.) If I were to build a permanent version of this antenna, I would change the lengths of the spokes slightly to move the high SWR portions of the antenna outside the amateur bands of interest. I believe a similar antenna could be constructed for general multiband use.

Column E reports the SWR profile on an antenna with three spokes in the disk and six spokes in the cone. This antenna would also be useful over the same frequency range.

Both *The ARRL Antenna Book* and the *Radio Handbook* describe HF skeleton discones in which the tips of the disk and cone spokes are connected by conducting wire. In order to determine what effect this would have on the antenna, I constructed a version of the discone with eight spokes in both the cone and the disk. All spoke tips were connected by circular wire rings.

This proved to be a mixed blessing. I did not run a complete SWR profile, but performed spot checks on several frequencies instead. The low-frequency cutoff dropped to about 77 MHz, but the SWR profile was generally higher. By utilizing the connecting rings, a physically smaller antenna could be made for the same low-frequency cutoff point. However, this approach may cause a slightly higher SWR at some frequencies and additional construction problems.

I tried a four-spoke cone/three-spoke disk version with circular rings connecting the spoke tips. It was totally unacceptable. The SWR throughout most of its range was between 2:1 and 3:1 with occasional dips and peaks.

Armed with the information outlined above, I constructed an HF version of the skeleton discone. I used four spokes, each six feet long, for the disk. I used four wires, each eighteen feet long, to simulate the cone. The top of the antenna was mounted on a push-up mast at about twenty feet. The wires were anchored to the ground with strings at an angle of about 60 degrees. The antenna was fed with 60 feet of RG8X coax. The SWR profile of this antenna is given in Table 2.

**Table 2**
**SWR Profile of the HF Skeleton Discone with Four Spokes in the Disk and Cone**

| *Frequency (MHz)* | *Antenna SWR* |
|---|---|
| 18.07 | 2.5 |
| 18.16 | 2.4 |
| 21.01 | 1.4 |
| 21.1 | 1.3 |
| 21.2 | 1.3 |
| 21.3 | 1.3 |
| 21.4 | 1.3 |
| 24.93 | 2.4 |
| 28.1 | 2.3 |
| 28.3 | 2.2 |
| 28.5 | 2.2 |
| 28.7 | 2.3 |
| 28.9 | 2.3 |
| 29.1 | 2.3 |
| 29.3 | 2.3 |
| 29.5 | 2.3 |
| 29.7 | 2.3 |
| 144 | 1.7 |
| 145 | 1.8 |
| 146 | 1.4 |
| 147 | 1.2 |
| 148 | 1.2 |

The SWR is certainly higher than desired in some bands. It does not, however, make the antenna totally unusable. For testing purposes I used a Kenwood TS-830S transceiver which features 6146 tubes as final amplifiers. It loaded easily on all HF amateur bands from 18 MHz and up without a tuner. The antenna also worked on 2 meters with both a mobile and a hand-held rig. I suspect that a solid-state HF rig would require a tuner to avoid SWR shutdown.

SWR alone is not a measure of good antenna performance. After all, my dummy load has an excellent SWR curve well into the VHF region, but it is a lousy antenna! Some verticals have a good SWR only because of significant ground loses. In order to determine how well the VHF version of the skeleton discone antenna performed on transmission and reception, it was compared with a more traditional discone antenna.

A traditional discone was constructed using hardware cloth covered with aluminum foil. The slant height was about 22.5 inches giving a low-frequency cutoff well below 144 MHz. All of my tests indicated that it performed as expected.

With my traditional discone as a reference, I made field-strength measurements on each of the five skeleton models outlined in Table 1. At a distance of twenty feet (roughly 3 λ) there was no measurable difference in radiated field strength between the skeleton discones and the traditional discone on 2 meters. Two distant repeaters with marginal signals at my QTH were used for reception testing. I found no measurable difference in received signal strength between the traditional discone and the skeleton discones. All of these measurements were made at the same location in my yard with the disks approximately seven feet above the ground.

The HF version of the skeleton discone could not be conveniently measured against a similar solid discone. I could only compare it with more conventional antennas. Obviously, my triband beam at 41 feet outperformed the discone. In its favored direction the beam is about two to four S units better. Off the sides and back of the beam, however, the discone is frequently better. Marginal signals are much easier to copy with the beam. This is to be expected since the beam is twice as high and has substantially greater gain.

I desired a fairer comparison for the HF skeleton discone so I constructed some temporary ¼-λ verticals with ground planes about six feet above the ground. Both field-strength and reception measurements indicated that the skeleton discone was as good as the verticals—and often better. Any signal that was readable on one antenna was also readable on the other. When monitoring signals from the United States, South America and Europe, I noticed that the skeleton discone had an advantage of one to two S units on 15 meters. Fewer signals were heard on 17 meters, but the results were similar. On 12 and 10 meters not many signals were available when I was checking the antennas. With the few signals that I did monitor, the discone performed as well as the verticals. This included one signal from the Azores on 10 meters.

I measured transmitted field strength from the verticals and the skeleton discone at a distance of 175 feet. The meter had a vertically polarized pickup antenna about 6 feet long. On 17, 15 and 12 meters the skeleton discone had a 2- or 3-dB advantage. On 10 meters the vertical-antenna signal strength was about 1 dB stronger.

## Conclusions and Suggestions

I arrived at several conclusions as a result of these experiments:

1) Experiments can be conducted by amateurs with simple and inexpensive instruments and still achieve meaningful results.

2) The skeleton discone is a practical antenna that could serve as a general-purpose, wide-band antenna. It certainly will not perform as well as a single-band gain antenna adjusted for optimum performance, but it should be quite useful for many amateur applications. Element lengths are not critical if the builder makes them sufficiently long while maintaining the disk diameter at about 67-70% of the total cone spoke length.

3) As the number of spokes in the skeleton is decreased, the SWR curve has more variations. This indicates that one should use as many spokes as possible. An HF antenna with just a few spokes should work, however. The number of spokes in the cone appears to be more important than the number in the disk. A simple Transmatch could reduce the SWR so that even the most sensitive solid-state rigs would be usable. The additional line loss that is present with an SWR in the 2:1 range is not generally a problem at HF if a very simple skeleton is desired. For VHF I would recommend a version with eight or more spokes to obtain a flatter SWR curve.

4) A skeleton discone that does not utilize connecting rings at the tips of the spokes has a low-frequency cutoff point where the length of the cone spokes approaches one-third wavelength. When designing a skeleton discone, a length slightly longer than one-third wavelength ($330/f_{feet}$ or $100/f_{meters}$) at the lowest operating frequency should be chosen for the cone spokes. The length of the disk spokes (measured from center to tip) should be about 0.34 times the length of the cone spokes. This will give a disk diameter of just over two thirds the slant height of the cone.

5) A skeleton discone constructed with connecting rings requires more spokes for acceptable performance. A skeleton discone with eight or more connected spokes has a low-frequency cutoff point where the spoke length of the wire simulating the cone is about a quarter wavelength ($246/f_{feet}$ or $75/f_{meters}$). Therefore, the length of the elements for a given frequency are shorter, but *more* elements are required.

6) Because there are peaks and valleys in the SWR curve of a skeleton discone, it should be possible to design an antenna with low SWR points within the amateur bands. This would require some cut-and-try testing, but could produce an excellent antenna for general use.

The author would like to hear from others who have experimented with this sort of antenna. An SASE would be appreciated if a reply is desired.

### Notes

[1]G. Kandoian, "Three New Antenna Types and Their Applications," *Proc IRE*, Volume 34, Feb 1946, pp 70w-75w.

[2]*The ARRL Antenna Book*, 15th edition (The American Radio Relay League, 1988), pp 9-7 through 9-12.

[3]W. I. Orr, *Radio Handbook*, 22nd edition (Howard W. Sams & Company, Inc, 1981), pp 27.24-27.26.

# The T-L DX Antenna

By Robert Wilson, AL7KK
Box 110955
Anchorage, AK 99511

I had an opportunity to operate out in the bush on short notice and I needed a DX antenna that was cheap and easy to build. It had to be theoretically sound and capable of being assembled in one afternoon. It also had to fit between two spruce trees at a reasonable height. The result was the "T-Lambda" antenna (see Fig 1). It looks like the letter T standing on top of the Greek letter lambda (Λ). I simply call it the "T-L."

My experience as an advance planner for a large shortwave broadcast network showed me that only the lower 15° of an antenna pattern is valuable for DX. The easy way to get low angle radiation is with a half-wave vertical, but there are a number of practical problems with this approach. First of all, it is difficult to hoist a wire high enough for an effective half wave vertical. In addition, feeding a tall vertical is a problem because the antenna field is often disturbed by the coax. This makes matching a real headache. Taking these factors into consideration, I drew upon my commercial shortwave background for some unique ideas to overcome these problems.

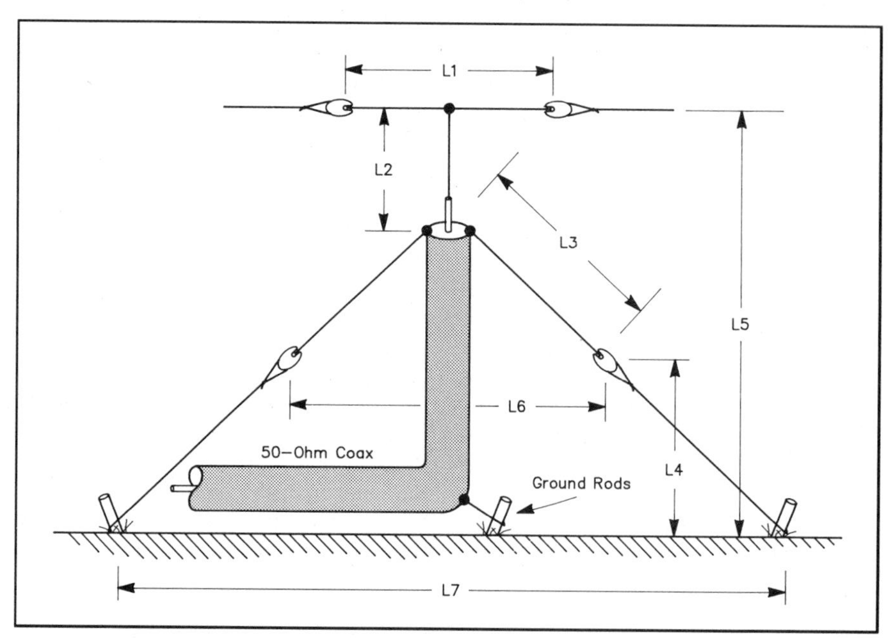

Fig 1—The complete T-L DX antenna design.

### The T-L Antenna Design

The T-L was designed to directly match a 50-ohm coax. Also, I wanted to string it like a horizontal dipole between two trees while still retaining a vertical radiation pattern. I configured the upper half like an old-fashioned top-loaded T antenna in an effort to shorten the height. (T antennas have been used for more than 70 years, but not this way!)

The conventional T vertical needs numerous ground wires, but this is not the case with the T-L. It has only two quarter-wave wires above the ground with an apex angle of 90°. Remarkable as it may seem, only *two* wires were required to give a good pattern. This is highly desirable where dry soil, sand or permafrost offer poor ground conductivity. The elevated T-L configuration places the radiation center higher above the ground and improves the low-angle radiation pattern characteristics.

If one were to slice through a typical discone antenna, it would look much like a T-L. The vertical leg of the T is longer than you might expect, but the additional length is required to match the 50-ohm line. One similarity to the discone is that the T-L is fairly broadbanded. While it is not quite as broad as a discone, it offers a bandwidth superior to most other antennas. This is an important consideration when attempting to use solid-state transceivers that automatically reduce power output in the presence of a high SWR.

The following formulas are required to calculate the dimensions of a T-L antenna. Please note that the bottom should be mounted at a minimum distance above ground to give a proper 50-ohm feed impedance at the desired frequency. Raising the antenna does not present a serious problem, but the pattern may suffer when the feed point is more than about ⅜ λ above ground. Only a four-function calculator is required to determine all dimensions in either feet or meters, for any frequency.

Table 1 shows the dimensions for the amateur bands in feet. Table 2 shows the dimensions for general purpose SWL antennas to cover the entire high frequency spectrum. For shortwave broadcast reception the T-L antenna may be optimized to the band of choice, but it will also receive very well on a much wider range of frequencies.

The horizontal bar on top of the T is L1. This is the wire length from one insulator to the opposite insulator. Values can be calculated for either feet or meters.

**Table 1**
**T-L DX Antenna Data for the Amateur Bands**

| *Band* | *Freq.* | *L1 Top Bar* | *L2 Vert Bar* | *L3 V Leg* | *L4 Bot. Ht.* | *L5 Top Ht.* | *L6 Width* | *Bandwidth, MHz* |
|---|---|---|---|---|---|---|---|---|
| 160 | 1.9 | 91.8 | 68.7 | 115.0 | 36.8 | 187.4 | 162.6 | 1.7–2.1 |
| 80 | 3.7 | 47.2 | 35.3 | 59.1 | 18.9 | 96.2 | 83.5 | 3.3–4.1 |
| 75 | 3.9 | 44.7 | 33.5 | 56.0 | 17.9 | 91.3 | 79.2 | 3.5–4.3 |
| 40 | 7.2 | 24.4 | 18.3 | 30.6 | 9.8 | 49.8 | 43.2 | 6.4–7.9 |
| 30 | 10.1 | 17.2 | 12.9 | 21.6 | 6.9 | 35.2 | 30.5 | 9.1–11.1 |
| 20 | 14.2 | 12.3 | 9.2 | 15.4 | 4.9 | 25.1 | 21.8 | 12.8–15.6 |
| 17 | 18.1 | 9.6 | 7.2 | 12.1 | 3.9 | 19.7 | 17.1 | 16.3–19.9 |
| 15 | 21.2 | 8.2 | 6.1 | 10.3 | 3.3 | 16.8 | 14.6 | 19.1–23.3 |
| 12 | 24.9 | 7.0 | 5.2 | 8.8 | 2.8 | 14.3 | 12.4 | 22.4–27.4 |
| 10 | 28.5 | 6.1 | 4.6 | 7.7 | 2.5 | 12.5 | 10.8 | 25.7–31.4 |
| 6 | 50.1 | 3.5 | 2.6 | 4.4 | 1.4 | 7.1 | 6.2 | 45.1–55.1 |
| 2 | 146.0 | 1.2 | 0.9 | 1.5 | 0.5 | 2.4 | 2.1 | 131.4–160.6 |

Note: All dimensions are in feet.

$$L1 = \frac{174.5}{f_{MHz}} \text{ feet}$$

$$L1 = \frac{53.2}{f_{MHz}} \text{ meters}$$

The vertical part of the T is L2. This runs from the center of L1 down to the top of an insulator where it is connected to the center wire of the 50-ohm coax.

$$L2 = \frac{130.5}{f_{MHz}} \text{ feet}$$

$$L2 = \frac{39.8}{f_{MHz}} \text{ meters}$$

Each half of the inverted V ground system is calculated by L3. These wires run from the lower part of the center insulator to two insulators located near the ground. The apex is connected to the shield of the 50-ohm coax. The coax must run vertically up the center line of the antenna from the ground. The angle of the apex is 90°.

$$L3 = \frac{218.5}{f_{MHz}} \text{ feet}$$

$$L3 = \frac{66.6}{f_{MHz}} \text{ meters}$$

A 10-meter T-L mounted on the roof of the author's backyard shed.

**Table 2**
**T-L DX Antenna Data for Selected SWL Bands**

| *Band* | *Freq.* | *L1 Top Bar* | *L2 Vert Bar* | *L3 V Leg* | *L4 Bot. Ht.* | *L5 Top Ht.* | *L6 Width* | *Bandwidth, MHz* |
|---|---|---|---|---|---|---|---|---|
| 1 | 1.0 | 174.5 | 130.5 | 218.5 | 69.9 | 356.0 | 309.0 | 0.9–1.1 |
| 2 | 1.2 | 145.4 | 108.7 | 182.1 | 58.3 | 296.7 | 257.5 | 1.1–1.3 |
| 3 | 1.4 | 121.2 | 90.6 | 151.7 | 48.5 | 247.2 | 214.6 | 1.3–1.6 |
| 4 | 1.7 | 100.9 | 75.4 | 126.3 | 40.4 | 205.8 | 178.6 | 1.6–1.9 |
| 5 | 2.1 | 84.3 | 63.0 | 105.6 | 33.8 | 172.0 | 149.3 | 1.9–2.3 |
| 6 | 2.5 | 70.1 | 52.4 | 87.8 | 28.1 | 143.0 | 124.1 | 2.2–2.7 |
| 7 | 3.0 | 58.4 | 43.6 | 73.1 | 23.4 | 119.1 | 103.3 | 2.7–3.3 |
| 8 | 3.6 | 48.7 | 36.5 | 61.0 | 19.5 | 99.4 | 86.3 | 3.2–3.9 |
| 9 | 4.3 | 40.6 | 30.3 | 50.8 | 16.3 | 82.8 | 71.9 | 3.9–4.7 |
| 10 | 5.2 | 33.8 | 25.3 | 42.3 | 13.5 | 69.0 | 59.9 | 4.6–5.7 |
| 11 | 6.2 | 28.2 | 21.1 | 35.3 | 11.3 | 57.5 | 49.9 | 5.6–6.8 |
| 12 | 7.4 | 23.5 | 17.6 | 29.4 | 9.4 | 47.9 | 41.6 | 6.7–8.2 |
| 13 | 8.9 | 19.6 | 14.6 | 24.5 | 7.8 | 39.9 | 34.6 | 8.0–9.8 |
| 14 | 10.7 | 16.3 | 12.2 | 20.4 | 6.5 | 33.3 | 28.9 | 9.6–11.8 |
| 15 | 12.8 | 13.6 | 10.2 | 17.0 | 5.4 | 27.7 | 24.1 | 11.6–14.1 |
| 16 | 15.4 | 11.3 | 8.5 | 14.2 | 4.5 | 23.1 | 20.1 | 13.9–17.0 |
| 17 | 18.5 | 9.4 | 7.1 | 11.8 | 3.8 | 19.3 | 16.7 | 16.6–20.3 |
| 18 | 22.2 | 7.9 | 5.9 | 9.8 | 3.2 | 16.0 | 13.9 | 20.0–24.4 |
| 19 | 26.6 | 6.6 | 4.9 | 8.2 | 2.6 | 13.4 | 11.6 | 24.0–29.3 |
| 20 | 32.0 | 5.5 | 4.1 | 6.8 | 2.2 | 11.1 | 9.7 | 28.8–35.1 |

Note: All dimensions are in feet.

The minimum height of the lower insulators above ground is L4. This value is not very critical, but to get optimum results over soils with high conductivity it is best to get reasonably close.

$$L4 = \frac{69.9}{f_{MHz}} \text{ feet}$$

$$L4 = \frac{21.3}{f_{MHz}} \text{ meters}$$

The height of the top wire above ground is given by L5. Supporting poles or trees should be at this height or perhaps just a bit higher, particularly if the wire sags excessively.

$$L5 = \frac{356.0}{f_{MHz}} \text{ feet}$$

$$L5 = \frac{108.4}{f_{MHz}} \text{ meters}$$

Bottom insulator to bottom insulator distance will be L6. This is the dimension that makes the inverted V a 90° angle.

$$L6 = \frac{309.0}{f_{MHz}} \text{ feet}$$

$$L6 = \frac{94.2}{f_{MHz}} \text{ meters}$$

Ground anchors will be needed and their separation distance can be estimated as L7. The actual antenna height above ground can influence this variable, so use it only as a reference point.

$$L7 = \frac{448.7}{f_{MHz}} \text{ feet}$$

$$L7 = \frac{136.8}{f_{MHz}} \text{ meters}$$

## Construction

When constructing the antenna it is a good idea to solder *every* joint. I also prefer to use Copperweld wire because it resists stretching. Make sure that there are no projecting wire ends on the antenna. They will promote corona discharge and will greatly reduce efficiency.

Ceramic strain insulators, also known as egg insulators, will work well in the T-L design. You will need at least five. Very small egg insulators may require two in series to get a sufficient voltage rating for rainy day operating. Electric power line strain insulators (known as number 502 insulators) are good up to about 1 kW.

My first 10-meter T-L antenna was built with copper pipe and standard plumbing fittings. All copper-to-copper joints were cleaned with crocus cloth and then soldered with hard tin-silver solder using a small gas torch. The center insulator was a piece of plastic pipe with an inner diameter that accommodated the outside diameter of the copper pipe. Stainless steel sheet metal screws held the copper and plastic pipes together so they would not rotate. Finally, I sealed any potentially leaky joints with silicone glue to waterproof the finished product.

A wire version of the antenna was also built for 15 meters and the entire assembly process was completed within a single hour. The cost of the wire version was essentially nothing since I used some wire left over from other antenna projects.

The 10- and 15-meter T-Ls were pulled into big spruce trees with nylon parachute cord. The results were excellent with superb DX performance. I also managed to obtain exact matches to their 50-ohm coaxial feed lines. I was even able to get my transmitter to load on the adjacent ham bands with little difficulty.

### Installation Tips

When installing the T-L try to keep it away from other metal objects. Mounting the antenna near metal pipes, power lines or telephone lines will cause the impedance and tuning to suffer. (Remember that there are pipes and wires inside most houses.) Roof mounting seems satisfactory, but only your own SWR checks will tell. If the SWR is 1.3:1 or better, it's fine. Try using L5 as a rule-of-thumb distance for spacing from large metal objects.

Lightning protection will be enhanced by using copper-coated ground rods as counterpoise anchors. Bring the coax directly down from the antenna if possible and ground the shield with an eight-foot ground rod. Bury the coax or at least run it close to the surface of the ground between the antenna and your shack. Just before it enters the building, make a lightning coil of about 20 turns of coax (approximately one foot in diameter and one foot long) and bury it in the ground. These techniques will tend to force the lightning to dissipate in the ground system rather than in your receiver or transmitter!

# A Beginner's Guide to Using Computer Antenna Modeling Programs

By L. B. Cebik, W4RNL
1434 High Mesa Dr
Knoxville, TN 37938-4443

When MININEC became available to antenna experimenters as an antenna modeling computer program, they absorbed it with relish. It saved them hours, if not days, of futile construction effort on designs that would not improve performance. Now the program is available in at least three versions for the IBM PC and compatible computers (MININEC3, MN, and ELNEC[1]) to the average ham at reasonable costs. Depending upon the version, it will run on PCs with or without math coprocessors. Two versions (MN and ELNEC) produce excellent screen graphics of antenna patterns, along with documentation of the design, source impedance factors, and current distribution.

Are these programs really useful to the beginning and moderately experienced ham? The answer is a resounding YES! When used within their limitations, these programs can go far beyond textbooks in teaching us why our antennas act the way they do.[2] They can also help us make better decisions on what antennas to build or buy and how to mount them. However, after a brief period of sampling the test designs included with the program, the antenna modeling program may end up in a disk file box. The reason for discarding these valuable programs is that most of us fail to understand all they can tell us, and that—in turn—is because we do not set up procedures to squeeze meaningful information out of the program.

The purpose of this article is to show the beginner how to start using an antenna modeling program effectively. A good beginning requires three areas of effort: (1) setting up certain program basics, (2) setting up consistent modeling conventions, and (3) developing a baseline of information about basic antennas located on one's own property. The first step permits us to focus on and master the essentials of the program, saving advanced features for later. Step 2 allows us to model accurately and confidently, with minimal error in comparing one design with another. Step 3 allows us to interpret intelligently the patterns that emerge from new designs we try. Once we have mastered both the program and what it can tell us, we can expand our knowledge by using its advanced features.

The suggestions presented here are no substitute for mastering the instructions that come with the program; instead, they are designed to supplement those instructions. The aim here is to make the program and its procedures as useful and instructive as possible, even for the beginning antenna modeler. The program I use is ELNEC, but the steps suggested here can be translated for use with nearly any MININEC-based program.

### Setting Up the Program

Advanced users of ELNEC and similar programs require considerable flexibility, so the programs offer many options to the user. Often these options inhibit the beginning student of antennas by offering choices among which the user cannot decide. Therefore, the first step in getting the most out of the antenna modeling program is to make decisions, even if they are initially made for weak reasons. For the new user, convenience may be the best reason available.

ELNEC offers a menu with many options, only a few of which we need for the purpose of getting used to the program. The "Wires" entry is for describing the antenna we shall model. The "Sources" entry is also crucial to each model, telling the program where the antenna is fed. Most of the basic antennas we shall start with—dipoles, verticals, Yagis, and the like—use only one source. Until we start modeling trap dipoles and beams, we shall likely have no use for the "Loads" entry (resistances and reactances as part of the antenna), so we may leave it at 0. Other fields to leave at default initially are the "Analysis Resolution" entry (1°), the "Step Size" entry (5°), and the "Field(s) to Plot" entry (total field only). After we become more advanced or develop special interests in the program, we may wish to alter those entries.

Two entries where we must make an initial decision are "Units" and "Ground Type." For US hams, feet and inches are the most commonly encountered antenna measurements. For HF work, start and stick with measurements in feet. This will ease the problem of calculating specific dimensions of antennas by sending us through the same steps with each antenna.

The selection of the type of ground to use is a bit more complex to decide. The program offers free space, perfect ground, and real-ground choices. My personal preference is for real ground. The default real-ground description uses average soil (as explained in the manual). Unless there is good reason to change this entry, initially stick with it. However, if you know the electrical characteristics of your soil and surrounding terrain (and water bodies, if present), it may pay to go through the setup procedures in the manual for establishing a detailed ground description. Remember, however, that the program does not account for yard clutter that may affect an antenna. Later you can compare real-ground patterns with patterns over perfect ground to see what differing soil conditions can do to an antenna's gain and pattern. You may also wish to compare your real-ground patterns with the free-space patterns so common in textbooks. Those exercises will be enlightening, but for comparing one antenna with another, stay with a consistent description of the ground. There are program limitations that the selection of real ground cannot overcome; we shall note the most significant one for the beginning modeler later.

Only two more menu entries require comment at this point. For the antenna "Title," choose your words and abbreviations carefully to pack in as much information as possible. That will allow you to easily distinguish one antenna design from another. "Dipole" is not a very good choice for a title, but "10 M Wire Dipole, 30 Ft Up" might be. The "Frequency" entry also requires care. For each band on which you compare different antennas, use the same frequency or set of frequencies. Do not shift from one end of a band to the other when changing antenna designs (unless you have a special reason for doing so); your comparisons may not be valid. If necessary, evaluate an antenna design at two or three frequencies within a band, using the same frequencies for all antenna designs.

Having now decided most of the menu entries in advance, we have reduced the remaining entries to a manageable few. To display and print out our results, we shall deal with decisions involving "Plot Type" and "Azimuth/Elevation Angle" later on. First we have to model our antenna design in order to enter it into the computer, and that takes some special forethought.

### Setting Up Modeling Conventions

Getting good results from an antenna program begins with pencil and paper. Whether you are modeling your own antenna, an idea from a handbook, or an experimental design, you will have to put the figures on paper and change their form to what the program wants to see. Therefore, you will want to develop a standard notebook page for each antenna you model. Here are the preliminary items that should go on the page.

*Item 1*

Make a neat sketch of the antenna, including all dimensions. This includes any changes in element dimensions, a common occurrence for beams using aluminum tubing.

*Item 2*

Tabulate the details of each element, including the total length, the length of each piece or section of wire or tubing that makes up the element, and their diameters or wire sizes. Also include the spacing between elements for multielement antennas. At this point, it may be good to place the tabulated data in a column on the left side of the notebook page, since you will want to manipulate that data before entering it into the program.

All MININEC programs require entry of the data in the form of x, y, and z coordinates for the ends of each "wire" or element section, where x is the standard left-right axis, y is the standard front-back (on the computer screen, up-down) axis, and z is the height of the element above ground. Each element section made from a different diameter wire or tube has two entries, corresponding to the two ends of the section. It makes no difference to the program whether you model the antenna using x for the element length and y for the spacing between elements or vice versa. The convention you choose depends on which system is most convenient.

Since the chief problem I have in setting up models for the computer is entry errors, I have chosen x for the antenna element lengths and y for the spacing. This puts the coordinates most likely to be numerous or to change (values for x) in the left-hand column of the entry readout, where I can survey them easily for errors. And hard-to-see errors do occur, as when I forgot a decimal point and made part of an antenna element 75 inches in diameter rather than 0.75 inches. Adopting this convention will require that we take our elevation plots at an angle of 90° rather than at 0°, as noted later on.

Let's go a step further. Although you can enter any set of x and y coordinates, so long as the element sections line up, you will probably make fewer errors if you set up your antenna symmetrically around x = 0 and y = 0. Since y is the antenna element spacing dimension, set the driven element at y = 0. A simple dipole, of course, will have no other element. A Yagi might have either or both a director and a reflector. Set the reflector behind the driven element by giving it a negative y sign. A reflector 6 feet behind a driven element shows up as y = –6 on the chart. Any directors receive positive spacing. This convention allows you to identify elements in your setup chart.

Set up your antenna elements symmetrically around x = 0. For a single-wire element, this means taking the total length and dividing it by two. The ends of the wire then have identical x entries, but one is negative while the other is positive. For example, a 66-ft dipole would have x entries of 33 and –33 for its two ends. Multisection elements of different diameters are only slightly more complex. I have a portable 10-meter dipole made from 4-foot sections of ¼-inch diameter aluminum rod with end pieces of 3⁄16-inch rod. The overall length is 16.6 feet. We can consider the center rods as one 8-foot piece, which gives us –4 and +4 for the pair of x coordinates. Each end piece is 4.3 feet long. The x coordinates for the outside ends will be –8.3 and +8.3, with the other ends of the end rods having the same x coordinates as the center rod.

There are two advantages to using this system. First, you can spot errors more easily on the element entry screen or printout. Just look for numbers that are supposed to be the same except for a sign change. (And be sure to survey all the signs for correct negative or positive values while you are at it.) Your notebook page can have compact entries in preparation for computer work.

*Item 3*

Enter in columns next to each element section the arithmetic you did to determine element coordinates. For example, the end element sections of the dipole might show 16.6 ft – 8 ft = 8.6/2 = 4.3 ft ±4 ft = ±8.3 ft.

*Item 4*

Enter the final x coordinates for each element section. If you are careful, you can use abbreviated notations. You can put sections 1 and 3 together as ±8.3 and 4.0, with section 2 as ±4.0. This means that section 1 goes from –8.3 to –4.0, section 2 from –4.0 to +4.0, and section 3 from 4.0 to 8.3. If you are not comfortable with abbreviated listings, then list each piece or section separately. Note that what we have called "pieces" or "sections" are called "wires" in the program. What the program calls "wires" may actually be wire conductors, or they may be lengths of metal tubing or rod.

Figs 1 and 2 illustrate two notebook pages, one for the simple dipole, the other for a 3-element Yagi. Notice that all the entries are in feet and decimal parts of feet. Therefore, add another item to the pencil and paper you need in order to plan your entries: a calculator. You will often encounter dimensions like 9 feet 9½ inches, which a calculator easily converts to 9.7917 feet.

| Band | Antenna Type | # El. | Height | Ground | Date | Run ? |
|---|---|---|---|---|---|---|
| 1Ø M | Dipole (portable) | 1 | 2Ø' | R | Ø1 / 12 / 1991 | x |

L. B. Cebik, W4RNL

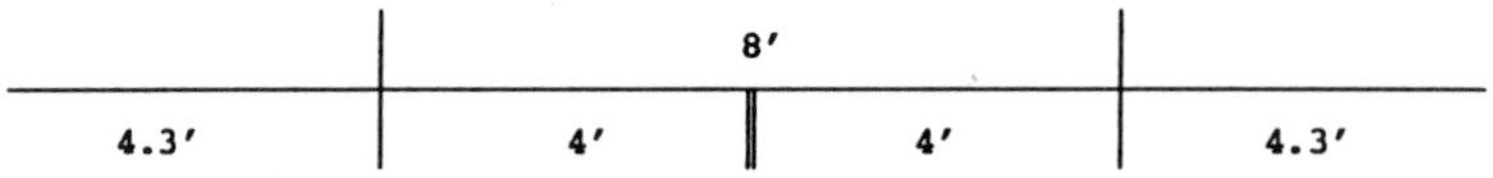

8'

4.3' 4' 4' 4.3'

| El. # | Lgth feet | Sp to DrEl | Calculations | Wire # | E1 x | y | z | E2 x | y | z | Dia. | Segs |
|---|---|---|---|---|---|---|---|---|---|---|---|---|
| 1 | 16.6 | ----- | 16.6-8=8.6/2=4.3 | 1 | -8.3 | Ø | 2Ø | -4 | Ø | 2Ø | .1875 | 2 |
| | | | 4.3+/-4.Ø=+/-8.3 | 2 | -4 | Ø | 2Ø | 4 | Ø | 2Ø | .25 | 8 |
| | | | | 3 | 4 | Ø | 2Ø | 8.3 | Ø | 2Ø | .1875 | 2 |

Portable dipole, disassembles and stores inside 2 5-foot PVC tubes used as mast.

Fig 1—Example notebook page for a 10-meter dipole.

| Band | Antenna Type | # El. | Height | Ground | Date | Run ? |
|---|---|---|---|---|---|---|
| 10 M | Yagi (Ant. Bk. 15th Ed p. 11-11) | 3 | 30 | R | 01 / 12 / 1991 | x |

L. B. Cebik, W4RNL

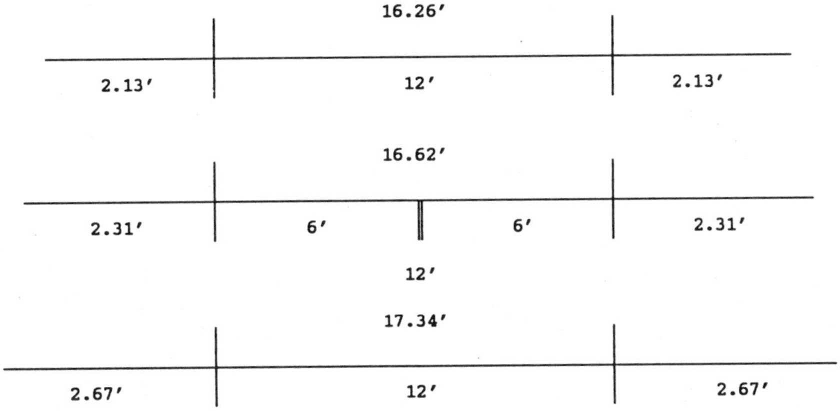

16.26′

2.13′ | 12′ | 2.13′

16.62′

2.31′ | 6′ | 6′ | 2.31′

12′

17.34′

2.67′ | 12′ | 2.67′

| El. # | Lgth feet | Sp to DrEl | Calculations | Wire # | E1 x | y | z | E2 x | y | z | Dia. | Segs |
|---|---|---|---|---|---|---|---|---|---|---|---|---|
| 1 | 16.62 | --- | 16.62-12=4.62/2=2.31 | 1 | -8.31 | 0 | 30 | -6 | 0 | 30 | .875 | 2 |
| | | | 2.31+/-6=+/-8.31 | 2 | -6 | 0 | 30 | 6 | 0 | 30 | 1.0 | 8 |
| | | | | 3 | 6 | 0 | 30 | 8.31 | 0 | 30 | .875 | 2 |
| 2 | 17.34 | 5.23 | 17.34-12=5.34/2=2.67 | 4 | -8.67 | -5.23 | 30 | -6 | -5.23 | 30 | .875 | 2 |
| | | | 2.67+/-6=+/-8.67 | 5 | -6 | -5.23 | 30 | 6 | -5.23 | 30 | 1.0 | 8 |
| | | | | 6 | 6 | -5.23 | 30 | 8.67 | -5.23 | 30 | .875 | 2 |
| 3 | 16.26 | 3.49 | 16.26-12=4.26/2=2.13 | 7 | -8.13 | 3.49 | 30 | -6 | 3.49 | 30 | .875 | 2 |
| | | | 2.13+/-6=+/-8.13 | 8 | -6 | 3.49 | 30 | 6 | 3.49 | 30 | 1.0 | 8 |
| | | | | 9 | 6 | 3.49 | 30 | 8.13 | 3.49 | 30 | .875 | 2 |

Antenna Book design used for program test purposes.

Fig 2—Example notebook page for a 3-element Yagi.

Too, you will be using fractionally dimensioned tubing (for example, 7/8 inch diameter) that you need to enter as a decimal (0.875). And, of course, the calculator serves as a check on addition and subtraction.

*Item 5*

Multielement antennas remind us to add two more listings. For each element, list the y and z coordinates, the spacing from the driven element and the height above ground. For all single-wire antennas, such as the dipole, set y = 0. Yagis will have other elements separated from the driven element. Each element will be the same distance above ground. Remember to set the source in the center of the driven element for each of the antennas noted here. In order to ensure that the program places your source at the exact center, use an even number of wire segments for the driven element. Otherwise, your source may be offset.

Use the height that you actually anticipate the antenna will be. If you model quads, of course the vertical sections will have changing z coordinates. Once you get used to setting up a few dipoles and Yagis, the changes for vertical sections will come naturally. Start by dividing the vertical section symmetrically and add or subtract from the height of the boom or hub, which you can determine from the real or anticipated tower height.

This completes the notebook page, except for any reminders you may wish to add. I generally enter a note that tells why I modeled the antenna. That refreshes my memory weeks later when I run across the page and wonder why I ever spent computer time on that crazy design.

**Building a Baseline of Antenna Information**

Let's begin this part of our work with a little scenario. You model a 2-element X beam, using some guesswork and intuition for the dimensions. The program produces the patterns shown in Fig 3. Now, what have we learned? Not much, perhaps, beyond the fact that this is probably not an antenna we'd want to actually build. But you can learn much more by practicing with basic antennas over the soil and land around your own QTH.

This is the reason why, in the program setup, I suggested following the instruction manual and selecting ground conditions that most closely correspond to your own land. I chose average soil for all the examples here because everything indicates that my hilltop QTH on what once was Tennessee farmland is just that: average.

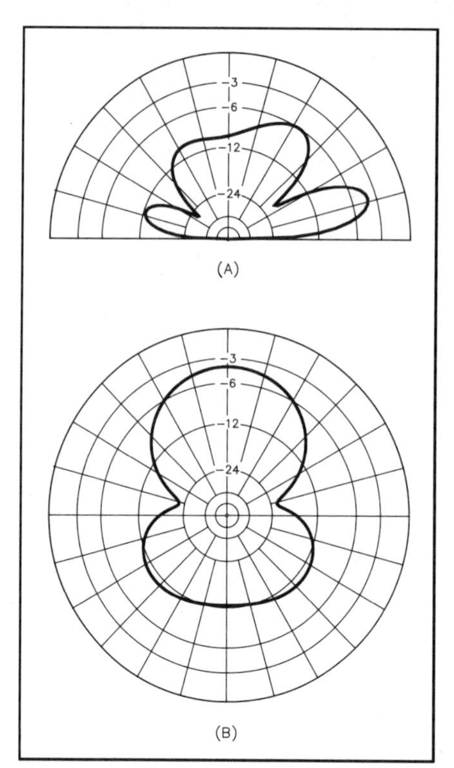


Fig 3—Elevation and azimuth patterns for a 10-meter X beam design, 30 feet high. The maximum gain is calculated to be 9.1 dBi. The azimuth pattern is taken at an elevation angle of 16°. **All patterns in this article were calculated with ELNEC 2.20 for "average" earth having a conductivity of 5 millisiemens per meter and a dielectric constant of 13. In all pattern plots, the 0-db reference (outer ring) is 13 dBi, and all azimuth patterns are taken at the elevation angle of maximum forward gain.**

Once you select and model your ground conditions, stick with them for all models, or your designs may not be directly comparable. If you discover that your soil, rock, or sand differs from your original estimates, then revise and rerun all the antennas that make up your baseline of information.

The next step is to model some basic antennas. Which antennas? You may wish to start with a dipole, a 2-element Yagi, and a 3-element Yagi, as these are all well-established antenna designs with fairly well developed characteristics. Yagi designs are available in many handbooks and ham magazines.

None of these antennas requires more than a single source, and they introduce no loads. Save those complexities until you have mastered the program basics (and then review the manual on how to use them). Moreover, all of these antennas conform to the general idea of avoiding program limitations that may produce misleading results. There are no acute angles for which special techniques are required. Likewise, none of these designs use close-spaced wires, as with folded dipoles. They, too, require special techniques to give accurate results. And none of these basic designs require three or more wires to join at one point.[3] When you are fully comfortable modeling basic designs, then you can add other techniques to your repertoire, one at a time.

Which frequency or frequencies should you use? Select the band in which you are most interested. Later you can expand your baseline to include other bands, especially if you decide to develop multiband antennas. The examples here use 28.2 MHz.

What antenna height or heights should you model? Use realistic heights relative to your situation. For convenience, I modeled the example antennas at 20 and 30 feet, since the lower height is where my temporary dipole is, and the higher elevation is about the size tower I plan to erect. Although two levels are sufficient to illustrate why the baseline data is important, expanding the baseline to 40, 50, and 60-foot heights is not unrealistic.

Heights of at least 20 feet work well for modeling at 10-meter frequencies. However, at lower frequencies (for example, 3.5 MHz), 20 feet would place these horizontal antennas well below the 0.2-λ minimum height for good modeling.[4] Table 1 provides a quick reference to the 0.2-λ height for the HF bands. Below these heights, results are likely to show incorrect

**Table 1**
**0.2 Wavelength at the Lower End of Low-Frequency Amateur Bands**

| *Frequency* | *0.2 Wavelength* |
|---|---|
| 1.8 MHz | 109 feet |
| 3.5 | 56 |
| 7.0 | 28 |
| 10.1 | 19.5 |
| 14.0 | 14 |
| 18.068 | 10.9 |
| 21.0 | 9.4* |
| 24.89 | 7.9 |
| 28.0 | 7 |

*Heights below this point are not generally useful for modeling, as yard clutter may affect patterns more than program inaccuracies. Mobile, portable, and experimental antenna designs may be exceptions.

impedances and excessive gain. Always keep in mind that MININEC-based programs calculate impedances over a perfect ground, whether you select a perfect or real ground. The results are usable for real ground above the 0.2-λ limit, but may not be usable for less height. Under some circumstances, you may have to erect an antenna below the modeling limit; the antenna may work, but the program may not accurately model its operation.

The three chosen antennas at two elevations each require only six runs of the program. Using constant-diameter elements for the baseline simplifies the design work and lets the computer race through the calculations. More complex antenna construction, of course, gives you more element sections and inevitably longer calculation times.

My early experiences have sold me on the idea of beginning with an elevation pattern. (Elevation is simply the vertical radiation pattern, and azimuth is the horizontal radiation pattern.) Since the x coordinate is in line with the antenna elements, the elevation pattern requires a 90° orientation angle to catch the main lobe. The elevation pattern then tells what angle to use for the azimuth pattern to catch the perimeter of maximum radiation. You can check other azimuth patterns at other angles, but for initial information, a basic elevation and azimuth pattern combination will be instructive. For the purposes of comparison, all the antennas use the same value for the scale on which they are plotted. This permits the omission of detailed analysis information, although you may want that information on the plots you take for your own use.

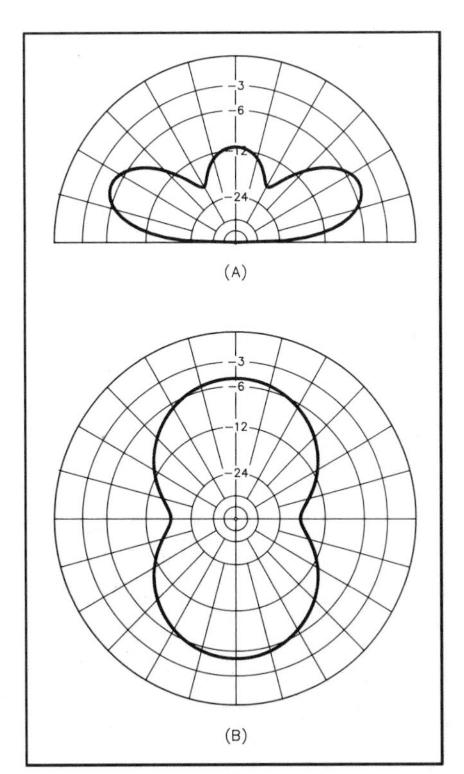

Fig 4—Patterns for a 10-meter dipole, 20 feet high. Azimuth pattern at 24° elevation; maximum gain 8.1 dBi.

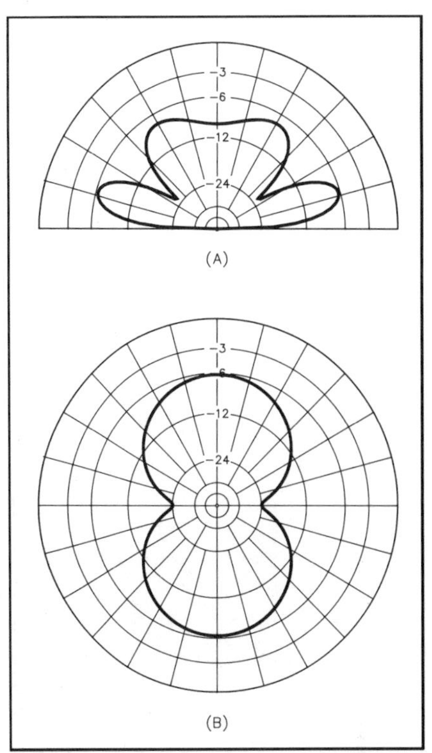

Fig 5—Patterns for a 10-meter dipole, 30 feet high. Azimuth pattern at 16° elevation; maximum gain 6.9 dBi.

Figs 4 and 5 illustrate the 10-meter dipole performance at 20- and 30-foot heights. The difference is revealing. The 30-foot-high dipole channels significant energy into high radiation angles, while the 20-foot high dipole appears to concentrate more energy at lower radiation angles. The patterns may come as a surprise if the only patterns you have viewed are free space types. Moreover, the patterns seem to supply an answer to my wondering about the excellent reports from Europe on my 20-foot high 10-meter dipole (without nearby trees). Figs 4B and 5B provide the azimuth patterns of the dipole. They show definite reductions off the ends of the antenna, but not the classic pinch-waisted "figure 8" of free-space patterns. Note that these patterns apply to 10 meters and not necessarily to other bands, because patterns will vary with (among other factors) the percentage of a wavelength the antenna is placed above ground.[5]

It is tempting to compare the 30-foot-high dipole patterns of Fig 5 with Fig 3, the intuitive X beam at 30 feet, and notice the variation of the elevation pattern from the dipole. However, before we can evaluate whether or not the variation—which gives some gain and front-to-back ratio—is good, bad, or indifferent, let's look at the Yagis.

Figs 6 through 9 present patterns for the Yagis at 20 and 30 feet. The 2-element Yagi comes from a design by Bill Orr, W6SAI,[6] while the 3-element Yagi is adapted from *The ARRL Antenna Book.*[7] By comparing the various patterns at equivalent heights, we can see the evolution of the dipole pattern into something with gain and front-to-back ratio. Had we skipped the 2-element design, we might have missed the continuity of pattern development, which shows with special clarity at 30 feet.

The patterns reveal a number of other factors of relevance. First, the Yagis provide significant, but not overwhelming gain relative to real dipoles. (Note: ELNEC provides gain in dBi, gain over an isotropic source. Our interest is in comparing gains of real antennas over real property. To do that, we need only use the *difference* in gain figures to see if we are making significant improvements, and how much.) Only when we combine the gain with the front-to-back ratio do we find the real merits of a beam. The 3-element Yagi shows the most significant increase in front-to-back ratio among the antennas used here for baseline data. A complete analysis of our individual antenna situations would require more standard models, but we have enough here to begin generating expectations. Rational expectations are what good baseline data are designed to give us.

For example, any beam we might propose to build at selected heights should show patterns that compete with the Yagis. We can now see that the intuitive X beam at 30 feet does not do the job. Its pattern is little more than a barely deflected dipole pattern. It does not match even the 2-element Yagi for gain and front-to-back ratio. That does not mean there are not good X beam designs; rather, I came up with a bad beam antenna design and used it for this example. What told me this is a bad design was not a textbook or the program manual, but the baseline information I collected from the antenna modeling program.

## The Next Step: Your Own Antennas

Your antenna situation differs from mine, which means that many of the pat-

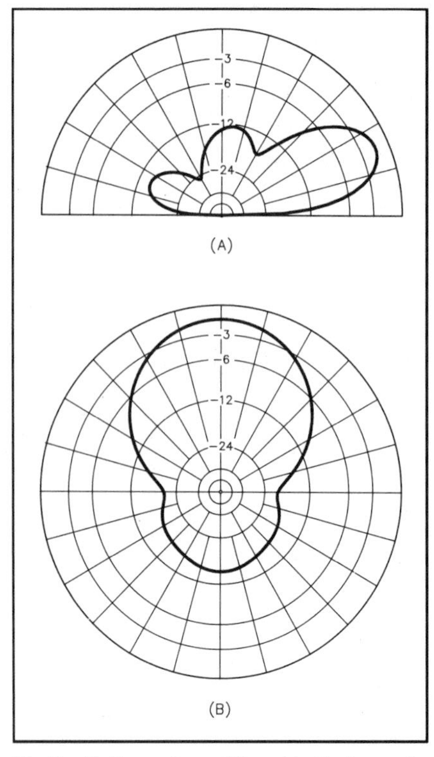


Fig 6—Patterns for a 10-meter 2-element Yagi, 20 feet high. Azimuth pattern at 23° elevation; maximum gain 11.6 dBi.

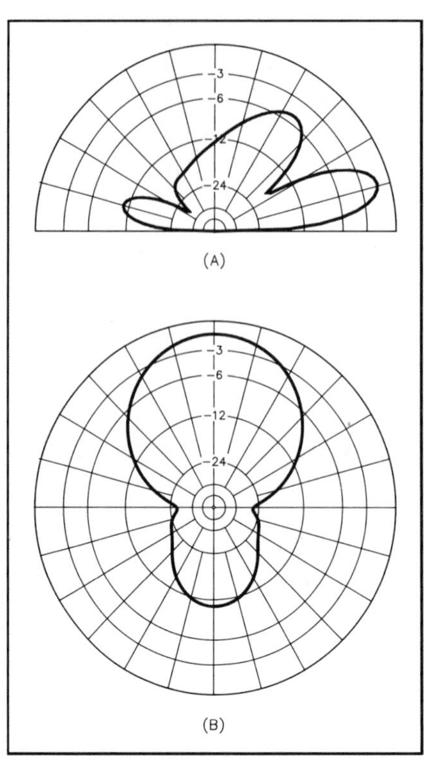


Fig 7—Patterns for a 10-meter 2-element Yagi, 30 feet high. Azimuth pattern at 16° elevation; maximum gain 11.8 dBi.

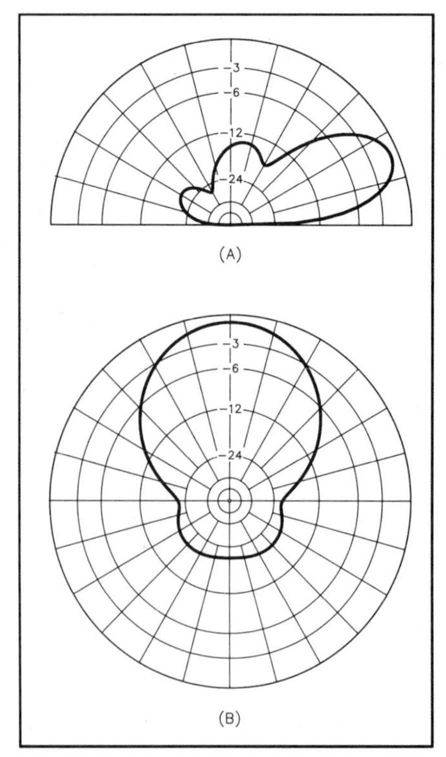


Fig 8—Patterns for a 10-meter 3-element Yagi, 20 feet high. Azimuth pattern taken at 23° elevation; maximum gain 12.3 dBi.

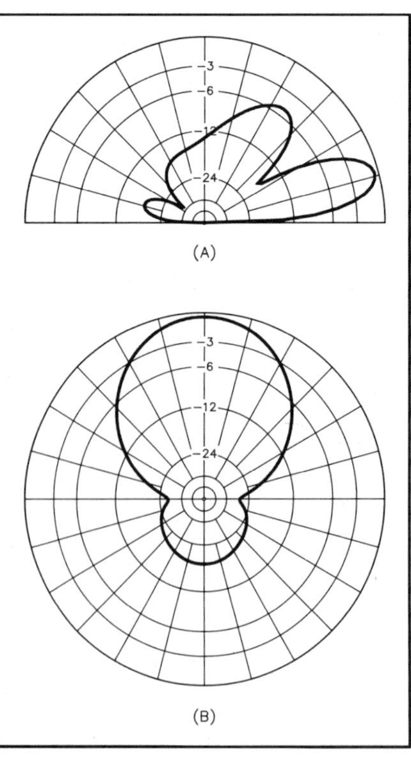


Fig 9—Patterns for a 10-meter 3-element Yagi, 30 feet high. Azimuth pattern taken at 16° elevation; maximum gain 12.6 dBi.

terns shown here may be irrelevant to you. However, they do illustrate the techniques needed by beginners to get the most out of a computer modeling program. The program can be like a textbook written for one person: you. From it, you can put theory into practice without undergoing the expense and time required to build every antenna design you encounter. You can also develop enough models to help you decide where to place your antenna dollar most wisely. Your antenna dollar includes not only the antenna itself, but the mast and tower as well. So even if you practice using the program by modeling the antennas shown here, the next step is to develop a set of models specific to your own needs. A good place to begin is with the antennas you are actually using (unless they contain traps or other complexities that may require advanced techniques to model). The assembly manuals or the calculations you performed to roll your own probably provide almost all the data you need.

Whether you choose your existing antennas or a set of proposed antennas to evaluate with the antenna modeling program, there are a few additional suggestions that may make your effort more profitable. For any antenna, be certain to evaluate it across the entire frequency band of interest. What the band of interest is depends upon your operating interests. If you use the entire band, from the lowest CW segment to the highest phone segment, then you will want antennas that are broadbanded. Their patterns may not show the most gain among your models, but they will show good gain and proper takeoff angles (and, for beams, reasonable front-to-back ratio) over the entire band. If your interests are confined to a narrow portion of any given ham band, then you should check every design at selected spots within that spectrum slice. When making comparisons between designs, use the same set of frequencies for each run.

At this point, the "Source" information becomes important. You will have to be able to match a real antenna to the feed line over the entire band of interest. A high-performance design that will not accept power is not a good antenna for you. Of course, there are many feed-point matching schemes, but choosing one goes outside the program. As you can see, an antenna modeling program is one important part—but not the only part—of an antenna design and construction project or program.

There are many other facets of using antenna modeling programs like ELNEC. The instruction manual is a good guide to them. Working with vertical antennas and ground planes or radial systems, for example, is a topic unto itself with many attached cautions.[8] Accurately modeling real ground requires reference to the program instructions and an understanding of the effects of ground upon antennas.[9] The information presented here is intended to get the beginning antenna modeler started in using the program to learn about antennas, even if he or she never builds one from scratch. Besides all these practical benefits, the programs are fun to use—especially on a rainy day when the bands are closed.

### Notes

[1] MININEC3 by Naval Ocean Systems Center, San Diego, is available from National Technical Information Services, Springfield, VA 22161, order no. ADA181681. This is public domain software, but a generous fee is charged for the diskette and documentation. MN by Brian Beezley, K6STI, is available commercially from Brian at 507-½ Taylor St, Vista, CA 92084. ELNEC by Roy Lewallen, W7EL, is available commercially from Roy at PO Box 6658, Beaverton, OR 97007. (MN and ELNEC are enhanced versions of MININEC3.) The ARRL in no way warrants these offers.

[2] For a "must-read" description of MININEC program limitations, see R. Lewallen, "MININEC: The Other Edge of the Sword," *QST*, Feb 1991, pp 18-22. (Photo copies are available for a nominal fee from ARRL HQ, 225 Main St, Newington, CT 06111-1494.)

[3] See the reference of note 2, pp 19-20.

[4] See the reference of note 2, p 21.

[5] [*Editor's note:* Always remember that the gain of an antenna is a function not only of its design, but also its height and the electrical characteristics of the earth. Comparing gain figures of antennas at different heights above ground can be misleading, as Figs 4 and 5 clearly show: The maximum gains differ by 1.2 dB, *for the same antenna*! Of course an antenna cannot have gain over itself. Similarly, front-to-back ratios may change with height at the wave angle of maximum forward response, as Figs 8 and 9 indicate. (The difference for these two heights is 2.1 dB.) As you gain experience with the modeling program, you may want to eliminate any height effects by initially modeling the antennas to be compared in free space (ground-mounted vertical systems excepted). *Then* model your optimized designs over ground to observe their performance in a real-life environment.]

[6] B. Orr, "A Compact 2-Element Yagi for 10 Meters," *CQ*, Dec 1990, pp 83-84.

[7] G. Hall, Ed, *The ARRL Antenna Book*, 15th or 16th eds (Newington: ARRL, 1988 or 1991), design charts on p 11-11.

[8] See the reference of note 2, pp 20-22.

[9] See the reference of note 7, Chapters 3 and 8.

# Modeling HF Antennas with MININEC—Guidelines and Tips from a Code User's Notebook

**By Dr. John S. Belrose, VE2CV, ARRL TA**
**17 Tadoussac Dr**
**Aylmer, QC J9J 1G1, Canada**

MININEC,[1] a computer program for the analysis of wire antennas using an IBM compatible microcomputer, has opened the door to radio amateurs interested in the exciting field of computational electromagnetics for antenna design and performance evaluation/optimization. The application of this code effectively eliminates the "black magic" component of antenna analysis—or does it? The user is cautioned that there still is a required skill to set up the problem and to interpret the result, no matter how user-friendly the program may seem to be.

An antenna design and performance study is much more easily done by numerical modeling than by constructing antennas and measuring performance in the real world. The ARRL technical staff and a number of radio amateurs are using this antenna modeling code, or versions of it designed for amateur application,[2-4] and clearly we are going to see an increasing number of antenna articles by hams using this new computational tool. However, users of this program, and the readers of amateur magazines, should be aware that MININEC is based on certain assumptions that limit its applicability and its accuracy. While this subject has already been addressed in *QST* by Roy Lewallen[5] and Rus Healy,[6] much more can be said on the subject.

Let me comment on the continuing development of antenna modeling codes. First I will tell you a story. In 1983 at an AGARD lecture, "The Performance of Antennas in their Operational Environments," speaker Belrose remarked that the need for more accurate design and the availability of powerful digital computers have combined to guide antenna design along a path that is providing a continuous transition from an empirical art to a mathematical science. But one of the fellow speakers at this lecture course, Edmund Miller, who had worked with Jerry Burke and others at the Lawrence Livermore National Laboratory, Livermore, CA during the development of NEC, cautioned that antenna modeling was still an art. There still is a required skill to set up the problem and to interpret the result.

A computer model for the antenna is specified to the MININEC code by giving the x-y-z coordinates of the start and end point of each wire, and the diameter of each wire. A structure actually constructed from wires, such as a dipole, a loop antenna, a Yagi or a log-periodic antenna, is readily coded into an input file for MININEC. Assumptions made in deriving the Pocklington's Integral Equation and in solving it with the moment method put restrictions on the geometry of interconnections of wires which may be solved by MININEC. A set of "modeling guidelines" is used, representing the code developer's and user's experience in designing models that will obtain the best results. MININEC itself does not give the user any warning that the input geometry violates the modeling guidelines. MININEC goes ahead and computes physically meaningless currents and fields for such structures.

The information presented in this article includes (1) modeling guidelines for users of the code, and (2) the experience of the author and his colleagues in developing models of HF wire antennas. Some of the guidelines discussed have been arrived at through research using NEC-3,[7] however, Rockway et al[8] have shown that MININEC has limitations that are in close accord with NEC.

**Coordinate System for Numerically Modeling Antennas**

MININEC models antennas as a set of connected straight wires that simulate the antenna configuration. Additional wires, ground connected or insulated, can be included as required to numerically simulate parasitic elements, as well as the effects of support towers, isolated towers, guys, and other wire structures in the operational environment of the antenna.

As mentioned, the numerical modeling code MININEC and versions of it employ rectangular x-y-z coordinates, for the purposes of constructing the wire model of the antenna, and for reference when plotting the antenna radiation patterns. See Fig 1. The arrow heads mark the positive directions for these axes. The direction of the +x axis is the 0° azimuth direction, and the azimuth angle $\phi$ is defined by a radial vector rotating clockwise from this direction. Thus, $\phi = 90°$ corresponds to the +y direction, $\phi = 180°$ corresponds to the −x direction, and $\phi = 270°$ corresponds to the −y direction. To clarify, think of a normal x-y coordinate graph. The abscissa is the x axis, the ordinate is the y axis, and a radial vector that rotates from the +x-axis direction to the +y-axis direction rotates in the clockwise direction.

For antennas the radiation pattern is sometimes viewed as a 3-dimensional

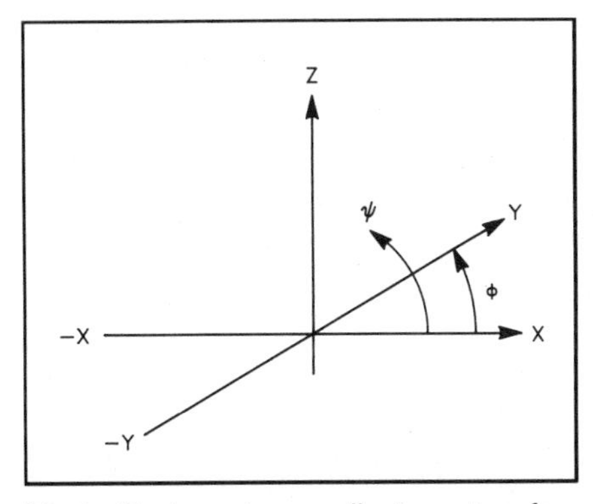


Fig 1—Rectangular coordinate system for constructing the MININEC wire model of the antenna and plotting antenna radiation patterns.

(volumetric) pattern, however for MININEC the most useful patterns are the vertical-plane pattern at the azimuth angle (ϕ) of maximum gain and the azimuthal pattern at the takeoff angle (ψ) of maximum gain. The latter is itself the principle plane pattern.

**Modeling Guidelines**

*Thin Wires*

For analysis by MININEC the wires of the model must be "thin" because the current is assumed to flow axially on the wire with no circumferential component. Various definitions of thickness are given, for example, that the wire diameter $d/\lambda << 1$, where d is the wire diameter. Wires are considered to fully satisfy the guideline if they are thinner than λ/50. As a rule of thumb, wires fatter than λ/15 are too fat to be considered "thin."

*Segment Length*

The modeling guidelines state that each wire should be subdivided into segments comparable to λ/10. There is a loss in accuracy if segments are allowed to be as long as λ/5. Segments as short as λ/20 may be required "in critical regions of the antenna." Wires must not be so finely subdivided that the segment length is less than λ/1000, but this is rarely a problem; an exception might be the modeling of transmission lines.

The shortest permissible segment is usually determined by the wire diameter. For an accurate solution, the segment length should be $\ell > 4d$.

When the segment length on straight wires is changed, for example to model transmission lines, it is desirable to progressively reduce the segment length as one approaches the source(s). Adjoining segments should not differ in length by more than a factor of 2.

Wires joining at an acute angle must have shorter segments than single straight wires joining in a line. Instead of making the entire wires out of short segments, the segments can be made short near the junction, tapering to a longer length away from the junction. ELNEC Version 2.01 and higher automates the process of segment tapering. The basic procedure is to replace the original wire by several wires of different lengths. The new wire nearest the junction is made very short (say λ/400) and with one segment. The second wire is made twice the length of the first, also with one segment. This procedure is repeated until the segment length is about λ/20, and the remainder of the original wire is made up of a multisegment wire of approximately this length.

If one wants to calculate the pattern of a large antenna, some versions of MININEC and MININEC itself have a limit on the highest frequency for which the program can be used. The reason is that the segment length must be $\ell \leq \lambda/10$, and the program will handle only a certain number of segments (300 for MININEC, 127 for ELNEC, and up to 260 for ELNEC with MaxP).

As discussed above, a computer model is specified to the MININEC code by giving the x-y-z coordinates of the beginning and end point of each wire (end 1 and end 2 of the wire) that is used to form the wire structure. It does not matter in what order the wires are specified, or which end is which. However, if more than one current (or voltage) source is used, such as to phase an antenna system or to model a transmission line, some precautions must be taken, because the phase of the resulting currents can be 180° from what was intended. Reversing the direction of numbering is equivalent to reversing the direction of current flow on the wire (or the phase of the current on the wire). When modeling transmission lines, for example, the wires that make up a two-conductor transmission line must be numbered in the same direction as one proceeds from the source(s). When placing a source in series with a wire, the source is always placed so the positive terminal faces end 2 of the wire.

*Wire Spacing*

Royer made a detailed study of the problem presented to NEC by closely spaced wires, a problem encountered when modeling a balanced transmission line as a part of the antenna system.[9] He compared the characteristic impedance of a two-wire transmission line, as computed by NEC, with that given by analytical equations. The modeling guideline that emerged from his investigations is that the ratio of the diameter of the wire to the wire spacing, d/s, should be less than approximately 0.2. If the two closely spaced wires have different diameters $\sqrt{d1 \times d2}/s$ should be < 0.2.

*Interconnections of Wires*

It is sometimes necessary to connect a thin wire to a thick wire (a thin wire to a tower for example). The diameter ratio of the two wires at the junction should be less than 5 and never greater than 10.

*Interpreting the Result*

If the resulting antenna gain seems intuitively to be too high, if the pattern is nonsymmetrical for antennas that have symmetry, if the antenna impedance or the pattern changes in an unexpected way with a small change in frequency or a change in the number of segments that make up the antenna, then look carefully at other parameters calculated by MININEC. Examine the currents on the wires and the antenna source impedance. Ridiculous values of gain and/or impedance, particularly of reactance, and in some cases a negative resistance,[10] are indicative of problems. Some example case studies that require particular attention are discussed below.

**Some Illustrative Case Studies From the Author's Files**

Loops, multiwire antennas, nonsymmetrical antennas such as loops with unequal sides (particularly corner-fed triangular loops near the ground), off-center-fed dipoles, and antennas with tuned feeders can give difficulty. When modeling such antennas, the number of segments for each wire or side should be chosen in proportion with the wire length, so the segment length does not change in going from one wire to the next.

**Selecting the Axis to Construct the Model**

Some consideration should be given to the selection of the axis to set up the wire model. In my view, the axis should be chosen to correspond with the expected pattern symmetry of the antenna. If this consideration is ignored, it doesn't necessarily mean that the computational results are wrong, but it could, if for example the antenna's symmetry is used to reduce the number of segments needed to model the antenna. However, the azimuthal plot is "neater" if 0° is the direction of maximum directivity, or if 90° corresponds to the broadside direction for a dipole. Thus, the x-z plane for the 160-meter full-wave hori-

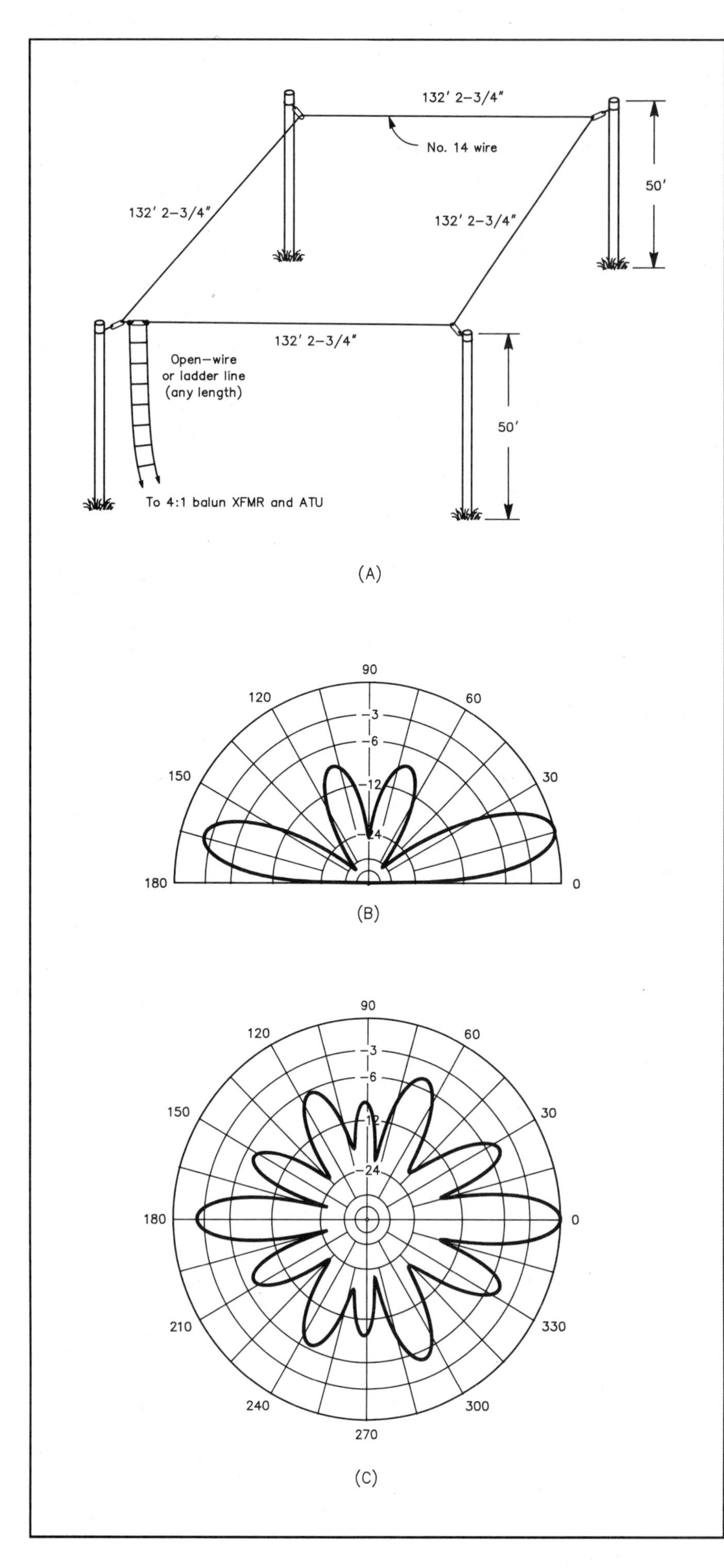


← Fig 2—At A, the W1FB full-wave 1.9-MHz loop (see Note 11). With a Transmatch, this antenna may be used for multiband operation. At B and C, elevation-plane and azimuthal patterns for the loop. The corners of the loop containing the feed and the opposite corner are in the 0-180° plane. The direction away from the feed is the direction of the 0° azimuth. The patterns have been calculated for average ground; the azimuth pattern is taken at 15.5° elevation.

zontal diamond loop of Fig 2A should be the plane that contains the feed and opposite corner, as the antenna is symmetrical with relation to this plane, as opposed to selecting the x-z plane parallel to one of the sides.[11]

With the axis chosen accordingly, the radiation patterns were calculated using ELNEC for average ground ($\sigma$ = 3 mS/m and $\varepsilon$ = 13). B and C in Fig 2 show the predicted patterns for 14.2 MHz. Note that with the axis chosen as we have, the antenna radiation pattern has symmetry in the x-y plane. And the gain is a maximum in the direction away from the feed toward the opposite corner, in the 0° azimuth direction.

### Modeling Ground

MININEC assumes a perfect ground for impedance and current calculations. The "real" ground description is used only for determining the shape and strength of the far-field pattern. This means that one cannot determine the efficiency of ground radial systems with MININEC, and the resistive component of impedance and hence absolute gain for low horizontal antennas won't be correct. The code predicts the resonant frequency of the antenna quite well, because the antenna reactance is not a strong function of ground conductivity.[12]

If we visualize the antenna above ground together with its image in the ground, we can physically describe and interpret the ground-loss effect (which MININEC does not take account of). The antenna and its image are closely coupled, and because of the finite conductivity of the ground beneath the antenna, a coupled loss resistance appears at the driven terminals of the antenna. The antenna resistance is the vector sum of radiation resistance and loss resistance. This is not a fictitious resistance; it is a real resistance. It can be measured, and it can be predicted by the more sophisticated code NEC-3. The magnitude of the antenna resistance is a strong function of ground conductivity, height of the antenna above ground in wavelengths, and polarization.[13]

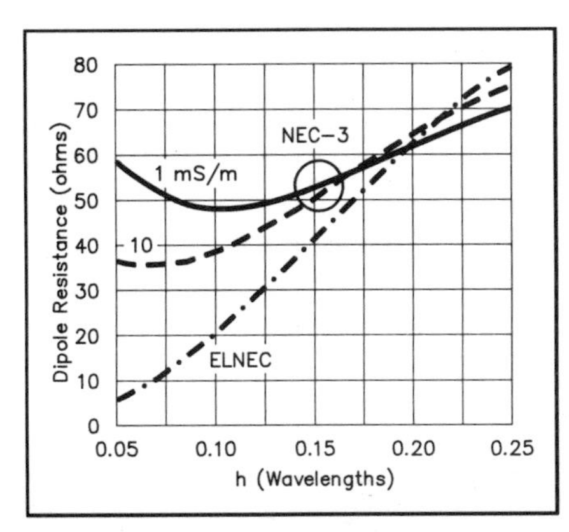


Fig 3—Antenna resistance for a 4-MHz horizontal dipole versus dipole height in wavelengths for two ground conductivities calculated by NEC-3, compared with MININEC3 (ELNEC version).

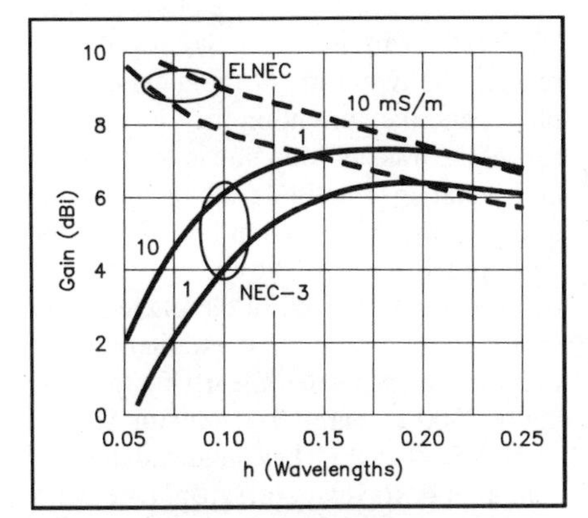


Fig 4—Gain at 90° elevation for a 4-MHz horizontal dipole versus dipole height for two ground conductivities calculated by NEC-3, compared with MININEC3 (ELNEC version).

Ground loss becomes an important parameter for horizontal antennas at low heights (h < 0.2 λ), and poor conductivity. Fig 3 compares the antenna resistance for a horizontal dipole over ground with two conductivities predicted by NEC-3, compared with MININEC3 (ELNEC version). NEC-3 is considered to be the industry standard, and the complexity of problems that can be solved by the program is continually being enhanced.[14] The frequency for the comparison in Fig 3 was 4 MHz, however for HF the dominant factor is the dipole height above the ground in wavelengths. Fig 4 compares the antenna gain predicted by NEC-3, compared with MININEC3.

Because MININEC assumes perfect ground for impedance and current calculations, vertical monopoles are modeled as if they employed an extensive radial ground screen (say 120 radials). Therefore the predicted gain will be high with regard to practical antennas employed by radio amateurs, who typically employ a few radials or only a ground stake. The gain is rather well predicted for elevated vertical antennas with elevated radials, as would be expected from the study by Al Christman, KB8I,[15] who showed that a monopole with three or four elevated radials had a gain equivalent to that of a ground-based monopole with 120 buried radials. The gain for center-fed vertical dipoles is also predicted rather well by MININEC. Table 1 compares the gain for two groundplane type antennas: a ground-mounted monopole, and an elevated monopole with elevated radials. For the ground-mounted monopole the comparison is made between a 3.8-MHz ¼-λ monopole with 120 ¼-λ long buried radials computed by Christman using NEC-2 and a monopole with no radials computed by the author using ELNEC.

**Table 1**
**Calculated Power Gain for the NEC-2 and MININEC3 Models of a Vertical ¼-λ Monopole Antenna**

Calculations are for average ground, f = 3.8 MHz. Values are dBi.

**Ground Mounted Monopole**

| *Elevation angle, degr.* | *NEC-2** | *MININEC3*† |
|---|---|---|
| 0 | −∞ | −38.68 |
| 5 | −6.15 | −6.03 |
| 10 | −2.41 | −2.36 |
| 15 | −0.87 | −0.84 |
| 20 | −0.18 | −0.16 |
| 25 | +0.04 | +0.06 |
| 30 | −0.03 | −0.02 |
| 40 | −0.84 | −0.83 |
| 50 | −2.37 | −2.36 |
| 60 | −4.69 | −4.68 |
| 70 | −8.14 | −8.13 |
| 80 | −14.14 | −14.13 |
| 90 | −158.44 | −99.99 |

**Four ¼-λ Radials, Elevated 3 Meters**

| *Elevation angle, degr.* | *NEC-2* | *MININEC3* |
|---|---|---|
| 0 | −∞ | −99.99 |
| 5 | −6.01 | −6.04 |
| 10 | −2.32 | −2.35 |
| 15 | −0.87 | −0.89 |
| 20 | −0.30 | −0.31 |
| 25 | −0.21 | −0.23 |
| 30 | −0.46 | −0.47 |
| 40 | −1.17 | −1.70 |
| 50 | −3.74 | −3.73 |
| 60 | −6.58 | −6.56 |
| 70 | −10.50 | −10.48 |
| 80 | −16.84 | −16.82 |
| 90 | −152.72 | −95.97 |

*With 120 ¼-λ long buried radials
†No radials, but connected to ground

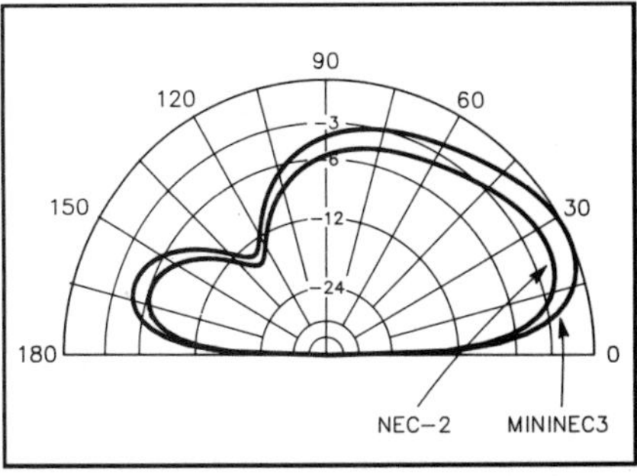


Fig 5—Elevation-plane radiation patterns for a one-radial elevated 3.8-MHz vertical monopole isolated from the support mast (height 4.6 m) calculated by NEC-2 and by MININEC3. The pattern has been calculated for average ground.

Recall that MININEC assumes perfect ground for impedance calculations. MININEC does have an option to include radials, but application of this option over-exaggerates the gain. A comparison between the patterns calculated by these two codes for a "VE2CV Field Day Special" antenna, a ¼-λ 3.8-MHz vertical monopole with one ¼-λ elevated radial, can be seen in Fig 5. The one radial is pointed in the 0° azimuth direction.

While the ground beneath the antenna affects its impedance and gain, the ground in front of the antenna affects its radiation pattern. The shape of the antenna pattern is rather well predicted using MININEC3, because the radiation pattern is calculated using Fresnel reflection coefficients for finitely conducting ground.

### Modeling Transmission Lines

The input impedance of an antenna fed by a parallel-conductor transmission line can be determined by calculating the impedance of the antenna, with the source on the antenna (for example, no feeder). Then, using standard transmission-line equations (or a Smith Chart), one can calculate the input impedance to the transmission line. But this only works for center-fed symmetrical dipoles. For some antennas the feeder is a part of the radiating system, and the feeder must therefore be a part of the numerical model. Examples are off-center-fed dipoles and end-fed Zepp antennas. Parallel-conductor transmission lines can be modeled as follows.

*Two-Source Model*

In this model the feeder is extended to

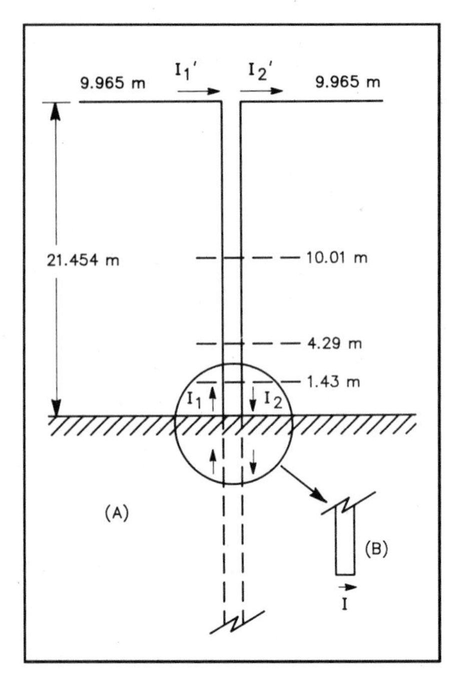


Fig 6—A balanced dipole with tuned feeders. At A, preferred method for numerically modeling a dipole and its transmission-line feeder. At B, an alternative method of modeling that can be used, with caution, where the length of the transmission-line feeder is shorter than the dipole height.

and terminated in the ground plane; if a longer feeder is to be modeled, slope the feeder. Place voltage or current sources with a 0-180° phase relation in series with each conductor, on the segment nearest to ground. This is equivalent to connecting a feeder to a balanced center-tapped voltage or current source. The total input impedance is the sum of the two source impedances. This method avoids the problem of having an abrupt end to the transmission line. In fact, considering the antenna and its image, the feeder in effect "connects through" the ground plane to the image of the antenna below the ground plane (see Fig 6A).

*Single-Source Model*

A disadvantage of the two-source model is that transmission lines shorter than the height of the dipole cannot be modeled. An alternative method is to place a single source (current or voltage) at the center of a short wire connected to the ends of the feeder, Fig 6B. While this method can give good results provided the impedance of the transmission line and hence the conductor spacing is not too small, this method must be used with caution. Not only does the wire on which the source is placed make right-angle intersections with the conductor wires of the transmission line (note the guideline above for wires joining at an angle), but the segment lengths on such a short wire are indeed very short.

A problem with the single-source method is that very short segments must be used near the fed end of the line, comparable with the conductor spacing. Rather than subdividing the full length of a straight parallel-conductor transmission line into many very short segments, the procedure I adopted is to replace each full-length wire by several wires of different lengths. The new wire near the source is the shortest. The second wire is made twice the length of the first (see Fig 6A), and so on until the end of the wire is reached. Each wire is subdivided into the same number of segments, and hence the segment length on which the source is placed can be made very short. I have used four such wires to model transmission lines. The wires connected to the dipole have the same segment length as the segments on the wires of the arms of the dipole, and all lower wires have the same number of segments as the top transmission-line wires.

A full-length feeder yields results that are closely identical with the sectionalized transmission line, providing that a sufficient number of segments are used on the feeder wires and the dipole, that the wire spacing for the transmission line is not too small, and that the two-source model is used.

### Center-Fed Dipole

Let us consider first a dipole with equal arm lengths, fed by a balanced transmission line, Fig 6A. I have modeled numerous such antennas, including the G5RV, in general using current sources. But for a balanced antenna it matters naught whether voltage or current sources are used, whether the feeder is grounded or not, or whether the dipole is horizontal or drooping. The transmission line is nonradiating so long as the arms of the dipole are symmetrical with regard to the feeder. The transmission-line input currents $I_1 = I_2$, and the impedances to these sources $Z_1 = Z_2$. The terminal-to-terminal balanced input impedance is equal to the sum of the source impedances $Z_1 + Z_2$. Also, $I_1' = I_2'$, that is, the currents into each arm of the dipole are equal.

In a case study I modeled a 40-meter dipole (7.15 MHz) fed with a parallel-conductor transmission line. The dipole height was ½ λ above the ground, so that a ½-λ transmission line could be used to feed the dipole. The current distribution on this antenna system (antenna and feeder) calculated by ELNEC is shown in Fig 7.

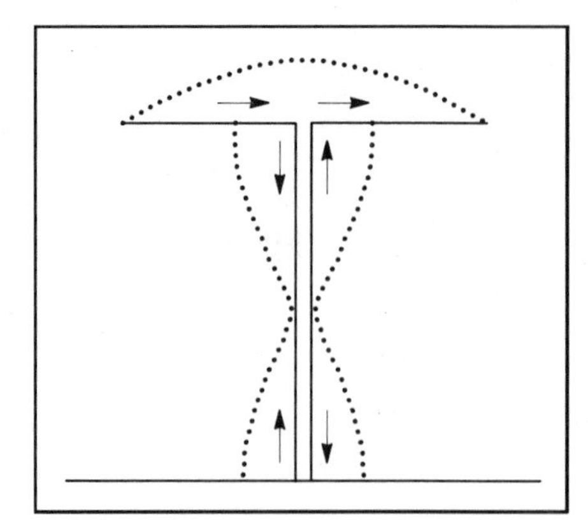

Fig 7—Current distribution on a ½-λ dipole with a ½-λ feeder.

The phase of the current is indicated by the arrows since everywhere the phase is either 0° or 180°, within a few degrees. The current on the two arms of the antenna is in phase, and the current on the two conductors of the transmission line is out of phase and equal in amplitude along the entire length of the transmission line. The radiation pattern, not shown, is entirely horizontally polarized in the plane broadside to the dipole, that is, there is no radiation by the feeder. Clearly for this case study (600-Ω parallel-conductor feeder) MININEC/ELNEC has modeled the antenna and its transmission line very accurately and correctly.

With a half-wavelength feeder the terminating impedance (the dipole impedance) should be reflected to the input, whatever the characteristic impedance of the feeder. For this model a dipole length of 19.93 m and a dipole height of 21.454 m was chosen. The input impedance of the dipole with no feeder, according to ELNEC, is 60 – *j*54.2 ohms. With a 600-Ω parallel-conductor transmission line, conductor spacing 149.25 mm for 2 mm diameter (no. 12 wire), the input impedance for the two-source model is 59.4 – *j*59.9 ohms, according to ELNEC. However the height of the dipole, and hence the feeder length, was selected to provide this agreement, but the feeder length is about right (0.51 λ, which is about a half wavelength).

Recall that one usually assumes a velocity of propagation for open-wire line equal to 0.975, in which case the feeder length should have been 0.4875 λ. But real transmission lines are spaced by insulating spreaders. The input impedance calculated according to the single-source method is 59.2 – *j*57.4 ohms.[16] So far so good!

I repeated the calculation for a 450-Ω feeder (wire spacing 42.7 mm) and for a 300-Ω feeder (wire spacing 12.2 mm); see

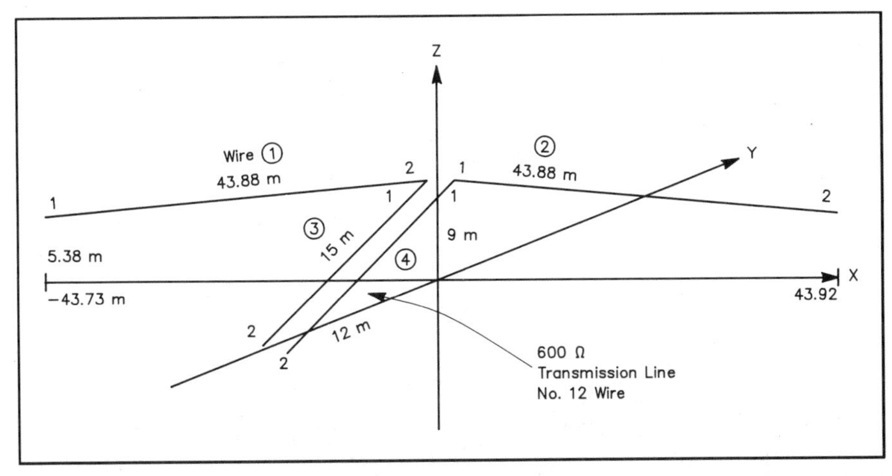


Fig 8—A drooping dipole with tuned feeders.

**Table 2**
**Input Impedance at 7.15 MHz for a Dipole at a Height of ½ λ with a ½-λ Transmission Line**

The dipole impedance (no feeder) is 60 − *j* 54.2 ohms, so a half-wave transmission line should reflect this impedance to its input.

| *Characteristic Impedance of Feeder* | *Two-Source Method* | *Single-Source Method* |
|---|---|---|
| 600 Ω | 59.4 − *j* 59.9 | 59.2 − *j* 57.4 |
| 450 Ω | 55.3 − *j* 99.6 | 55.3 − *j* 101 |
| 300 Ω | 64.0 − *j* 231 | Ridiculous value |

Table 2. The antenna resistance is transferred quite accurately, but the agreement is less satisfactory for the antenna reactance. There is good agreement between the two-source method and the single-source method, except for the closely spaced parallel-conductor transmission line (300-Ω characteristic impedance) where the single-source method failed to give a sensible result. An increasing capacitive reactance with decreasing characteristic impedance of the transmission line (wire spacing) suggests that MININEC has not quite correctly calculated the electrical length of the feeder. (Lengthening the line will reduce the capacitive reactance). While the input impedances to transmission lines having low characteristic impedance are not quite right, this will have little effect on the pattern, and so the transmission line can indeed be included as a part of the model. The single-source method fails (calculated impedance meaningless) when the wire spacing is too small. However the guideline is not respected for such a short end wire.

Finally, while the analysis discussed above seems to be self consistent as judged by using the program itself, we need to validate that MININEC is indeed correctly taking account of the effect of the feeders. A drooping dipole with tuned feeders, Fig 8, has been accurately modeled using NEC-3 for perfect and imperfect ground. Insofar as impedance is concerned, MININEC can model antennas only over perfect ground. Fig 9 compares the input impedance of this dipole with feeders calculated by NEC-3 (for perfect ground)[17] and by MININEC (the ELNEC version). The agreement is good. The two-source model was used for this comparison. Therefore, so long as the dipole is not too close to the ground (height greater than 0.2 λ), its input impedance with tuned feeders is correctly predicted.[18]

### Modeling Off-Center-Fed Dipoles

We will now consider the case where the arms of the dipole are unequal, such as an off-center-fed dipole, and we will model first an ideal OCFD. The arms of the dipole are horizontal, and the dipole height is ½ λ so that half-wave feeders can be used for comparison with the center-fed dipole (Figs 6 and 7). The OCFD is an antenna that is more complicated than one might realize. The current on the antenna system (the arms of the dipole and the transmission line), the antenna impedance (at the input to the transmission line) and the radiation pattern depend on (1) the dipole configuration (drooping or horizontal), (2) the method of support (tree or tower), (3) the length and type of the transmission line (parallel conductor or coax), and (4) the type of feed (voltage or current).

The model assumes a balanced transmission line feeding the antenna. The two sources inject equal voltages or currents having opposite phase (0° and 180°) into the two conductors of the transmission line. The current distribution on the antenna calculated by ELNEC is shown in Fig 10A for current feed, and in Fig 10B for voltage feed. Compare with the current distribution of a center-fed dipole (Fig 7). Even though with voltage feed the input voltages are equal and differ in phase by 180°, the input currents are unequal (Fig 10B), and the phase difference (not shown) is more nearly in phase rather than out of phase (only 17° apart). Normal transmission-line currents should differ in phase by 180°. Also, quite different from the center-fed dipole case, the current maximum on the right-hand feeder wire corresponds with a current minimum on the left hand wire. The currents into each arm of the OCFD are also unequal, and there is a large phase difference (163°). Ideally, the currents into each arm of the dipole should be equal, in both amplitude and phase.

For current feed, Fig 10A, the input currents are equal, and the phase difference is 180° (this was specified to the program for current sources). The currents into each arm of the dipole are approximately equal,

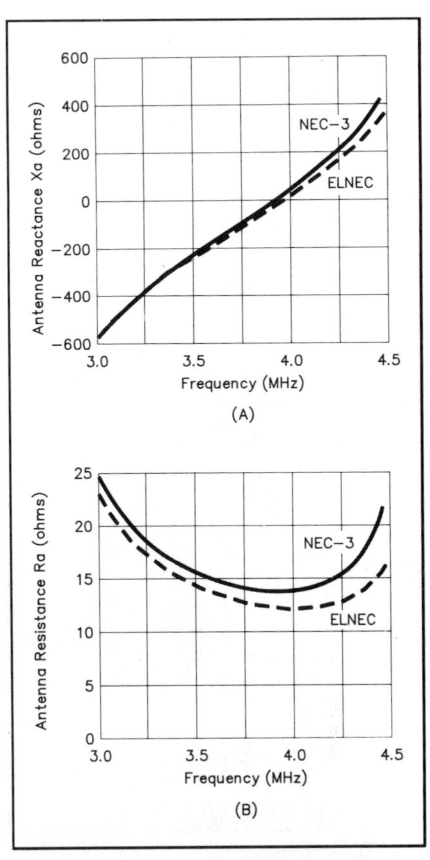


Fig 9—Input impedance of the dipole with tuned feeders (Fig 7), calculated by NEC-3 (for perfect ground) and by ELNEC.

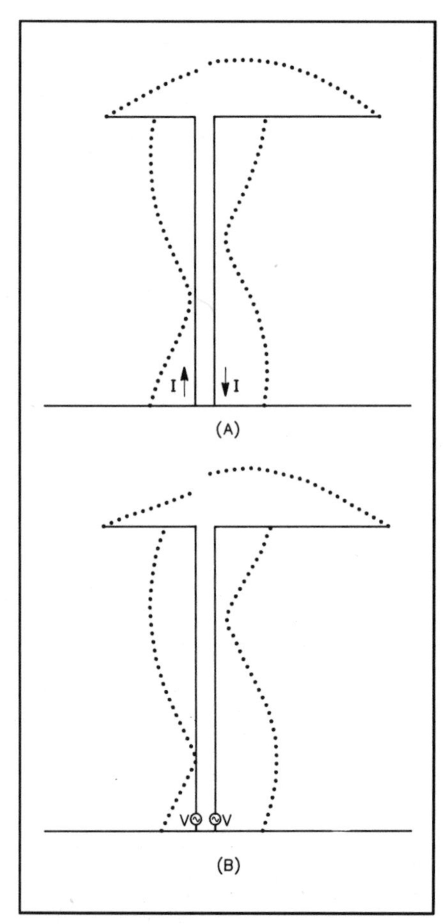

Fig 10—Current distribution on a ½-λ off-center-fed dipole with a ½-λ feeder. At A, with current sources, and B, with voltage sources.

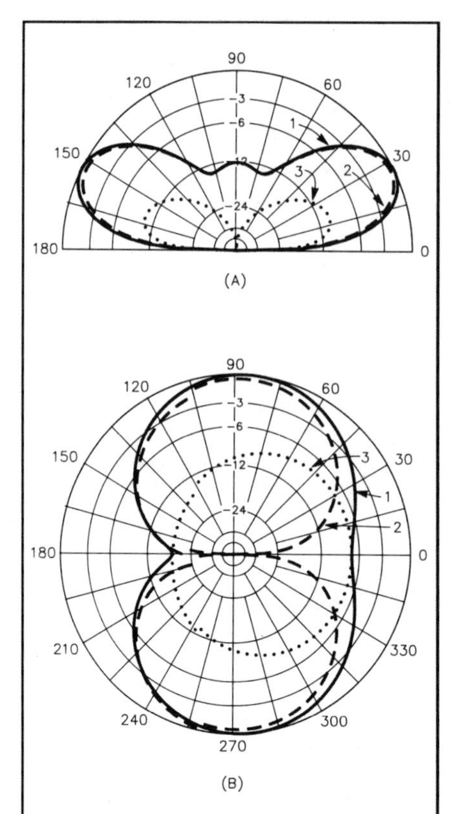

Fig 11—Elevation and azimuth pattern (at 26.5° elevation angle) for the antenna of Fig 10A, with current sources. Curve 1, total field; curve 2, horizontally polarized component, and curve 3, vertically polarized component of field.

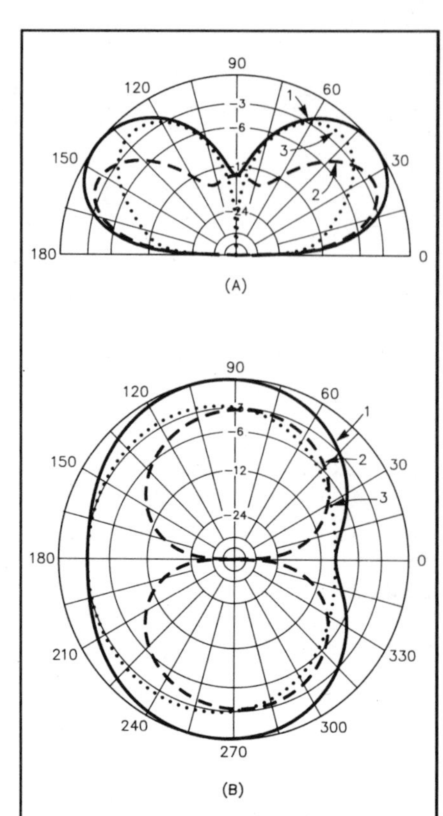

Fig 12—Elevation and azimuth pattern (at 37.5° elevation angle) for the antenna of Fig 10 with voltage sources. Curve 1, total field; curve 2, horizontally polarized component, and curve 3, vertically polarized component of field.

and there is only a small phase difference (about 10°). But this is not always the case; see below.

The input impedances for a 600-Ω transmission line for current feed are

$Z1 = 34 - j295\ \Omega$
$Z2 = 86 + j247\ \Omega$
$Z_{in} = 120 - j48\ \Omega$

and for voltage feed are

$Z1 = 12 - j327\ \Omega$
$Z2 = 63 + j229\ \Omega$
$Z_{in} = 75 - j126\ \Omega$

The impedance of the OCFD with no feeder is 76 – $j$98. ohms. Recall that a nonradiating ½-λ feeder, regardless of its impedance, reflects the antenna impedance to its input. But this is not the case irrespective of the method of feed, current or voltage. The feeder is a part of the radiating system.

Figs 11 and 12 compare the vertical-plane patterns and the azimuthal patterns at the elevation angle of maximum gain for current and voltage feed respectively. There is a marked difference. Feeder radiation is markedly less when the antenna is driven by a current source. The vertically polarized component in the plane broadside to the dipole is 10 dB less than the horizontally polarized component, and the angle of fire is lower by about 8°. With voltage feed the two components in the broadside plane are comparable.

Some readers may say, so what! The radiation of a vertically polarized component, resulting in mixed polarization, may be useful under certain signal-fading conditions. However, it is generally desirable to keep radiation from the line to a minimum. Parallel currents and radiation from the feeder portion of the antenna can give trouble, by inducing currents to flow on other vertical metal structures, for example the neighbor's coaxial TV downlead.

Let us now consider a drooping-dipole configuration of the OCFD. While this is a good practical arrangement that I have used,[19] in retrospect it is a worse case with regard to asymmetry. The arms of the dipole are not symmetrical with regard to to the transmission-line feeder, nor with regard to ground.

In a case study, the dipole arm lengths were 13.8 m and 27.7 m (for 3.5 MHz and higher operation). The apex height was 15.24 m, and the angle between the dipole arms 127°; see Fig 13. The center support is a tree or wooden pole. This study revealed an unexpected result. One of the source resistances calculated by the two-source model with current feed is negative. For example, for this dipole, fed with a 600-Ω transmission line at 7.15 MHz, the impedances are

$Z1 = -185 + j126$ ohms
$Z2 = 1880 - j1141$ ohms, and
$Z_{in} = 1694 - j1014$ ohms

At first this seemed to be an incorrect solution—a negative resistance! However the solution is valid. NEC-3 gives a similar result.[20] If we force equal currents into each conductor of the parallel-conductor line, we may have to take power from one side of the antenna system in order to feed the other conductor with an equal current. This is a tough job for a current balun! The

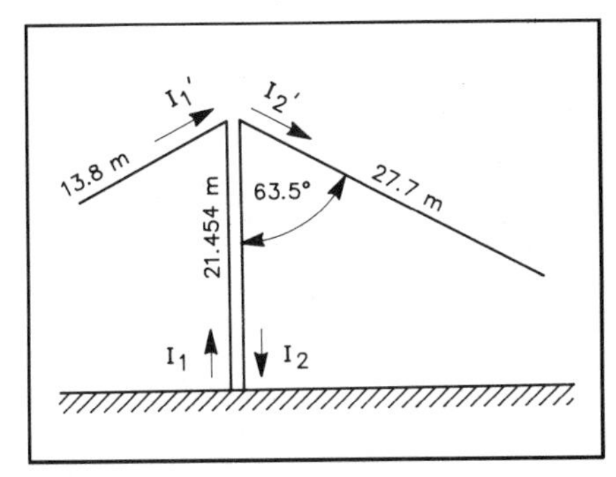


Fig 13—An off-center-fed drooping dipole.

antenna and its transmission line are a coupled system. An additional surprise that emerged from this study was that even for equal input currents I1 and I2, I1′ and I2′ can be unequal (for this case study I1′/I2′ = 2.5 /28°).

With voltage feed the input currents are naturally unbalanced, since the impedance seen by each source is quite different. The input currents are not "forced" to be equal and of opposite phase. The feature where one of the source resistances (always the one on the side of the feeder connected to the short arm of the dipole) can have a negative value has not been found with voltage feed for antenna configurations and frequencies modeled.

The impedance and azimuthal pattern for the "confounded" OCFD depends on the dipole configuration (drooping or horizontal), on the method of feed (parallel-conductor transmission line, or balun and coaxial cable), the length of the feeder, the type of balun employed (voltage or current balun), and the method of support (tree or tower).

A coaxial feeder with a perfect current balun at the antenna can be modeled as follows. Place a current source on the dipole; drop a conductor having the outside diameter of the coax to the ground, or configured in the way of a practical installation, with connection to ground at the transceiver. Insulate this conductor from the dipole. If one does this it becomes very clear that the coaxial feeder is a part of the antenna system: Because this conductor carries a current, it will radiate. The magnitude of the current depends on whether the feeder is grounded or not, and whether it is a resonant length or not.

## Transmission-Line Stubs

Short-circuited transmission line stubs are sometimes used for tuning and phasing antenna elements. For example, consider the 2-element quad beam shown in Fig 14A. The perimeter of the reflector needs to be about 5% greater than the perimeter of the driven element. One way to do this is to make the reflector element physically larger. Another arrangement is to employ a reflector that is the same physical size as the driven element, but in effect to lengthen its perimeter with a shorted transmission-line stub, as suggested in ARRL publications.[21] This is a convenient arrangement because the length of the tuning stub can be adjusted for maximum forward gain, or for maximum front-to-back ratio. Can MININEC model this short-circuited stub as a part of the wire structure, that is, by adding three more wires to the menu of wires? Yes, MININEC correctly models the antenna system including the short-circuited stub.

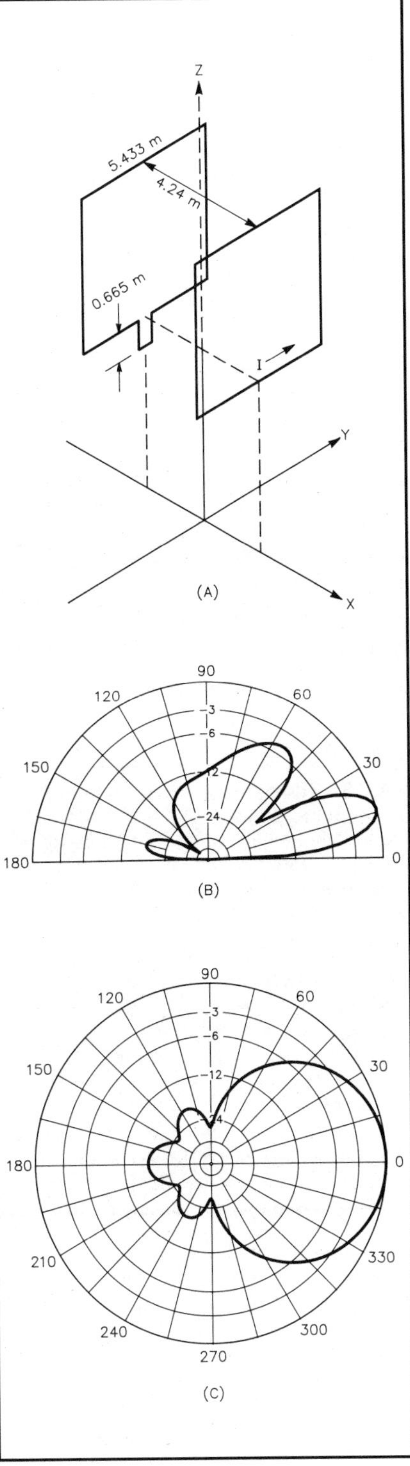


Fig 14—Modeling a short-circuited transmission line as part of the antenna system. At A, a 2-element 20-m quad beam with a short-circuited transmission-line stub to tune and phase the reflector element. At B and C, vertical and azimuthal radiation patterns for this antenna calculated by ELNEC. (The azimuth pattern is taken at 16° elevation.) The director-to-reflector spacing for this case study was 0.2 λ, the height of the quad at its center 0.85 λ (17.96 m). The patterns have been calculated for average ground.

In B and C of Fig 14 are the vertical-plane pattern and principle-plane azimuthal pattern for this quad. That MININEC has correctly modeled this quad beam with its tuning stub can be judged by the fact that the expected pattern is actually predicted,[22] and by favorable comparison with a model employing a physically larger reflector element (not shown).

## Multiwire Fan Monopoles

A multiwire fan monopole was also modeled; see Fig 15A. The impedance versus frequency for this antenna was calculated using ELNEC. Note from Fig 15B that the antenna is resonant and antiresonant at frequencies of 1.8 and 5.5 MHz (when its height is 0.15 λ and 0.45 λ). Above this latter frequency the variation of impedance with frequency is rather erratic. In particular there is a sharp change near 6 MHz. Inspection of the currents on the wires provided the clue to this behavior. The loops formed by the wire structure result in resonances at this frequency. Is this a real effect, or a nuance of the calculation? With the addition of a horizontal wire connected to all wires forming the fan, Fig 16A, we break up these resonances, and the impedance versus frequency is well behaved. See Fig 16B. In this case, this was a real effect. One has confidence in the analysis because, although the currents on the wires are different and the difference changes with frequency, the current distribution on the wires is symmetrical. The current distribution on the two outside wires is the same, as is the current distribution on the two wires next to the outside wires, and so forth.

## Concluding Remarks

In conclusion, MININEC is a powerful tool that can be used to predict antenna performance, providing the antenna problem is correctly modeled and the results of

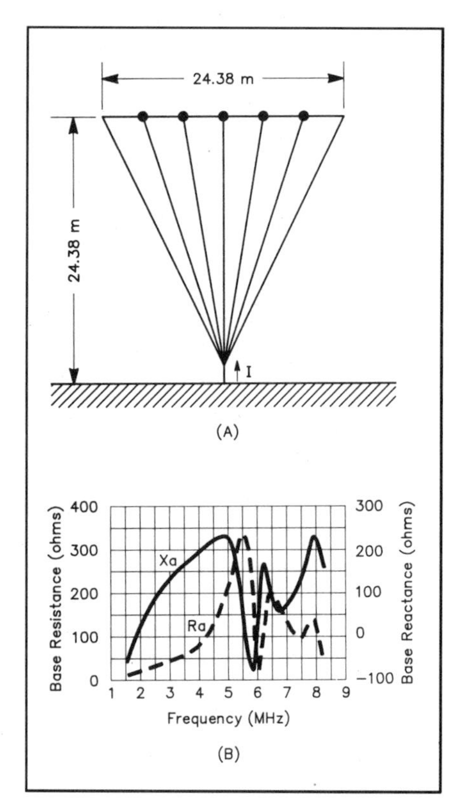


Fig 15—At A, a fan monopole, and at B, its impedance versus frequency, as calculated by ELNEC.

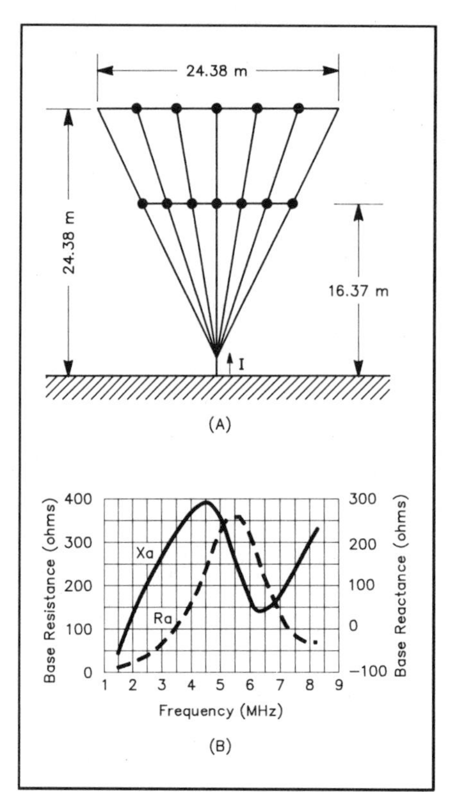


Fig 16—The same antenna as in Fig 15, but with the addition of a horizontal wire connecting all the wires that make up the fan. This added wire breaks up resonances on the structure.

the analysis can be plausibly interpreted.

One has to use many segments when wires are bent to form loops, particularly triangular loops, to model the correct current distribution on the antenna in the vicinity of the corners of the triangle. The calculation of the pattern for such an antenna made up of many segments (say 127) using ELNEC, takes about 1 hour on an IBM compatible PC type 286/4 (4-MHz clock rate), 15 minutes on a 286/12, 2 to 3 minutes on a 386/25 and 1 to 1.5 minutes on a 386/33. Each machine has its own matched math coprocessor.

## Acknowledgment

The author works at the Communications Research Centre, Ottawa, Ontario. He acknowledges discussions with his colleague Max Royer, and with Prof Stanley Kubina, Concordia Univ, Montreal. He particularly thanks Roy Lewellan, W7EL, for the opportunity to test numerous prototypes of ELNEC (beginning with early before-release versions), and for the numerous enjoyable telephone conversations and correspondence.

## Notes

[1] J. W. Rockway, J. C. Logan, D. W. S. Tam and S. T. Li, *The MININEC System: Microcomputer Analysis of Wire Antennas* (Norwood, MA: Artech House, Inc, 1988).

[2] R. P. Haviland, *Practical Antenna Design and Analysis*, a MININEC PC package designed for the Amateur Radio fraternity. Available from Minilab Books, PO Box 21086, Daytona Beach, FL 32019-1086. The ARRL in no way warrants this offer.

[3] MN by Brian Beezley, K6STI. Available from Brian Beezley, 507 ½ Taylor St, Vista, CA 92084. See Product Review by Rus Healy, "MN and Yagi Optimizer Antenna-Analysis Software," *QST*, Aug 1990, pp 41-42.

[4] ELNEC by Roy Lewallen, W7EL. Available from Roy Lewallen, PO Box 6658, Beaverton, OR 97007.

[5] R. Lewallen, "MININEC: The Other Edge of the Sword," *QST*, Feb 1991, pp 18-22.

[6] J. W. Healy, "Correction: May 1990 Antenna Radiation Patterns," Technical Correspondence, *QST*, Sep 1990, pp 38-39.

[7] C. W. Trueman and S. J. Kubina, "The Calculation of Radar-Cross-Section in the HF Band by Wire Grid Modeling," *Report TN-EMC-90-01*, Univ of Concordia, Montreal, QC, 1990.

[8] See Note 1.

[9] Private communications, 1980.

[10] Negative radiation resistances are not a figment of the imagination; they can be real. A negative resistance can be found in the case of multielement phased array antenna systems of particular design, and in case studies of a transmission line feeding an off-center-fed dipole (further discussed in the text). A negative resistance means that one has to take power from, rather than deliver power to, the element having a negative radiation resistance.

[11] The antenna of Fig 2A originally appeared in an article by D. DeMaw (W1FB), "A Closer Look at Horizontal Loop Antennas," *QST*, May 1990, pp 28-29, 35.

[12] J. S. Belrose, G. M. Royer and L. E. Petrie, "HF Wire Antennas over Real Ground: Computer Simulation and Measurement," *AGARD LS 165, Modern Antenna Design Using Computers and Measurement: Application to Antenna Problems of Military Interest*, Sep 1989, pp 6-1 to 6-30. Available from NTIS, Springfield, VA (Ref NASA Acces. No. N90-17932).

[13] See Note 12.

[14] G. J. Burke, "Recent advances to NEC: Application and Validation," *AGARD LS 165* (see Note 12), pp 3-1 to 3-25.

[15] A. Christman, "Elevated Vertical Antenna Systems," *QST*, Aug 1988, pp 35-42.

[16] An 8-wire transmission line was used for the model; each conductor was broken into 4-wires. The wire closest to the source was the shortest. The second wire was twice the length of the first, and so on. Each wire was subdivided into the same number of segments (12), and hence the shortest segment length was 119 mm. For the single-source method, the fed end is short-circuited by a wire that contains the source. This wire was subdivided into four segments. If this source wire is subdivided into two segments, the calculated reactance is ridiculous (for whatever reason).

[17] G. M. Royer, "The Characteristics for a Broadband (2-10 MHz) Switched Drooping Dipole," *Technical Memorandum DRC 87-03*, Communications Research Centre, Department of Communications, Ottawa, May 1987.

[18] Center-fed multiband dipoles with tuned feeders are used by many radio amateurs. The feeders are made resonant by the conjugate matching action of the antenna tuner, and a knowledge of the antenna input impedance is required to design matching circuitry, or to determine if a tuner will match the antenna. For a discussion of this topic see W. Maxwell, *Reflections—Transmission Lines and Antennas* (Newington: ARRL, 1990), pp 20-11 to 20-22.

[19] J. S. Belrose and P. Bouliane, "The Off-Center-Fed Dipole Revisited: A Broadband, Multiband Antenna," *QST*, Aug 1990, pp 28-34.

[20] Burke, private communications, 1991.

[21] *The ARRL 1991 Handbook*, p 17-17, Fig 38.

[22] This quad was dimensioned for resonance at about 14 MHz. In this case study it is interesting to note (not shown) that with an increase or decrease in frequency from antenna-system resonance, the gain decreases, but slowly. However, as the frequency is increased, the F/B ratio increases, from about 8 dB at 14 MHz to 20 dB at 14.35 MHz. Clearly, F/B ratio and gain can be traded off. But the purpose here is not to optimize a quad beam, but to illustrate the applicability of MININEC to model the transmission-line stub used as part of the reflector.

# Smith Chart Impedance Matching on your PC

By Lynn A. Gerig, WA9GFR
RR1, 6417 Morgan Rd
Monroeville, IN 46773

Everyone who does any design of RF circuits sooner or later gets involved with impedance matching. Whether attempting to match the output of one circuit to the input impedance of the following amplifier or needing to match that new antenna design to a 50-Ω coaxial line, impedance matching can be a frustrating experience. Now you can use your IBM compatible PC to provide impedance results for any type of matching network in tabular form or plotted on a Smith Chart in high resolution graphics.

If you have done much impedance matching, you have probably been exposed to the Smith Chart. Although computers are routinely used to perform much number crunching for a variety of applications, many RF types still prefer to use the graphical approach of a Smith Chart which, to those who understand it, provides a visual means of analysis which gives a "feel" not available from viewing long lists of tabulated impedances from a computer printout.

But using a printed Smith Chart form can be frustrating and tedious, with questions arising such as, "Do I follow impedance or admittance curves for a shunt element, and do I multiply or divide by $Z_0$ (the characteristic impedance of the system)?" My MS-DOS computer program SCHART will now put those frustrations behind you. Just input a new matching element and let the computer do the crunching and plotting for you.

It isn't necessary to understand Smith Charts to use this program. Load impedances are entered at each frequency of interest. Next you select a matching element, and the program performs the necessary calculations and lists the resulting impedances at each frequency in tabular form. If your computer has a graphics card, you can optionally plot the Smith Chart results. If Smith Charts are foreign to you, review the references listed at the end of this article for more information.

Various matching network elements (inductors, capacitors, transformers, series transmission lines, open or shorted stubs etc) can be cascaded in any combination. Each time a new element is added, the new resulting impedances are tabulated and can be optionally plotted.

### Using the Program

The program is completely menu driven for ease of use. After an introductory title screen is displayed you must select a graphics option appropriate to your computer hardware. Choices are (1) text only, (2) CGA graphics, or (3) Hercules graphics. You must then input the number of frequencies and the starting impedance (resistive and reactive components) at each frequency.

**Table 1**
**Matching Network Selection Menu**

```
    CHOOSE TYPE OF MATCHING SECTION

 1 SERIES C
 2 SERIES L
 3 SERIES TUNED (SERIES L-C)
 4 SERIES TUNED (PARALLEL L-C)
 5 SERIES TRANSMISSION LINE
 6 SHUNT  C
 7 SHUNT  L
 8 SHUNT  TUNED (SERIES L-C)
 9 SHUNT  TUNED (PARALLEL L-C)
10 SHUNT  TRANSMISSION LINE
11 TRANSFORMER
12 SERIES R
13 SHUNT  R

14 STOP ADDING SECTIONS

15 CALCULATE MODE (L-C-LINE VALUES)

16 REVIEW LAST GRAPHICS SCREEN

WHAT IS YOUR CHOICE (1-16)?
```

After being given the option to print the tabular inputs to your printer and/or plot the initial (load) impedances on a Smith Chart (unless you chose the text only option previously), you are presented with the menu shown in Table 1.

Options 1 through 13 lead to appropriate questions about the component values. Select 14 to exit the program. If you wish to review your last graphics screen, selecting 16 will give you a look at the most recently plotted Smith Chart.

Option 15 places you in a mini calculator mode. A typical matching situation may require a series inductor having a reactance of 75 Ω, but what value of inductance do you need? Or perhaps you want to add a 20° length of transmission line, but how many inches do you input? Option 15 will give you those answers and then return you to this menu.

After each new element value is entered, a tabular listing of the resulting impedances is shown on the screen, then the menu of choices shown in Table 2 is presented. Plotting options are omitted if you did not previously choose a graphics option.

The tabulated impedance values can optionally be plotted on the last Smith Chart viewed or on a new clean chart. Constant SWR circles can be superimposed on the plot to let you quickly determine whether or not your results are within a particular SWR range.

The results for each matching element selected are only "temporarily" stored in memory until Option 1 is selected. If you don't like the most recent "trial" network, selecting Option 3 will cause the trial values to be discarded, and you will be returned to the menu shown in Table 1 to select a new circuit or component value.

Once Option 1 is selected, the current

**Table 2**
**Trial Value Selection Menu**

```
1 --> GOOD VALUE:  KEEP & PROCEED
2 --> PRINT LATEST ITERATION & VALUES ON PRINTER
3 --> BAD VALUE:  DISCARD & TRY A NEW CIRCUIT
4 --> PLOT ON CLEAN CHART
5 --> PLOT ON LAST CHART
6 --> PLOT CONSTANT VSWR CIRCLES
```

impedance values become the new impedance to match. You are then returned to the menu shown in Table 1 where you can select a new matching element to be cascaded to your previous network.

Option 2 will cause the tabulated impedances to be sent to your printer along with the current element type and value shown at the top of the screen. To print the graphical Smith Chart, you must press the <SHIFT>+<PRT-SCRN> keys while viewing the chart (see "Printing Results" below).

Two examples shown below demonstrate the use of the program.

### Matching Examples

When I erected my 60-foot tower, I chose to feed the top set of guy wires as 40-meter slopers. Although insulators were inserted appropriately to make the radiators the proper length, they were not resonant because of coupling to the tower. The following series of typical screens provide a working "walk through" of the procedure I used to provide a good 50-Ω match. The starting impedances measured at the shack end of the coax feed line were first input as follows.

```
WHAT IS THE CHARACTERISTIC IMPEDANCE (IN OHMS)? 50

HOW MANY FREQUENCIES (1-10)? 4

INPUT FREQUENCY 1 IN MHZ? 7.0
INPUT RS, XS  OF LOAD AT  7 MHZ? 52,-110
INPUT FREQUENCY 2 IN MHZ? 7.1
INPUT RS, XS  OF LOAD AT  7.1 MHZ? 57,-100
INPUT FREQUENCY 3 IN MHZ? 7.2
INPUT RS, XS  OF LOAD AT  7.2 MHZ? 65,-90
INPUT FREQUENCY 4 IN MHZ? 7.3
INPUT RS, XS  OF LOAD AT  7.3 MHZ? 70,-80
```

After I input the load impedances at each frequency, the following screen listed them in tabular format for me, along with presenting me with the option to start over, print them, and/or plot them.

THESE WERE YOUR LOAD IMPEDANCE INPUTS:

| FREQ (MHZ) | RS (OHMS) | XS (OHMS) |
|---|---|---|
| 7.000 | 52.000 | -110.000 |
| 7.100 | 57.000 | -100.000 |
| 7.200 | 65.000 | -90.000 |
| 7.300 | 70.000 | -80.000 |

```
ARE YOU SATISFIED (Y=YES)? y
PRINT LOAD VALUES ON PRINTER (Y=YES)? n
PLOT LOAD IMPEDANCE (Y=YES)? y
```

Although I could have obtained a reasonable match with a series inductor, I wanted to use leftover lengths of coaxial cable to gain experience in matching with lines/stubs. I selected a series line as follows.

```
ADD SERIES TRANSMISSION LINE
WHAT IS LINE IMPEDANCE (OHMS)? 50
WHAT IS VELOCITY FACTOR? .67
WHAT IS LENGTH (IN INCHES)? 130
```

| FREQ (MHZ) | RS (OHMS) | XS (OHMS) |
|---|---|---|
| 7.000 | 9.747 | -25.467 |
| 7.100 | 11.601 | -23.891 |
| 7.200 | 14.152 | -22.968 |
| 7.300 | 16.492 | -21.892 |

```
1 --> GOOD VALUE:  KEEP & PROCEED
2 --> PRINT LATEST ITERATION & VALUES ON PRINTER
3 --> BAD VALUE:  DISCARD & TRY A NEW CIRCUIT
4 --> PLOT ON CLEAN CHART
5 --> PLOT ON LAST CHART
6 --> PLOT CONSTANT VSWR CIRCLES
```

I plotted the values on the previous Smith Chart, and then selected Option 1, which permitted me to match the values shown above with a shunt shorted stub as shown below.

```
ADD SHUNT TRANSMISSION LINE
WHAT IS LINE IMPEDANCE (OHMS)? 50
WHAT IS LINE VELOCITY FACTOR? .67
WHAT IS LENGTH (IN INCHES)? 100
OPEN (O) OR SHORTED (S) STUB? S
```

| FREQ (MHZ) | RS (OHMS) | XS (OHMS) |
|---|---|---|
| 7.000 | 74.465 | -11.641 |
| 7.100 | 59.703 | -8.105 |
| 7.200 | 51.398 | -1.240 |
| 7.300 | 45.365 | 2.915 |

I plotted the resulting impedances on the last chart, then I plotted a constant SWR circle of value 2. A quick look at the chart showed that the results were well within an SWR of 2:1 across the band. Fig 1 is a screen dump showing the 2:1 SWR circle with the final, intermediate, and initial impedances all superimposed on the same Smith Chart.

I am presently using this network built from RG-58 cables. The results, as measured with an HP Vector Impedance Meter, are within 5% of the calculated values.

A few years ago I was asked to broadband a "rubber duckie" antenna on a manpack radio across the 110-160 MHz military communications band. That was an interesting challenge because the scenario was equivalent to extending the range of a 2-meter handheld (for which the antenna looks like about 70 Ω at resonance, including ground losses) to cover the aircraft and public service bands. The results of that challenge are shown below.

THESE WERE YOUR LOAD IMPEDANCE INPUTS

| FREQ (MHZ) | RS (OHMS) | XS (OHMS) |
|---|---|---|
| 110.000 | 45.000 | -90.000 |
| 120.000 | 50.000 | -50.000 |
| 130.000 | 70.000 | 0.000 |
| 146.000 | 80.000 | 60.000 |
| 160.000 | 140.000 | 90.000 |

ADDED SHUNT INDUCTOR OF .06 UH

| FREQ (MHZ) | RS (OHMS) | XS (OHMS) |
|---|---|---|
| 110.000 | 17.667 | 60.522 |
| 120.000 | 40.563 | 49.101 |
| 130.000 | 23.026 | 32.888 |
| 146.000 | 12.344 | 37.291 |
| 160.000 | 12.072 | 47.357 |

ADDED SERIES CAPACITOR OF 60 PF

| FREQ (MHZ) | RS (OHMS) | XS (OHMS) |
|---|---|---|
| 110.000 | 17.667 | 36.408 |
| 120.000 | 40.563 | 26.997 |
| 130.000 | 23.026 | 12.483 |
| 146.000 | 12.344 | 19.122 |
| 160.000 | 12.072 | 30.779 |

ADDED SHUNT CAPACITOR OF 30 PF

| FREQ (MHZ) | RS (OHMS) | XS (OHMS) |
|---|---|---|
| 110.000 | 90.944 | 12.622 |
| 120.000 | 40.831 | -26.883 |
| 130.000 | 28.777 | -5.408 |
| 146.000 | 36.322 | 14.319 |
| 160.000 | 87.670 | -15.883 |

Fig 2 shows a Smith Chart plot where I chose to include only the initial and final impedances. A 2:1 SWR circle was also plotted showing that the results were within a reasonable range.

### Printing Results

All text is sent to your computer's default printer port as standard ASCII characters using the BASIC LPRINT command. This straight text will provide tabular impedance listings on any type of printer—dot matrix, laser printer, or daisy wheel.

For high-resolution screen dumps of Smith Chart plots, you will need a printer capable of printing graphics (no daisy wheels). It will be necessary to run the appropriate high resolution screen dump utility file for your printer from DOS before running the SCHART program. For most users this will be the file named GRAPHICS.COM, one of the many files which came with your DOS (disk operating system). There are some exceptions. Some Zenith users will find programs such as PSCMX80.COM (for Epson compatible printers) furnished with Zenith's own version of DOS. If you have a Hercules compatible video system, use the appropriate software furnished with your graphics card. If you have an HP compatible LaserPrinter, look to your local bulletin board or

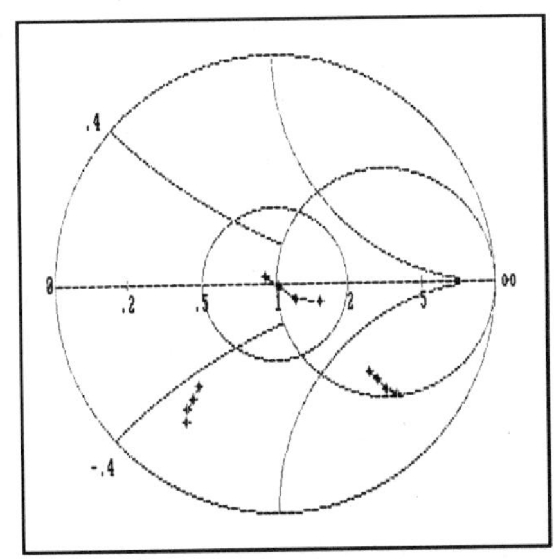


Fig 1—Plot from matching 40-meter sloper antennas. The larger circle tangent to the outer rim of the chart at the right is the 1.0 resistance circle, and the arcs touching the outer rim at top and bottom center are the 1.0 reactance circles. Three impedance groups are shown, as discussed in the text. The smaller circle centered on chart center represents a constant 2:1 SWR, and nicely encloses the impedance group obtained from matching.

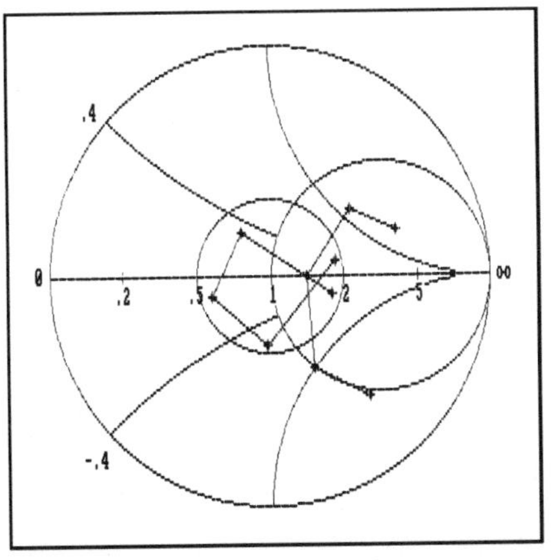


Fig 2—Plot from broadbanding a 2-meter "rubber duckie" type antenna; see text. The initial impedance group is to the right of center, almost contained in the 1.0 resistance circle, while the matched-load impedances are enclosed by the 2:1 SWR circle. Intermediate impedance groups from the matching procedure are not plotted.

other source of shareware programs for a utility such as LG.COM. The GRAPHICS.COM utility provided with DOS 5 supports laser printers.

With the appropriate program in memory (waiting in the background), pressing <SHIFT>+<PRT-SCRN> while viewing the Smith Chart will cause the graphical impedance plot to be transferred to your printer in a special graphics mode. Because of slight differences between screen aspect ratios and printer aspect ratios (horizontal versus vertical dots per inch) the Smith Chart might not print as a perfect circle, but it will be close.

**Program Description**

The source code for my program is listed as Program 1 (at the end of this article), and it is included on the companion diskette as the file SCHART.BAS. (See **Diskette Availability** on an early page of this book.) The various sections of the program are described below. Note that the source code does not use line numbers and therefore will not run under interpretive BASIC (BASICA or GWBASIC). If you have a modern BASIC compiler you can load and run the program in that environment (see "For Programmers" below). However, a compiled version of the program, which will run directly in your DOS environment, is included on the companion disk as the file SCHART.EXE.

Equations for each type of matching network accommodated by the program are included in the Appendix. If you don't have a computer which runs MS-DOS, you can use that information to develop a program for your own system. The equations are valid for use with any calculator. As a matter of interest, my first version of this program was developed for the HP-9830 Desktop Calculator.

The first few program lines are remarks documenting the program history. The section beginning at the label "START:" prints the initial title screen where the user selects the desired text/graphics option. If CGA or Hercules graphics are chosen, the next lines draw a blank Smith Chart and store it in memory (the GET...BLANK# statement). When using the program, the existing chart is flashed onto the screen when needed—it is not redrawn each time.

The lines following the label "INPUTS:" request user inputs of the working frequencies and impedances. The section after the label "MENU:" prints the menu shown earlier as Table 1 and requests user inputs.

The next 13 sections of the program contain equations for the 13 different types of matching elements which can be selected. The label at the beginning of each section is an abbreviated description of the component selected. For example: SERC for series capacitor, SHL for shunt inductor, SHLINE for shunt transmission line, etc. The equations in each section correspond to those in the Appendix on an element by element basis.

The section beginning at "CALC:" is the mini-calculator for determining component values for a desired impedance or line length.

The code associated with the menu shown earlier in Table 2 begins at the label "MENU2:" where plotting, printing and good/bad component values are selected. If Option 3 is selected, you are returned to the main menu for a new trial network. If Option 1 is selected indicating that you wish to keep the chosen component and proceed further, the code beginning at the label "GOODVAL:" converts the "temporary" impedances resulting from the trial to the new load impedance which you must match. Then you are returned to the main menu. PRINTING, PLOTTING, and SWR circle selections branch to the associated code beginning with labels bearing those names.

**For Programmers**

This program was written using Borland's Turbo BASIC (and Spectra's newer Power BASIC) compiler. The code is broken into modules with appropriate labels to help identify the various sections. (Note: This program does not use line numbers and will not run under interpretive BASICA or GWBASIC). I have compiled and run this code on many different types of computers including an IBM PC XT, PC's Limited Turbo XT, Zenith 158, Compaq, and several newer 386 and 486 machines. Smith Charts have been successfully plotted on a variety of CGA, Hercules, EGA, and VGA graphics systems. Hundreds of copies of a previous version are in use with no known problems. The only incompatibility I know of is that the program won't run on an IBM PC Jr.

Feel free to modify the program for your own purposes. A couple suggestions for expansion are adding graphics routines for higher resolution VGA modes or adding an option to save intermediate data to a disk file for later recall.

Experienced programmers may question my use of long lines of code containing multiple statements and my limited use of remarks/documentation. My programming style is a carry-over from earlier days of programming computers with less than 64k of available memory where each line number and each remark in interpretive BASIC stole several bytes of precious memory, and "compaction" was required to permit programs to fit into memory. In fact, much of this code was imported directly (as a file transfer) from my Commodore 64 version of this program. Another reason is that "publication" was in mind when I generated this program. Using several statements per line and eliminating comments drastically reduces the number of lines to be printed; this saves on the number of published pages which reduces publishers' costs, and it saves trees.

**QuickBASIC Users**

If you compile this program with Microsoft QuickBASIC, a few changes will be necessary. At two places in the program I print a temporary message on the screen, delay 2 seconds, then erase the message. Turbo BASIC and Power BASIC use the "DELAY" command ("DELAY 2" causes the computer to wait 2 seconds before executing the next command). QuickBASIC does not recognize that command but uses the SLEEP command; you must change DELAY to SLEEP.

QuickBASIC V4.5 will run the modified source code, but it does not appear to utilize memory as efficiently as Turbo BASIC. If you attempt to compile and save an executable EXE file to disk with the "Full Menus" option selected and have other programs in background memory, it may report that two equations are too complex to compile. If you encounter this anomaly, turn OFF the Full Menus option and no error messages should be encountered. As an alternative, you can break the two equations into two smaller parts. For example, a line of code in the format

```
X = (A + B) / (C + D)
```

could be broken into the following two lines.

```
X = (A + B)

X = X / (C + D)
```

Another difference is the way the different compilers handle Hercules graphics cards. The source code provided makes a SCREEN 2 call. Turbo BASIC and Power BASIC check your hardware at that point. If you have a CGA card (EGA and VGA are compatible) you are placed in the 640 by 200 pixel graphics mode just as if you had made a SCREEN 2 call from BASICA or from GWBASIC. If you have a Hercules card, appropriate action is taken to plot in the 720 by 348 pixel Hercules graphics mode.

Microsoft QuickBASIC works quite differently. If you have a Hercules card, you must change the SCREEN 2 command to SCREEN 3. In addition, you must run the DOS program MSHERC.COM (which is provided with QuickBASIC) before running the SCHART program on a Hercules system.

**References**

J. R. Fisk, W1HR, "How To Use the Smith Chart," *Ham Radio*, Mar 1978, p 92.

W. I. Orr, W6SAI, *Radio Handbook* (Indianapolis: Howard Sams), 1987, pp 21-13 to 21-17.

P. H. Smith, *Electronic Applications of the Smith Chart* (New York: McGraw-Hill), 1969.

Any recent edition, *The ARRL Antenna Book* (Newington: ARRL).

## Appendix

The convention I used for the various matching elements and impedances is shown below. The equations used for each type of reactive matching element follow.

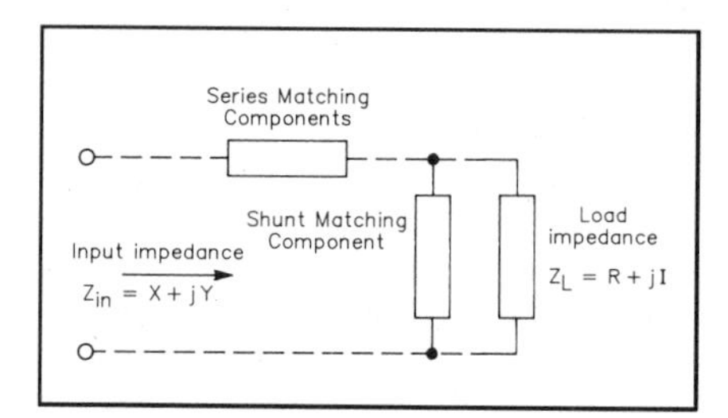


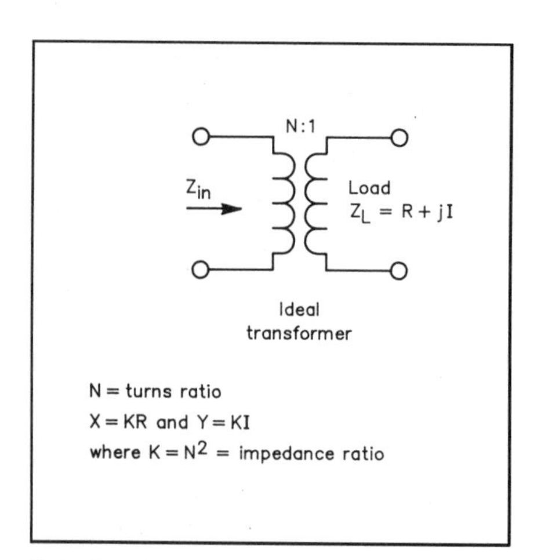


Transformer

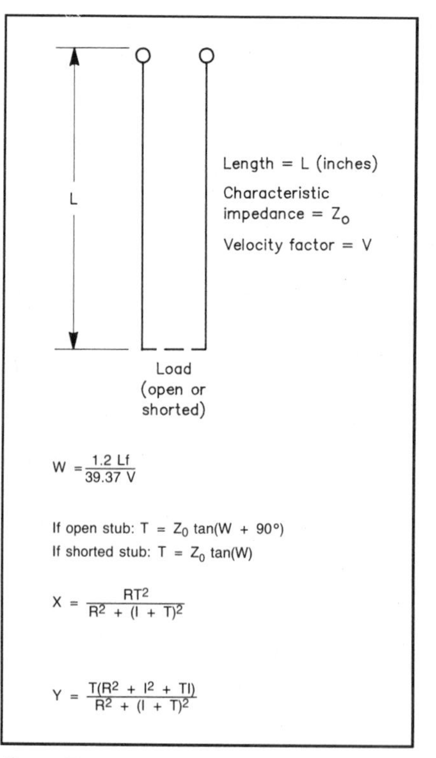


Shunt Transmission Line

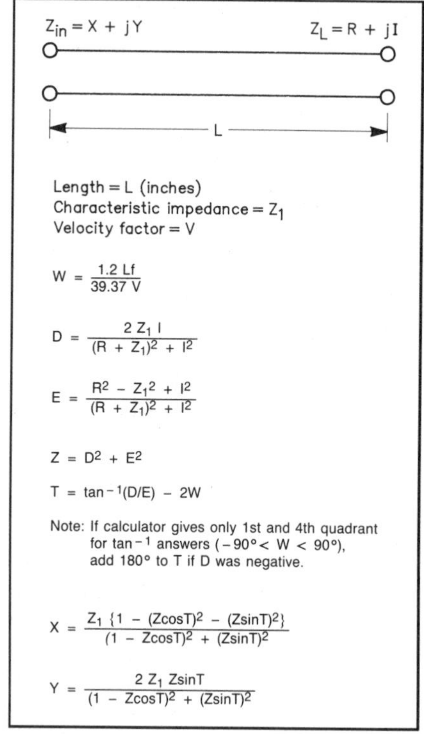


Series Transmission Line

## Appendix (continued)

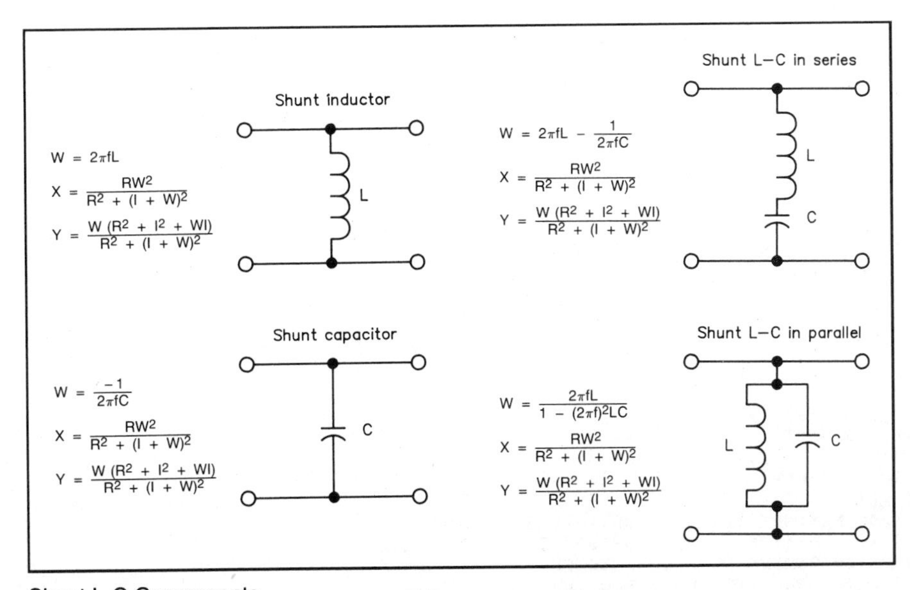


Shunt L-C Components.

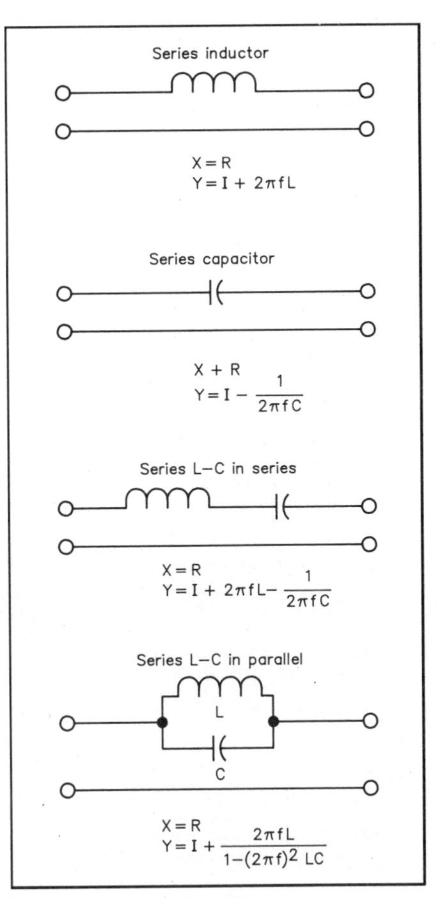


Series L-C Components.

## Program 1
## Source Code for SCHART.BAS

The code is written for the Turbo BASIC (Power BASIC) compiler. It is usable with QuickBASIC with a few modifications (see text) but will not run with interpretive BASIC programs such as GWBASIC and BASICA. This program is available on a companion diskette; see **Diskette Availability** on an early page of this book. Also included on the companion diskette is SCHART.EXE, a compiled version that runs directly from DOS.

```
' ****************************************************************************
' ** PROGRAM BY:     L.A. GERIG, 6417 MORGAN RD., MONROEVILLE, IND  46773   **
' ****************************************************************************
' **                 JAN '82 RF DESIGN MAGAZINE (HP-9830 DESKTOP CALCULATOR)**
' ** DOCUMENTATION   OCT '84 HAM RADIO MAGAZINE (COMMODORE 64 VERSION)      **
' ** HISTORY         JUN '85 RF DESIGN           (COMMORORE 64 VERSION)      **
' **                 MAY '87 COMMUNICATIONS (MS-DOS GW-BASIC CGA GRAPHICS)  **
' **                 ARRL ANTENNA COMPENDIUM, VOLUME 3 (THIS VERSION 2.3)    **
' ****************************************************************************

ON ERROR GOTO ERRORS:KEY OFF

START:
    PI=3.141593:CL$=CHR$(12):DN$=CHR$(31):PT$="U2D4U2R4L8R4E1G2E1H1F2H1
    PRINT CL$DN$,"IMPEDANCE MATCHING PROGRAM WITH SMITH CHART DISPLAY
    PRINT DN$,"          FOR THE IBM-PC OR COMPATIBLES
    PRINT DN$,"             V2.3 COPYRIGHT (C) 1991":PRINT DN$,,"          BY
    PRINT DN$,,"    LYNN A. GERIG":PRINT,," R.R.#1, MONROEVILLE
    PRINT ,,"    IN 46773 USA
    PRINT DN$DN$DN$DN$"SELECT YOUR HARDWARE CONFIGURATION FROM THE FOLLOWING MENU:
    PRINT:PRINT"     1 --> TEXT ONLY (SUCH AS MONOCHROME TEXT CARD)
    PRINT"     2 --> IBM COLOR/GRAPHICS EMULATION--CGA GRAPHICS
    PRINT"     3 --> HERCULES GRAPHICS CARD":PRINT
  INGR:
    INPUT GR
    IF GR<>1 AND GR<>2 AND GR<>3 THEN GOTO INGR
```

```
    IF GR=1 THEN GOTO INPUTS
    IF GR=2 THEN XC=32Ø:YC=1ØØ:XR=217:AR=.4166:XL=1Ø3:XRT=537:YT=1Ø:YB=19Ø
    IF GR=2 THEN YR=9Ø:X1=429:A=175:B=248:C=392:D=465:AA=98:AB=1Ø2
    IF GR=3 THEN XC=36Ø:YC=175:XR=24Ø:AR=.6667:XL=12Ø:XRT=6ØØ:YT=15:YB=335
    IF GR=3 THEN YR=16Ø:X1=48Ø:A=2ØØ:B=28Ø:C=44Ø:D=52Ø:AA=171:AB=179
    SCREEN 2:CLS:PRINT"THANKS FOR WAITING WHILE I DRAW THE CHART."
    CIRCLE (XC,YC),XR,,,,AR
    LINE (XL,YC)-(XRT,YC)
    CIRCLE (XRT,YT),XR,,PI,1.45*PI,AR
    CIRCLE (XRT,YB),XR,,.55*PI,PI,AR
    CIRCLE (X1,YC),XR/2,,,,AR
    LOCATE 13,12:PRINT"Ø"
    LOCATE 13,69:PRINT"oo":IF GR=2 THEN PSET (55Ø,1ØØ):PSET (551,1ØØ)
    IF GR=3 THEN PSET(62Ø,175):PSET(621,175):PSET(62Ø,176):PSET(621,176)
    LOCATE 5,18:PRINT".4":LOCATE 22,18:PRINT"-.4
    FOR J=AA TO AB:PSET (A,J):PSET (B,J):PSET (C,J):PSET (D,J): NEXT J
    LOCATE 14,22:PRINT".2"
    LOCATE 14,31:PRINT".5"
    LOCATE 14,41:PRINT"1"
    LOCATE 14,5Ø:PRINT"2"
    LOCATE 14,59:PRINT"5"
    IF GR=2 THEN CIRCLE (537,325),544,,.626*PI,.742*PI
    IF GR=3 THEN CIRCLE (6ØØ,575),6ØØ,,.626*PI,.742*PI,.6667
    IF GR=2 THEN CIRCLE (537,-125),544,,1.258*PI,1.374*PI
    IF GR=3 THEN CIRCLE (6ØØ,-225),6ØØ,,1.258*PI,1.374*PI,.6667
    DIM BLANK#(27ØØ),FULL#(27ØØ),TEXT#(75Ø)
    IF GR=2 THEN GET (8Ø,9)-(56Ø,191),BLANK#:GET (8Ø,9)-(56Ø,191),FULL#:CLS
    IF GR=3 THEN GET (1ØØ,14)-(63Ø,335),BLANK#:GET (1ØØ,14)-(63Ø,335),FULL#:CLS
    PRINT"TO ENABLE CAPABILITY OF GRAPHICS SCREEN DUMP TO PRINTER (USING THE
    PRINT"<PRINT SCREEN> KEY FROM THE KEYBOARD), YOU MUST RUN THE APPROPRIATE
    PRINT"HI-RES SCREEN DUMP UTILITY FILE FOR YOUR HARDWARE AT THE DOS LEVEL
    PRINT"BEFORE RUNNING THIS PROGRAM."DN$DN$
    PRINT"DO YOU WISH TO PLOT DISCRETE FREQUENCIES?      ( +    +     +)"DN$
    PRINT"OR DO YOU DESIRE A SMOOTH GRAPHICS PLOT?       ( +---+-----+)"DN$
  INPL:
    INPUT"INPUT D OR S";D$
    IF D$="d" THEN D$="D"
    IF D$="s" THEN D$="S"
    IF D$<>"D" AND D$<>"S" THEN GOTO INPL

INPUTS:
   CLS:INPUT"WHAT IS THE CHARACTERISTIC IMPEDANCE (IN OHMS)";ZØ
   PRINT:INPUT"HOW MANY FREQUENCIES (1-1Ø)";N:PRINT
   IF N<1 OR N>1Ø THEN GOTO INPUTS
   FOR J=1 TO N
   PRINT"INPUT FREQUENCY"J;:INPUT"IN MHZ";F(J)
   PRINT"INPUT RS, XS  OF LOAD AT "F(J)"MHZ";:INPUT R(J),I(J)
   X(J)=R(J):Y(J)=I(J):NEXT J:CLS:PRINT
   PRINT"      THESE WERE YOUR LOAD IMPEDANCE INPUTS:":PRINT
   PRINT"      FREQ (MHZ)        RS (OHMS)       XS (OHMS)":PRINT:FOR J=1 TO N
   PRINT USING "    #####.###";F(J);:PRINT USING "       #####.###";R(J);
   PRINT USING "          #####.###";I(J):NEXT J
   LOCATE 18:INPUT"ARE YOU SATISFIED (Y=YES)";A$
   IF A$<>"Y" AND A$<>"y" THEN GOTO INPUTS
   INPUT"PRINT LOAD VALUES ON PRINTER (Y=YES)";A$
   IF A$="Y" OR A$="y" THEN GOSUB PRINTER
   IF GR<>1 THEN INPUT"PLOT LOAD IMPEDANCE (Y=YES)";P$
   IF P$="Y" OR P$="y" THEN GOSUB PLOT

MENU:
   PRINT CL$"    CHOOSE TYPE OF MATCHING SECTION";DN$DN$
   PRINT" 1 SERIES C
   PRINT" 2 SERIES L
   PRINT" 3 SERIES TUNED (SERIES L-C)
   PRINT" 4 SERIES TUNED (PARALLEL L-C)
   PRINT" 5 SERIES TRANSMISSION LINE
   PRINT" 6 SHUNT  C
   PRINT" 7 SHUNT  L
   PRINT" 8 SHUNT  TUNED (SERIES L-C)
   PRINT" 9 SHUNT  TUNED (PARALLEL L-C)
   PRINT"1Ø SHUNT  TRANSMISSION LINE
   PRINT"11 TRANSFORMER
   PRINT"12 SERIES R
   PRINT"13 SHUNT  R
   PRINT DN$"14 STOP ADDING SECTIONS"
   PRINT DN$"15 CALCULATE MODE (L-C-LINE VALUES)"DN$
   IF GR<>1 THEN PRINT"16 REVIEW LAST GRAPHICS SCREEN"DN$
 INGR1:
   IF GR=1 THEN INPUT"WHAT IS YOUR CHOICE (1-15)";M:IF M<1 OR M>15 THEN GOTO INGR1
 INGR2:
   IF GR>1 THEN INPUT"WHAT IS YOUR CHOICE (1-16)";M:IF M<1 OR M>16 THEN GOTO INGR2
   ON M GOTO SERC,SERL,SERS,SERP,SERLINE,SHC,SHL,SHS,SHP,SHLINE
   ON (M-1Ø) GOTO XFMR,SERR,SHR,QUIT,CALC,REVIEW

SERC:
   PRINT CL$;"ADD SERIES CAPACITOR
```

```
   INPUT"WHAT IS VALUE (IN PF)";C
   FOR J=1 TO N
   X(J)=R(J)
   Y(J)=I(J)-1/(2*PI*F(J)*C*.000001)
   NEXT J: GOTO MENU2

SERL:
   PRINT CL$;"ADD SERIES INDUCTOR
   INPUT"WHAT IS VALUE (IN UH)";L
   FOR J=1 TO N:X(J)=R(J)
   Y(J)=I(J)+2*PI*F(J)*L:NEXT J:GOTO MENU2

SERS:
   PRINT CL$;"ADD SERIES TUNED (SERIES L-C)
   INPUT"WHAT IS VALUE OF C (IN PF)";C
   INPUT"WHAT IS VALUE OF L (IN UH)";L
   FOR J=1 TO N: X(J)=R(J)
   Y(J)=I(J)+2*PI*F(J)*L-1/(2*PI*F(J)*C*.000001):NEXT J:GOTO MENU2

SERP:
   PRINT CL$;"ADD SERIES TUNED (PARALLEL L-C)
   INPUT"WHAT IS VALUE OF C (IN PF)";C
   INPUT"WHAT IS VALUE OF L (IN UH)";L
   FOR J=1 TO N: X(J)=R(J)
   Y(J)=I(J)+(2*PI*F(J)*L)/(1-((2*PI*F(J))^2)*L*C*.000001):NEXT J:GOTO MENU2

SERLINE:
   PRINT CL$;"ADD SERIES TRANSMISSION LINE
   INPUT"WHAT IS LINE IMPEDANCE (OHMS)";Z1
   INPUT"WHAT IS VELOCITY FACTOR";V
   INPUT"WHAT IS LENGTH (IN INCHES)";LL
   FOR J=1 TO N
   T=1.2*LL*F(J)/39.37/V
   D=(R(J)+Z1)^2+I(J)^2
   R=(R(J)^2-Z1^2+I(J)^2)/D
   I=2*Z1*I(J)/D
   Z=SQR(R*R+I*I)
   T=180/PI*ATN(I/(R+1E-30))-2*T+180*(R<0)
   R=Z*COS(T*PI/180)
   I=Z*SIN(T*PI/180)
   D=(1-R)^2+I^2
   X(J)=Z1*(1-R^2-I^2)/D
   Y(J)=2*Z1*I/D
   NEXT J:GOTO MENU2

SHC:
   PRINT CL$;"ADD SHUNT CAPACITOR
   INPUT"WHAT IS VALUE OF C (IN PF)";C
   FOR J=1 TO N: W=2*PI*C*.000001
   D=(1-W*F(J)*I(J))^2+(R(J)*W*F(J))^2
   X(J)=R(J)/D
   Y(J)=(I(J)*(1-W*F(J)*I(J))-R(J)^2*W*F(J))/D:NEXT J:GOTO MENU2

SHL:
   PRINT CL$;"ADD SHUNT INDUCTOR
   INPUT"WHAT IS VALUE OF L (IN UH)";L
   FOR J=1 TO N: W=2*PI*F(J)*L
   D=R(J)^2+(I(J)+W)^2
   X(J)=R(J)*W^2/D
   Y(J)=W*(R(J)^2+I(J)^2+W*I(J))/D
   NEXT J:GOTO MENU2

SHS:
   PRINT CL$;"ADD SHUNT TUNED (SERIES L-C)
   INPUT"WHAT IS VALUE OF C (IN PF)";C
   INPUT"WHAT IS VALUE OF L (IN UH)";L
   FOR J=1 TO N
   W=2*PI*F(J)*L-(1000000!)/(2*PI*F(J)*C)
   D=R(J)^2+(I(J)+W)^2
   X(J)=R(J)*W^2/D
   Y(J)=W*(R(J)^2+I(J)^2+W*I(J))/D
   NEXT J:GOTO MENU2

SHP:
   PRINT CL$;"ADD SHUNT TUNED (PARALLEL L-C)
   INPUT"WHAT IS VALUE OF C (IN PF)";C
   INPUT"WHAT IS VALUE OF L (IN UH)";L
   FOR J=1 TO N
   W=(2*PI*F(J)*L)/(1-((2*PI*F(J))^2)*L*C*.000001)
   D=R(J)^2+(I(J)+W)^2
   X(J)=R(J)*W^2/D
   Y(J)=W*(R(J)^2+I(J)^2+W*I(J))/D
   NEXT J:GOTO MENU2

SHLINE:
   PRINT CL$;"ADD SHUNT TRANSMISSION LINE
   INPUT"WHAT IS LINE IMPEDANCE (OHMS)";Z1
```

```
    INPUT"WHAT IS LINE VELOCITY FACTOR";V
    INPUT"WHAT IS LENGTH (IN INCHES)";LL
    LOCATE 4,40:PRINT"OPEN (O) OR SHORTED (S) STUB";
   INS:
    INPUT S$
    IF S$="s" THEN S$="S"
    IF S$="o" THEN S$="O"
    IF S$<>"O" AND S$<>"S" THEN GOTO INS
    FOR J=1 TO N
    T=LL*F(J)*1.2/39.37/V
    IF S$="S" THEN W=Z1*TAN(T*PI/180)
    IF S$="O" THEN W=Z1*TAN((T+90)*PI/180)
    D=R(J)^2+(I(J)+W)^2
    X(J)=R(J)*W^2/D
    Y(J)=W*(R(J)^2+I(J)^2+W*I(J))/D
    NEXT J:GOTO MENU2

XFMR:
    PRINT CL$;"ADD TRANSFORMER
    INPUT"STEP UP OR DOWN (U OR D)";T$
    IF T$<>"U" AND T$<>"D" AND T$<>"u" AND T$<>"d" THEN GOTO XFMR
    INPUT"WHAT IMPEDANCE RATIO";W
    IF T$="D" OR T$="d" THEN W=1/W
    FOR J=1 TO N: X(J)=W*R(J)
    Y(J)=W*I(J): NEXT J: GOTO MENU2

SERR:
    PRINT CL$;"ADD SERIES RESISTOR
    INPUT"WHAT VALUE OF R";RS
    FOR J=1 TO N: X(J)=R(J)+RS
    Y(J)=I(J):NEXT J: GOTO MENU2

SHR:
    PRINT CL$;"ADD SHUNT RESISTOR
    INPUT"WHAT VALUE OF R";RS
    FOR J=1 TO N
    D=(R(J)+RS)^2+I(J)^2
    X(J)=RS*(R(J)^2+RS*R(J)+I(J)^2)/D
    Y(J)=I(J)*RS^2/D: NEXT J: GOTO MENU2

QUIT:
    CLS:INPUT "WANT TO RUN ANOTHER ONE (Y=YES)";A$
    IF LEFT$(A$,1)<>"Y" AND LEFT$(A$,1)<>"y" THEN END
    IF GR=1 THEN GOTO INPUTS
    CLS:IF GR=2 THEN PUT (80,9),BLANK#:GET (80,9)-(560,191),FULL#:GOTO INPUTS
    IF GR=3 THEN PUT (100,14),BLANK#:GET (100,14)-(630,335),FULL#:GOTO INPUTS

PLOT:
    IF GR=2 THEN GET (0,0)-(639,30),TEXT#:CLS
    IF GR=3 THEN GET (0,0)-(700,55),TEXT#:CLS
    IF X=4 AND GR=2 THEN PUT (80,9),BLANK#:GOTO PLOTVAL
    IF X=4 AND GR=3 THEN PUT (100,14),BLANK#:GOTO PLOTVAL
    IF GR=2 THEN PUT (80,9),FULL#
    IF GR=3 THEN PUT (100,14),FULL#
    IF M=16 THEN PRINT"PRESS ANY KEY TO CONTINUE":GOTO INDUMMY
   PLOTVAL:
    FOR J=1 TO N
    D=(X(J)+Z0)^2+Y(J)^2
    PX(J)=((X(J)-Z0)*(X(J)+Z0)+Y(J)^2)/D*XR
    PY(J)=-2*Y(J)*Z0/D*YR
    PSET (PX(J)+XC,PY(J)+YC):DRAW PT$
    IF D$="S" AND J>1 THEN LINE (XC+PX(J-1),YC+PY(J-1))-(XC+PX(J),YC+PY(J))
    NEXT J
   MESSAGE1:
    PRINT"<PRT SCN> FOR SCREEN DUMP TO PRINTER OR ANY KEY TO CONTINUE"
    delay 2:LOCATE 1:PRINT SPACE$(79)
   INDUMMY:
    I$=INKEY$:IF I$="" THEN GOTO INDUMMY
    IF GR=2 THEN GET (80,9)-(560,191),FULL#
    IF GR=3 THEN GET (100,14)-(630,335),FULL#
    CLS:PUT (0,0),TEXT#
RETURN

VSWR:
    LOCATE 18:FOR J=1 TO 6:PRINT SPACE$(79):NEXT J
   BADVSWR:
    LOCATE 21,1:INPUT"WHAT VALUE OF VSWR CIRCLE DO YOU WANT TO PLOT";VS
    IF VS<1 THEN LOCATE 19:PRINT"VSWR > 1.0":GOTO BADVSWR
    IF GR=2 THEN GET (0,0)-(639,30),TEXT#:CLS
    IF GR=3 THEN GET (0,0)-(700,55),TEXT#:CLS
    IF GR=2 THEN PUT (80,9),FULL#
    IF GR=3 THEN PUT (100,14),FULL#
    VR=XR*(VS-1)/(VS+1)
    CIRCLE (XC,YC),VR,,,,AR
    GOTO MESSAGE1
```

```
MENU2:
   LOCATE 6:PRINT"     FREQ (MHZ)        RS (OHMS)       XS (OHMS)"DN$
   FOR J=1 TO N
   PRINT USING "    #####.###";F(J);:PRINT USING "        #####.###";X(J);
   PRINT USING "      #####.###";Y(J):NEXT J
MENU2A:
   LOCATE 18
   PRINT"1 --> GOOD VALUE:  KEEP & PROCEED
   PRINT"2 --> PRINT LATEST ITERATION & VALUES ON PRINTER
   PRINT"3 --> BAD VALUE:  DISCARD & TRY A NEW CIRCUIT
   IF GR=1 THEN GOTO NOGRAPH
   PRINT"4 --> PLOT ON CLEAN CHART
   PRINT"5 --> PLOT ON LAST CHART
   PRINT"6 --> PLOT CONSTANT VSWR CIRCLES                    ";
  NOGRAPH:
   INPUT X
   IF X=1 THEN GOTO GOODVAL
   IF X=2 THEN GOSUB PRINTER:GOTO MENU2A
   IF X=3 THEN GOTO MENU
   IF X=4 AND GR<>1 THEN GOSUB PLOT:GOTO MENU2
   IF X=5 AND GR<>1 THEN GOSUB PLOT:GOTO MENU2
   IF X=6 AND GR<>1 THEN GOSUB VSWR:GOTO MENU2
   GOTO NOGRAPH
  GOODVAL:
   FOR J=1 TO N:R(J)=X(J)
   I(J)=Y(J):NEXT J
   LOCATE 18:FOR J=1 TO 6:PRINT SPACE$(79):NEXT J
   LOCATE 18:PRINT"YOU MUST NOW MATCH THE IMPEDANCES LISTED ABOVE.";
   delay 2:GOTO MENU

CALC:
   PRINT CL$"DO YOU WISH TO CALCULATE  (1) L-C VALUES
   PRINT DN$"                      OR  (2) LINE LENGTHS"DN$
   INPUT "WHAT IS YOUR CHOICE (1 OR 2)";CH
   IF CH<>1 AND CH<>2 THEN GOTO CALC
   ON CH GOTO LCVAL,LINEVAL
  LCVAL:
   PRINT DN$DN$:INPUT"INPUT OPERATING FREQUENCY (MHZ)";OF
  GETX:
   INPUT "INPUT DESIRED REACTANCE (OHMS)";DX
   IF DX=0 THEN GOTO GETX
   IF DX<0 THEN DX=-DX
   IND=DX/(2*PI*OF)
   CAP=1000000!/(2*PI*DX*OF)
   PRINT DN$"USE INDUCTANCE OF ";:PRINT USING "#####.###";IND;:PRINT" UH ";
   PRINT "OR CAPACITANCE OF ";:PRINT USING "#######.##";CAP;:PRINT" PF
   PRINT DN$"FOR REACTANCE OF ";:PRINT USING "#######.##";DX;:PRINT" OHMS AT ";:
        PRINT USING "#####.###";OF;:PRINT" MHZ
   GOTO ENDCALC
  LINEVAL:
   PRINT DN$DN$:INPUT"INPUT OPERATING FREQUENCY (MHZ)";OF
   INPUT "INPUT VELOCITY FACTOR OF YOUR LINE";V
  GETLEN:
   INPUT "INPUT ELECTRICAL LENGTH DESIRED IN DEGREES (0-360)";EL
   IF EL<0 OR EL>360 THEN GOTO GETLEN
   WL=300*39.37/OF
   PRINT DN$"ONE WAVELENGTH IN AIR =";:PRINT USING "######.##";WL;:PRINT" INCHES AT ";:
        PRINT USING "#####.###";OF;:PRINT" MHZ
   L1=INT(100*WL*EL/360*V+.5)/100
   PRINT DN$;:PRINT USING "###.##";EL;:PRINT" DEGREES WITH VEL FACTOR OF ";:
        PRINT USING "##.###";V;:PRINT" = ";:PRINT USING "######.##";L1;:PRINT" INCHES
  ENDCALC:
   PRINT DN$DN$"PRESS ANY KEY TO RETURN TO MATCHING NETWORK MENU"
   I$=INPUT$(1):GOTO MENU

REVIEW:
   CLS:GOSUB PLOT:GOTO MENU

PRINTER:
    LPL$=STR$(L):QQ=INSTR(LPL$,"."):IF QQ>0 THEN LPL$=LEFT$(LPL$,QQ+2)
    LPC$=STR$(C):QQ=INSTR(LPC$,"."):IF QQ>0 THEN LPC$=LEFT$(LPC$,QQ+2)
    IF S$="O" THEN LPS$="OPEN STUB SHUNT TRANSMISSION LINE
    IF S$="S" THEN LPS$="SHORTED STUB SHUNT TRANSMISSION LINE
    LPW$=STR$(W):QQ=INSTR(LPW$,"."):IF QQ>0 THEN LPW$=LEFT$(LPW$,QQ+2)
    LPR$=STR$(RS):QQ=INSTR(LPR$,"."):IF QQ>0 THEN LPR$=LEFT$(LPR$,QQ+2)
    LPV$=STR$(V):QQ=INSTR(LPV$,"."):IF QQ>0 THEN LPV$=LEFT$(LPV$,QQ+2)
    LPLL$=STR$(LL):QQ=INSTR(LPLL$,"."):IF QQ>0 THEN LPLL$=LEFT$(LPLL$,QQ+2)
    LPZ1$=STR$(Z1):QQ=INSTR(LPZ1$,"."):IF QQ>0 THEN LPZ1$=LEFT$(LPZ1$,QQ+2)
    IF T$="U" OR T$="u" THEN LPT$="STEP UP TRANSFORMER OF IMPEDANCE RATIO"
    IF T$="D" OR T$="d" THEN LPT$="STEP DOWN TRANSFORMER OF IMPEDANCE RATIO"
    IF M=0 THEN LPRINT"THESE WERE YOUR LOAD IMPEDANCE INPUTS"
    IF M=1 THEN LPRINT"ADDED SERIES CAPACITOR OF";LPC$;" PF
    IF M=2 THEN LPRINT"ADDED SERIES INDUCTOR OF"LPL$" UH
    IF M=3 THEN LPRINT"ADDED SERIES TUNED (SERIES L-C)
    IF M=3 THEN LPRINT"L="LPL$" UH AND C="LPC$" PF
    IF M=4 THEN LPRINT"ADDED SERIES TUNED (PARALLEL L-C)
    IF M=4 THEN LPRINT"L="LPL$" UH AND C="LPC$" PF
```

```
    IF M=5 THEN LPRINT"ADDED"LPLL$" INCHES OF SERIES TRANSMISSION LINE
    IF M=5 THEN LPRINT"OF"LPZ1$" OHM IMPEDANCE AND VELOCITY FACTOR OF"LPV$
    IF M=6 THEN LPRINT"ADDED SHUNT CAPACITOR OF"LPC$" PF
    IF M=7 THEN LPRINT"ADDED SHUNT INDUCTOR OF"LPL$" UH
    IF M=8 THEN LPRINT"ADDED SHUNT TUNED (SERIES L-C)
    IF M=8 THEN LPRINT"L="LPL$" UH AND C="LPC$" PF
    IF M=9 THEN LPRINT"ADDED SHUNT TUNED (PARALLEL L-C)
    IF M=9 THEN LPRINT"L="LPL$" UH AND C="LPC$" PF
    IF M=1Ø THEN LPRINT"ADDED"LPLL$" INCHES OF "LPS$
    IF M=1Ø THEN LPRINT"OF"LPZ1$" OHM IMPEDANCE AND VELOCITY FACTOR OF"LPV$
    IF M=11 THEN LPRINT"ADDED "LPT$;LPW$
    IF M=12 THEN LPRINT"ADDED SERIES RESISTOR OF"LPR$" OHMS
    IF M=13 THEN LPRINT"ADDED SHUNT RESISTOR OF"LPR$" OHMS
    LPRINT:LPRINT"     FREQ (MHZ)        RS (OHMS)        XS (OHMS)":LPRINT
    FOR J=1 TO N
    LPRINT USING "    #####.###";F(J);:LPRINT USING "       #####.###";X(J);
    LPRINT USING "       #####.###";Y(J):NEXT J:LPRINT:LPRINT:LPRINT
RETURN

ERRORS:
    BEEP:IF ERR<>25 AND ERR<>27 AND ERR<>57 THEN GOTO ERROR2
    PRINT:PRINT"PRINTER NOT READY--PRESS ANY KEY WHEN PRINTER IS READY"
    PPP$=INPUT$(1):RESUME
ERROR2:
    ON ERROR GOTO Ø
```

# Horizontal Antennas and the Compound Reflection Coefficient

By Charles J. Michaels, W7XC
13431 N 24th Ave
Phoenix, AZ 85029

One night on 40 meters there was a discussion on certain details of the pattern of an Extended Double Zepp over real earth. Analysis required the ability to plot patterns in the vicinity of the nulls and secondary lobes to a fine resolution. It appeared that although current antenna articles contain many real earth patterns at any required azimuth and wave angle, such patterns are hardly ever seen in older articles.

Since the method of calculating such real earth patterns is conceptually rather obvious, the lack of them prior to the advent of NEC and MININEC would appear to be primarily due to the tedious computations involved. However, the computations are very reasonable with a minimal personal computer or even a handheld programmable scientific calculator with program loading means such as the Hewlett-Packard HP41C. Many hams have computers which can't handle MININEC but can handle field calculations using algebraic-trigonometric solutions for sinusoidal current distributions.

Equations for fields of horizontal antennas above real earth in the plane broadside to the antenna, where polarization is purely horizontal, and in the vertical plane containing the antenna, where polarization is purely vertical, are found in some texts. All that seemed lacking was equations for the fields at azimuth and wave angles between the principal planes where the field is a mixture of horizontally and vertically polarized waves. A search for such equations yielded some interesting findings.

### The Search

Textbooks on hand and others borrowed from the library or friends were consulted. A representative group treating horizontal antennas over real earth was selected.[1-8]

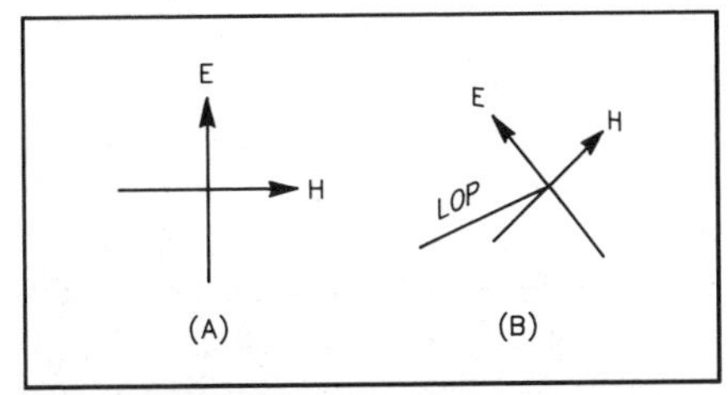


Fig 1—(A) Electromagnetic wave coming out of page. E and H fields are perpendicular to each other. (B) H and E fields are also perpendicular to the line of propagation (LOP) coming out of page at angle and to the reader's right.

All discuss the effects of perfect and real earth to various degrees. There are differences among them, most of which are correct in context. Some treat only pattern shapes—others include field strength calculations. Some present them only in context of half-wave dipoles. Some devote separate sections to patterns for some particular antenna lengths when a general equation could cover many, leaving the intermediate lengths untreated. Some contain misleading or ambiguous statements—some, outright errors.

In general, treatments start with an equation for free-space patterns of infinitely thin antennas then present a multiplying expression to yield patterns over perfect earth—most stating, "*in the plane perpendicular to the antenna*," and rarely, "*in the vertical plane containing the antenna*." The reasonable reader would probably assume this to exclude other azimuth and wave angles; actually, it doesn't.

The simplifying assumption of an infinitely thin antenna is acceptable since the radiation resistance and loop currents involved are very close to those at the loop(s) of typical amateur MF and HF antennas of finite diameter as opposed to those at the feed point which may vary widely with antenna diameter-to-length ratio. Of course, for some antennas such as the center-fed half wave, the feed point *is* at a current loop.

None provide an explicit equation for patterns between the principal planes where the field is a mixture of horizontally and vertically polarized waves for the real earth case and only Terman[9] appears to state one as applying to the perfect-earth case.

Finally, taking the bull by the horns, an effort was undertaken to derive the required equations. A study of the perfect-earth case provides a basis for that effort.

### Defining Polarization

A clear understanding of polarization is vital to understanding the formation of antenna patterns. An electromagnetic wave comprises an oscillating Electric Vector, E, and an oscillating Magnetic Vector, H, perpendicular to each other and to the "line of propagation" (LOP) of the wave. See Fig 1.

Referring to Fig 2A—consider a wave incident on and reflecting from a surface oriented such that the E vector is parallel to the surface. The LOP of this incident wave and that of the reflected wave lie in a plane perpendicular to the surface and at an angle $\psi$ (the wave angle) to the surface. This wave is "horizontally polarized" or has "horizontal polarization." The H vector, which is always perpendicular to the E vector, lies in the perpendicular plane and at an angle of 90° – $\psi$ to the surface. The angle of reflection is the same as the angle of incidence.

Referring to Fig 2B—now consider a wave incident on and reflecting from a surface oriented such that the H vector is parallel to the surface. Again the LOP for

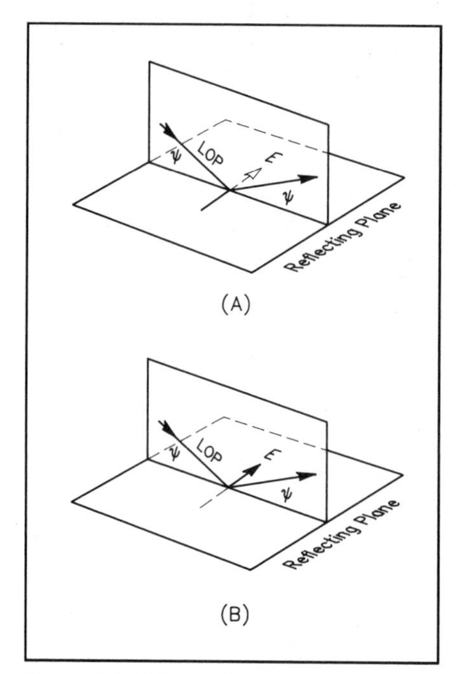


Fig 2—(A) EM wave incident on a reflecting plane at wave angle ψ. The LOP of the incident and reflected waves lie in the vertical plane. The E vector lies in the reflecting plane. This is horizontal polarization. The H vector, not shown, is in the vertical plane and perpendicular to the LOP and the E vector. (B) EM wave incident on a reflecting plane at wave angle ψ. The LOP of the incident and reflected waves lie in the vertical plane. The E vector is in the vertical plane and at an angle to the reflecting plane. The H vector, not shown, is in the horizontal plane perpendicular to the LOP and the E vector. This is vertical polarization. Note: the E vector is not perpendicular to the reflecting plane except when ψ is 0.

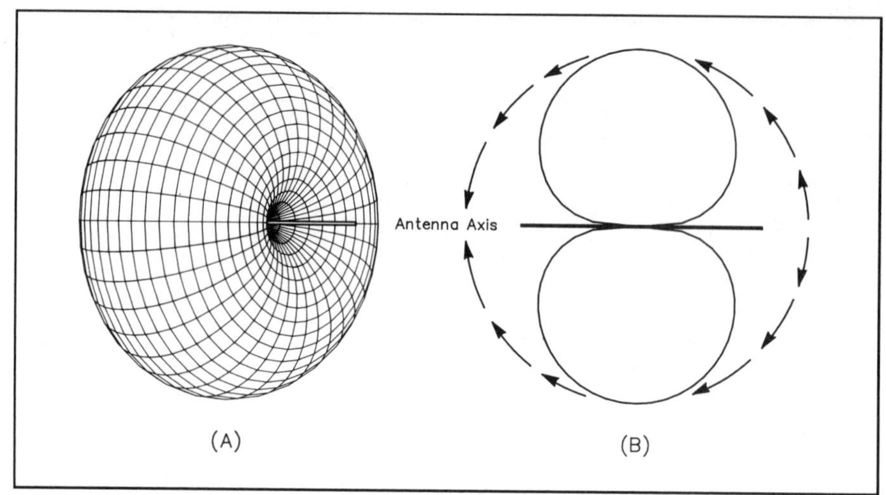


Fig 3—(A) Field of half-wave dipole in free space. (B) Pattern in any plane containing antenna forms typical figure 8. Note that the electric vector directions are symmetrical about the antenna axis.

the incident wave and the reflected wave lie in the plane perpendicular to the surface. This wave is "vertically polarized" or has "vertical polarization." The E vector, which is always perpendicular to the H vector, lies in the perpendicular plane at an angle of 90° – ψ to the surface.

It is important to note that in the vertically polarized case, contrary to some common misconceptions and texts,[10] the E vector is *not* perpendicular to the surface except when the wave angle is zero.

If the wave angle is 90°, both the E and H vectors lie in the surface and the wave may be considered as having either polarization. Although either the E or the H vectors may be used in calculations, the E vector is usually used.

### Reflection in the Perfect-Earth Case

The reflected waves of horizontally and vertically polarized waves from a perfectly conducting surface are very different. A perfectly conducting surface cannot have a tangential electric force. Nature takes care of this by making the simultaneously present horizontally polarized reflected E vector reverse its phase relative to the incident wave so that they cancel in the surface. Therefore:

- The reflected horizontally polarized wave, $E_r$, is of the same amplitude as the incident E vector but its phase is changed by –180°.
- The reflected vertically polarized wave, $E_r$, is equal in amplitude and of the same phase as the E vector of the incident wave.

The ratio of a reflected wave, $E_r$, to the incident wave, $E_i$, is called the *reflection coefficient*. The reflection coefficients are of the form $R_h \underline{/\, r_h}$ and $R_v \underline{/\, r_v}$, where $R_h$ and $R_v$ are the attenuation constants and $r_h$ and $r_v$ are the respective associated phase angles. Thus, for this perfect-earth case, since there is no earth loss:

$E_r = E_i \times 1 \underline{/\, -180^\circ}$ for horizontally polarized waves

$E_r = E_i \times 1 \underline{/\, 0^\circ}$ for vertically polarized waves

### The Free-Space Center-Fed Antenna

Eq 1 may be used to calculate the three-dimensional pattern of center-fed antennas of arbitrary length in free space having sinusoidal current distribution. For simplicity, a half-wave dipole pattern is shown in Fig 3A. It has the classic "doughnut" shape. Because dipole free-space patterns are symmetrical about the wire axis, the half-wave dipole example pattern can be shown as in Fig 3B for any plane in which the antenna lies. Note the symmetry of the direction of the E vectors. This will be of importance later.

$$E = 37.282\, I_o \left\| \left[ \frac{\cos((L/2)\cos\theta) - \cos(L/2)}{\sin\theta} \right] \right\| \quad \text{(Eq 1)}$$

where

$E$ = absolute value of the electric vector of field, mV/meter at one mile distance. Absolute value indicated by the vertical bars.
$L$ = total angular length of antenna, ie, 180° for a half-wave dipole
$\theta$ = angle of the LOP of a wave relative to antenna axis
$I_o$ = current at the current loop(s), amperes

The constant 37.282 in Eq 1 causes it to yield the field strength in millivolts per meter at a distance of one mile. The "one mile" is a completely arbitrary heritage of the BC antenna industry. For many amateur purposes there is no real mathematical reason to state the field in these terms but it gives a sense of physical significance to the field values. The bracketed expression is the *shape factor*. The field strength is proportional to $I_o$ and $I_o = \sqrt{P/R_r}$, where P = the radiated power and $R_r$ = the radiation resistance at the current loop(s). Derivation of some version of Eq 1 can be found in many textbooks.

Unless anticipated, when θ is zero a calculator or computer would stop with an error display—division by zero being forbidden. Repetitive tests for zero division in a program are time consuming. Addition of $10^{-6}$ degrees to the value of θ ensures that barring the unlikely use of $-10^{-6}$ degrees for θ, the ordinary rounding and display of results eliminates the very small effect on calculated fields.

The trigonometry of Eq 1 causes the sign of the bracketed expression to be negative for some θ for some lengths. This indicates the phase change between lobes on some patterns and is of no significance for our purposes. The negative signs are ignored and the absolute value used.

## Referencing to Azimuth and Wave Angles

Eq 1 is in terms of θ. By some elementary trigonometry, Eq 2 yields θ in terms of azimuth angle φ and wave angle ψ. φ = 0° is perpendicular to the antenna and increases counterclockwise looking down on the antenna. While this is contrary to the conventional compass directions, it is almost universally used to match the customary polar diagram directions.

$$\theta = \arccos(\cos\phi \cos\psi) \qquad \text{(Eq 2)}$$

where
- θ = angle of LOP relative to the antenna axis
- φ = azimuth angle (0 broadside to the antenna)
- ψ = wave angle (0 on the horizon)

## The Antenna above a Perfectly Conducting Surface

Fig 4 shows the antenna over a perfectly conducting flat surface—close enough such that only the phase changes due to reflection are obvious and the phase changes due to path length differences are not yet very large.

Fig 4A is the end-on view of the antenna. $E_d$ is the E vector of a direct wave propagating toward some distant point in a plane perpendicular to the antenna. Waves in this plane are purely horizontally polarized. $E_d$ is shown as an open circle indicating an arrow directed into the page.

$E_i$ is the E vector of a wave in the same plane as $E_d$ propagating toward and incident on the surface from which it is reflected to become $E_r$, at the same wave angle as $E_d$ and propagating toward the same point as $E_d$, with which it will be combined. $E_r$ is shown by a dot in a circle indicating an arrow pointing out of the page due to the 180° phase change associated with the reflection of a horizontally polarized wave from a perfectly conducting surface.

Fig 4B is a view from broadside to the antenna. The E vectors shown are in the vertical plane that contains the antenna.

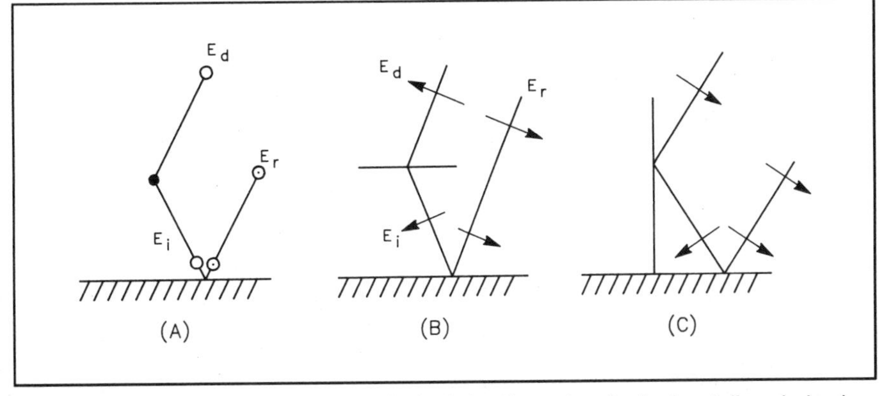


Fig 4—For horizontal antennas, the reflected electric vectors for horizontally polarized waves broadside to the antenna (A), and those for the vertically polarized waves in the vertical plane containing the antenna (B), behave in the same fashion. That is, they are in opposite directions to the direct wave. Contrast this with vertically polarized waves from a short vertical antenna (C), where they are in the same direction as the direct wave. This is due to the symmetry of electric vectors about antennas, as shown in Fig 3.

Since the E vectors are in that plane they are shown as ordinary arrows. The waves in this plane are purely vertically polarized and hence suffer no phase change when reflected from the perfectly conducting surface. Notice that the direction of the $E_i$ and $E_d$ arrows is to the left because the field is symmetrical about the axis of the antenna. Looking along the LOP in the direction of propagation of $E_i$, the arrow points to the right of the LOP so looking along the LOP of $E_r$, it also points to the right of the LOP there being no phase change on reflection.

Again $E_d$ is propagating to the distance point, $E_i$ toward the reflecting plane and $E_r$ toward the same distance point to be combined with $E_d$. *The arrows are in opposition just as they were in the end view containing the horizontally polarized wave.* Contrast this with the situation for a short vertical antenna on perfect earth shown in Fig 4C where the reflected wave is in the same direction as the direct wave.

In other words—with respect to its interaction with the direct waves, the reflected vertically polarized waves from a horizontal antenna behave as if the sign of the reflected E vector is reversed, that is, phased −180°. This makes the reflected $E_r$ of the vertically polarized waves the same as the $E_r$ of the horizontally polarized waves. Therefore, in this perfect-earth case the vertically polarized and horizontally polarized waves behave the same in this respect and there is no need to differentiate between them—even at azimuth angles containing waves polarized at intermediate angles between vertical and horizontal. In the real-earth case, the sign of the reflected vertically polarized component of such waves must be reversed, that is, phased an additional −180°.[23]

## The Multiplying Factor

The path difference for $E_d$ and $E_r$ causes phase differences such that at some antenna heights, and azimuth and wave angles, $E_r$ reinforces $E_d$ and at others partially or totally cancels $E_d$, thus forming patterns that depend on antenna height. Many texts present some version of Eq 3 to convert the free-space patterns of Eq 1 to the pattern above a perfect earth.

$$E_p = 2 \sin [(360\, h/\lambda) \sin \psi]\, E \qquad \text{(Eq 3)}$$

where
- $E_p$ = total field, mV/meter at 1 mile
- E = free-space antenna field from Eq 1, mV/meter at 1 mile
- h = height above earth, same units as λ
- λ = wavelength, same units as h
- ψ = wave angle above horizon, degrees

The 2 in Eq 3 represents the "6-dB reflection gain" ($20 \log_{10} 2 = 6$ dB), the same as the 6 dB to be added to the "reflection factors" mentioned in the Figure 11 caption on page 3-8 of *The ARRL Antenna Book*, 15th Edition. Those perfect-earth "pattern factors" are plots of the remainder of Eq 3 with unity at the 0-dBi circle.

Of all the texts previously cited, only Terman's *Radio Engineers Handbook*[9] presents the multiplying factor as simply, "For horizontal antennas ..."—no restrictions. Others say, for example, "For the plane perpendicular to the antenna." True, but not exclusive. Terman is right—the perfect-earth multiplying factor applies to all

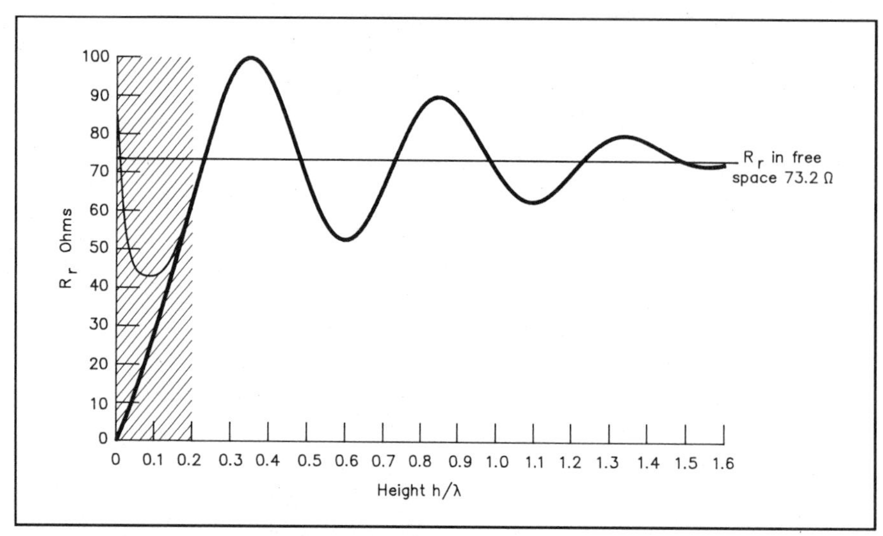


Fig 5—Radiation resistance, $R_r$, of a half-wave dipole as a function of height above ground. Heights below 0.2 wavelength will not yield reliable results because of local induction field earth losses coupling into the antenna and the finite thickness of the earth current. At greater heights, these effects are considered negligible. Below 0.2 wavelength, the input impedance of the half-wave antenna will be approximately as shown by the dotted line, depending on earth parameters and frequency.

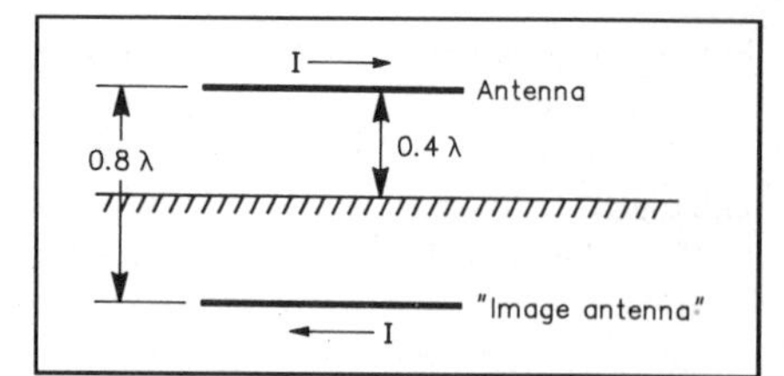


Fig 6—For purposes of calculations for mutual resistance, $R_m$, the mathematical "image antenna" is separated from the real antenna by twice the height of the real antenna above earth.

azimuth and wave angles.

At this point the *shape* of patterns of horizontal antennas above perfect earth may be plotted for any combination of selected azimuth and wave angles. Enter the selected azimuth and wave angles in Eq 2 to yield θ. Enter θ in the *shape factor* (the absolute value of the portion within the brackets of Eq 1) with the antenna length. Then multiply the resulting E by the number obtained from Eq 3 for the selected height h/λ and wave angle. The results, however, cannot be used to compare the gain of antennas of differing lengths or heights. Some texts neglect to mention this.

### Radiation Resistance

Because $R_r$ differs with antenna length and height above earth, the gain of different antennas cannot be compared unless $R_r$, permitting calculation of $I_o$ for equal radiated power from each antenna, is known.

Almost all antenna texts present the curve of Fig 5 showing $R_r$ of a thin half-wave antenna as a function of height above earth in wavelengths. (One reference text[11] has an error of a factor of 2 in the height scale of this plot).

The curve of Fig 5 is generated by the "method of images." $R_r$ of the "real antenna" is changed from its free-space value by the resistance component, $R_m$, of the mutual impedance, $Z_m$, between it and its mathematical "image antenna." See Fig 6.

The free-space $R_r$ of this half-wave dipole (180° length) is 73.2 ohms. At a height of 0.4 wavelength above earth as in Fig 6, separation between it and the image is 0.8 wavelength. $Z_m$ and its associated phase angle for a 90° monopole (one half of the 180° dipole) can be obtained from curves of $Z_m$ versus separation found in many texts. Multiplying the $Z_m$ by the cosine of the phase angle yields –9.25 ohms. Doubling this for the 180° antenna makes $R_m$ –18.5 ohms. Eq 4 applies and the $R_r$ at this height is 73.2 – (–18.5) = 91.5 ohms. The negative sign in Eq 4 reflects the 180° phase of the image of horizontal antennas relative to the real antenna.

$$R_r = R_{fs} - R_m \qquad \text{(Eq 4)}$$

where
$R_r$ = radiation resistance of real antenna above earth, ohms
$R_m$ = resistance component of mutual impedance, ohms
$R_{fs}$ = radiation resistance of antenna in free space, ohms

Repeating this calculation for various heights of a half-wave antenna yields the curve of Fig 5. From 0 height, where the two antennas would cancel each other, the curve rises to a peak then looks like a damped sine wave about the free-space $R_r$ value and approaches the free-space $R_r$ value at great heights.

These curves usually carry an indication that they are valid only for heights exceeding 0.2 wavelength because below that height, over real earth, local induced earth loss and the finite thickness of the earth current flow become significant factors difficult to analyze. The same problem occurs with current versions of MININEC.

$I_o$ for a selected power to the antenna is given by Eq 5.

$$I_o = \sqrt{P/R_r} \qquad \text{(Eq 5)}$$

where
$I_o$ = current at antenna loop(s), amperes
P = power to antenna, watts
$R_r$ = radiation resistance of antenna at current loop as corrected for height above earth, ohms

If a power of 100 watts is assumed in this case, $I_o = \sqrt{100/91.5} = 1.045$ amperes. Many texts, having explained the half-wave case this way, then fail to warn the reader of the difficulties of applying the method to antennas of other lengths.

Because $I_o$ will differ for the same power level in antennas of different heights because of the variation of $R_r$ with height, in some height ranges gain will be enhanced while in other ranges gain will be diminished. Hence the often quoted 2.15 dB gain of a half wave in free space over an isotropic source does not translate to 8.15 dBi except for certain lobes and for those heights for which the radiation resistance is the same as that of the antenna in free space over perfect earth and at chance combinations of height and earth parameters over real earth.

### $R_r$ for Other Antenna Lengths

Free-space $R_r$ and image $R_m$ are needed to calculate $R_r$ of an antenna above earth. Free-space $R_r$ can be obtained from some texts[12], but unfortunately, the curves in most texts for $Z_m$ or $R_m$[13] for antennas other than the 90° monopole (for 180° dipole) and the 230° monopole (for the Extended Double Zepp 460° dipole) are difficult to use since they are usually plotted for odd lengths of particular interest to the broadcast industry. Interpolation between curves is difficult to impossible because the curves overlap with varying intervals due to their cyclic nature. Therefore, a means to calculate $R_m$ is needed.

The resistance component of $Z_m$ at the current loop of a pair of monopoles (double for the dipole) may be calculated by the equations in Appendix 2. With a suitable subroutine for generating the sine integral, $S_I$, and cosine integrals involved (see Appendix 3), a personal computer can calculate $R_m$ in a blink of an eye. Free-space $R_r$, although available in several texts as a curve, can be similarly calculated by the equations of Appendix 3. In fact, I have programmed a hand-held programmable scientific calculator (HP41C) to run these and all the other calculations of this paper. While running time in the HP41CV for $Z_m$ and $R_r$ is several minutes, it still beats looking them up in textbooks.

For antennas shorter than 180° dipoles, the current loop cannot actually physically exist on the antenna itself. Nevertheless the values of $R_r$ and $I_o$ resulting are proper for use in these calculations.

When calculating $R_m$ by the means of Appendix 2, calculators or computers will run "out of range" for some height depending on the length of the antenna due to the large values reached by the factorial numbers in the dividend of the series expansion of Si and Ci. This limits the height range over which $R_m$ may be calculated. For most MF/HF horizontal antennas this limit provides for practical antenna heights.

**Summary of Calculations for Perfect Earth**

Perfect-earth patterns are often used simply because the calculations are simpler than those for real earth but also serve as a reference to permit identification of the effects of real earth.

Using Eqs 1, 2, and 3 with $I_o$ calculated for the same power levels by Eq 5 and $R_r$ appropriate to the length and height of each antenna, patterns of antennas of differing length and/or height above perfect earth may be compared to determine effects on patterns such as number, size and wave angle of lobes and nulls or minima.

**Real-Earth Case**

Real earth acts like a lossy capacitance. A portion of the incident wave is lost in the reflection process and the reflected portion undergoes a phase change differing from that of the perfect-earth case depending on the dielectric constant and conductivity of the real earth in the earth areas from which the reflections take place.

As an example, Fig 7A shows curves for $R_h$ and $r_h$ (horizontal polarization) for an average earth at 21 MHz. As the wave angle varies from 0° (on the horizon) to 90° (straight up) the fraction reflected, $R_h$, varies from 1 to 0.6 while the phase angle doesn't get very different from –180°, not greatly different from the perfect-earth case.

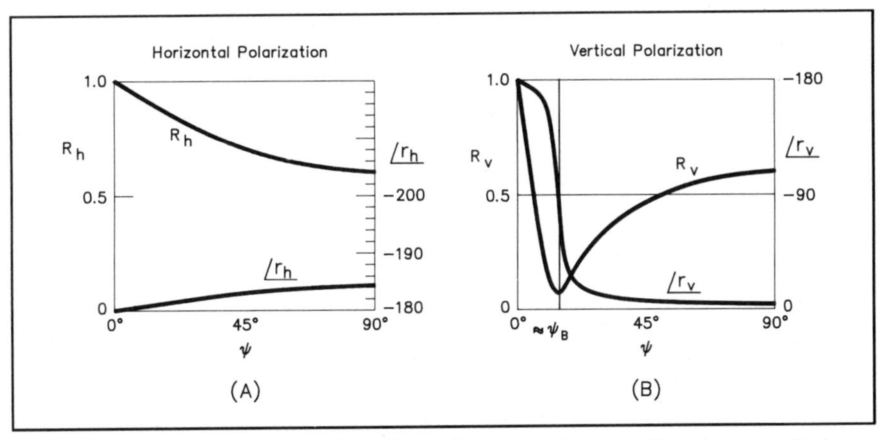


Fig 7—(A) Reflection coefficient for horizontally polarized waves. The phase angle, $r_h$, as calculated by Eq 6 may be from –180° to +160° for the –180° to –200° range shown. (Some computers and calculators convert all angles more negative than –180° to the equivalent, but still proper, positive angles.) (B) Reflection coefficient for vertically polarized waves. Both curves for $e_r = 15$, $\sigma = 5$ mS/meter (average sort of earth) at 21 MHz.

Fig 7B shows curves for $R_v$ and $r_v$ (vertical polarization) for that same earth and frequency. The values vary widely. At high angles $R_v$ is close to $R_h$ and $r_v$ is close to the 0° of the perfect-earth case. But at lower angles $R_v$ drops to a low value then rises rapidly for low angles to 1 at 0° (the horizon), while $R_v$ goes quickly from a small negative angle to –180 on the horizon. The minimum of $R_v$ occurs very close to the wave angle at which $R_v$ is –90°. This is the "Pseudo-Brewster Angle" when this reflection coefficient is used with vertical antennas.[14,20]

Aside from the violent behavior of the vertical polarization reflection coefficient, the salient feature of vertically polarized waves from a horizontal antenna is that the phase of the reflected vertically polarized E vector is reversed by –180° in calculations because of the symmetry of the E fields as shown in the perfect-earth analysis. This reverses the effect of the vertically polarized reflected waves from real earth such that at low angles, instead of opposing the direct wave, as it does for vertical antennas over real earth, it reinforces the direct wave.[23]

Since all antennas have a null directly end-on, that null still exists, but depending on the earth parameters and frequency, the field in the immediate vicinity of the null-cone may rise more rapidly with wave angle thus enhancing the low angle radiation off the ends at these low angles. This effect fades for the intermediate angles and at high angles the vertically polarized radiation off the ends approaches the value of the horizontally polarized field broadside. At 90° they are exactly equal. The method by which the direct and reflected waves are combined is treated later.

Eq 6 is the horizontal polarization coefficient. Eq 7 is the vertical polarization coefficient.

$$R_h \,/\underline{r_h} = \frac{\sin\psi - \sqrt{(e_r - jx) - \cos^2\psi}}{\sin\psi + \sqrt{(e_r - jx) - \cos^2\psi}} \quad \text{(Eq 6)}$$

$$R_v \,/\underline{r_v} = \frac{(e_r - jx)\sin\psi - \sqrt{(e_r - jx) - \cos^2\psi}}{(e_r - jx)\sin\psi + \sqrt{(e_r - jx) - \cos^2\psi}} \quad \text{(Eq 7)}$$

where

$x = 1.8 \times 10^4\,\sigma/f$

$R_h$ = attenuation constant of the horizontal polarization coefficient

$r_h$ = phase angle of the horizontal polarization coefficient

$R_v$ = attenuation constant of the vertical polarization coefficient

$r_v$ = phase angle of the vertical polarization coefficient

$\psi$ = wave angle (0° on the horizon)

$e_r$ = dielectric constant of the earth relative to air = 1

$\sigma$ = conductivity of earth, siemens per meter (S/m)

f = frequency, MHz

j = complex operator

Values of $e_r$ and $\sigma$ may be obtained from *The ARRL Antenna Book*, 15th or 16th Editions, page 3-3, Table 1 or from text listings.[21] Values of earth conductivity shown on maps usually relate to the 0.5- to 1.6-MHz BC band and are not necessarily relevant to HF conditions where depth of penetration of earth currents is much shallower.

The textbooks can lead one on a confusing chase in obtaining Eqs 6 and 7. The expressions are given in a variety of equivalent but differing dimensioning systems. For example, some[21] use earth conductivity in the physicist's emu. The emu is multiplied by $10^{11}$ to obtain the practical unit of siemens/meter. In the practical system $5 \times 10^{-14}$ emu would be 0.005 siemens/meter or 5 millisiemens/meter. Some use earth resistivity, which is the reciprocal of the conductivity and would be 200 ohm-meters.

Some derive the coefficients for use in the "image antenna" concept. Others do not and hence include the 180° phase of the image antenna in the coefficient. Thus Terman's[16] $r_h$ differs from that of Jordan and Balmain[17] by 180° (some numerator signs reversed). This paper uses those of Jordan and Balmain, as do most texts. Both, however, are correct as used in their respective texts.

Laport's *Radio Antenna Engineering*[18] has an error in that the radicals in both the numerator and denominator extend over the entire expression. Johnson and Jasic's *Antenna Engineering Handbook*[19] has a sign error in the denominator of the horizontal polarization coefficient and the authors of that chapter erroneously apply the horizontal reflection coefficient at all azimuth angles.

Many of the texts cited show how to apply the horizontal polarization coefficient in the plane perpendicular to the horizontal antenna and some show how to apply the vertical polarization coefficient in the vertical plane containing the antenna.[23] None of these, however, provide an explicit equation that will apply to all azimuth and wave angles.

### The Compound Reflection Coefficient

Fig 8A requires a bit of imagination to see it as a three-dimensional diagram. A horizontal antenna is shown lying in a vertical plane (azimuth angle $\phi = 90°$ and 270°) at height $h/\lambda$ above the plane of the earth. There is a vertical plane (azimuth angle $\phi$ = 0° and 180°) at right angles to the first plane and passing through the center of the antenna. All LOP will lie in planes containing the antenna and have electric vectors perpendicular to them in those planes.

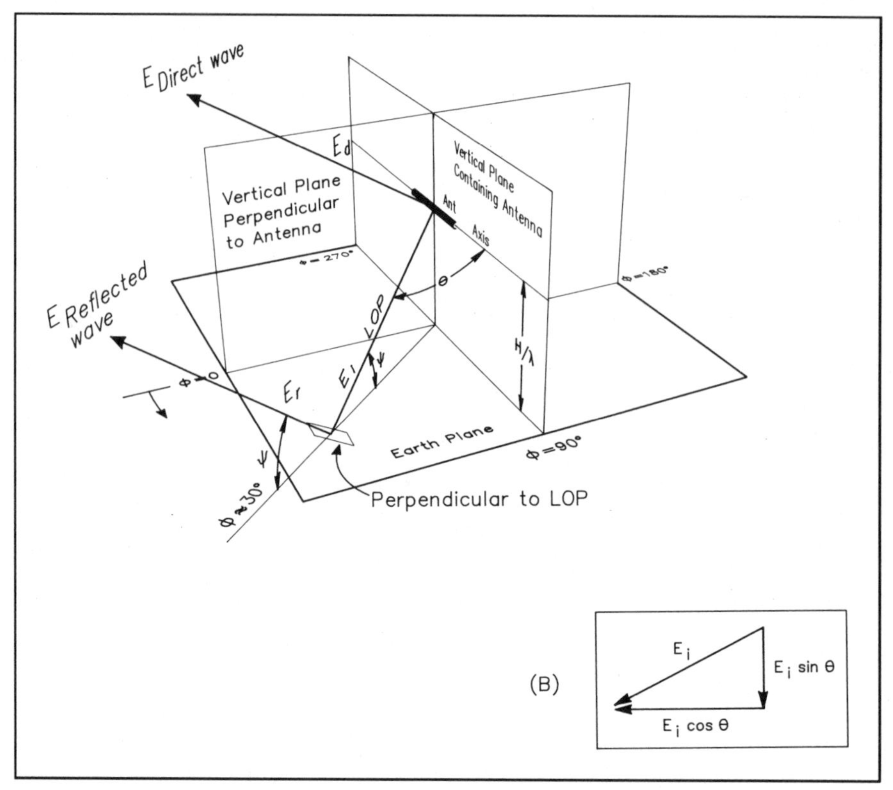


Fig 8—(A) A horizontal antenna $h/\lambda$ above an earth plane. (B) The small tilted plane of A at $\phi \approx 30°$ is shown enlarged. The reflected wave, $E_r$, and the direct wave, $E_d$, are parallel and combine at a distant point to form the field of E vectors composing the pattern. Both should be seen as coming out of the page and to the reader's left.

The electric vectors of the LOP lying in the $\phi$ = 0 plane are purely horizontally polarized and hence parallel to the earth plane.

The electric vectors of the LOP lying in the $\phi$ = 90° plane are purely vertically polarized and are at various angles to the earth plane.

For each wave incident on and reflected from the earth there is a corresponding direct wave at the same azimuth and wave angle with which each reflected wave will ultimately be combined.

Observe the LOP at about $\phi$ = 30° shown at wave angle $\psi$. Its E vector is of course in a slanted plane containing the antenna and the LOP, perpendicular to the LOP and at an angle to the earth plane. It lies in the little tilted plane shown at the point of earth incidence which is shown enlarged in Fig 8B.

The electric vector in Fig 8B can be resolved into two components as shown. One component lies in the earth plane (horizontally polarized) and is equal to $E_i \cos\phi$. The other component is at an angle to the earth plane and in a vertical plane containing the LOP (vertically polarized) and is equal to $E_i \sin\phi$.

The incident wave has now been resolved into a purely horizontally polarized component and a purely vertically polarized component. Each can be multiplied by the reflection coefficient appropriate to its polarization for its wave angle to obtain $E_r$(Hor) and $E_r$(Vert). The two resulting reflected waves can then be combined into a compound reflected wave by the square root of the sum of the squares of the two components. By algebraic and trigonometric manipulation a result can be derived that will be the compound reflection coefficient, $R_c \,/\underline{r_c}$. Remember that the vertically polarized E vector phase must be reversed by −180°.*

Reflected horizontal component

$$E_r\text{Hor} = E_i \cos\phi \, R_h \,/\underline{r_h}$$

Reflected vertical component

$$E_r\text{Vert} = E_i \,/\underline{-180} \sin\phi \, R_v \,/\underline{r_v}$$

$$= E_i \sin\phi \, R_v \,/\underline{r_v - 180}$$

*While it would appear that using a minus sign would be equivalent to the −180 phase change, the minus sign would be lost in the squaring process.

Forming the square root of the sum of the squares

$$E_r = [(E_i \cos\phi\, R_h \angle r_h)^2 + (E_i \sin\phi\, R_v \angle r_v - 180)^2]^{1/2}$$

Expanding the squares. Noting that $(y \angle x)^2 = y^2 \angle 2x$

$$E_r = [E_i^2 \cos^2\phi \bullet R_h^2 \angle 2r_h + E_i^2 \sin^2\phi\, R_v^2 \angle 2r_v - 360]^{1/2}$$

Extracting $E_i$ outside the square root brackets

$$E_r = E_i\,[\cos^2\phi\, R_h^2 \angle 2r_h + \sin^2\phi\, R_v^2 \angle 2r_v - 360]^{1/2}$$

Dividing both sides by $E_i$

$$E_r/E_i = [\cos^2\phi\, R_h^2 \angle 2r_h + \sin^2\phi\, R_v^2 \angle 2r_v - 360]^{1/2}$$

Converting each term to rectangular coordinates

$$E_r/E_i = [\cos^2\phi\, R_h^2 (\cos 2r_h + j \sin 2r_h) + \sin^2\phi\, R_v^2 (\cos(2r_v - 360) + j \sin(2r_v - 360))]^{1/2}$$

Let $\cos^2 R_h^2 = A$ (Eq 8a)

Let $\sin^2 R_v^2 = B$ (Eq 8b)

Substituting A and B for convenience of manipulation and calculation

$$E_r/E_i = [A(\cos 2r_h + j \sin 2r_h) + B(\cos(2r_v - 360) + j \sin(2r_v - 360))]^{1/2}$$

Expanding into individual terms

$$E_r/E_i = [A \cos 2r_h + j A \sin 2r_h + B \cos(2r_v - 360) + j B \sin(2r_v - 360)]^{1/2}$$

Collecting real and imaginary parts

$$E_r/E_i = [A \cos 2r_h + B \cos(2r_v - 360) + j(A \sin 2r_h + B \sin(2r_v - 360))]^{1/2}$$

Let $A \cos 2r_h + B \cos(2r_v - 360) = C$ (Eq 8c)

Let $A \sin 2r_h + B \sin(2r_v - 360) = D$ (Eq 8d)

Substituting C and D for convenience of manipulation

$$E_r/E_i = [C + j D]^{1/2}$$

The definition of the A reflection coefficient is the ratio of $E_r$ to $E_i$, therefore

$$E_r/E_i = R_c \angle r_c = [C + j D]^{1/2} \qquad \text{(Eq 8e)}$$

Converting to polar format and taking the square root, noting that

$$\angle x^{1/2} = \angle x/2$$

$$R_c \angle r_c = [\,|M| \angle \Delta - 360\,]^{1/2} \qquad \text{(Eq 8f)}$$

where

$$|M| = \sqrt{C^2 + D^2}$$

and

$$\frac{\Delta - 360}{2} = \frac{\arctan \frac{D}{C}}{2} - 180^\circ$$

Hence

$$R_c \angle r_c = \sqrt{C^2 + D^2}\ \angle\ \frac{\arctan \frac{D}{C}}{2} - 180^\circ \qquad \text{(Eq 8)}$$

This is the *compound reflection coefficient*, a function of the horizontal polarization coefficient and the vertical polarization coefficient which are functions of the wave angle, azimuth angle, frequency and earth parameters.

Calculate A and B of Eqs 8a and 8b. Enter them into Eqs 8c and 8d to form C and D, then C and D into Eq 8e. Converting to polar form by Eq 8f yields the compound reflection coefficient in Eq 8. This reflection coefficient is valid for any azimuth and wave angle used in its computation.

**Combining the Direct and Reflected Waves**

The method of images may be used to derive the phase delay of $E_r$ relative to $E_d$ due to the height of the antenna above earth to be added to the phase angle of the reflection coefficient. See Appendix 1.

$E_d$ and $E_r$ combine vectorially to form $E_p$ (the field at a distant point). Two vectors, a and b, at an angle $\Gamma$ may be added by a relative of the "Law of Cosines," $c = (a^2 + b^2 + 2ab \cos\Gamma)^{1/2}$, applied here as:

$$E_p = E\left[1 + R_c^2 + 2R_c \cos\left(r_c + 720\frac{h}{\lambda}\sin\psi\right)\right]^{1/2} \qquad \text{(Eq 9)}$$

where

$E_p$ = vector sum of $E_d$ and $E_r$ (total field), mV/meter at 1 mile
$E$ = free-space field E from Eq 1, mV/meter at 1 mile
$R_c$ = attenuation constant of compound reflection coefficient
$r_c$ = phase angle of compound reflection coefficient
$h$ = height above earth, same units as $\lambda$
$\lambda$ = wavelength, same units as h
$\psi$ = wave angle, 0° on horizon†

Terman[22] presents an equation similar to Eq 9 but uses a minus sign before the 2 R cos term when the equation is applied to horizontally polarized waves. This provides for the 180° image phase not included in his version of the reflection coefficient for horizontally polarized waves.

**Summary of Real-Earth Field Calculations**

1) Calculate $R_r$ for the antenna length and height where given by reference to $R_m$ and $R_r$ values or by means of Appendices 2 and 3 and Eq 4.

2) Select a current $I_O$ corresponding to the selected power level using Eq 5.

3) Calculate the direct field at the selected azimuth and wave angles using Eq 1 and 2.

4) Calculate the Reflection Coefficient for the selected earth parameters using Eq 6 and 7.

5) Calculate the Complex Reflection Coefficient of Eq 8 using Eq 8a through 8f.

6) Calculate the total field strength using Eq 9.

7) Repeat steps 3 through 6 for each azimuth and wave angle of the pattern.

†It can be shown that Eq 9 can be reduced to Eq 2 if $R_c = 1$ and $r_c = 180°$ (perfect earth conditions) and the trigonometric identity, $\sin(x) = ((1 - \cos 2x)/2)^{1/2}$, applied.

## Gain Calculations

Gain calculations must include losses in the antenna system. However, the word "gain" may be correctly applied here in that typical amateur MF/HF antennas at heights more than 0.2 wavelength above earth show negligible earth loss and copper loss in the loop impedance and earth-reflection loss is included in the calculations. Since the patterns for horizontal antennas below 0.2 wavelength in height vary very little in shape over this range, such results may be useful but the gain calculations will be less accurate because of the local induction field induced earth loss.

A popular reference for gain properly called "dBd" is the maximum of a half-wave dipole in free space radiating the same power as the subject antenna. However, it sometimes causes confusion in attempts to reference the subject antenna to a half-wave dipole at the same height and radiated power as the subject antenna over perfect earth or the same earth as the subject antenna. The field of this half-wave dipole includes "reflection gain." The confusion is caused by erroneous use of the designation "dBd" for this reference, too. More confusion results if reference is intended or misunderstood to be over the entire field of the reference dipole rather than to only the maximum. In the case of antennas such as loops the height to be used for the reference half wave may be difficult to select, or at least controversial. A more convenient reference is the "free-space isotropic antenna" used in almost all contemporary antenna literature.

An "isotropic antenna" cannot physically exist. It is a mathematical device which is assumed to radiate equally in all directions. Discussion has been in terms of field strength at one mile. It requires only determination of the field of an isotropic free-space antenna at a distance of one mile at the same power as the subject antenna to establish the reference over the entire pattern. See Appendix 4 for derivation of Eq 10.

$$E_{iso} = 3.4034\sqrt{P} \qquad \text{(Eq 10)}$$

where

$E_{iso}$ = field of isotropic antenna at 1 mile with P watts, mV/meter

P = power into antenna being referenced, watts

$P = I_o^2 R_r$, $I_o$ and $R_r$ being those of the subject antenna

As each point in the pattern is calculated it is referenced to the isotropic antenna to obtain the pattern in terms of dBi using Eq 11.

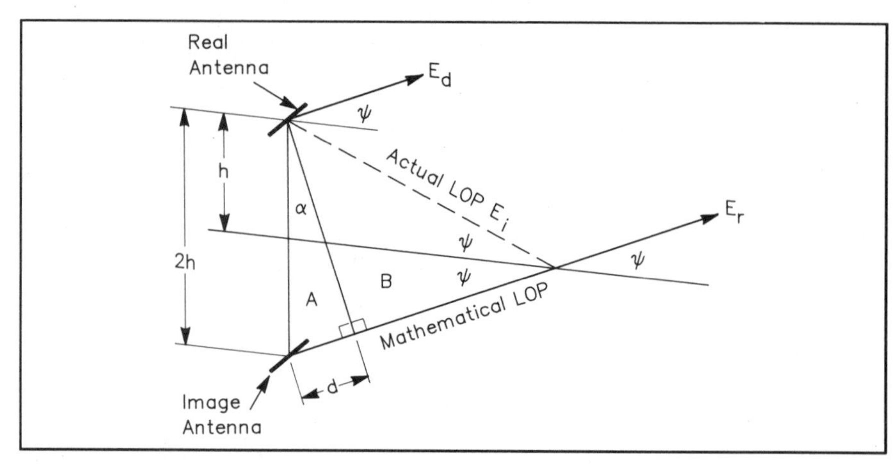


Fig A, Appendix 1

$$dBi = 20\log_{10}(E_p/E_{iso}) \qquad \text{(Eq 11)}$$

where

dBi = field strength of antenna versus isotropic antenna, decibels

$E_p$ = field strength of subject antenna, same field strength and distance units as $E_{iso}$

$E_{iso}$ = field strength of isotropic antenna at same distance, power level and field-strength units as $E_p$

## Plotting Patterns

When plotting on standard ARRL polar grids with –dBi circles to the 0 dBi outer circle, a note must accompany most plots indicating the value of the 0 dBi circle. The patterns are often normalized to place the maximum lobe of the plot on the 0 dBi circle. Since this may make each plot to a different scale, it becomes difficult to compare patterns by eye, often even in various azimuth and wave angle cuts in the same pattern. While there must be cases of extreme ranges of dBi where this might be necessary, in most cases a much better plot provides dB circles outside the 0 dBi circle to about +20 dBi, sufficient for most amateur patterns. Such plots using the log periodicity of 0.89 per 2 dB factor of the ARRL grids, which is an excellent choice, can be drafted and reproduced on a copy machine. See *The ARRL Antenna Book*, 15th or 16th Editions, page 2-12 for details on other pattern grids.

## How About that Double Extended Zepp...

The method permits calculation of fields at any azimuth and wave angle over perfect or real earth for horizontal antennas with sinusoidal current distribution for plotting of patterns.

By running plots throughout the MF and HF range with various earth dielectric and conductivity values, I have found that the earth dielectric constant is of relatively minor importance below 30 MHz but of increasing importance as frequency increases.

Getting back to the question of the Double Extended Zepp—nulls in the free-space pattern produced by Eq 1 are preserved in both the perfect-earth and real-earth patterns. Nulls and minima resulting from earth reflections are increasingly "filled in" as the earth quality becomes poorer.

## Appendix 1
## Phase Delay in $E_r$ and Combining Direct and Reflected Waves

See Fig A for the following.

d = physical path length difference $E_r$ relative to $E_d$ over perfect earth

$d/\lambda$ = path length difference in fractions of wavelength

$\frac{2h}{\lambda}$ = hypotenuse of triangle A in fractional wavelength

$$\frac{\frac{d}{\lambda}}{2\frac{h}{\lambda}} = \sin\alpha$$

$$\text{Phase delay in } d = 2\pi\frac{d}{\lambda} = D \text{ radians}$$

$$\text{Distance } 2\frac{h}{\lambda} = 2\pi\, 2\frac{h}{\lambda} = 4\pi\frac{h}{\lambda} \text{ radians}$$

$$\frac{D}{4\pi \frac{h}{\lambda}} = \sin \alpha$$

But $\alpha = \psi$, since triangles A and B are "similar triangles." So

$$\frac{D}{4\pi \frac{h}{\lambda}} = \sin \psi \text{ and } D = 4\pi \frac{h}{\lambda} \sin \psi$$

or in degrees $D = 720 \frac{h}{\lambda} \sin \psi$

But reflection adds $r_c$, the phase angle of the reflection coefficient. Therefore

$$\Gamma = r_c + 720 \frac{h}{\lambda} \sin \psi$$

where $\Gamma$ is the total phase between $E_d$ and $E_r$.

*Combining $E_d$ and $E_r$ in real-earth case*

A "relative" of the cosine law is used to add vectors.

$$c = [a^2 + b^2 + 2\, a\, b \cos \Gamma]^{1/2}$$

Let $a = E_d$ but $E_d = E_i$ and $E_i = E$

Let $b = E_r$ but $E_r = E_i R_c = ER_c$ ($r_c$ is in $\Gamma$). So

$$E_p = [E_d^2 + E_r^2 + 2\, E_r\, E_d \cos \Gamma]^{1/2}$$

$$= E^2 + E^2 R_c^2 + 2\, E\, E\, R_c \cos \Gamma]^{1/2}$$

$$= E\, [1 + R_c^2 + 2\, R_c \cos \Gamma]^{1/2}$$

## Appendix 2
## Calculation of Mutual Resistance between Monopoles

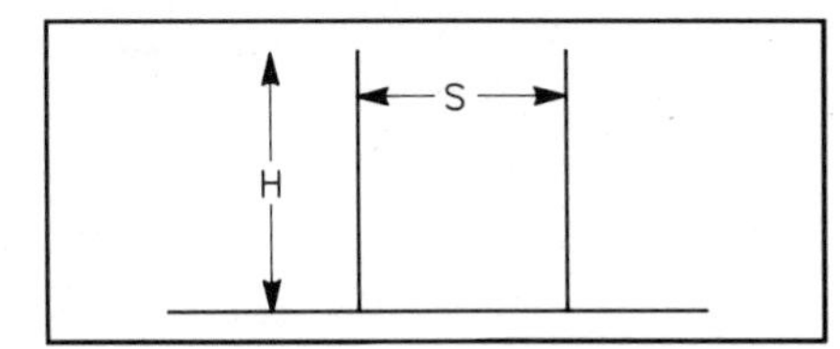


Fig B, Appendix 2

These equations produce the mutual resistance component, $R_m$, at the current loop(s) between two monopoles of angular height H and angular separation S over perfect earth. Multiply by 2 for $R_m$ of dipoles in free space. Application to dipoles:

$S = 4\, \pi\, h/\lambda$ radians

$H = 0.01745\, L/2$ radians

L = dipole length, degrees

H, S, U and V are in radians

$$R_{21} = 30\, \{\sin H \cos H\, [Si(U_2) - Si(V_2) - 2\, Si(V_1) + 2\, Si(U_1)] - (\cos 2\, H)/2\, [2\, Ci(U_1) - 2\, Ci(V_1) - Ci(U_2) - Ci(V_2)] - [Ci(U_1) - 2\, Ci(U_0) + Ci(V_1)]\}$$

where

$U_0 = S$

$U_1 = \sqrt{S^2 + H^2} - H$

$V_1 = \sqrt{S^2 + H^2} + H$

$U_2 = \sqrt{S^2 + 4\, H^2} + 2\, H$

$V_2 = \sqrt{S^2 + 4\, H^2} - 2\, H$

Si (X) = sine integral of X

Ci (X) = cosine integral of X

$R_{21}$ = resistive component of $Z_m$ between antennas 1 and 2

See Appendix 3 for series to calculate Si (X) and Ci (X).

This function based on Jordan and Balmain, *Electromagnetic Waves and Radiation Systems*, 2nd ed, Prentice-Hall, 1968, page 540. From P. S. Carter, *Proc IRE*, 20, p 1004, 1932.

## Appendix 3
## Calculation of Radiation Resistance of a Thin Antenna

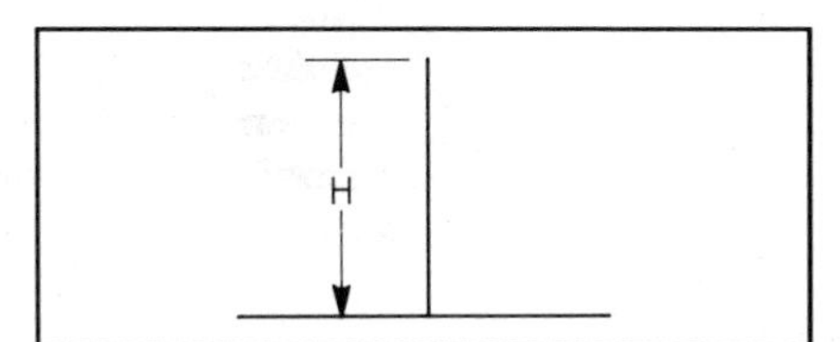


Fig C, Appendix 3

These equations are for the radiation resistance at current loop(s) of a monopole of length H above a perfect earth. Multiply by 2 for $R_r$ of dipole of 2H in free space.

$$R_r = 30\, [((\cos 2H)/2)\, (S1(4H)) + (1 + \cos 2H)\, (S1\, (2H)) + \sin 2H\, ((Si\, (4H))/2) - (Si\, (2H))]$$

where

$$Si\,(X) = \int_0^X \frac{\sin X}{X}\, dx$$

$$= X - \frac{X^3}{3 \times 3!} + \frac{X^5}{5 \times 5!} \cdots \frac{X^n}{n \times n!}$$

$$S1\,(X) = \int_0^X \frac{1 - \cos X}{X}\, dx$$

$$= \frac{X^2}{2 \times 2!} - \frac{X^4}{4 \times 4!} + \frac{X^6}{6 \times 6!} \cdots \frac{X^n}{n \times n!}$$

$$Ci\,(X) = \ln_e X + 0.5772157 - S1\,(X)$$

0.5772157 is Euler's constant

Si and S1 may be calculated by a reiterative subroutine that tests for some preset minimum term value to stop the series calculation and summing. A small difference between relatively large terms in the equation requires good accuracy for Si and S1, particularly for short antennas. A limit of $10^{-4}$ is recommended.

A curve labeled "$b/\lambda = 0$" corresponding to this $R_r$ function will be found in Terman, *Radio Engineer's Handbook*, 1943, page 793, Fig 26.

## Appendix 4
## Field of Isotropic Antenna at 1 Mile Distance with Radiated Power, P

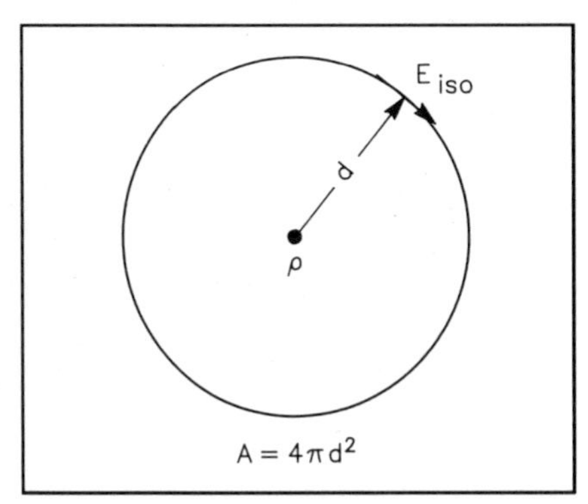


Fig D, Appendix 4

The radiated power is distributed uniformly over the area of a sphere having a radius of one mile (1 mile = 1609.3 meters)

$E_{iso}$ = field at surface of sphere mV/meter

P = power at isotropic antenna, watts

d = radius of sphere, meters

A = area of sphere, square meters

Just as $E = \sqrt{P R}$ in circuits, in a field, $E = \sqrt{D Z}$

where

D = power density at surface of sphere, watts per square meter

Z = impedance of free space, ohms ($120\,\pi$, generallyexpressed as rounded to 377 ohms)

$$D = \frac{P}{A} = \frac{P}{4\pi d^2}$$

$$E_{iso} = \sqrt{(P\ 120\pi) / (4\pi d^2)}$$

$$= \sqrt{(P\ 30) / d^2}$$

$$= \sqrt{P\ (30 / 1609.36^2)}$$

$$= \sqrt{P}\ \sqrt{1.15828 \times 10^5}$$

$$= 0.0034034\ \sqrt{P} \text{ volts per meter}$$

$$= 3.4034\ \sqrt{P} \text{ mV/meter}$$

$$E_{iso} = 3.4034\ \sqrt{P} \text{ mV/meter at 1 mile}$$

For example, a half-wave dipole 0.4 wavelength above perfect earth has $R_r$ of 91.5 ohms. With 100 watts power, it produces a field of 77.95 mV/meter at 1 mile at azimuth 0, wave angle 38.75° (a lobe). $E_{iso} = 3.4034\sqrt{100} = 34.03$ mV/meter; dBi $= 20 \log_{10} 77.95/34.03 = 7.2$ dBi. Of this, 6 dB is "reflection gain"; the remainder, 1.2 dB, is due to particular height above earth (quite different from the customary assumed 2.15 dB).

At a height near 0.5 wavelength, where the $R_r$ is the same as the free-space value, the broadside lobe at 33.17° shows gain of 8.15 dBi. Of this, 6 dB is "reflection gain" leaving the often quoted free-space gain of 2.15 dB of a half wave over isotropic. However, this is unique to heights where $R_r$ is the same as the free-space value and then only to certain lobes and over perfect earth.

### References


[1]Terman, *Radio Engineer's Handbook*, McGraw-Hill, 1943.

[2]Shelkunoff, *Electromagnetic Waves*, D. Nostrand, 1943.

[3]Shelkunoff and Friis, *Antennas, Theory and Practice*, John Wiley, 1952.

[4]Johnson and Jasik, *Antenna Engineering Handbook*, McGraw-Hill, 1984.

[5]Kraus, *Antennas*, McGraw-Hill, 1988.

[6]Jordan and Balmain, *Electromagnetic Waves and Radiating Systems*, Prentice-Hall, 1968.

[7]Laport, *Antenna Engineering*, McGraw-Hill, 1952.

[8]*ARRL Antenna Book*, 15th Edition, 1988.

[9]Ref 1, p 788-789.

[10]Ref 7, Preface, p VII.

[11]Ref 3, p 412 Fig 13.7.

[12]Ref 6, p 541 Fig 14-3.

[13]Ref 1, p 793 Fig 26 for $b/\lambda = 0$.

[14]C. J. Michaels, "Some Reflections on Verticals," Jul 1987 *QST*.

[15]*ARRL Antenna Book*, 15th (1988) or 16th (1991) Edition, p 2-13.

[16]Ref 1, p 696, Eq 27b.

[17]Ref 6, p 631, Eq 16-7.

[18]Ref 7, p 233.

[19]Ref 4, p 26-6.

[20]Ref 8, "The Effects of the Earth," pp 3-2 to 3-4.

[21]Ref 1, p 709, Table 1.

[22]Ref 1, p 791.

[23]Ref 6, p 643 and top p 399.

# Quadro-Line: The Ideal UHF Transmission Line

By Karol Dillnberger, DL7SS
Seesener Str 33
1000 Berlin 31
Germany

This article discusses the advantages of an unjustly neglected type of transmission line, called *Quadro-Line*, abbreviated Q-L. Q-L is a low-loss, four-conductor transmission line that you can build, especially suited for UHF and microwave applications.

Transmission-line attenuation results from radiation, and resistive and dielectric losses. Attenuation in a given section of transmission line usually increases with frequency. At VHF and above, it is necessary to make every effort to reduce dielectric and "skin-effect" losses. Examples of low-loss VHF-UHF-microwave transmission lines are rigid coax, known as "hardline," and open-wire. While hardline is often available at little or no cost from sources in the cable-TV industry, connectors are very expensive. Open-wire line is a poor choice for long runs because nearby objects affect its characteristic impedance, it tends to twist, and its characteristic impedance ($Z_0$) is high (450 to 600 ohms). A high-$Z_0$ line is hard to interface with low-Z equipment.

The ideal solution is open-wire line with no dielectric and thick conductors. The line described in this article comprises four separate parallel conductors, which are held at uniform separation on the periphery of circular insulating disks. The conductors also may be supported at four corners of a square plate. These disks, made of flat plastic material (crossed spacers may be used as well), play the same role as the spacers of twin-lead feeders, ie, holding all four conductors at a constant distance. The ends of opposite conductors are connected (ie, A1 with A2 and B1 with B2, Fig 1). This design determines the electrical properties of the line. The distance between the conductors (between 5 and 15 cm) for the line in Fig 1 is the same as for open-wire line. Quadruple lines of this type have lower characteristic impedance than simple twin-conductor lines ($Z_0$ from 180 Ω to 280 Ω). The characteristic impedance may be computed from Eq 1.

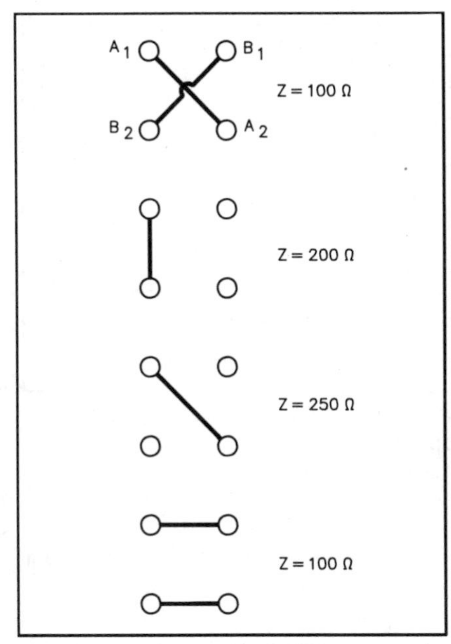


Fig 1—Possible connections of transmission-line conductors and their characteristic impedances for a conductor diameter of 3 mm and spacing of 12 mm. Quadro-Line is shown at the top.

$$Z_0 = 138 \log s/r \quad \text{(Eq 1)}$$

where
s = distance between adjacent conductors
r = conductor radius

This line provides for an ultimate balance and low radiation loss. Additionally, this line is not as sensitive to external influence as the twin-conductor line.

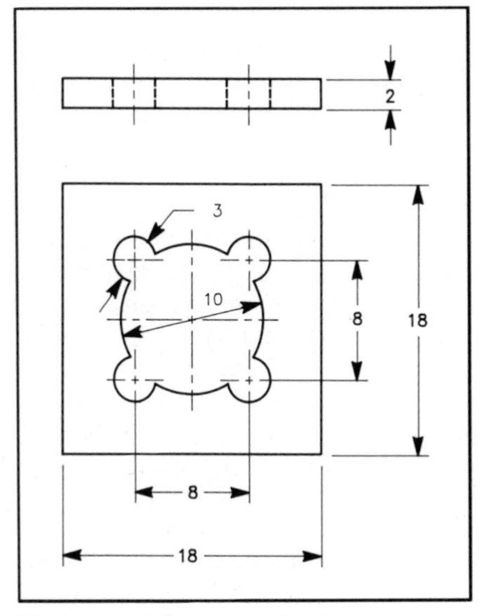


Fig 2—A suitable spacer for four conductors of 3 mm diameter.

If quadruple-conductor line is so good, why haven't you heard of it already? Unfortunately, unlike common transmission lines, quadruple line is not easily acquired or installed. These difficulties inspired my design, which allows smaller conductor separation, and minimizes the effects of nearby objects.

### Construction Materials

I chose 3-mm-diameter aluminum rod for the conductors. At VHF and above, the difference in conductivity between copper and aluminum is slight, and aluminum won because of its lighter weight. For shorter lines used at higher frequencies, copper-plated steel welding rods (3-mm diameter) can be used if you polish the copper plat-

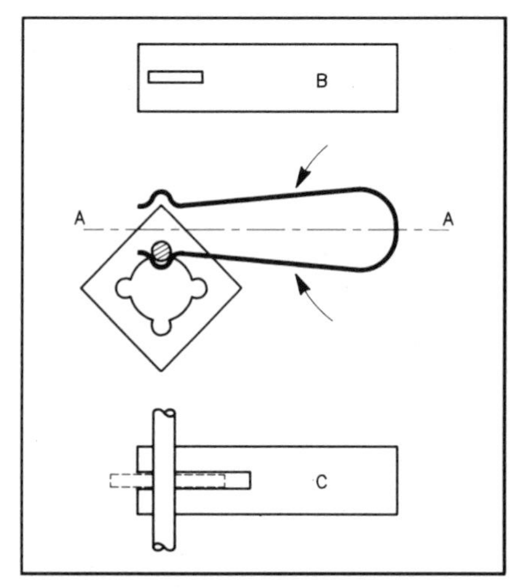


Fig 3—A tool made by the author to ease snapping conductors into spacers.

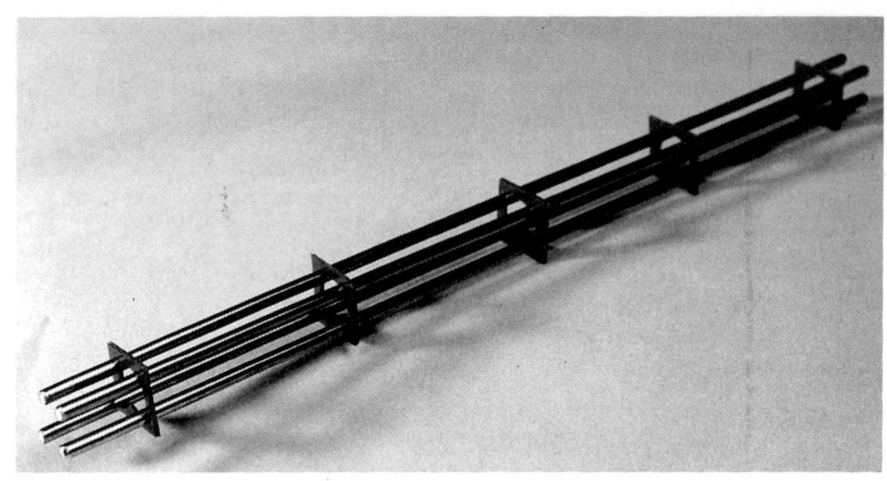

Fig 4—A completed section of Quadro-Line.

ing. Clamping the rod in a lathe or drill press allows faster polishing. Use the finest emery cloth and finish with a rubbing compound (used to restore the finish of faded automobile paint). Silver plating would improve performance, but the plating doesn't last long outdoors. Aluminum rod comes in coils, and must be stretched to straighten it before assembling the line.

Fig 2 gives the dimensions for a square spacer used with 3-mm-diameter rods to produce a Q-L with $Z_0$ = 100 Ω. Lexan plastic or other lossless material about 2-mm thick is suitable. Each spacer is 18 mm square, providing 11.7-mm diagonal separation between rods. Corner holes are bored slightly undersize, so the rods can be snapped into place. Fig 3 shows a flexible, U-shaped tool I made to ease pressing the rods into place, precisely and without excessive force. Fig 4 shows a section of completed line.

**Connecting To Quadro-Line**

There are three ways to connect to Q-L:

1) Direct connection from the rods to the load or source.

2) Connection via a transition to coaxial cable or other type of line.

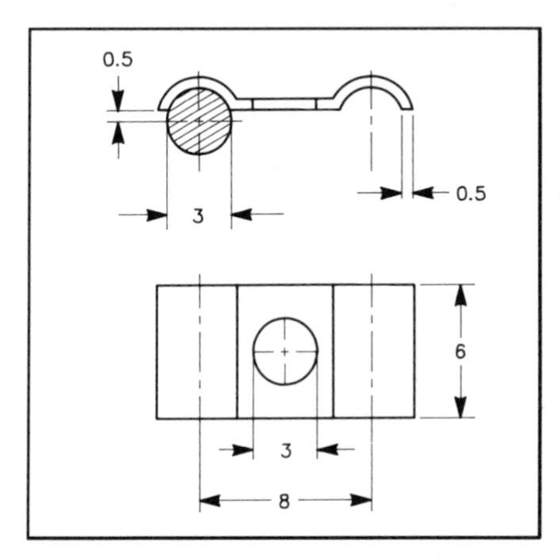


Fig 5—Metal clamp for 8-mm conductor spacing.

3) Attaching a suitable connector for transition to another type of line.

Fig 5 shows a suitable clamp for connecting adjacent rods (8-mm separation), while Fig 6 shows a similar clamp for connecting diagonally separated rods (12-mm separation). The metal used for the clamps should not cause an electrolytic reaction with the aluminum rods. Suitable metals include steel (or stainless steel), zinc and cadmium. The M2 screws and nuts used to tighten the connecting clamps

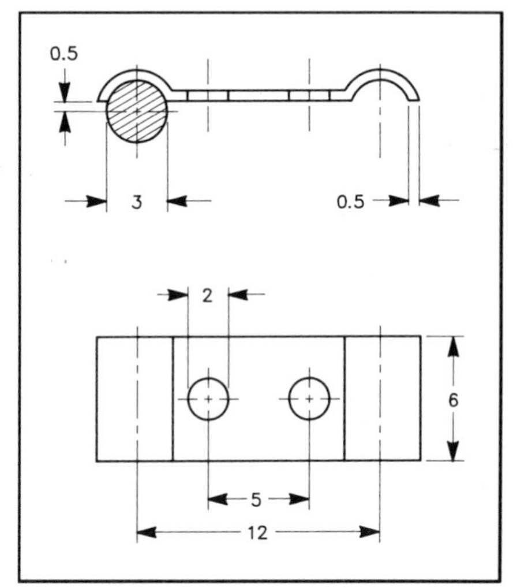


Fig 6—Metal clamp for 12-mm conductor spacing.

do not directly contact the aluminum rods, so they may be made of any metal. The connections with the ends of the Q-L conductors may vary, depending on the application.

One way to make an impedance-transformation is to use coaxial-cable braid, slid over the aluminum rods. A piece of steel spring wire wrapped around the netting

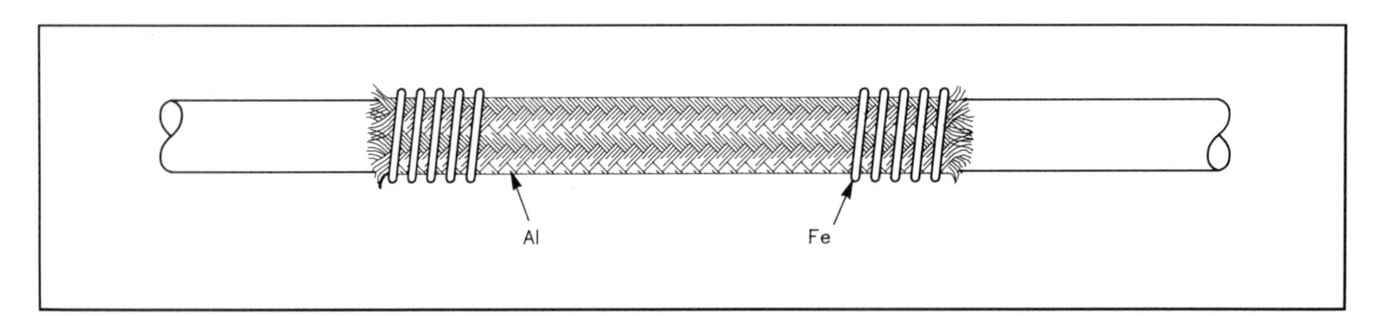


Fig 7—Coaxial-cable braid slid over the conductors provides a means of connecting to them.

Fig 8—A coaxial cable connector can be clamped to the line when a transition is needed. Details of the clamp are shown in Fig 9.

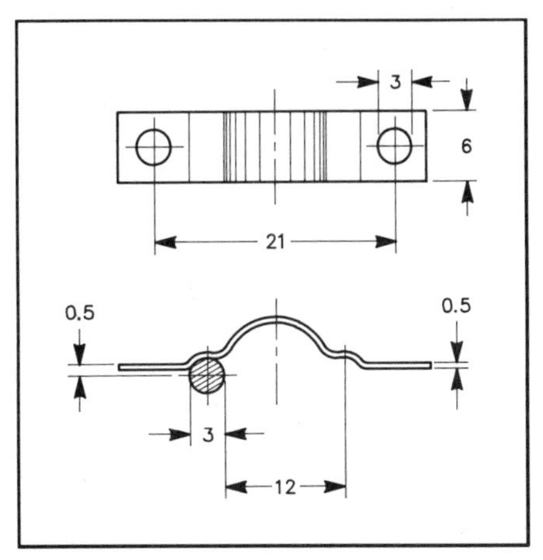


Fig 9—Coaxial-cable-connector clamp details.

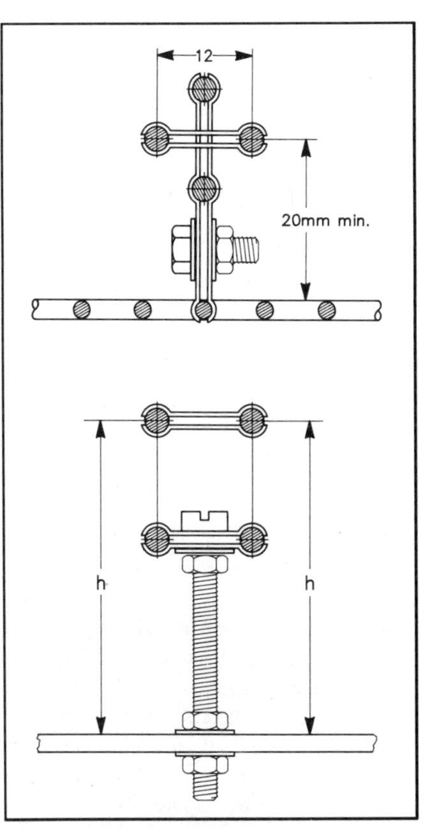


Fig 10—Quadro-Line can be mounted to a wall or metal surface such as a tower using the methods shown here. The author recommends a minimum spacing of 20 mm.

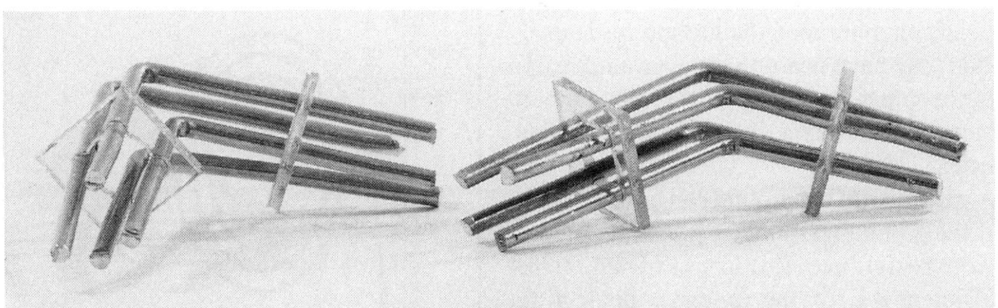

Fig 11—Although Quadro-Line is rigid, it can be bent when necessary. It's probably easiest to make the bend a separate section, then secure it to the straight runs on either side.

holds it in place (Fig 7). You can also use heat-shrink tubing to secure the braid and seal the junction.

Fig 8 shows the termination to a type-F connector using the clamps shown in Fig 9. Their internal thread corresponds to the plug, comprising two half-nuts, between which the aluminum rod ends are tightly clamped.

### Protection From The Elements

Q-L itself is not affected by moisture, but the transitions to other types of feed line may be, if they aren't suitably protected. If you mount the line vertically, use plastic collars to prevent water from running into the connectors. You can also use the traditional methods of sealing outdoor connections, such as with silicone grease and CoaxSeal.

### Mounting The Line

Fig 10 shows the two methods of securing the line to a wall. Mount the line at least 20 mm from any conductive surface, to minimize mistuning. Although Q-L line is not flexible, by planning ahead you can install the line with very few corners. Fig 11 shows practical 45- and 90-degree corners. Spacers should be closer together near the corners, or additional clamps (Figs 5 and 6) may be installed. Maintain symmetry between the "hot" lines and nearby objects. The "cold" lines may be used to support the line anywhere along its length.

### Conclusion

Quadro-Line is a low-loss transmission line suitable for home construction. In some stations, the transmission line may be the only commercial part. That need no longer be the case! Quadro-Line is relatively easy to construct and performs well. You can select a characteristic impedance almost at will, and even taper the line to accomplish impedance transformations. I hope other experimenters will do advanced work with Quadro-Line.

# Efficient Feed For Sky Hooks

By Stan Gibilisco, W1GV
2301 Collins Ave, #A-632
Miami Beach, FL 33139

The antipodes of Field Day mean long hours of darkness instead of light, and being indoors rather than out. One dreads solar storms if not thunderstorms. Broadcast stations at 910 kHz are anathema. Vacuum cleaners, heating pads, electric blankets and hair dryers lurk, their spark-makers poised, waiting to wreck your fun. Old cordless phones spew their chatter for miles. Yes, it's the 160-meter contest. I anticipate this contest at least as much as Field Day, for this is when I test new antenna ideas on one of our most radical and challenging bands.

I fly balloons and kites, weather permitting, with the antenna as a tether. I have found that the "sky hook," in the true Marconi tradition, is a 160-meter contest antenna that really does the job—unless you can have a 200-foot tower. Compromise antennas will work, and it's better to get on the air with a clothes hanger than to stay QRT. But how much more fun it is to be told "BIG SIG HR OM" when you're only running 85 watts! And oh, the pleasure of being able to "run" stations without commandeering the clothes dryer outlet or setting off burglar alarms!

I have written other articles about kites and balloons as antenna supports, and the mechanics (and the essential safety precautions) involved when flying them.[1,2] This paper concerns an equally important point: how best to make these antennas radiate. The feed method described here is useful with end-fed wires on any band, but it applies especially to the low bands, 160 and 80 meters. After much theorizing and testing, I believe this is the most elegant, efficient and easy-to-tune arrangement for low and moderate power levels, especially at 1.8 MHz.

### The Antenna

An end-fed wire needs a good ground, or so we are told. This is true in a strict theoretical sense. Any unbalanced antenna must work against ground, and an end-fed wire is an unbalanced antenna. But the efficiency of such an antenna system depends on more than the ground resistance. You may have heard that the ground resistance must be low for an unbalanced antenna to function well. This is not always true.

The loss in the ground can be calculated if the ground resistance $R_G$, and the radiation resistance $R_R$, are known. The efficiency E of the antenna, neglecting other possible losses, is

$$E = \frac{100R_R}{R_G + R_R} \text{ percent}$$

As an example, suppose the antenna presents a radiation resistance of 36 Ω, typical for a full-size, ¼-λ vertical. Also assume a ground resistance of 100 Ω, quite possible if the soil is dry and you have only a ground rod at the base of the antenna. Then the efficiency is:

$$E = \frac{100 \times 36}{100 + 36} = \frac{3600}{136} = 26 \text{ percent}$$

This means that almost three-fourths of the power that you put into the antenna only heats up the earth. You're losing 5.8 dB in the soil, even if your standing wave ratio (SWR) is 1:1. If you pump 1500 watts into the thing, you'll radiate only 400 watts. This could be your reward for erecting a pole 130 feet high for use at 1.8 MHz!

Shortened antennas can perform even more poorly, but not if we pay some attention to the ratio $R_G/R_R$. We want this ratio to be as low as possible. Perhaps better stated, we want the ratio $R_R/R_G$ as high as possible. The two resistances are in series electrically. Power divides between them, so the larger resistance gets most of the power. The greater the difference in resistance, the greater the proportion of power dissipated (or radiated) by the larger resistance. Ohmic values are less important than the ratio between them. Ideally, $R_R$ will be at least several times as large as $R_G$.

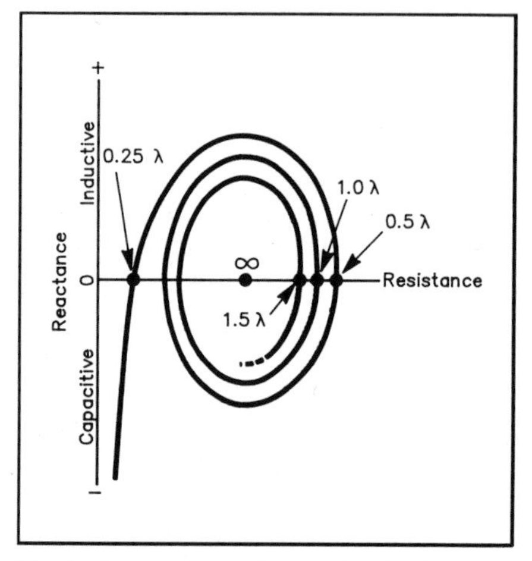


Fig 1—Impedance of an end-fed wire versus length when operated against a perfectly conducting ground. Reactance is zero at all integral multiples of ¼ λ. The curve spirals inward with increasing wire length.

In some installations, notably with short, inductively loaded verticals having low $R_R$, getting a good ratio is difficult. Many ground radials are required to lower $R_G$. For a longer wire antenna however, especially one supported by a kite or balloon, the problem is far easier to solve, because $R_R$ can be made extremely large. If the wire length is a multiple of ½ λ, $R_R$ is very high. In fact, the resistance may be so high that, at 1.8 MHz especially, many Transmatches must be modified to cope with it. Fortunately, the modification is often simple.

Fig 1 shows the resistive and reactive components of impedance for an end-fed wire, operated against a perfect ground ($R_G = 0$). When the wire is short, the resistance is low and there is capacitive reactance. Resistance $R_R$ rises to about 36 Ω and the reactance disappears at a length of ¼ λ. This represents quarter-wave resonance, and is often used to advantage as a so-called Marconi antenna. As the length is increased farther, inductive reactance appears and resistance $R_R$ continues to rise. The reactance reaches a maximum at about 3⁄8 λ, and then declines to zero again at ½ λ. Marconi would have done better at this length! By now, the resistance has become very large. If the wire is thin and no conducting objects are too close to it, $R_R$ will be thousands of ohms at the feed point when the length is ½ λ.

The highest possible feed-point resistance exists when the wire measures ½ λ. At 1.8 MHz this is 260 feet (468/1.8). High values also exist at small integral multiples of this length: 1, 1.5 and 2 λ. To get the lengths in feet, add 273 feet for each extra half wavelength beyond the first half wavelength (492/1.8). Any such length gives a resonant antenna. What's more important, the antenna accepts power and radiates it efficiently. Of course, we have to modify our tuning networks to handle this large resistive impedance. The high $R_R$ practically guarantees that the ratio $R_R/R_G$ will be large, and the efficiency high, if we have a halfway decent RF ground.

A ½-λ wire antenna, flown by a balloon or a kite, is essentially a free-space antenna, and approaches the ideal case. We may assume that $R_R$ is at least 3000 Ω; it may be much higher. (It is hard to predict $R_R$ accurately; even atmospheric humidity may affect it. But variations are easily dealt with by the right kind of tuning network.) If the ground loss is 100 Ω or less, the efficiency is at least

$$E = \frac{100 \times 3000}{100 + 3000} = \frac{300{,}000}{3100} = 97 \text{ percent}$$

and the loss in dB is negligible.

An end-fed wire needs a ground, but it doesn't have to be elaborate. A couple of ¼-λ wires (135 feet at 1.8 MHz) laid out on the surface will suffice for an end-fed, ½-λ wire antenna. This ground system works over any type of soil—even sand or rock. The effective RF resistance is probably under 100 Ω at worst, giving a worst-case efficiency of 97%.

This concludes the drawing-board work. The only remaining hurdle is the Transmatch, which may not tune a ½-λ, end-fed wire directly. Put away the pencil and calculator, and take out the soldering gun and wire cutters. There are a couple of alternatives to rewiring the innards of the Transmatch, and I'll mention them in case they interest you. I rejected both of these external options however, as inelegant, not versatile enough, and not optimally efficient.

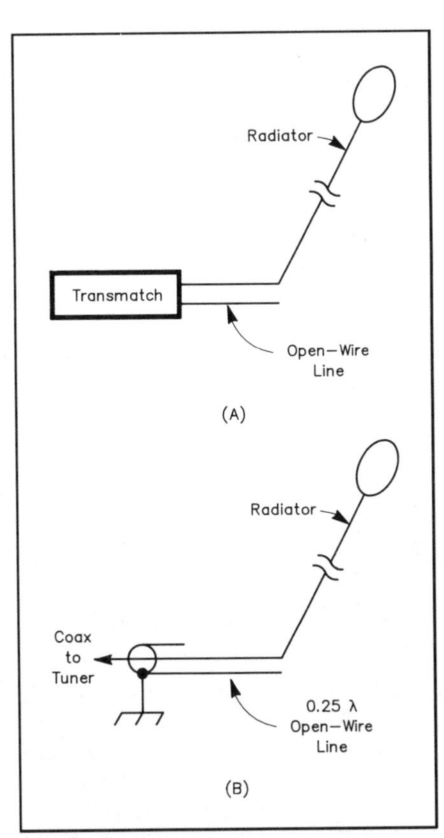


Fig 2—At A, Zepp feed for an end-fed, ½-λ wire. At B, a J-pole style of feeding a ½-λ wire. The section of transmission line at A should be at least 40-feet long; at B, the open-wire section should be ¼ electrical wavelength or about 130 feet at 1.8 MHz.

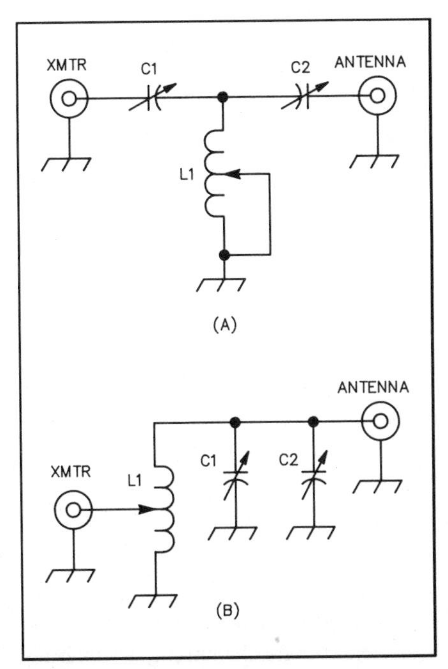


Fig 3—At A, my Transmatch circuit before rewiring; at B, after rewiring. The circuit at B allows matching extremely high, resistive impedances.

## The Matching System

One way to get a ½-λ wire to "look decent" to a Transmatch is to Zepp feed it, as shown in Fig 2A. This can be done if the Transmatch has a balanced output. I found that 40 feet of homebrew open-wire line was enough to bring the antenna impedance within the range of my tuner. This introduces reactance into the system, but the Transmatch is designed to tune that out. The Zepp feed line radiates, since Zepp feed cannot have perfect balance. This may not be a real problem though, especially if the feed line can be suspended in line with the wire. If the feed line is above the ground, its radiation is useful.

The feed line must have low loss, because the SWR is high. This mandates the use of open-wire line, not television twin lead. Finally, it is naive to hope that the Transmatch balun won't waste power. I cracked a Transmatch balun core with just 300 watts, after taking its kilowatt rating seriously. (I guess it was at a current loop. It must have been pretty hot! This heat can only have come out of the RF that should have gone to the antenna.) In general, I think there are too many wires, dielectrics and core materials in the system of Fig 2A. I wanted something simpler, with fewer places for power to be wasted as heat.

Another method is the scheme shown in Fig 2B. The ¼-λ line again should be of the open-wire type; television ribbon is too lossy. The matching section can be flown as part of the antenna, if the kite or balloon has enough lift. This makes the overall length ¾ λ. This feed method can be inconvenient to adjust, and may not offer a good match over a very wide band of frequencies. I discarded this idea for these reasons, and also because it was close to contest time, and I had no ladder line. Instead, I took the solder gun and wire cutters to my Transmatch. There went the warranty!

The original Transmatch circuit (without the balun) is shown in Fig 3A. In my opinion, this is not an optimum design. It's versatile, as the inductance can be decreased to zero; but it can be inefficient for certain antenna lengths. In some cases it won't provide a match to 50 Ω. My particular unit could not match a ½-λ wire on 160 meters. The antenna had to be at least 2/3 λ to obtain a 1:1 SWR. This defeated my main objective. The value of $R_R$ would be too low to suit me if the antenna were that long. Besides, I noticed coil warming at power levels over 200 watts output. (The unit is supposed to handle 1500 watts.)

I rewired the Transmatch as shown in

Fig 3B. This is a classic circuit for matching high, resistive impedances to 50 Ω. The roller inductor is put to better use in my design, as compared with the manufacturer's circuit. It serves as a variable transformer, rather than as variable inductor. We need such a device to manage the high $R_R$ of an end-fed, ½-λ, free-space wire antenna.

My modification limits the Transmatch to 160 and 80 meters, since the tank circuit always contains the entire coil. This is too much inductance at 7 MHz and higher frequencies. But it matches a ½-λ, end-fed 160-meter or 80-meter wire perfectly, and with plenty of leeway. My interest is only in these two bands anyway, at least for end-fed, kite- and balloon-supported wires.

You can make this modification to any Transmatch that has a roller inductor. It involves only a little cutting, desoldering and resoldering. Be sure you get the wiring right. In my case, the two identical capacitors are wired in parallel. This ensures that the unit can be tuned to resonance at 1.8 MHz. I set one capacitor at minimum and adjust the other one. If that's not enough, the first one can be adjusted after the other is fully meshed.

I found this unit forgiving of errors in antenna measurement, which came as a pleasant surprise. When I tested it further, I discovered it tunes practically any wire at 1.8 MHz, no matter what its length. It tuned a six-foot wire without coil heating at 85 watts intermittent (CW) power. In fact, it even "tuned" a piece of wet spaghetti, though there was arcing across the capacitor at only 75 watts output. I suppose the short ground leads were getting more RF than the spaghetti. But the coil still did not get warm, even after calling CQ at 70 watts output for a half hour.

### Observations and Modifications

While the coil has too much inductance to work at 7 MHz and above, it should be easy enough to place inductances in parallel with the main coil. Doing so will raise the resonant frequency while retaining all the advantages of the roller inductor. An example is shown in Fig 4. L2 is a tapped toroidal inductor. Total inductance of the parallel combination of L1 and L2 is suitable for the desired band. I didn't actually try this, as my interest is in 160 and 80 meters. I have used the technique with active ferrite loopstick receiving antennas though, and it works fine.

Voltage across the tuned circuit is large when the wire is short, as proved by the capacitor arcing at power levels of 75 watts

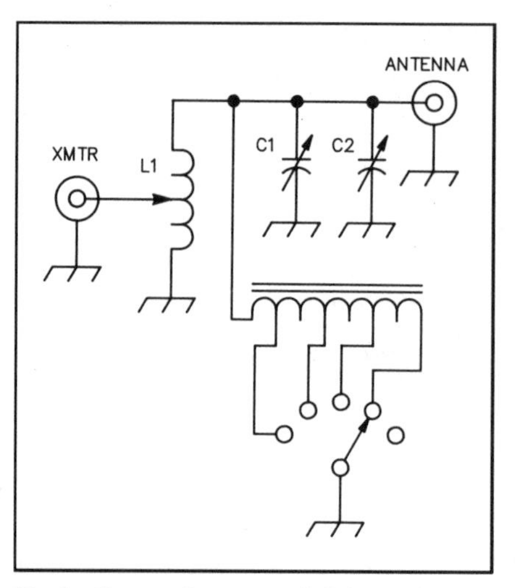


Fig 4—Connecting a parallel, tapped inductance to the main transformer coil may allow tuner operation on 7 MHz and above. L2 is a tapped toroidal inductor.

or so. The original Transmatch is supposed to handle the legal power limit. My modification subjects it to voltages larger than those in the original circuit. I guess RF just leaks off wherever it gets a chance.

This Transmatch arrangement should interest anyone who wants to work on 160 meters in limited space. But with short wires—less than ¼ λ or 130 feet—a good ground becomes important. Take pains to minimize $R_G$.

### How Well Does It Work?

When I get on the radio I want to take it easy. I'm not much of a contester. I'll let the contest awards go to those who labor to get them. To try out this system, I made a few contacts and held the frequency a few times. A station in California told me that a station in Hawaii (KH6) was calling me, but I couldn't hear the other station. This made me think, as my ½-λ wire dangled above the Minnesota tundra, why not hang a parasitic element from another balloon nearby? Say, a director west of my ½-λ wire? If I'd had one when the KH6 called me, I might have heard him.

Let's see. It would have to be about 240 feet long, and about 50 feet away. And also about 52 weeks in the future, if the weather is good. I did get as far as inflating a second five-foot balloon for the purpose, but when it popped in my face, I gave up.

### Using the 160-Meter Antenna on 80 Meters

The ½-λ wire, carefully measured to 260 feet according to the classic dipole formula, is close to a full wavelength at 3.5 MHz. The coil tap position is similar on both bands: about 15% of the way from ground to the "hot" end; the setting change is minor. The capacitance needs some adjustment, of course, to change from 160 to 80 meters.

I also worked numerous stations on 80 meters when I got bored with 160. The 1-λ wire presents a high-resistance $R_R$, and all the advantages that go along with it. It doesn't have to be exactly a multiple of ½ λ; the Transmatch tunes out reactance well enough that an approximate length is all right. On neither band, incidentally, did I suffer from "RF in the shack," even though the wire came right down to the station window. Of course, I wasn't running much power. With high power, you could have arcing problems with this arrangement unless you use a capacitor with wide plate spacing. This is why we have hamfests—so we can get these oddball components. If you want to use this scheme with high power, I recommend a major metropolitan hamfest. You might also have to locate the feed point, and hence the Transmatch, some distance away from the station, and use coax between the transmitter and tuner, to keep the RF out of station accessories and house wiring.

I used to dream of working worldwide DX on 80 meters, back when I was a kid trying to force-feed a 40-meter dipole with my crystal rig on 3.714 MHz. I thought it was a miracle when I worked Ohio that way. Not many years later, albeit with slightly more power, I was easily logging European and South American stations like they were locals. Yes, my Elmer had assured me, these things can and do happen.

During the contest, several stations commented on my loud signal. Other stations were routinely coming in at S9 plus 20 or 30 dB. Some, not local, were even louder. When I called a station, I almost always got through on the first try. The coil never got warm, even after torrid contesting at 85 watts for a half hour at a stretch.

### Bibliography

[1] S. Gibilisco, "Balloons as Antenna Supports," *The ARRL Antenna Compendium, Volume 2* (Newington: ARRL, 1989), pp 142-144.

[2] S. Gibilisco, "Balloon and Kite Supported Antennas," *The ARRL Antenna Compendium, Volume 2* (Newington: ARRL, 1989), pp 145-149.

# The Z-Match Coupler — Revisited and Revised

By Charles A. Lofgren, W6JJZ
1934 Rosemount Avenue
Claremont, CA 91711

At one time or another, almost every amateur working the HF bands needs a Transmatch, commonly known as an antenna coupler or tuner. A flexible and convenient design for balanced feed lines is the multiband Z-Match antenna coupler, originally described by Allen King, W1CJL, in 1955[1] and subsequently appearing with minor variations in the *Handbook* and elsewhere.[2] While continuing to draw attention in Great Britain,[3] it seems to have fallen from sight on this side of the Atlantic.

The purpose of this article is twofold. First, it reviews the original Z-Match, examines its operation, and offers some revisions that make it even more useful. Second, it suggests how the Z-Match may be modified and used in a variation of the L-Match that I call the "L-Z Match." My intention is not to provide how-to-do-it detail (although suggestions are included), but to offer ideas that I have used in actual practice.

### The Original Z-Match

The original Z-Match, which covers 80 through 10 meters, is shown in Fig 1A. According to King's measurements, adjustment of C1 and C2 allows the coupler to handle "reactive and nonreactive loads ranging from 10 to 2,500 ohms,"[4] with no band switching other than changing between links L1 and L4. To explain the multiband and matching capabilities of the Z-Match, two aspects need attention.

Multiband operation results from the tank circuit formed by C2, L2, and L3. This simultaneously tunes two ranges of frequencies, as illustrated by Figs 1B and 1C. On the lower range (approximately 3.5-10.5 MHz), L3 has a relatively low reactance and tends to appear as a short circuit, which places the two sections of C2 in parallel across L2 (see Fig 1B). On the higher range (approximately 10-30 MHz), L2 has a relatively high reactance and tends to disappear (or act as an RF choke), which places the two sections of C2 in series across L3 (see Fig 1C).[5]

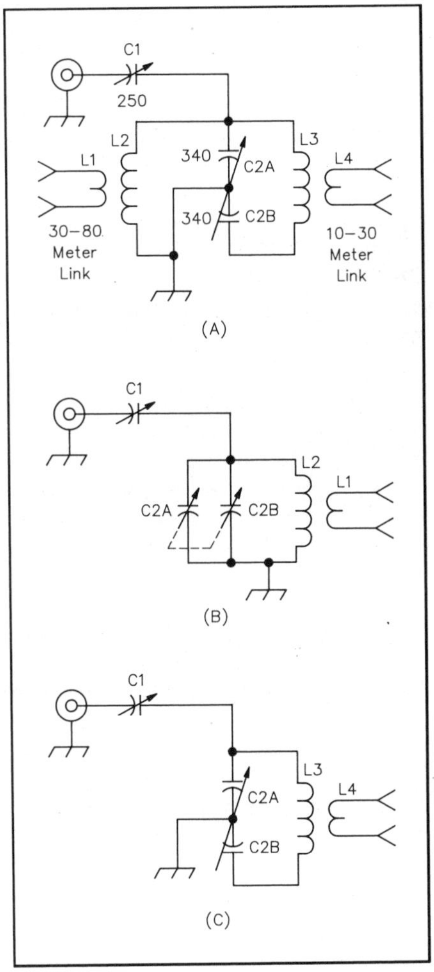

Fig 1— Schematic diagrams of the original Z-Match Coupler. At A, the complete circuit. At B, the equivalent circuit for approximately 3.5- to 10.5-MHz operation. At C, the equivalent circuit for approximately 10 to 30 MHz.

The key to the Z-Match's impedance-matching capability is its function as an L network.[6] The evolution of the circuit is shown in Fig 2 and may be explained as follows: First, consider an L network, as shown in Fig 2A. It has an inductive reactance in the series arm and a capacitive reactance in the shunt (or parallel) arm. Second, reverse the reactances, as in Fig 2B. For impedance-matching purposes, the reversal does not affect operation of the network, provided the absolute values of reactance in each arm remain the same. (In conventional notation, the signs of the reactances in each arm are now reversed.) Third, substitute a parallel L-C circuit in the shunt arm of the L network, as shown in Fig 2C. When the L-C circuit is tuned to the high side of resonance, it appears as an inductive reactance and is equivalent to the circuit in Fig 2B. Finally, add a tightly coupled link, as shown in 2D. This is the equivalent of the Z-Match in Fig 1A, with the exception that in the Z-Match a multiband circuit replaces the simple L-C circuit in the shunt arm.

### The Refined Z-Match

The Z-Match is not quite "switchless." As already indicated, changing between the high and low HF bands requires a switch to select either L1 or L4 as the output link. In addition, my experience with the circuit indicates that for some antenna and feed-line combinations—especially those presenting low impedances to the coupler—two other refinements are useful to avoid the necessity of changing the length of the feed line.

One refinement is a means of varying the number of turns in the output links. This can be done by tapping down from the top

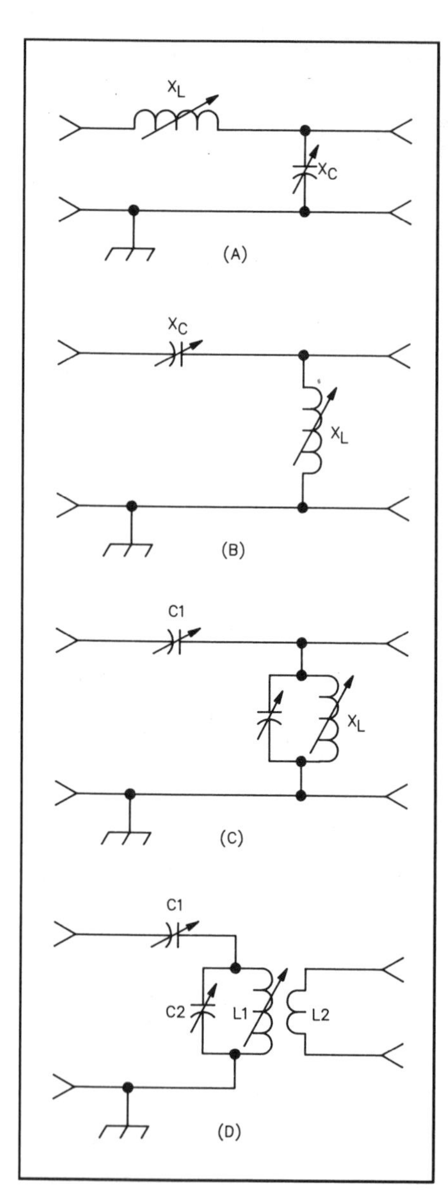


Fig 2— Evolution of the Z-Match. At A, a conventional L network. At B, the L network with reactances transposed. The circuit at C has a detuned L-C network in the shunt arm. D shows an L network with link-coupled output.

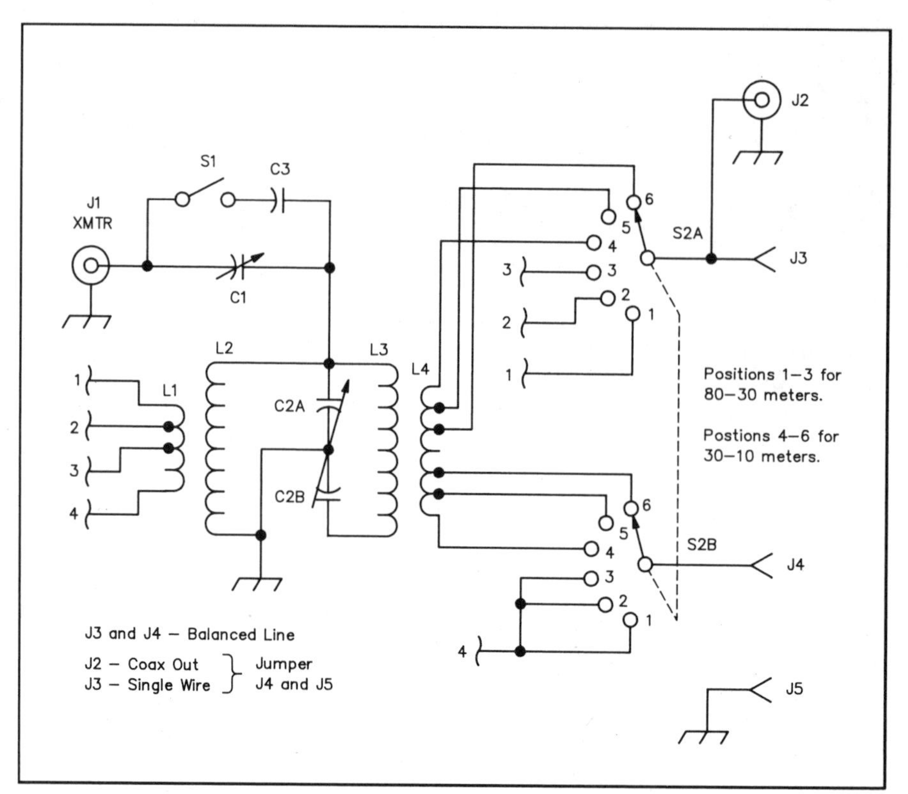


Fig 3—The "refined" Z-Match. Refinements include tapped output links and the provision to switch parallel capacitance across C1.

C1—325 pF or greater.

C2—325 pF or greater per section (see text).

C3—300 pF.

L1—3.2 µH; 9 turns no. 16, 2 inches diameter, 1.5 inches long (6 turns/inch). Tap at 3 and 5 turns from top. Mount concentric with and inside L2.

L2—3.1 µH; 8 turns no. 12, 2.5 inches diameter, 2 inches long (4 turns/inch).

L3—1.15 µH; 4 turns no. 12, 2.5 inches diameter, 1 inch long (4 turns/inch).

L4—1.8 µH; 6 turns no. 16, 2 inches diameter, 1 inch long (6 turns/inch), tapped 1 and 2.25 turns from each end. Mount concentric with and inside L3.

on L1 and in from each end of L4. The effect is to change the transformation ratio *between* the L network and the antenna feed line. If the switch used to select the output link is a 2-pole switch with 4 or 6 positions per pole, it will allow for *both* changing the links *and* selecting from 2 or 3 taps on each link.

A second refinement is to allow for adding additional capacitance in parallel with C1. This is particularly useful in handling loads sometimes encountered on the lower frequency bands.

The Z-Match with refinements is shown is Fig 3. A few points may be noted. Links L1 and L4 should be placed over (or inside) L2 and L3, respectively, to ensure tight coupling. Rather than being wound directly over (on top of) the associated tank coils, they should be larger (or smaller) in diameter than L2 and L3 by about ¼ to ½ inch. This lessens the capacitive coupling between the tank circuits and the links, which helps with harmonic reduction and in maintaining current balance in the output. Suitable coil dimensions are included in Fig 3, but may be varied to suit the capacitor used at C2. Also, the number of taps on each link may be changed from that shown in Fig 3, depending on the positions available on S2. The two coils, L2 and L3, should be placed at right angles to each other, to minimize coupling between them.

An important consideration is wire size. Depending on the reactances being tuned out, the circulating current in the tank circuit formed by C2 and either L2 or L3 may be quite high, owing to the unusually high C/L ratio. This means that to reduce $I^2R$ losses (heat), L2 and L3 should be wound with large diameter wire or, for high power, with tubing. For the same reason, the wiper contacts for the rotor plates in capacitor C2 should be able to carry high current. (Even at QRP levels, the need to mitigate losses in these components is important.) To obtain adequate overlap between the low- and high-frequency coverages, and hence to ensure reliable coverage of 30 meters, it is desirable to have a capacitor of at least 325-pF per section at C2, and larger is still better. (At the same time, however, a low minimum capacitance per section is also desirable.)[7]

As for the resulting feed-line balance, measurements with an RF ammeter on various antenna and feed-line combinations indicate that current balance in the feed line is within about 10 percent. As with most Transmatches, it tends to be better when the feed-point impedance is relatively low.[8] It is better, too, when out-

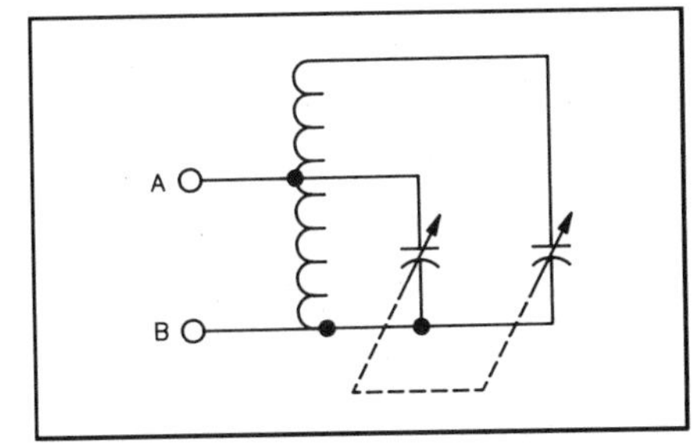


Fig 4— Multiband tank circuit with single coil.

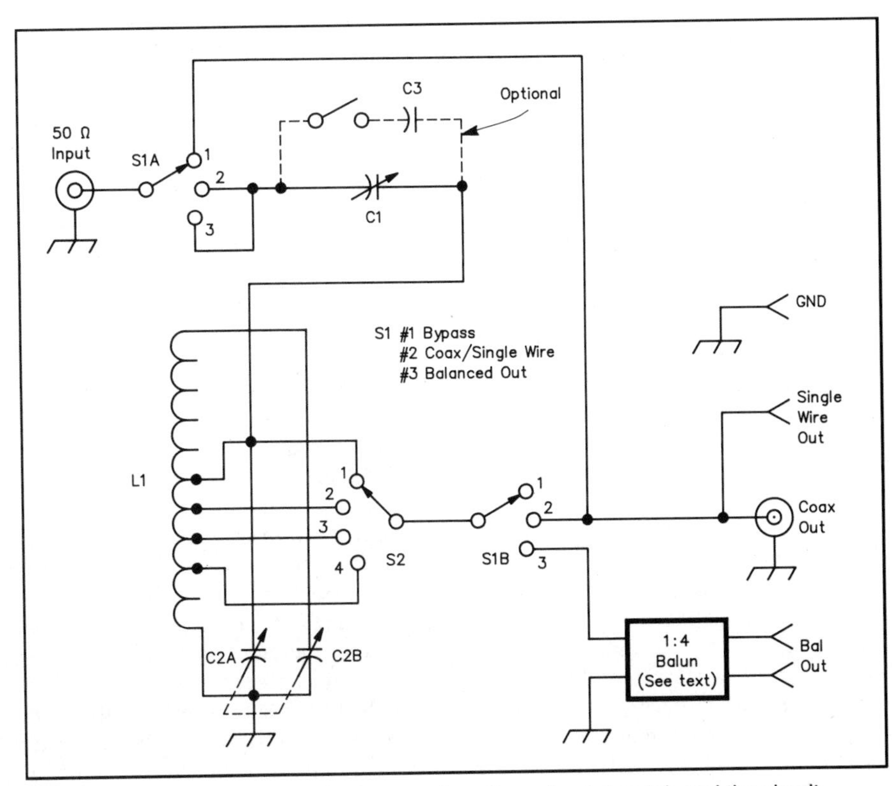


Fig 5— The L-Z Match combines features from the refined Z match and the circuit shown in Fig 4. Values for C1, C2 and C3 are the same as given in Fig 3.

L1—22 turns no. 16 tinned or enameled wire on a T-157-6 core. Tap at 11, 9, 6 and 3 turns from the bottom.

put is from the high-band link (L4), as would be expected from examining the equivalent circuit in Fig 1C.

## An Adaptation: The L-Z Match

The heart of the Z-Match is its multiband tank circuit. A similar circuit will function as the shunt element in a standard single-ended L-Match, as indicated earlier in Fig 2. Such an arrangement allows for continuous adjustment of the inductive reactance without resorting to a roller inductor. Before turning to the details of this version, let me explain the variation of the multiband circuit that it uses.

Back in the days when the multiband tank circuit was enjoying its popularity in tube-type transmitters, R. W. Johnson, W6MUR, described a multiband tuner that employed only one coil. This is shown in Fig 4. Anyone interested in the derivation of the design should consult Johnson's informative 1954 article.[9] Suffice to say that a single center-tapped inductor of the appropriate dimensions may be tuned to cover two simultaneous ranges of frequencies. Between terminals A and B, the L-C circuit in Fig 4 appears as a parallel-resonant circuit with two resonant frequencies.[10]

Coverage of the two frequency ranges depends on three variables: the value of L1, the coupling between the two halves of L1 (which is a function of its length-to-diameter ratio), and the value of each section of C1. To cover the full range from 3.5 to 30 MHz, with adequate overlap in the vicinity of 30 meters, C1 should have 325 pF *or more* per section, and the coil should have an inductance of approximately 4.8 μH with a length-to-diameter ratio of about 1:1. An example is a coil of 12 turns, 2 inches in diameter, and 6 turns per inch (making its length 2 inches). (Johnson's article contains graphs as well as formulas for calculating the coil configuration, but the easiest design approach is to do some experimenting with a dip meter.)

The multiband tank shown in Fig 4 may be used as the shunt element in an L network, with the necessary inductive reactance being produced by detuning the circuit to the high side of resonance.[11] If the circuit is used in this fashion, the matching range will depend on whether the shunt element is on the input or the output side of the network. If on the input side, the network will match to impedances lower than the input impedance. If on the output side, the network will match to impedances higher than the output impedance.

A practical application is the L-Z Match in Fig 5. It uses a powdered-iron toroid in place of an air-core coil at L1. Rather than reversing the input and output ports in order to accommodate the full range of impedances equal to, above, and below the input impedance, the design includes taps on the lower half of L1. This allows a simpler switching arrangement. In practice, two or three taps distributed below the coil midpoint should give an adequate low-impedance range. Keep in mind, however, the desirability of intermediate options between the center-tap and the lowest tap. (For highest efficiency, the tap selected with S2 should be no lower than necessary.) If the antenna and feed-line combination presents a high impedance or voltage loop at the coupler, it should not be necessary to move S2 from the first (coil center-tap) position.) The tuning of C2 is often quite sharp, and for this reason a vernier drive is useful.

Compared to the standard Z-Match, the design in Fig 5 is yet more susceptible to circulating currents in the tank circuit under certain tuning conditions. For this reason, large diameter wire is used for L1. If space permits, a T-200-6 core, wound with no. 12 or no. 14 wire, would be better than the T-157-6 core and no. 16 wire specified in Fig 5. If the substitution is made, the number of turns can probably be left unchanged.

S1 provides selection of tuner or bypassed operation, coax/single-wire output via the tuner, or balanced output via the tuner. A more elaborate switching arrangement could incorporate additional options.

For the balun in Fig 5, I have found that a 1:4 "voltage" balun with a ferrite core (μ=125) is quite satisfactory at the low power levels recommended for the coupler (see below). Core saturation is not a problem, and measurements with an RF ammeter show good feed-line balance. Nonetheless, purists may wish to try a current balun. In any case, I recommend the 1:4 ratio rather than a 1:1 ratio. While the reactances present in a complex feed-line impedance may affect the balun's actual impedance transformation, the 1:4 ratio seems to provide better performance at high feed-line impedances, while not compromising the low-impedance performance of the coupler.

Even with heavy gauge wire for L1, I would not recommend the design for power levels above 25 watts. For QRP levels, however, the L-Z Match makes a convenient and physically small coupler. Without the bulk of a roller inductor, its multiband tank circuit allows continuous adjustment of the inductive arm of the L network, which is an advantage over the usual tapped-coil alternative.

### Operation

For either the Z-Match or its L-Z offshoot, tuning is simple. Initially set C1 at its midpoint and tune C2 for an indication of a dip in SWR between the Transmatch and the rig. It may be necessary to try another setting of C1 before a dip is observed. (Peaking on receiver noise is another option for initial tuning.) Then fine tune both C1 and C2 for a 1:1 SWR. The two adjustments will interact to some extent. With the Z-Match, of course, be sure to select the correct link. If no match proves possible, the feed line is probably presenting a low impedance at the coupler. If this is the case, select one of the low-impedance taps with S2. (It should be noted that usually you can also "tune" a high-impedance feed point when S2 is set on a low-impedance tap. Efficiency may suffer, however, so don't use a tap any lower than necessary.)

Coverage of 30 meters may fall at either the low or the high capacitance end of C2, or at each end. Where it actually occurs will depend on the exact component values in the circuit and on the feed-line reactance seen at the coupler. When possible, tune the 30-meter band at the low-capacitance end of C2, and, for the Z-Match, use the low-band link. In theory, this gives marginally better efficiency, owing to the lower C/L ratio in the multiband tank circuit.

The Z-Match and its variations are not cure-alls, but they probably have good harmonic-reduction qualities.[12] At the very least, they offer convenient adjustment, including ease in band changing.

### A Note on Feed-Line Imbalance

In most circumstances, the ability of a Transmatch to produce balanced currents in the feed line is really tested when it feeds the line at a high-impedance point. This is true for the Z-Match, for other link-coupled tuners, and also for today's more common solution of using the combination of an unbalanced or single-sided matching network along with a balun on either the input or the output end. The reason is that at high impedances, even a high reactance offers leakage coupling to ground. This offending reactance may take the form of a small capacitive coupling between key components (especially links) and ground, or the inductive reactance (or capacitive reactance) to ground in a balun.

The problem may be present even when the balun is used on the *input* end of a Transmatch, as sometimes recommended. In this instance, when a single-sided matching network is used (for example, the popular T network or an L network), the balun also presents a reactance between one side of the *output* and ground, although the reactance does not appear as such in the circuit diagram. In other words, it is only partly correct to say that placing a balun at the input of the Transmatch, in a low-impedance and purely resistive circuit, avoids the problem of using it in a situation for which it is not designed, that is, in a reactive, high-impedance circuit. A portion of one side of the output current of the Transmatch still may flow through the balun to ground *if* the balun's reactance to ground is low *relative* to the antenna feed-line impedance that appears at the Transmatch. (The balun now acts as an RF choke. No one would put an RF choke of only several hundred ohms reactance between ground and one side of a Transmatch output feeding a high impedance line!)

The imbalance develops primarily when the impedance being matched is high, however, and detecting it is thus a problem. At a high-impedance point, the feed-line current is low and often close to zero, which makes it difficult to measure with any accuracy using a standard RF ammeter. The solution is to measure the current balance a quarter wavelength from the Transmatch, *toward* the antenna. The same result may be obtained by temporarily inserting an extra half wavelength of feed line between the Transmatch and the regular feed line, with RF ammeters in the middle of the temporary half-wave segment. (In making the measurements, two ammeters may be used, one on each side of the transmission line, or a single meter may be switched from one side to the other.)

### Notes

[1]A. King, "The 'Z-Match' Antenna Coupler," *QST*, May 1955, p 11.

[2]For example, *The Radio Amateur's Handbook*, 1957 ed., pp 340-341; *Understanding Amateur Radio*, 1963 ed., pp 288-290. At least two commercial versions also appeared in the 1950s. Harvey-Wells Electronics produced the coupler originally described by King, "The 'Z-Match' Antenna Coupler," Note 1 above, and World Radio Laboratories manufactured another version.

[3]For example, see Radio Society of Great Britain, *Radio Communication Handbook*, 5th ed. (1976), pp 12.50-12.51; L. Varney, "An Improved Z-Match ASTU," *Radio Communication*, Oct. 1985, pp 770-71, 776; J. Heys, *Practical Wire Antennas* (RSGB, 1989), pp 88-89. Heys' presentation is based on Varney's.

[4]King, "The 'Z-Match' Antenna Coupler," Note 1 above, p 11. Of course, there was no attempt in those days to cover 30 meters. Care in selection of component values allows King's design to handle the band, which appears in what can be called the "transition zone" between the coupler's two ranges.

[5]The "unused" inductors do have a slight effect on the tuning of the tank circuit, but in practice this may be ignored. For a fuller description of the circuit's operation, see C. Chambers, "Single-Ended Multiband Tuners," *QST*, July 1954, pp 23-24.

[6]The following analysis in this paragraph draws on the explanation in *The Radio Amateur's Handbook*, 1957 ed., pp 340-341. The *Handbook* explanation is particularly important in view of the criticisms of the original Z-Match suggested by L. Varney, "An Improved Z-Match ASTU," Note 3 above. Varney criticizes the design for feeding a 50-ohm line into the top or high-impedance end of a parallel-tuned L-C circuit. In actuality, this is not done in the Z-Match, as can be seen when the multiband circuit in the design is understood as the shunt arm in an L network, with the 50-ohm line feeding instead into the series arm. Varney's data nonetheless suggest the need to provide a tapping arrangement, and he offers another alternative to the one discussed in the present article.

[7]Fair Radio Sales Co, PO Box 1105, Lima, Ohio 45802, carries a 4-section variable capacitor, 20-500 pF per section (part number 4G-535). For C2 in Fig 3, use two of the sections. (I would guess that in the Z-Match, its plate spacing should handle the voltages developed by 100 watts of RF; I have had no difficulty using it at the 50-watt level. Its relatively high minimum capacitance may require "playing" with the inductance values.) For levels up to 25 watts or so, "receiver-type" variables with 365 or 420 pF per section are adequate, but look for ones which have separate rotor wiper contacts—that is, which do not depend only on the end bearings for electrical contact between the frame and rotor.

[8]For an explanation, see "A Note on Feed-Line Imbalance" in the text at the end of this article.

[9]Johnson, "Multiband Tuning Circuits," *QST*, July 1954, pp 25-28, 122.

[10]The L-C circuit is also series resonant midway between the two parallel-resonant frequencies. This resonance does not affect its operation in the matching network described in the text.

[11]Alternatively, the multiband circuit could be used in the series arm of the network, but the desirability of grounding the stators of the variable capacitor makes its use in the shunt arm more convenient. In principle, the cir-

cuit could also be used in a T network. However, the two capacitors in a T network usually allow fine adjustments even when a tapped rather than roller inductor is used in the shunt arm, so the advantage of using a continuously tuned multiband tank is not so great. Also, the T network is less efficient than the L network, and my tests indicate that the problem with circulating current in the tank circuit (mentioned in the text) is greater when the multiband circuit is used in a T network.

[12]I have not been able to check harmonic reduction with a spectrum analyzer, but in the Z-Match, the tank circuit provides attenuation, and the link coupling should reduce capacitive coupling of harmonics. In the L-Z Match, the capacitive reactance between the output and ground resembles the arrangement in the SPC version of the T-Match —and the tuning is similarly sharp. A potential problem is that the multiband circuit may possibly tune both a fundamental frequency and one of its harmonics at the same settings of the controls. For this reason, when using these circuits in transmitters, component values are carefully selected to avoid simultaneous harmonic resonance. In Transmatch operation, a complex of unpredictable feedline variables is "reflected" into the circuit, which means there is little point in trying to guard against harmonic resonance in the design itself. In years of using variations of the Z-Match, I have not encountered a situation where the Transmatch tunes two bands with the same control settings using the same antenna.

# Fixed Inductor Variable Impedance Matching Networks

By Robert F. White, W6PY
130 Heather Lane
Palo Alto, CA 94303

This paper focuses on the design, but more especially on the analysis by calculator program, of impedance matching networks which use a fixed (tapped) inductor (L) as the central network element and two variable capacitors ($C_A$ and $C_B$) as the end elements.

Such networks are intended to match a range of unknown and, in general, complex load impedances to a known and fixed resistive source impedance. The two commonly used arrangements treated herein are the high-pass T-matched system of Fig 1A and the low-pass pi-matched system of Fig 1B.

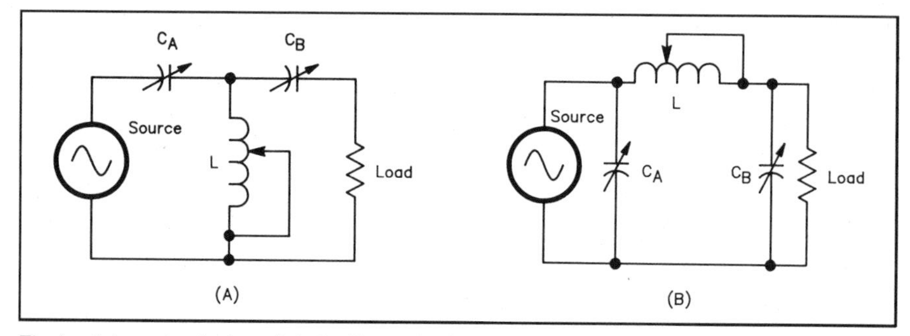


Fig 1—A is a physical model of a T-matched system. B is a physical model of a pi-matched system.

To analyze such systems, it is convenient to replace each physical model with a reactance model. In the model, L is replaced by a fixed positive reactance ($+X_0$), and $C_A$ and $C_B$ are replaced by two variable negative reactances ($-X_A$ and $-X_B$).

A reactance model will represent the corresponding physical model only at some specific but unstated design frequency ($f_0$). Ultimately, the fixed positive reactance can be converted to µH and the negative reactances to pF for some value of $f_0$.

The basic reactance models and some additional terminology used in this paper are given in Fig 2. The fixed parameters $R_A$, $R_0$, $X_0$ and $T_0$ in Fig 2A (for T networks) and Fig 2B (for pi networks) are so related that if any two are known or assumed, the other two are fixed and calculable by Eq 1 (the basic design equation) and either Eq 2A or Eq 2B as applicable.

The source impedance $R_A$ is almost always fixed and known. In most modern amateur stations, it is the 50-Ω unbalanced output of a solid-state, broadband transceiver and is independent of frequency. Unless otherwise stated, $R_A$ will be 50 Ω in this paper.

With $R_A$ fixed, there are two useful design choices: (1) to select a value for $R_0$ and use the relationship of Eq 1 to calculate $X_0$ or; (2) to select an arbitrary value for $X_0$ and use Eq 1 to calculate $R_0$. For the purposes of this article, it is most convenient to choose $R_0$ and derive $X_0$ and $T_0$.

### Defining the Load Range

In order to make an intelligent choice for $R_0$ it is desirable to estimate the range of load resistances that will need to be matched. In most modern stations the "load" is a coaxial cable (with a nominal $Z_0$ of 50 Ω) that is connected to an antenna of some sort at the far end. The range of impedances seen at the coax input depends only on $Z_0$ and the SWR of the antenna-to-$Z_0$ mismatch at the far end.

Smith Chart analysis shows that the range for the real parts of the input impedances is the same for either $R_{SE}$ or $R_{PE}$. For both cases, the range is given by:

$$\frac{Z_0}{SWR} \text{ to } Z_0 \times SWR$$

Table 1 shows the ranges for a $Z_0$ of 50 Ω with several SWRs. The last two rows are of special interest. Note that each of these ranges has geometric symmetry around $Z_0$, a point of some significance in system analysis.

**Table 1**
**$R_{SE}$ or $R_{PE}$ range versus SWR, for $Z_0$ = 50 Ω.**

| *SWR* | *Range For $R_{SE}$ or $R_{PE}$* | | |
|---|---|---|---|
| 2.0 | 25 | to | 100 |
| 2.5 | 20 | to | 125 |
| 3.0 | 16.67 | to | 150 |
| 3.5 | 14.29 | to | 175 |
| 4.0 | 12.5 | to | 200 |
| 4.419 | 11.315 | to | 220.95 |
| 5.050 | 9.901 | to | 252.5 |

### Criteria for a Satisfactory Match

Before getting into specific treatments, an additional item of information is needed. One of the desiderata in these systems is a reasonable loaded Q of the matched system. A maximum Q of 20 is often cited in the literature, and that value is used in this article.

The two necessary criteria for a system

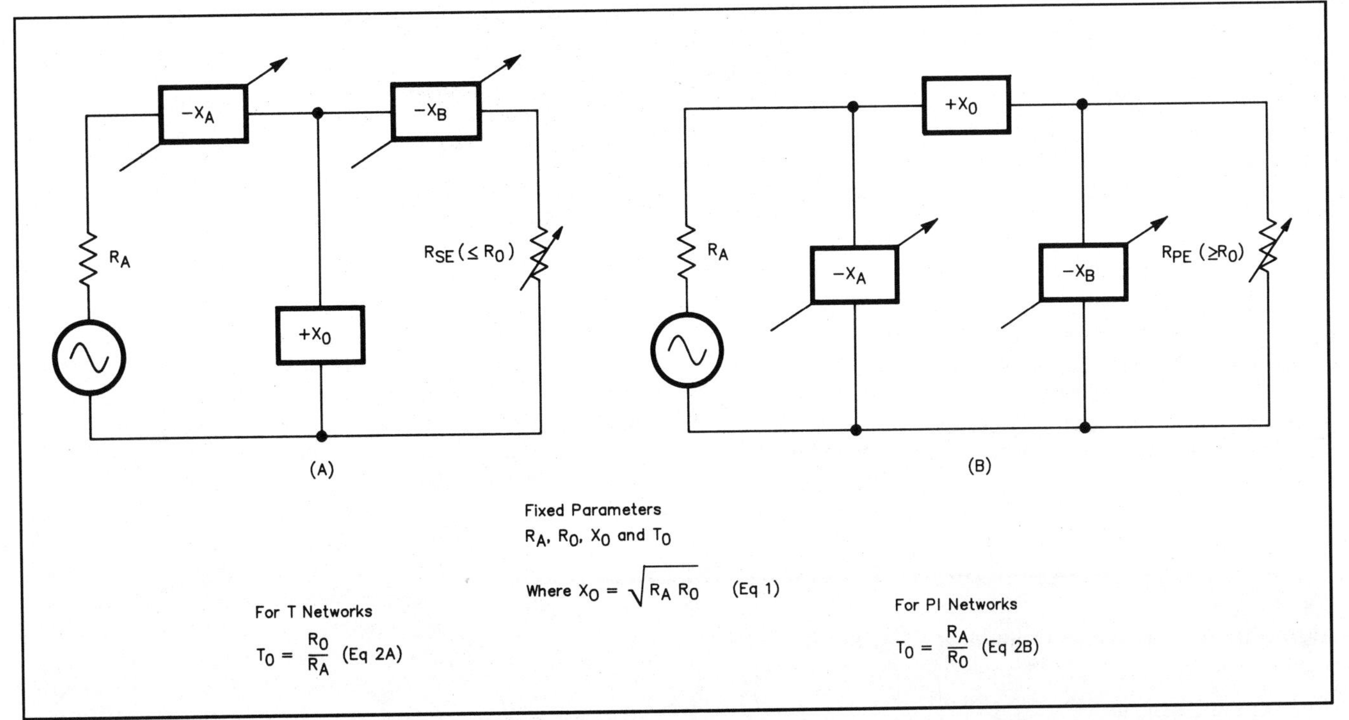


Fig 2—Basic reactance models. A is a high-pass T-matched system with fixed $X_O$ and variable $X_A$ and $X_B$ to match a real source impedance $R_A$ to the real part, $R_{SE}$ (≤ $R_O$), of a complex load impedance in series form. B is a low-pass pi-matched system with fixed $X_O$ and variable $X_A$ and $X_B$ to match a real source impedance $R_A$ to the real part, $R_{PE}$ (≥ $R_O$), of a complex load impedance in parallel-equivalent form. See the text for discussion of $R_O$, the parameter of choice in this article, and $T_O$ a dimensionless parameter of importance in system analysis.

to be satisfactory are therefore:

- The system must provide a mathematically correct match.
- The loaded Q of the system should not exceed 20.

The Q limitation has a profound effect on the analysis, since it provides an arbitrary lower bound on $R_{SE}$ in T systems and an arbitrary upper bound on $R_{PE}$ in pi systems. A higher value for Q would increase the "satisfactory" range in either system, and a lower Q requirement would have the opposite effect. All of this leads to a rather remarkable conclusion, discovered and verified by many hours of patient experimentation using the calculator program described later.

**Table 2**
**Simulated Worst-Case Load Set for a $Z_0$ of 50 Ω and an SWR of 4.419:1**

| | *Series Form* | | *Parallel Form* | |
|---|---|---|---|---|
| *Load* | *$R_{SE}$* | *$X_{SE}$* | *$R_{PE}$* | *$X_{PE}$* |
| 1 | 11.315 | 0 | 11.315 | ∞ |
| 2 and 12 | 12.083 | ±12.665 | 25.358 | ±24.193 |
| 3 and 11 | 14.833 | ±26.92 | 63.689 | ±35.093 |
| 4 and 10 | 21.527 | ±45.129 | 116.14 | ±55.398 |
| 5 and 9 | 39.232 | ±71.225 | 168.54 | ±92.835 |
| 6 and 8 | 98.589 | ±103.34 | 206.91 | ±197.40 |
| 7 | 220.95 | 0 | 220.95 | ∞ |

Note: Positive values of X correspond to load numbers 2 through 6, and negative values correspond to load numbers 8 through 12. Load 12 is the conjugate of load 2 and so on.

## The "4.419" Solution for the "Typical Station"

In the very special case where $R_A = Z_0 = 50$ Ω, a guaranteed-satisfactory solution meeting the above criteria is possible only when: (1) the SWR of the antenna-to-$Z_0$ mismatch is 4.419:1 or less, and (2) $R_O$ is so chosen that $T_O$ will be 4.419 as well. This means that $R_A$ must be 50 Ω and $R_O$ must be 220.95 Ω for a T network, and $R_A$ must be 50 Ω and $R_O$ must be 11.315 Ω for a pi network. For either network, the range of load resistances that can be matched is 11.315 to 220.95 Ω. This is identical to the range shown in Table 1 for an SWR of 4.419:1.

## Quantifying the Load Impedance

With either of these choices, it should be possible to find a satisfactory match for any load impedance with coordinates on or within the perimeter of a 4.419:1 SWR circle on a 50-Ω Smith Chart. To verify that assumption, and to establish the required ranges for the variable reactances $X_A$ and $X_B$, I've borrowed an idea (along with several others) from an excellent article by Earl Whyman, W2HB.[1]

For a given $Z_0$ and SWR, the Whyman approach uses 12 impedances evenly spaced around the perimeter of a Smith Chart SWR circle to create a simulated worst-case load set for a particular SWR. Table 2 shows such a set for a $Z_0$ of 50 Ω and an SWR of 4.419:1. The complex loads are shown in: (1) the usual series form ($R_{SE}$ and $X_{SE}$) required for T-network analysis, and (2) the rarer parallel-equivalent form

($R_{PE}$ and $X_{PE}$) required for pi-network analysis. The latter forms can be calculated by:

$$R_{PE} = \frac{(R_{SE}^2 + X_{SE}^2)}{R_{SE}}$$

$$X_{PE} = \frac{(R_{SE}^2 + X_{SE}^2)}{X_{SE}}$$

**System Analysis by Calculator Program**

The essential element in the approach of this paper is a rather simple calculator program. It allows rapid and accurate calculation and storage of all pertinent system parameters once given values of $R_A$, $R_0$ and a complex load impedance (for which a match is possible).

For discussion and illustration of the method, it is useful to describe two separate (but closely related) programs: program A for T-network analysis, and program B for pi-network analysis. List A gives the algorithm for Program A, and List B gives the algorithm for Program B. Each represents an actual program for a TI-59 (Texas Instruments) calculator.

It turns out, however, that a simple stratagem allows either program to be used for both T- and pi-network analysis. I've provided a coding list for Program A on the TI-59, along with instructions as to how it can be used directly for T-network analysis and indirectly for pi-network analysis. That information should make it relatively easy to duplicate the program on other calculators or computers.

The stratagem is: If the reciprocals of the pi-network input parameters ($R_A$, $R_0$, $R_{PE}$ and $X_{PE}$) are used as the inputs to Program A, the numbers calculated and stored by the program in registers <01>, <09>, <10> and <11> will be the reciprocals of $X_0$, $X_A$, $X_B$ and $X_B'$. The dimensionless parameters $T_0$, K, Q, V, P, $\alpha$ and $\beta$ will be unaffected.

Note that the first three inputs and the first ten outputs in each list produce the basic network reactances for the elements identified in Fig 2, while the fourth input, the eleventh output and the diagram (at the bottom of each list) show how the load-facing element $X_B$ is modified to absorb, or cancel, the imaginary part of a complex load. For such complex loads, the load-facing element becomes $X_B'$ instead of $X_B$. $X_B$ and $X_B'$ are identical when the load is a pure resistance.

To illustrate the method and provide benchmark examples for checking other programs, I've used Program A to do a complete post-choice analysis of a T network for a "typical station" ($R_A = Z_0 = 50\ \Omega$ and $R_0 = 220.95\ \Omega$). The maximum SWR is 4.419:1, and the "worst case load set" comprises the twelve $R_{SE}$, $X_{SE}$ pairs from Table 2. The results of the twelve program runs are shown in Table 3.

Similarly, I've used Program B to do a complete post-choice analysis for a pi network in a typical station ($R_A = Z_0 = 50$, $R_0 = 11.315$). The maximum SWR is again 4.419:1 and the "worst case load set" comprises the parallel-equivalent pairs from Table 2. The results are given in Table 4.

In effect, Table 3 tells us that the required T-network elements comprise a fixed inductor of +105.11 Ω (column <01>) and two variable capacitors each having a range of about –105 to –320 Ω (the minimum and maximum values found for $X_A$ and $X_B'$).

Similarly, Table 4 indicates that the required pi-network elements would comprise a fixed inductor with a reactance of +23.785 Ω and two variable capacitors each with a range from about –7.8 to –23.8 Ω.

For illustrative purposes, Fig 3A shows a complete reactance model for a T network with load 6 from Table 3. Fig 3B shows a similar model for a pi network with load 2 from Table 4. Note that the dimensionless parameters for the two are identical.

**List A**
**Algorithm for T-network Calculator Program A**

(Refer to Fig 2A and text.)

*Given: $R_A$, $R_0$, $R_{SE}$ AND $X_{SE}$, with $R_{SE} \le R_0$*

$$X_0 = \sqrt{R_A R_0} \quad \text{(Eq 1)}$$

$$T_0 = \frac{R_0}{R_A} \quad \text{(Eq 2A)*}$$

$$K = \frac{R_0}{R_{SE}} \quad \text{(Eq 3A)*}$$

$$Q = \sqrt{T_0} + \frac{K}{\sqrt{T_0}} + 2\sqrt{K-1} \quad \text{(Eq 4)*}$$

$$V = \sqrt{T_0 / K} \quad \text{(Eq 5)*}$$

$$P = 180 - \sin^{-1}\left(\frac{1}{K}\right) \quad \text{(Eq 6)*}$$

$$\alpha = 1 - \left(\frac{\cos P}{V}\right) \quad \text{(Eq 7)*}$$

$$\beta = 1 - V\cos(P) \quad \text{(Eq 8)*}$$

$$X_A = -X_0\,\alpha \quad \text{(Eq 9A)}$$

$$X_B = -X_0\,\beta \quad \text{(Eq 10A)}$$

$$X_B' = X_B - X_{SE} \quad \text{(Eq 11A)}$$

*The results of these equations are dimensionless parameters.

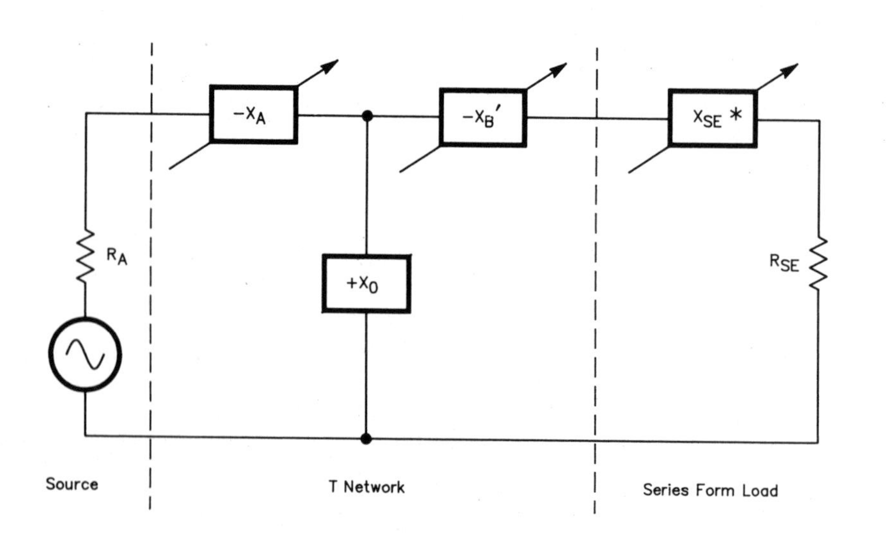


**The Final Step: L, $C_A$ and $C_B$**

The final steps in either analysis are: (1) Specify a frequency $f_0$, and (2) convert $X_0$,

$X_A$ and $X_B$ to the corresponding values for the physical models in Fig 1. The relationships are:

For any positive reactance:

$$L = \frac{0.159155 \times X}{f_0}$$

For any negative reactance:

$$C = \frac{159155}{f_0 \times |-X|}$$

where

L = inductance, in µH
C = capacitance, in pF
$f_0$ = frequency, in MHz

Using these formulas with the reactance data from Tables 3 and 4, I've calculated L, $C_A$ and $C_B$ for frequencies of 3.5, 7 and 14 MHz as shown in Table 5.

Technically speaking, the T and pi-network solutions for this special case seem to be precisely equivalent, but there is one difference that may help explain why (generally) only T networks seem to have been considered for this application. The difference is that variable capacitors to produce the very low reactances needed for the pi-network solution would be about 13.5 times larger than those needed for the T-network solution!

### Effects of a Change in Frequency

The design process described above produces a network with a fixed inductance of L µH, which will have a specific positive $X_0$ at a specific design frequency. For frequencies other than $f_0$, the reactance will change proportionately to the change in frequency. It can be shown (by a rather complicated analysis) that a system designed to handle the range of load impedances corresponding to an SWR of 4.419:1 at the design frequency can probably handle a range corresponding to an SWR of 4:1 for frequencies within about ±5% of $f_0$.

The bottom line is: Inductor taps should be closely spaced to get the maximum use out of a network with a fixed-tap inductor. The optimum solution, of course, is to use a roller inductor, which can be treated as a tapped inductor with an infinite number of taps.

### Generalizing the Approach: the Dimensionless Domain

When treating a specific example, as I have done so far, the focus is almost entirely on the R, X and $Z_0$ parameters, all of which are dimensioned in ohms. It would be possible, in theory at least, to derive a set of equations that would give the ohmic outputs directly, in terms of the ohmic inputs. The equations, however, would be very complicated and have no general meaning.

The simplicity and generality of the approach in this article may be better understood by means of the flow chart in Fig 4. It covers the basic elements of the two algorithms, namely the first three inputs and the first ten outputs in each case.

For those interested in theoretical analysis, the dimensionless parameters are more important than those in the ohmic domains. A complete analysis is far beyond the scope of this article, but I can point out a few things about Fig 4. Many of these points can be verified by simple inspection, and all of them can be verified by using the calculator program:

- It is not difficult to see why the stratagem of using Program A with the reciprocals of the pi-network inputs and outputs will work.
- It is not difficult to see that the reverse is also true: A stratagem of using Program B with the reciprocals of the T-network inputs and outputs will also work.
- It is not too difficult to see that multiplying all of the ohmic parameters of any

---

**List B**
**Algorithm for Pi-network Calculator Program B**

(Refer to Fig 2B and text.)

*Given: $R_A$, $R_0$, $R_{PE}$ AND $X_{PE}$, with $R_{PE} \geq R_0$*

$$X_0 = \sqrt{R_A R_0} \quad \text{(Eq 1)}$$

$$T_0 = \frac{R_A}{R_0} \quad \text{(Eq 2B)*}$$

$$K = \frac{R_{PE}}{R_0} \quad \text{(Eq 3B)*}$$

$$Q = \sqrt{T_0} + \frac{K}{\sqrt{T_0}} + 2\sqrt{K-1} \quad \text{(Eq 4)*}$$

$$V = \sqrt{T_0 / K} \quad \text{(Eq 5)*}$$

$$P = 180 - \sin^{-1}\left(\frac{1}{K}\right) \quad \text{(Eq 6)*}$$

$$\alpha = 1 - \left(\frac{\cos P}{V}\right) \quad \text{(Eq 7)*}$$

$$\beta = 1 - V\cos(P) \quad \text{(Eq 8)*}$$

$$X_A = \frac{-X_0}{\alpha} \quad \text{(Eq 9B)}$$

$$X_B = \frac{-X_0}{\beta} \quad \text{(Eq 10B)}$$

$$X_B' = \frac{1}{\frac{1}{X_B} - \frac{1}{X_{PE}}} \quad \text{(Eq 11A)}$$

*The results of these equations are dimensionless parameters.

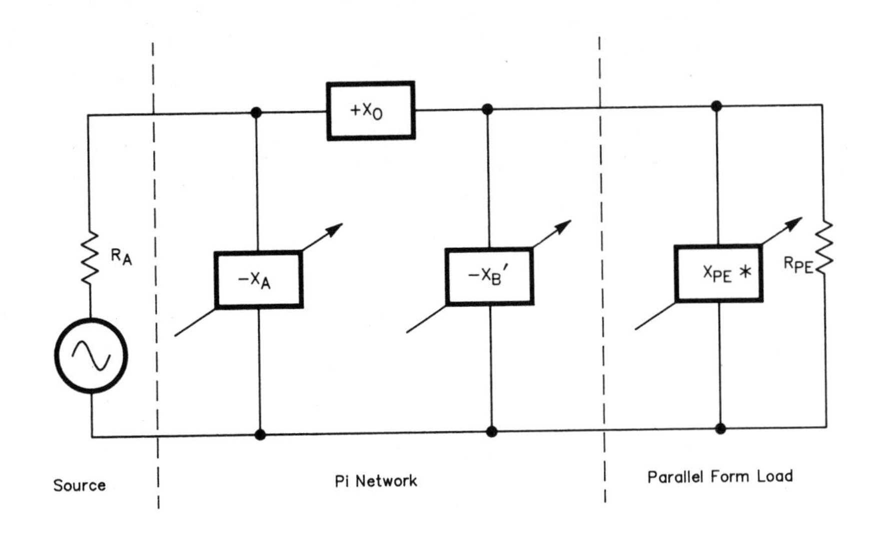

**Table 3**
**T-Net Parameters for the "Typical Station" Example**

Fixed inputs: $R_A$ = 50 Ω, stored in <21>: $R_0$ = 220.95 Ω, stored in <22>.

| | Memory Register Number | | | | | | | | | | | |
|---|---|---|---|---|---|---|---|---|---|---|---|---|
| | <23> | <24> | <01> | <02> | <03> | <04> | <06> | <07> | <08> | <09> | <10> | <11> |
| *Load* | $R_{SE}$ | $X_{SE}$ | $X_0$ | $T_0$ | $k$ | $Q$ | $P$ | $\alpha$ | $\beta$ | $X_A$ | $X_B$ | $X_{B'}$ |
| 7 | 220.950 | 0 | +105.11 | 4.419 | 1.00 | 2.58 | 90° | 1 | 1 | −105.11 | −105.11 | −105.11 |
| 6 or 8 | 98.589 | ±103.34 | +105.11 | 4.419 | 2.24 | 5.40 | 138.09° | 1.530 | 2.045 | −160.81 | −214.84 | −318.28 or −111.60 |
| 5 or 9 | 39.232 | ±71.225 | +105.11 | 4.419 | 5.63 | 9.09 | 155.08° | 2.024 | 1.803 | −212.72 | −189.54 | −260.77 or −118.32 |
| 4 or 10 | 21.527 | ±45.129 | +105.11 | 4.419 | 10.26 | 13.07 | 161.81° | 2.448 | 1.623 | −257.29 | −170.63 | −215.76 or −125.50 |
| 3 or 11 | 14.833 | ±26.92 | +105.11 | 4.419 | 14.90 | 16.64 | 164.98° | 2.773 | 1.526 | −291.49 | −160.40 | −187.32 or −133.48 |
| 2 or 12 | 12.083 | ±12.665 | +105.11 | 4.419 | 18.29 | 19.12 | 166.48° | 2.978 | 1.478 | −312.99 | −155.34 | −168.01 or −142.68 |
| 1 | 11.315 | 0 | +105.11 | 4.419 | 19.53 | 20.00 | 166.92° | 3.048 | 1.463 | −320.32 | −153.81 | −153.81 |

**Note**: Numbers on the left of column 11 correspond to positive values of $X_{SE}$ and numbers on the right to negative values of $X_{SE}$.

**Table 4**
**Pi-Net Parameters for the "Typical Station" Example**

Fixed inputs: $R_A$ = 50 Ω, stored in <21>: $R_0$ = 11.315, stored in <22>.

| | Memory Register Number | | | | | | | | | | | |
|---|---|---|---|---|---|---|---|---|---|---|---|---|
| | <23> | <24> | <01> | <02> | <03> | <04> | <06> | <07> | <08> | <09> | <10> | <11> |
| *Load* | $R_{PE}$ | $X_{PE}$ | $X_0$ | $T_0$ | $K$ | $Q$ | $P$ | $\alpha$ | $\beta$ | $X_A$ | $X_B$ | $X_{B'}$ |
| 1 | 11.315 | $\rightarrow\infty$ | +23.785 | 4.419 | 1.00 | 2.58 | 90° | 1.00 | 1 | −23.785 | −23.785 | −23.785 |
| 2 or 12 | 25.358 | ±24.193 | +23.785 | 4.419 | 2.24 | 5.40 | 138.09° | 1.530 | 2.045 | −15.546 | −11.631 | −7.85 or −22.401 |
| 3 or 11 | 63.689 | ±35.093 | +23.785 | 4.419 | 5.63 | 9.08 | 155.07° | 2.023 | 1.803 | −11.755 | −13.189 | −9.586 or −21.124 |
| 4 or 10 | 116.14 | ±55.398 | +23.785 | 4.419 | 10.26 | 13.07 | 161.81° | 2.448 | 1.623 | −9.717 | −14.652 | −11.587 or −19.921 |
| 5 or 9 | 168.54 | ±92.835 | +23.785 | 4.419 | 14.90 | 16.64 | 164.98° | 2.773 | 1.526 | −8.577 | −15.586 | −13.346 or −18.731 |
| 6 or 8 | 206.91 | ±197.40 | +23.785 | 4.419 | 18.29 | 19.12 | 166.48° | 2.978 | 1.478 | −7.937 | −160.94 | −14.880 or −17.522 |
| 7 | 220.95 | $\rightarrow\infty$ | +23.785 | 4.419 | 19.53 | 20.00 | 166.92° | 3.048 | 1.463 | −7.805 | −16.254 | −16.254 |

**Note:** Numbers on the left of column 11 correspond to the positive values of $X_{PE}$ and numbers on the right to negative values of $X_{PE}$.

matched system by any constant will produce a system which is also matched (the dimensionless parameters will be unchanged).

- Finally, it is not too difficult to see that changing the signs of all Xs in a high-pass matched system will produce a low-pass system (and vice versa) that is also matched (again, the dimensionless parameters won't be affected).

Obviously the characteristics and behavior of the dimensionless parameters have a significance far beyond that of any particular application.

**Table 5**
**Actual Component Values for the Networks in Tables 3 and 4**

| T Network | | | Pi Network | | |
|---|---|---|---|---|---|
| $F_0$ (MHz) | $L$ (μH) | $C_A$ or $C_B$ (pF) | $F_0$ (MHz) | $L$ (μH) | $C_A$ or $C_B$ (pF) |
| 3.5 | 4.78 | 142 – 433 | 3.5 | 1.08 | 1911 – 5830 |
| 7.0 | 2.39 | 71 – 216 | 7.0 | 0.54 | 955 – 2915 |
| 14.0 | 1.20 | 35.5 – 108 | 14.0 | 0.27 | 478 – 1458 |

## The Importance of $T_0$ and K

The key parameter in the entire process is $T_0$, which is defined as $R_0/R_A$ for T systems and $R_A/R_0$ for pi systems. $T_0$ becomes fixed as soon as $R_A$ and $R_0$ are chosen, and it can be considered the "signature" parameter for any system. In theory it could have any positive value between zero and infinity, but the practical range is much more limited.

The next most important parameter is K, defined as $R_0/R_{SE}$ for T systems and as $R_{PE}/R_0$ for pi systems. Values of K between 0 and 1 are forbidden, so K could (in theory) have any value between 1 and infinity. However, the requirement that $Q \leq 20$ imposes an upper limit on K, designated herein as K*.

K* depends only on $T_0$. For a given $T_0$, it can be found by patient experimentation using the calculator programs. I've done

## Table 6
## $K^*$ versus $T_0$ Based on a Maximum Q of 20

| $T_0$ | $K^*$ |
|---|---|
| 0.322 | 8.02 |
| 1.00 | 12.282 |
| 2.00 | 15.510 |
| 3.00 | 17.549 |
| 4.00 | 19.020 |
| 4.419 | 19.527 |
| 5.00 | 20.151 |
| 10.00 | 23.347 |
| 25.50 | 25.50 |
| 53 | 23.515 |
| 80 | 20.30 |
| 93 | 18.71 |
| 104 | 17.39 |
| 120 | 15.543 |
| 150 | 12.366 |
| 200 | 8.00 |

these calculations for a number of different $T_0$ values. The results are shown in Table 6. The dimensionless K range for each entry is from 1 to $K^*$.

Two of the entries in Table 6 are of special interest: In one, $K^*$ equals $T_0^2$. In the other, $K^*$ equals $T_0$.

When $K^*$ is 19.527, it is equal to $T_0^2$. This leads to the "4.419 solution" for which a complete analysis has already been given in Tables 3 and 4. As stated earlier, this is the optimum solution when the source impedance $R_A$ equals $Z_0$, as it does in the "typical" station. For this solution, the maximum tolerable SWR is 4.419:1.

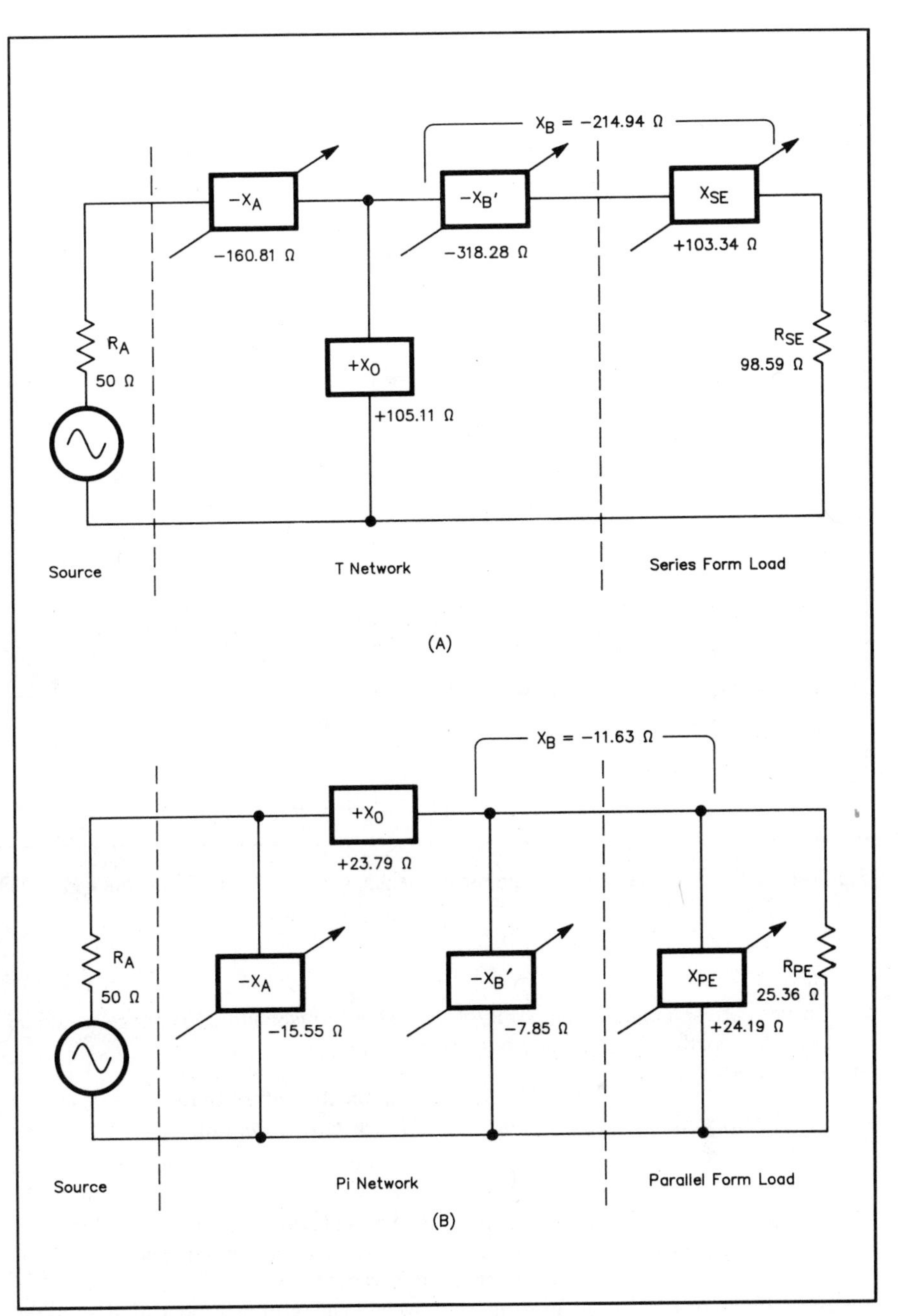


Fig 3—A is a T-network example, using Load 6 from Table 3. B is a pi-network example, using Load 2 from Table 4. The two examples have identical dimensionless parameters.

### A "Nontypical 5.050 Solution"

When $K^*$ and $T_0$ are equal to 25.5, there is a "5.050 solution." (The maximum tolerable SWR is the square root of $K^*$, namely, 5.050:1.) It is of little practical interest because it requires source-impedance ($R_A$) values that are unlikely to be available. Table 6 shows that 25.5 is the maximum possible value for $K^*$, so there is no possible solution that will accommodate an SWR greater than 5.050:1. The basic parameters for this optimum solution are:

| *T network* | *pi network* |
|---|---|
| $R_A = 9.901$ | $R_A = 252.5$ |
| $R_0 = 252.5$ | $R_0 = 9.901$ |
| $X_0 = 50$ | $X_0 = 50$ |
| $T_0 = 25.5$ | $T_0 = 25.5$ |

In both cases $K^*$ is 25.5, the maximum SWR is 5.050:1, and the $R_{SE}$ or $R_{PE}$ range is from 9.901 to 252.5 Ω. Since transceivers with an output impedance of 9.901 Ω (for a T network) or 252.5 Ω (for a pi network) are not readily available, this solution is mainly of theoretical interest. It represents the greatest possible matchable range for a $Z_0$ of 50 Ω and a maximum Q of 20.

Note however that both ("4.419" and "5.050") solutions depend on the underlying arbitrary selection of 20 as the maximum Q. If a different maximum Q is specified, the numbers change. For a fixed $Z_0$ and a given $Q_{max}$, the greatest possible matchable range will be achieved when $X_0$ is equal to $Z_0$ and $T_0$ is selected as:

$$T_0 = K^* = \left(\frac{Q_{max}}{4}\right)^2 + 0.5$$

A $Q_{max}$ of 20 gives the 25.5 value of Table 6. A $Q_{max}$ of 24 would produce a $T_0$ of 36.5 and a matchable range corresponding to an SWR of 6.04:1. A $Q_{max}$ of 16 would produce a $T_0$ of 16.5 and a match-

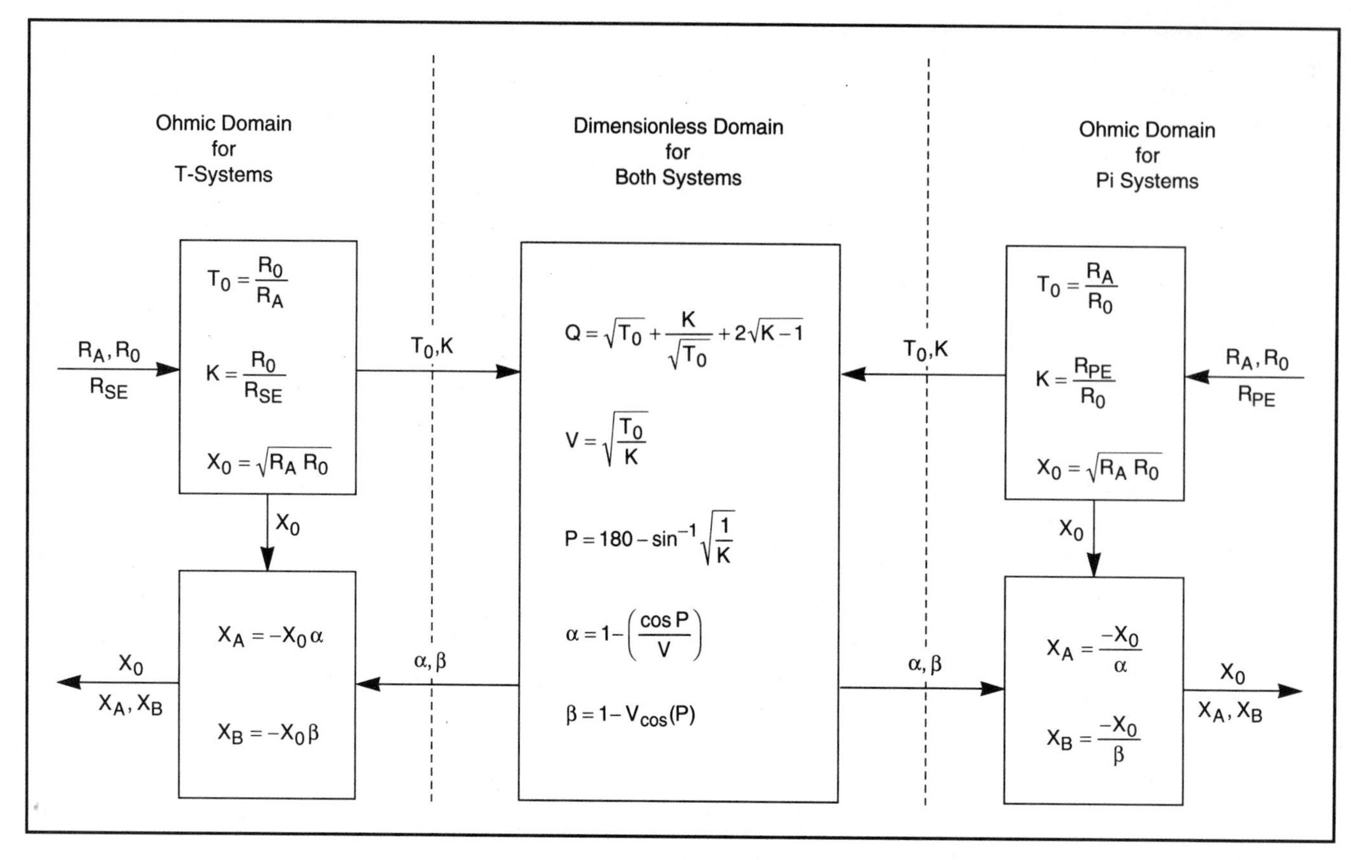


Fig 4—Flow chart for the basic algorithms. Restrictions: $R_{SE} \leq R_0$ for T networks, $R_{PE} \geq R_0$ for pi networks.

able range corresponding to an SWR of 4.06:1. The equation is empirical, but it appears nearly exact.

### A Special Application: Vacuum Tube Amplifiers

The low-pass variable pi networks used in the output stages of vacuum-tube amplifiers are a special case because they must convert a source impedance of several thousand ohms to a range of load resistances symmetrical around a $Z_0$ of 50 Ω. Thus, they could require values of $T_0$ approaching 200 (greater in rare cases). For example, an $R_A$ of 2500 Ω and an $R_0$ of 15.1515 Ω would give a $T_0$ of 165 and would accommodate a range of 15.15 to 165 Ω, which corresponds to a $Z_0$ of 50 Ω and a maximum SWR of 3.3:1. With higher values of $T_0$, the matchable range drops off very rapidly, requiring supplementary measures such as the pi-L system described by Whyman.[1]

High-pass T systems are never considered for this special application because they provide little or no harmonic suppression, and they are technically unsuitable. The $T_0$ values for such systems could be on the order of 0.06 or less, well below the minimum value in Table 6. The K range and the corresponding $R_{SE}$ range would be very limited.

### Graphical Analysis in the Dimensionless Domain

As shown in the flow chart of Fig 4, the dimensionless domain has two inputs ($T_0$ and K) and two principal outputs (α and β). $T_0$ is fixed by the initial design choice. K is a variable that can have any value between 1 and infinity in theory, but is limited to a range between 1 and K* in practice. A graph showing α and β versus K (for a given $T_0$) can be very revealing. I've prepared such graphs for four different values of $T_0$: 1, 4.419, 25.5 and 165 in Figs 5 through 8, respectively.

Examples for Figs 6, 7 and 8 have already been discussed earlier in the article. I've added an example for Fig 5 to give a more complete view of the overall picture. The conditions are:

T network: $R_A = R_0 = 175$ Ω

or

pi network: $R_A = R_0 = 14.286$ Ω

In either case, $T_0$ will be 1 and the range for either $R_{SE}$ or $R_{PE}$ will be 14.286 to 175 Ω. The range is symmetrical around 50 Ω and corresponds to a maximum SWR of about 3.5:1. This example is of little practical interest because of the unorthodox source impedances it would require.

The most striking feature of the four graphs is the behavior of β. For any value of $T_0$, β is a rising function of K as K increases from 1 to 2, then a descending function of K as K increases from 2 toward infinity. At infinity, β would again approach its initial value of 1.

On the other hand, α is single-valued. It is 1 when K is 1 and increases monotonically with increasing K. It approaches infinity as K approaches infinity.

Q is also single-valued. It has a minimum value of $T_0 + 1/T_0$ when K equals 1 and approaches infinity as K approaches infinity.

The double-valued nature of β divides the K range into major and minor sections, as shown on the four charts. The minor sections comprise the K range from 1 to 2 and a corresponding P (phase) range of 90° to 135°, which is equivalent to electrical lengths from ¼ λ to ⅜ λ for the network. The major sections comprise the theoretical K range from 2 to infinity and a

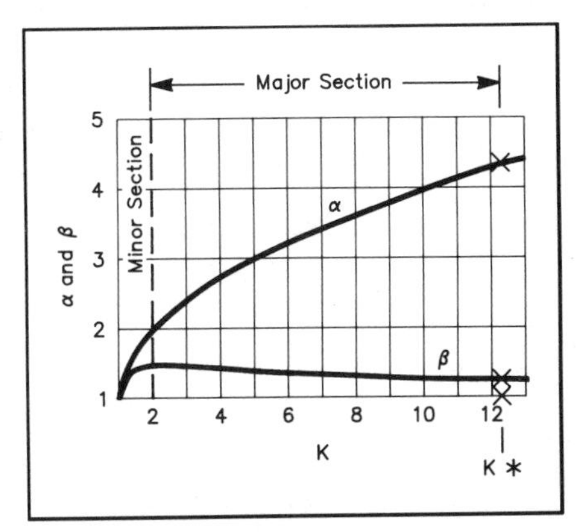


Fig 5—α and β versus K, for $T_0$ of 1. K ranges from 1 to 12.282.

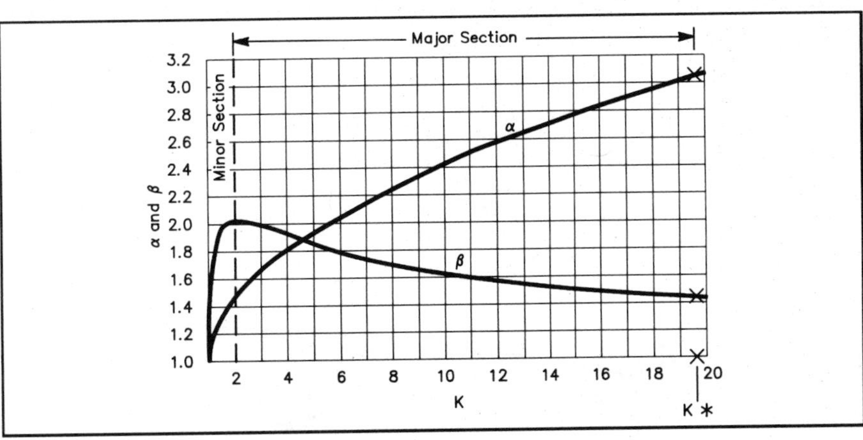


Fig 6—α and β versus K, for $T_0$ of 4.419. K ranges from 1 to 19.527.

corresponding P range from 135° to 180°, which is equivalent to electrical lengths from ⅜ λ to ½ λ for the network. (P is generated internally by the calculator program. It is not of direct interest in this application. In some contexts, it may be a useful bit of information. For example, see White.)[2]

In general, I'll let the four charts speak for themselves, but one point may need emphasis. For large values of $T_0$ (such as in Fig 8), the slope of β is so extreme for values of K very close to 1 that adjustments in that range may be difficult.

Here is a useful point to keep in mind when translating from the dimensionless domain to the ohmic domains: α and β act as multipliers to produce $X_A$ and $X_B$ for T networks, but the reciprocals (1/α and 1/β) act as multipliers to produce $X_A$ and $X_B$ for pi networks. Values for α and β will always be in the 1-to-infinity range, and the reciprocals will always be in the 1-to-zero range.

Hence, the magnitudes of $X_A$ and $X_B$ for T networks will always be equal to or greater than the magnitude of $X_0$, and the magnitudes of $X_A$ and $X_B$ for pi networks will always be equal to or less than the magnitude of $X_0$. The *signs* of $X_A$, $X_B$ (and $X_B'$) must always be negative and that of $X_0$ must be positive. (All reactances are assumed to be pure.)

Also keep in mind that $X_A$ is always the network element facing the source, and $X_B'$ is the network element facing the load.

### About the Calculator Program(s)

In order to show the close relationships between the T- and pi-network treatments, I've chosen to present and discuss them on a "side by side" basis. In using the methods, however, focus on one program and ignore the other completely. For my part, I have found it very convenient to use Program A for both systems (in the manner described later). Others may find it more convenient to use Program A for T networks and a program based on List B for pi networks.

*For T networks:* Once $R_A$ and $R_0$ have been chosen, Program A will provide a mathematically correct solution for any value of $R_{SE}$ from $R_0$ down to almost zero. No solution is possible for values of $R_{SE}$ greater than $R_0$. An arbitrary lower limit on $R_{SE}$ is established by the stipulation that Q is 20 or less.

*For pi networks:* Once $R_A$ and $R_0$ have been chosen, a program based on List B (or Program A using reciprocals) will provide a mathematically correct solution for any value of $R_{PE}$ from $R_0$ up to almost infinity. No solution is possible for values of $R_{PE}$ less than $R_0$. An arbitrary upper limit on $R_{PE}$ is established by the stipulation that Q is 20 or less.

The programs are somewhat "fail-safe" in that values of $R_{SE}$ greater than $R_0$, or values of $R_{PE}$ less than $R_0$, will make K less than one and cause the program to give an error signal when it tries to calculate Q.

### Instructions and Coding for TI-59 Program A

With Program A stored in memory, proceed as follows:

*For T-network analysis:*

Step 1: Fixed inputs (entered only at beginning).

Key in $R_A$ STO 21, then $R_0$ STO 22.

Step 2: Variable inputs (entered before each run)

Key in $R_{SE}$ STO 23, then $X_{SE}$ STO 24. ($R_{SE} \le R_0$, $X_{SE}$ positive for inductive loads, negative for capacitive loads or zero for purely resistive loads.)

Step 3: With all inputs entered and stored, press A to execute the program. (When the program halts, the 11 parameters calculated by Eqs 1 through 11 will be stored in data registers <01> through <11> and can be retrieved by RCL 01, RCL 02 and so on.)

See Table 3 for a "benchmark" set of calculations.

*For pi-network analysis:*

Step 1: Fixed inputs (entered only at beginning):

Key in $R_A$ 1/X STO 21, then $R_0$ 1/X STO 22.

Step 2: Variable inputs (entered before each run)

Key in $R_{PE}$ 1/X STO 23, then $X_{PE}$ 1/X STO 24. ($R_{PE} \ge R_0$, $X_{PE}$ positive for inductive loads, negative for capacitive loads or zero for purely resistive loads.)

Step 3: With all inputs entered and stored, press A to execute the program. (When the program halts, the reciprocals of all the ohmic parameters will be stored in the respective data registers. The ohmic parameters can be retrieved by RCL 01 followed by 1/X and so on. The dimensionless parameters can be retrieved directly.)

See Table 4 for a "benchmark" set of calculations.

**Program A**
**T- and Pi-network Analysis for the Texas Instruments TI-59 Programmable Calculator**

| *LOCATION* | *CODE* | *KEY* | *COMMENTS* |
|---|---|---|---|
| 000 | 76 | LBL | |
| 001 | 11 | A | |
| 002 | 25 | CLR | |
| 003 | 43 | RCL | |
| 004 | 21 | 21 | |
| 005 | 65 | × | |
| 006 | 43 | RCL | |
| 007 | 22 | 22 | |
| 008 | 95 | = | |
| 009 | 34 | √X | |
| 010 | 42 | STO | |
| 011 | 01 | 01 | $X_0$ |

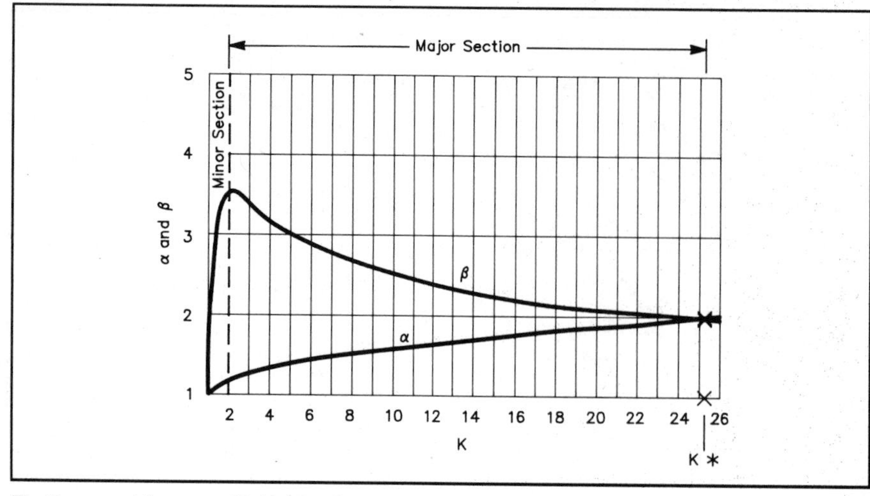


Fig 7—α and β versus K, for $T_0$ of 25.5. K ranges from 1 to 25.5.

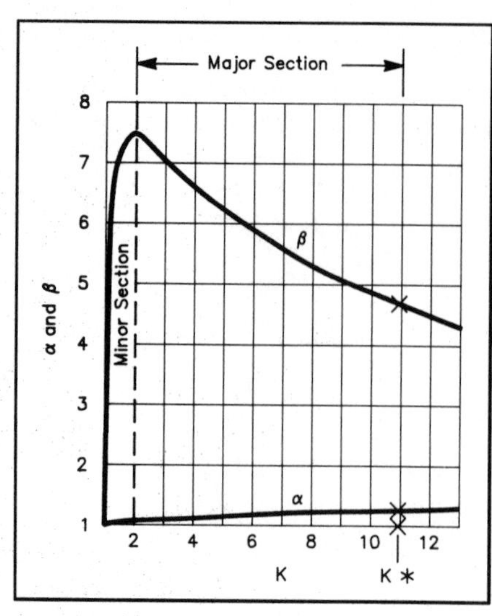


Fig 8—α and β versus K, for $T_0$ of 165. K ranges from 1 to 10.89.

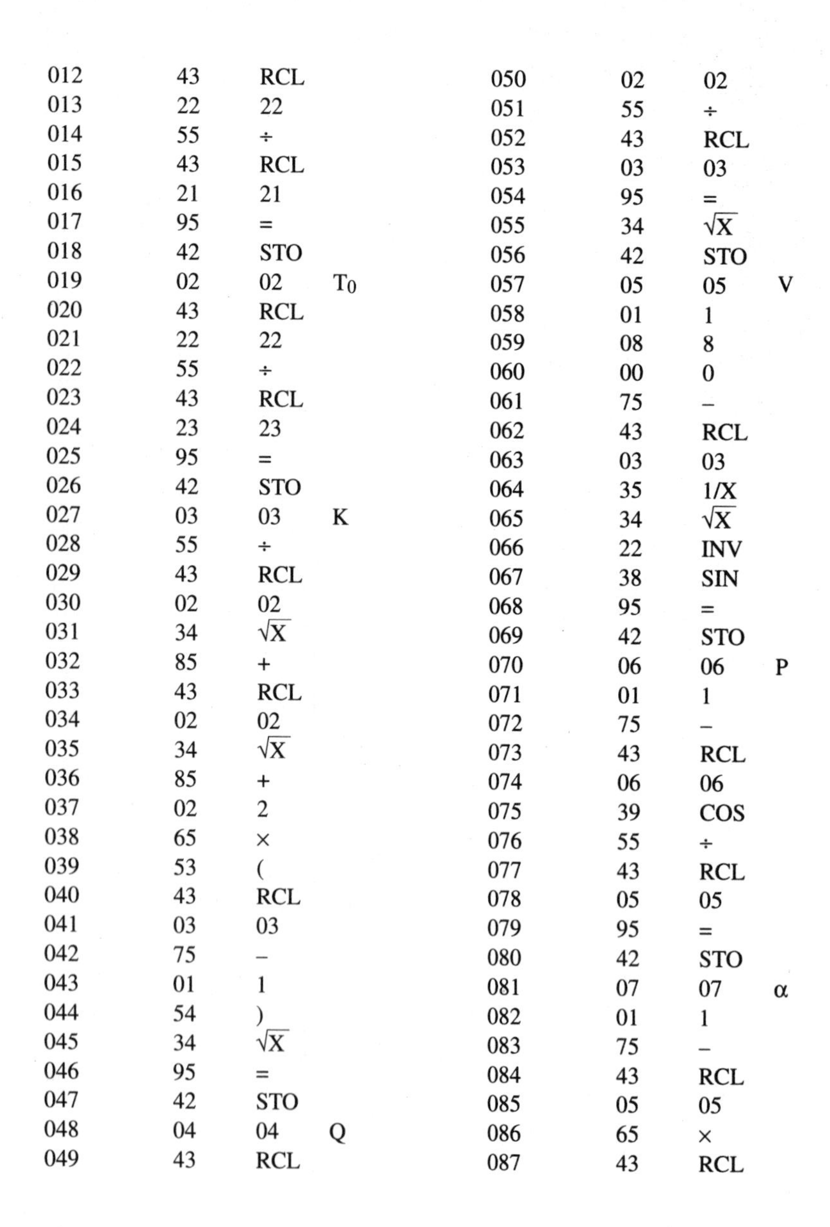

| | | | |
|---|---|---|---|
| 012 | 43 | RCL | |
| 013 | 22 | 22 | |
| 014 | 55 | ÷ | |
| 015 | 43 | RCL | |
| 016 | 21 | 21 | |
| 017 | 95 | = | |
| 018 | 42 | STO | |
| 019 | 02 | 02 | $T_0$ |
| 020 | 43 | RCL | |
| 021 | 22 | 22 | |
| 022 | 55 | ÷ | |
| 023 | 43 | RCL | |
| 024 | 23 | 23 | |
| 025 | 95 | = | |
| 026 | 42 | STO | |
| 027 | 03 | 03 | K |
| 028 | 55 | ÷ | |
| 029 | 43 | RCL | |
| 030 | 02 | 02 | |
| 031 | 34 | √X | |
| 032 | 85 | + | |
| 033 | 43 | RCL | |
| 034 | 02 | 02 | |
| 035 | 34 | √X | |
| 036 | 85 | + | |
| 037 | 02 | 2 | |
| 038 | 65 | × | |
| 039 | 53 | ( | |
| 040 | 43 | RCL | |
| 041 | 03 | 03 | |
| 042 | 75 | – | |
| 043 | 01 | 1 | |
| 044 | 54 | ) | |
| 045 | 34 | √X | |
| 046 | 95 | = | |
| 047 | 42 | STO | |
| 048 | 04 | 04 | Q |
| 049 | 43 | RCL | |
| 050 | 02 | 02 | |
| 051 | 55 | ÷ | |
| 052 | 43 | RCL | |
| 053 | 03 | 03 | |
| 054 | 95 | = | |
| 055 | 34 | √X | |
| 056 | 42 | STO | |
| 057 | 05 | 05 | V |
| 058 | 01 | 1 | |
| 059 | 08 | 8 | |
| 060 | 00 | 0 | |
| 061 | 75 | – | |
| 062 | 43 | RCL | |
| 063 | 03 | 03 | |
| 064 | 35 | 1/X | |
| 065 | 34 | √X | |
| 066 | 22 | INV | |
| 067 | 38 | SIN | |
| 068 | 95 | = | |
| 069 | 42 | STO | |
| 070 | 06 | 06 | P |
| 071 | 01 | 1 | |
| 072 | 75 | – | |
| 073 | 43 | RCL | |
| 074 | 06 | 06 | |
| 075 | 39 | COS | |
| 076 | 55 | ÷ | |
| 077 | 43 | RCL | |
| 078 | 05 | 05 | |
| 079 | 95 | = | |
| 080 | 42 | STO | |
| 081 | 07 | 07 | α |
| 082 | 01 | 1 | |
| 083 | 75 | – | |
| 084 | 43 | RCL | |
| 085 | 05 | 05 | |
| 086 | 65 | × | |
| 087 | 43 | RCL | |

| | | | |
|---|---|---|---|
| 088 | 06 | 06 | |
| 089 | 39 | COS | |
| 090 | 95 | = | |
| 091 | 42 | STO | |
| 092 | 08 | 08 | β |
| 093 | 43 | RCL | |
| 094 | 01 | 01 | |
| 095 | 65 | × | |
| 096 | 43 | RCL | |
| 097 | 07 | 07 | |
| 098 | 95 | = | |
| 099 | 94 | ± | |
| 100 | 42 | STO | |
| 101 | 09 | 09 | $X_A$ |
| 102 | 43 | RCL | |
| 103 | 01 | 01 | |
| 104 | 65 | × | |
| 105 | 43 | RCL | |
| 106 | 08 | 08 | |
| 107 | 95 | = | |
| 108 | 94 | ± | |
| 109 | 42 | STO | |
| 110 | 10 | 10 | $X_B$ |
| 111 | 75 | – | |
| 112 | 43 | RCL | |
| 113 | 24 | 24 | |
| 114 | 95 | = | |
| 115 | 42 | STO | |
| 116 | 11 | 11 | $X_B'$ |
| 117 | 91 | R/S | |

## Notes

[1]E. Whyman, "Pi-Network Design and Analysis," *Ham Radio*, Sep 1977, pp 30-39.

[2]R. White, "Phase-Shift Design of Pi, T and L networks," *The ARRL Antenna Compendium, Volume 2* (Newington, CT: ARRL, 1989), pp 187-196.

# A Precision Tuner For Antenna Couplers

By Jack Kuecken, KE2QJ
2 Round Trail
Pittsford, NY 14534

I developed the gadget described in this article over a period of years. I feel that it solves several problems that plague many hams. I have been using some variation of this scheme for several years, both professionally and as a ham. Your fellow hams might welcome the fact that you use it, too!

There are a great many reasons to use an antenna coupler, especially if you wish to operate on all 9 HF bands. You may not have enough room for 9 separate monoband antennas. The use of trap multiband antennas is common, but most trap antennas do not cover the wider bands in their entirety—80 meters in particular. Also, there is a certain economy of style associated with the use of a single antenna to cover all bands. For this reason, many hams use some form of antenna tuner or Transmatch.

There is no real reason (or perhaps excuse is more appropriate) for the number of strong "tuner uppers" to be heard on the ham bands. A final amplifier can be tuned into a dummy load if the antenna is matched. However, to tune up a Transmatch or an antenna coupler you must radiate some signal. The use of a noise bridge is no exception to this statement.

## The Noise Bridge

The noise bridge is simply an RF bridge that uses noise as a signal source, and employs a receiver as a null detector. To measure an antenna or to tune a coupler, the noise source is applied and the bridge controls adjusted to null the noise in the passband of the receiver. The same bridge could be excited with a signal generator instead of the noise generator, but the majority of hams do not own a signal generator. They do own a receiver, and the noise source makes a low-cost alternative to an all-band signal generator. Make no mistake about this noise source, however; it actually radiates a signal, a noise signal. Even when the bridge is nulled a noise signal is being radiated, not only at the null frequency but at all frequencies covered by the noise generator. Of course the antenna modifies the noise spectrum that is radiated. A saving grace is the fact that the noise signal is low in spectral density (meaning that in any given narrow passband the power is small); therefore, the radiated noise generally doesn't interfere with anyone at any distance. However it is not nonexistent. The navies of the world learned more than a half century ago that their direction finders could locate the radiated receiver noise of a ship operating in what it thought was "radio silence."

Unfortunately, the noise-bridge approach is not without its problems in implementation. The noise source is generally square-wave modulated (on and off) to help in identification. On busy, noisy bands it may prove difficult or impossible to hear the null in your noise signal when the bridge is balanced. The bridge does not null the incoming signals at the same setting that nulls the noise. As a matter of fact, the bridge usually scarcely attenuates incoming signals. Think about trying to tune something while listening for a null in the noise signal on 40 meters, when the foreign broadcasters are coming in at 10 dB over S9!

Of course, tuning with the aid of a conventional directional coupler radiates essentially the full power applied to the system. This is usually reduced power compared to full operating power for the amplifier, but it is a very strong unwanted signal nonetheless. The device described here uses the transmitter signal; however, it radiates only about 2 to 3 mW maximum.

A further advantage of this gadget is the fact that tuning is accomplished with the transmitter looking into a matched load. Solid-state nontuned amplifiers in modern transceivers are a bit fussy about the load impedances they see. The process of tuning an antenna coupler can produce some extreme mismatches, in an almost random order. Some amplifiers have unstable impedances which can send them into parasitic oscillation. On others, a directional coupler monitors the forward output and tries to hold this constant through the ALC (automatic level control). If a low impedance or a high reactance is encountered, the circuit cranks up the drive and increases the current until the overcurrent circuit shuts the system down, the fuse blows or something more drastic happens. The SWR of this system never exceeds about 1.1:1 during the wildest impedance gyrations of the tuner.

## Theory of Operation

The basic system is illustrated in Fig 1. In the RUN condition, S1 is in the position shown and the entire circuit is bypassed. When tuning is initiated, S1 is switched into the alternate position. S1A directs the transmitter output to a dummy load, which is shunted by an attenuating resistor and a fixed Wheatstone bridge. One leg of the bridge is formed by the antenna load through S1B. S1C turns on the power to a small solid-state dc amplifier. If the antenna load looks exactly like 50 Ω, there will be no net voltage between points B and C and the dc output of the detector formed by D1 will be zero.

A few words about this detector are in order. The transformer shown as T1 is actually just a bifilar balun. This was wound on a Ferronics 11-220-K core about ⅜-inch diameter and ³⁄₃₂-inch thick. About 8

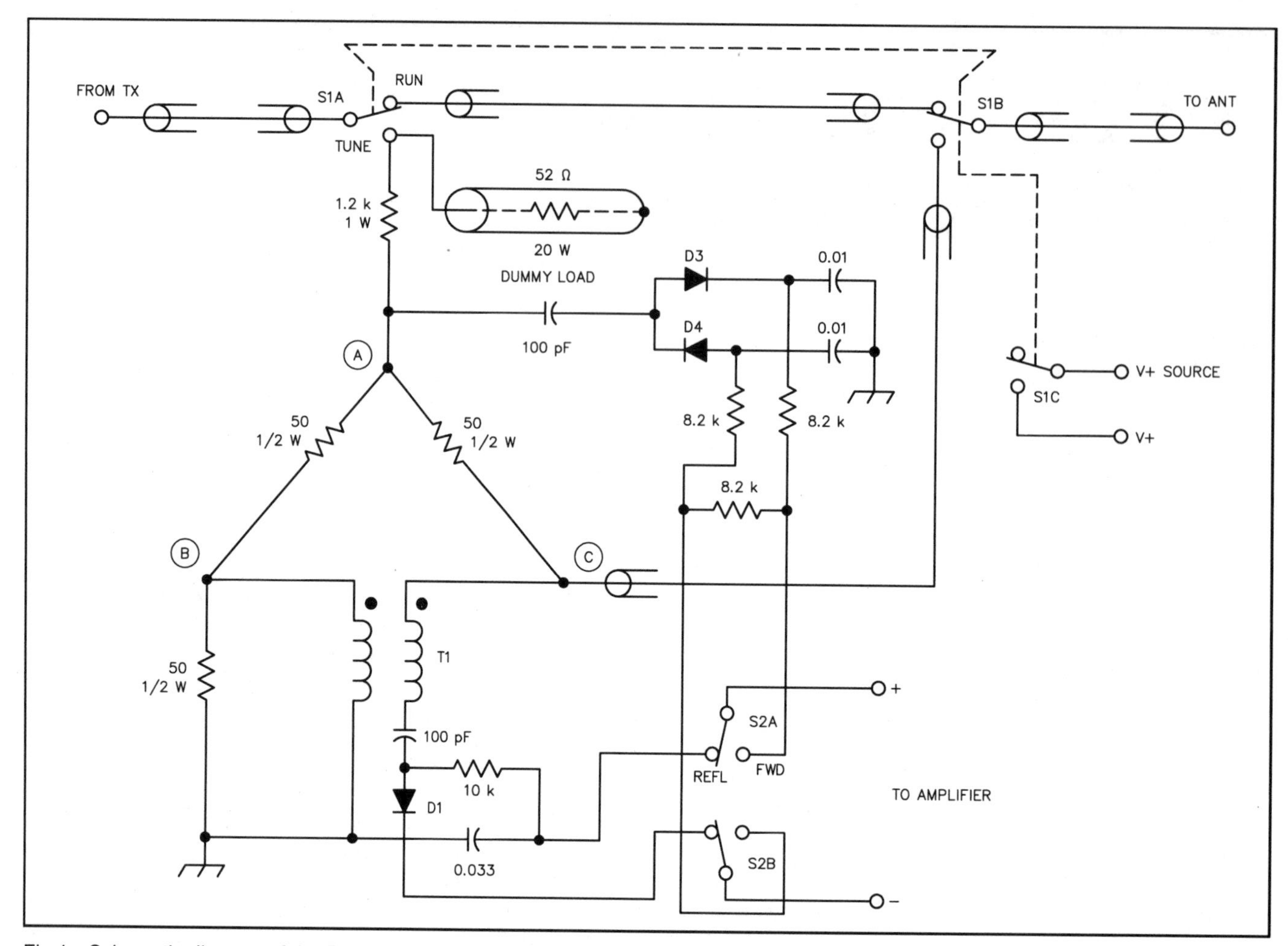


Fig 1—Schematic diagram of the Precision Tuner bridge. Resistors are 1/4-watt carbon or metal film unless otherwise specified. Capacitors are 50-volt disk ceramic. D1-D4 are 1N3666 germanium diodes.

feet of no. 30 enameled wire was doubled and then twisted with a hook held in the chuck of a hand drill, giving about 4 feet of twisted pair. The small core will hold about 40 turns of the twisted-pair wire. The K material has a mu of 125 and the *A* rating of the core is 55.2 nH per turn. Measured at 1 to 4 MHz, the coil represents about 74 μH of inductance. It is noteworthy that a measurement with a 1-kHz inductance bridge showed an inductance of only 12 μH. The inductance of these devices must be measured at RF! The capacitance of the between the wires of the winding measured 55 pF.

The transformer was installed as shown, in balun fashion; note that the dots are on the same end. One end of the bifilar winding was separated and used for the input and the other end used for the output. The detector proper consists of a single 1N3666 germanium diode. A 1N33 could also have served. The germanium diode has a lower forward drop than a silicon diode such as a 1N914. A voltage-doubler circuit like the D3-D4 detector could be used in this balanced bridge application to eliminate the balun, but it does not work as well. At very low signal levels the output of all diode detectors falls off, due to the forward drop of the diode. In the case of the doubler there are two forward drops so the detector does not respond well at low levels. This would tend to crowd the 2:1 SWR reading closer to the 1:1 end of the scale. The single-diode detector is much more linear at the low levels encountered when the bridge is balanced.

## How It Works

If the voltage at point A remained constant with changing loads, the voltage differential between B and C would be proportional to the reflection coefficient of the load at C. This isn't exactly what happens, however. The large (1.2-kΩ) resistor between the line and the bridge makes the current essentially constant. As the impedance at C varies, the impedance from A to ground changes, and the voltage across the bridge varies accordingly.

The bridge works as follows: If the impedance at C is a short circuit, the voltage at C is equal to zero and the voltage at B is just one-half of the voltage at A (neglecting the tiny current drawn by the rectifier).

Conversely, if C is open circuited, the voltage at C is equal to the voltage at A and the voltage at B is one-half of the voltage at A. Either of these cases corresponds to an infinite SWR or a reflection coefficient of 1.

The detector formed by D3 and D4 has been padded to approximately equal the output of the bridge detector for a reflection coefficient of 1. If you want to know the actual SWR it can be calculated by the following.

Voltage reflection coefficient = $\rho = V_{REFL}/V_{FWD}$

Then

$SWR = (1 + \rho)/(1 - \rho)$

In most cases you only need to know that the impedance match is good enough, and this can be seen from the reading of

$V_{REFL}$ alone. A reasonably designed antenna coupler will often permit reduction of $V_{REFL}$ to very near zero. As popular as directional couplers are, you might wonder why I chose to use a fixed bridge instead of one of the more common reflectometer designs. The first reason is, it is easier to build a bridge and detector circuit that operates at very low power. In a conventional HF reflectometer the detector operates on only a tiny sample of the reflected power. With a bridge, nearly half of the entire reflected power is applied to the detector. The detection system of a bridge is therefore at an advantage of 20 to 30 dB in sensitivity, compared to a reflectometer.

A second reason is that the bridge can usually be made more accurate than a directional coupler. A directional coupler that covers the range from 1.8-30 MHz with a directivity (the ability to discriminate between forward and reflected power) of 25 dB is exceptional. By comparison, I have built a number of these bridges, which function from a few hundred kHz to more than 250 MHz, with a balance of better than 40 dB over the entire range. A directivity of 25 dB on the directional coupler says that with 1 V forward an error voltage of 0.056 V would exist. This error voltage corresponds to an SWR of

$$\text{Error SWR} = (1 + 0.056)/(1 - 0.056) = 1.12$$

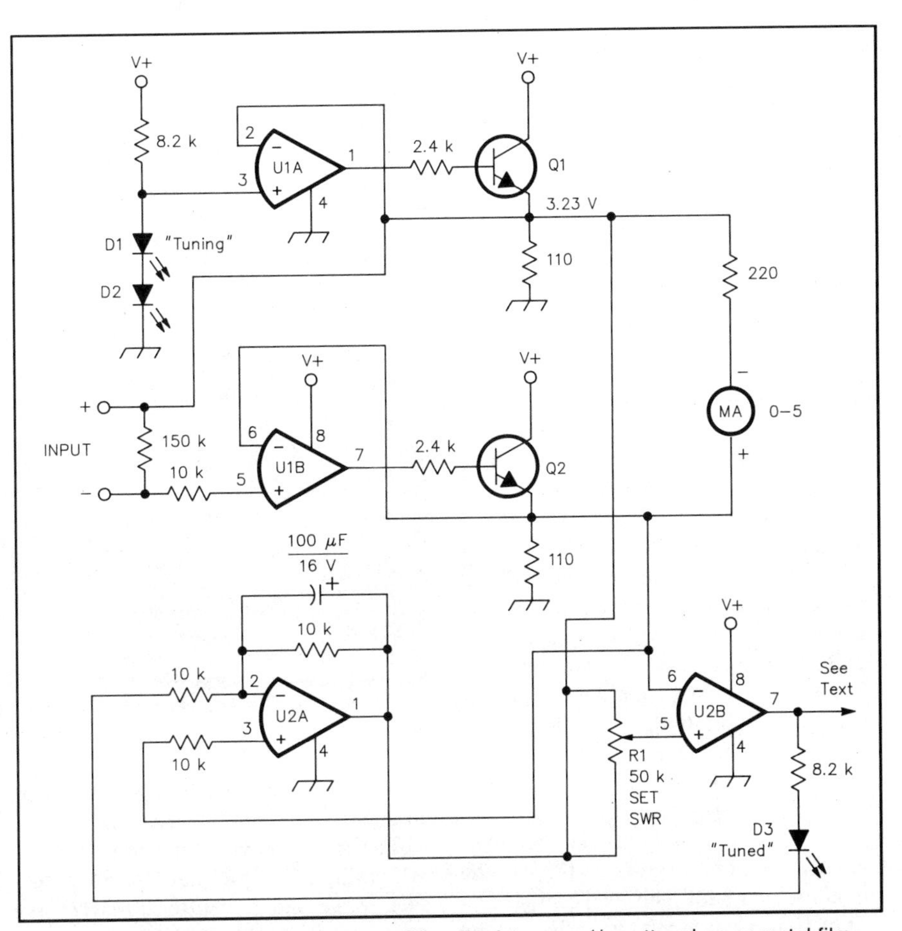


Fig 2—Schematic diagram of the dc amplifier. Resistors are 1/4-watt carbon or metal film, except R1, which is a 1/2-watt linear-taper potentiometer. Q1 and Q2 are 2N3904 or equivalent. U1 and U2 are LM1458 dual operational amplifier ICs. D1-D3 are Hewlett Packard HL4MP D155 LEDs or equivalent.

By comparison, the 40-dB balance yields an error in SWR of 1.02. In short, it is relatively simple to build a bridge with a vastly wider frequency range and a significantly better accuracy.

The first point is perhaps more telling. You will note that the bridge circuit has a significant attenuation; as a matter of fact it calculates and measures out to be 34 dB. If we excite the tuner with 50 watts, the radiated power from the antenna in a matched condition is 2 mW! At 75-watts input to the coupler, the antenna is radiating 3 mW. The same bridge can be used with a signal generator as a source applied at point A.

### Fast Or Automatic Tuning

People tune up antenna couplers using a variety of strategies; however, there is almost always an interaction between the inductor and capacitor. They both must reach the correct values for a match to take place. Neither one alone in the correct setting will produce a matched condition. In a remotely tuned antenna coupler this can become an even greater problem, since you cannot directly see the setting of the components. One strategy I have used is what I call an "onion slicer" technique. Suppose that we are using a coupler in which we have a shunt L (a roller coil) on the antenna side and a series C on the line side. If we set the C into continuous rotation, it will go through all of its possible values twice for each revolution. We can then tune the inductor at some relatively rapid rate, for example one revolution for every five revolutions of C. We know that the correct value of C to give a match will occur within a fifth to a tenth of a turn of the exact value required of L, and that somewhere along the line as L rolls out, a near perfect match will take place.

The one problem with this method is the fact that the antenna goes into and out of a match so rapidly that the meter needle cannot follow the match all the way down. With a little experience you can tell when you have hit the match by the way the needle is kicking. Another circuit still can be a significant help.

The amplifier circuit is shown in Fig 2. When S1 is put in the TUNE position, S1C closes and supplies V+ to the amplifier. This amplifier is designed to run with a single-ended supply like a car battery, and therefore it is necessary to establish a reference voltage. The voltage drop across LEDs D1 and D2 is used to establish a reference voltage of 3.23 V. I used Hewlett-Packard HLMP D155 LEDs, but other diodes should serve as well. I happen to like these LEDs because they are particularly bright. In the interest of economy, D1 is also mounted on the front panel and labeled TUNING, to indicate that you are in the tune mode. I have had good success using the constant-voltage characteristic of an LED in place of a Zener diode, in this and other applications.

U1A and Q1 simply serve to give a low-impedance reference at the 3.23 V. level. U1 and U2 are LM1458 dual op amps. U1B amplifies the output of the RF detector selected by S2. The potential difference between the emitters of Q1 and Q2 drives the meter. If a more sensitive meter, such as a 50-μA or 1-mA movement were used, it might be possible to eliminate Q1.

If you are redesigning this circuit to accommodate another current range on the meter, it is important that Q1 draw slightly more current than the full-scale meter current, to prevent the meter current from cutting off Q1. When a matched condition is reached, the output of U1B swings sharply down to 3.23 V. Whenever it falls below the set fraction, U2B switches to high output and lights D3 on the panel, indicating an impedance match. I usually set the pot to flash the lamp for any SWR below 1.5. This signal could also be used to stop the rotation of the tuner roller coil in an automatic setup. In the latter case however, the signal would have to ANDed with an RF PRESENT signal, since the lamp is lit when no RF is present at the bridge. Once the lamp flashes, you can shut off the L drive and stop C from scanning. C is then jogged into place at slow speed using the meter to optimize the tuning. A little jog of L might help to attain the absolute minimum SWR. You can usually get a better SWR with manual tuning than with fully automatic tuning. Also, with a little experience with the antenna and coupler, a skilled operator can usually jump in the right direction to respond to a change in frequency.

Fig 3—Inside the bridge. The dummy load is on the left. On the right is the bell crank that drives S1. The amplifier board is in the center.

**Construction**

One of the first items required is a dummy load. Fig 3 shows, at the left, the dummy load I built. This load consists of a set of 10 resistors, each nominally 2 watts and 510 Ω. A few extra resistors were purchased and 10 were selected to give 52 Ω at dc. The plates shown in the photo are single-sided circuit board. With the ten resistors, the load should have a continuous-duty rating of 20 watts; however, I have not experienced any overheating with inputs to 75 watts over any reasonable tuning cycle.

A BNC chassis connector (UG-290A) is mounted to the foil side of one PC-board rectangle. The resistors are in two rows of 5, with a spacing of ¾ inch between resistors in the row and 1½ inch between rows. A no. 55 drill was used for the resistor drill plan, and the top and middle boards drilled together. The resistor leads were then fed through the holes with the boards held foil side out. The boards were then clamped in place parallel to each other, with a ¼-inch spacing between the resistor end and the boards. Then the leads were soldered. A lead was added between the center conductor of the BNC connector and the back board. Note that this board is now at line potential. To minimize radiation leakage a shield board was cut ½ inch longer in each dimension than the resistor boards. This board was attached, foil side in, to the top board (the one with the BNC connector) using six heavy wire spacers bent to the right length and soldered to the foil at both ends. This makes a fairly sturdy, nonradiating load with reasonable SWR.

The load was measured from 1 to 30 MHz using a General Radio 1606A impedance bridge, and from 30 to 220 MHz using a General Radio 1602B UHF admittance meter. The load shows an increasing capacitive reactance, due to the capacitance of the driven plate. The corresponding SWR values run from 1.05 at 1 MHz to 1.11 at 30 MHz. Beyond this frequency, the impedance progressively rolls out, reaching an SWR of 1.77 at 144 MHz. Note that the accuracy of this load does not affect the accuracy of the tuning, which is established by the quality of the bridge. It only affects the SWR which the amplifier sees on tuning.

The next item is S1. I have actually used a Cutler-Hammer 5-A industrial toggle switch, with all lines going to the switch being made of coax cable, and the shields brought together behind the switch. However, the feedthrough is not as small as I would like and the residual SWR that this contributes to the tuning reaches about 1.15 at 30 MHz. You could use a commercial DPDT RF coaxial switch with very good results. Unfortunately, these switches are expensive, in the $50 to $75 range, and require quite a few expensive cable fittings for the various jumpers, etc. The cable fittings also tend to make the assembly bigger.

I won't go into too much detail concerning the construction of the switch, as you may have to modify the design slightly in order to use available materials. In essence this is simply a two-position rotary switch. You might want to modify a ceramic wafer switch for this purpose. Note, however, that the ground plane is maintained throughout the switch to try to maintain the characteristic impedance as close to 50 Ω as possible.

Fig 4 shows the switch plate and the fingers. The detailed design of the switch is largely controlled by the finger-stock dimensions.The switch plate is made of 0.063-inch single-sided printed-circuit board, with the heaviest copper laminate available. I have used both G-10 and XXXP boards. The switch lands can be etched, but I elected to carefully scribe the outlines and peel the foil off with a razor knife. After the insulating outlines are removed to separate the lands, the entire switch plate should be wire brushed and sanded to remove any sharp edges on the lands. A very fine steel wool or 400-grit wet-or-dry tungsten-carbide paper (used to refinish auto bodies) can be used to round the edges. Then wash the board carefully and apply an electroless silver plating.[1] Apply the powder (which looks like a kitchen cleanser) to a damp rag and rub the copper until a silver finish results.

The electroless silver helps keep the surface clean and nonoxidized. It can be dispensed with if you don't mind cleaning up the switch lands now and then. The fingers are the sort of beryllium copper stock used to seal shielded boxes and doors. The finger stock I used has 6 fingers per inch, and this seems about right. You can use anything that approximates this. If

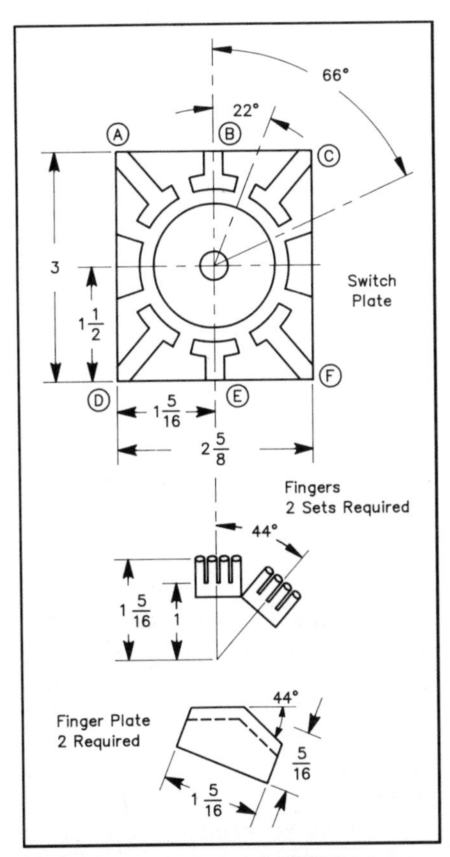


Fig 4—S1 is a homemade DPDT rotary switch. Stator contacts are cut on PC board; rotor contacts are made from finger stock. See text for details of assembly.

the fingers are too small, they are prone to bend with the wiping action of the switch. If they are too big, the switch must be made larger to obtain multiple contacts. For a good low-resistance connection I feel that three or four contact points per switch contact are required; particularly in a switch of this type, where the switch plate is not perfectly flat and some flexibility in the fingers is required. You can often find finger stock at a hamfest. This is the least expensive way to get it. Barring that, finger stock may be ordered by mail.[2]

The two finger plates also are made of single-sided PC board. The four finger sets are soldered to the foil up to the dotted lines, as shown in Fig 4. The stock I used had a tie band about 3⁄16 inch wide up to the roots of the fingers. This entire band was lapped onto the finger plate. Once assembled, the finger plates are epoxied to the rotor (epoxy cement on the board, not the foil side) in position so that the fingers ride about 3⁄16 inch upon the lands on the switch plate. The rotor can be a piece of nearly any plastic which is 2 × 15⁄16 × 1⁄4 inch. This should be a fairly strong plastic like XXXP or Bakelite, so that it can be drilled through on the center of the broad side for a 1⁄4-inch shaft. Do this on a drill press or a lathe, so the hole goes through the rotor as squarely as possible. The unit shown used a 1⁄4-inch ID collar with an arm soldered to it, to transmit torque to the rotor. The tip of the arm was screwed to the rotor, and the collar had a pair of 4-40 setscrews. This item is easily made, but I happened to have some of these pieces left over from some dismantled Command Sets. The items were used for a factory lock-in-place setting of C60 and C67 of the T19-ARC-5 or T22-ARC-5 (also BC-458A or BC-457A).

The center of the switch plate is drilled 3⁄8 inch for a standard panel bushing, which accepts a 1⁄4-inch shaft. This attaches with a nut clamping upon the plate, like a volume control. These bushings usually have a self-lubricating Oilite bearing for the 1⁄4-inch shaft to run on. The shaft should turn freely in the bushing, without too much wobble. A shield plate is made of single-sided PC board measuring 25⁄8 × 4 inches, with a 3⁄8-inch hole drilled in the center. This will be assembled with the foil side out, so that the switch plate and the shield plate will touch on the board, not the foil side. In other words, when they are assembled, there will be two thicknesses of the board dielectric between the foils. The extra thickness partially adjusts the characteristic impedance through the switch elements. The shield plate extends 1⁄2 inch above and below the switch ends. The switch plate and the shield plate can now be temporarily assembled, using the panel bushing to hold them together. The panel bushing has a flange about 1⁄8 inch thick, which should be on the switch-plate side. The length of the panel bushing extends through the shield plate. Next, the rotor should be assembled to a length of 1⁄4-inch shaft that runs freely in the panel bushing. A collar is then secured with setscrews to the shaft, on the shield-plate side of the board. The flange on the panel bushing should clear the finger plates by at least 1⁄8 inch. If it does not, remove the rotor and cut the finger plate back so that it does. This can be done with a Dremel tool or with a rotary file in a drill press. When this clearance is achieved, it should be possible to adjust the collar on the shield side so the rotor turns freely with a reasonable amount of sprung deflection of the fingers.

It should now be possible to rotate the rotor so that, for example, the fingers on one side short lands A and B together, and the fingers on the other side short lands E and F together, with no overlap of the fingers across the gaps between lands. Check for continuity with a good low-ohms meter. Any resistance beyond 0.1 Ω is unacceptable. It may be necessary to spring the fingers down somewhat to achieve a good low-resistance contact.

With a 44° rotation of the rotor, it should be possible to have one set of fingers short B and C and the other pair to short D and E. Apply the tests as above. The collar that holds the rotor in running position is a collar-plus-arm bell crank. Next, mark the location of the end of the bell crank in the two extreme switch positions. Eventually, a pair of stops fabricated from brass or double-sided PC board material can be soldered in place, to limit the rotor travel. In the position you intend to use for TUNE, you may next locate the position to attach the microswitch that forms S1C. I actually mounted the microswitch on a small block of plastic first. I then slid it into place so that it was actuated, and marked the location so that it could later be glued in place. Do not attach these items just yet.

Although the switch could have been constructed with flexible miniature coax such as RG-188, it is easier to build with semirigid miniature coax. The coax used has an OD of 0.140 inch. This line has a thin copper-tube outer conductor, Teflon insulation and a solid copper center conductor. The tubing is easily bent by hand without kinking, around a grooved form. The line is available from a number of sources.[3] It is widely used in microwave and radar applications.

A few notes are in order on the use of this coaxial tubing. To cut the tubing, score an even mark around the circumference of the outer conductor with a sharp knife. The tube will then break neatly if flexed at the score mark, exposing the insulation. This can be cut with the knife. The center conductor can be cut with a thin pair of diagonal cutters. For bending, either turn a form with a 0.140-inch round bottom groove or obtain a small V pulley. The bending radius should not be less than 0.75 inch. Place the tube in the groove and smooth the bend in with your fingers. If a lathe is available, brass adapters are easily turned so that nearly any RF connectors can be used. If a lathe is not available, most connectors intended for RG-188 cable can be adapted by reaming out a few of the parts with a drill.

The next step is to drill lands A through F to accept the coax center conductor. Remove the rotating parts but leave the bushing in place. With the switch plate and the shield board clamped together by the panel bushing, drill a hole through each land to accept the center conductor of the coax. With the 0.140-inch OD tubing, the hole should be made with a no. 60 drill (0.040

inch). Drill a hole in the center of each land 1/4 inch from the top or bottom edges of the switch plate. The hole should go through both PC boards and will serve for alignment. The bushing is now removed and the boards separated. Using the center conductor holes for a pilot, the shield board is drilled with a no. 28 or 9/64-inch drill, to accept the outer conductor of the coax. Do this carefully, to maintain the concentricity of the holes as much as possible.

Lands B and E of the switch represent the output and the input respectively. The path from land A to F represents the bypass path with the switch plunger all the way in. I preferred to arrange the switch on a pull-to-tune and push-to-run basis, so that an accidental bump of the switch actuator knob would have no effect while the system was running. Form a piece of coax to make the path from A to F, using the bending tool described earlier. Make sure the path clears switch S1C and the central shaft and hardware. In forming the tube, do not attempt to form the radius all the way to the end. Leave at least 1/2 inch of straight section beyond the bend.

This is the trickiest tube to make. It is worthwhile to bend up a piece of solder or soft wire into a dummy first to determine the required length. Start bending with the corner of the "L." Next, begin to roll in the end bends at 90° to the plane of the L. Begin these at the outer end of the cable and roll them in along the form. If the L is too big to fit the holes in the shield plate, simply roll the bend in a bit further and carefully straighten out the excess bend until a fit is obtained.

Next, cut the end of the tubes. Fig 5 shows the effect to be obtained. Score the coax outer conductor at the appropriate place and break it off. If you are removing no more than 5/8 inch you can grasp the outer conductor gently with a pair of pliers, and pull it off, leaving the insulator and center conductor protruding. Trim back the insulator, leaving a little less than 1/16 inch protruding beyond the outer conductor. *Take care not to score the center conductor in this step.*

Assemble the shield plate to the switch plate loosely, using the bushing to hold them together. Insert the L and go on to make the remaining tubes. The tubes attaching to lands B and E have only one bend, and take an RF fitting on the other end. The tube from D goes to the resistor that forms the bridge input. The tube from C comes from the bridge output. Form all of these tubes a little too long, so they can be trimmed after the bridge is assembled.

Next insert the remaining tubes temporarily, jiggling the shield plate for a best fit. Tighten the bushing to firmly clamp the plates together. If the plates are slightly bowed, clamp them together at the edges for a "flattest" condition. Now go around and, making sure that each coax tube is fully "home," solder first the center conductor and then the outer conductor. Make sure that you do not permit the solder to encroach on the area where the switch contacts ride. It is a good idea to do one on the ABC side and the next on the DEF side.

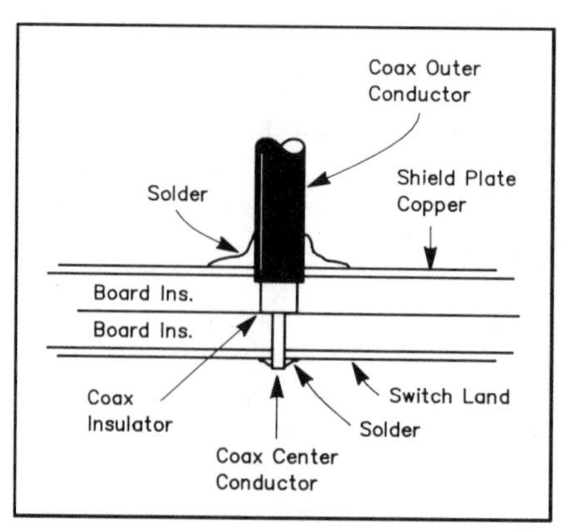


Fig 5—Details of connecting the semirigid coaxial cable to S1.

When this is complete, the travel stops may be soldered on, after which it is possible to cement the plastic block with S1C in place. Next assemble and install the rotor.

The fittings may be installed on tubes B and E. I used bulkhead-type connectors. These serve as part of the mounting for the switch. An angle bracket was attached to the shield plate and used for the third suspension point for the switch. This mounting is quite secure and sturdy.

This arrangement places the switch axis parallel to the panel. I could have used a pair of bevel gears and a rotary knob action, however I felt that the pull-push action would be simpler. A third bell crank arm is attached to the outer end of the switch shaft. This is drilled to take a push-pull rod. The rod was bent into an S shape to permit the knob to be placed halfway between the input and output fittings. A rubber grommet was installed in the panel for a guide bushing for the rod. This prevents vibration in mobile use. The portion of the rod bent to go through the hole in the bell crank is drilled and fitted with a washer and cotter pin, to secure it to the bell crank. The action of this linkage is very positive and has a good "feel."

The bridge and detector and the FWD detector could have been etched on double-sided board, but I used a somewhat simpler procedure. Single-sided boards were used for the circuits, with the components installed on the foil side of the board. The holes for the components were drilled from the foil side with a drill sized for the component leads. A 1/8-inch drill was then used to cut away the copper around each lead that was not intended to be grounded. Conductors for the other side were cut from 3M Scotch EMI/RFI copper shielding tape.[4]

The copper foil tape is adhesive. It can be cut to shape and stuck onto the board. Punch holes for the component leads with a scribe or needle. The component leads are soldered on in a conventional way but try to refrain from using too much heat. A hot, clean iron and a bit of polishing with steel wool on the component leads along with a drop of flux will usually ensure a good, fast joint with minimal heating.

### Results

This unit has been in use in my ham shack for several months now, and functions very effectively. The unit is placed between the transceiver and the linear amplifier. Tuning is done with the linear off and in the bypassed condition. I have two principal antennas, a vertical and a horizontal. Both are equipped with remote tunable couplers, with the couplers physically located at the antenna feed points. The horizontal covers 80-10 meters and the vertical covers 160-10 meters. Using this device, I can tune up without radiating more than 2 to 3 mW. In tests, people on the far side of town generally cannot copy my signal during tuning. The linear is tuned into a dummy load and simply switched in and out when required. Isolation between the two paths in the switch measures more than 58 dB at all frequencies below 148 MHz, and the insertion SWR in the bypassed condition measures less than 1.2:1 from 1-30 MHz.

As tuned with this bridge, the reflected power from the antenna couplers seldom exceeds 1 or 2 watts out of 100 forward, for an SWR of 1.33. Most of this is due to the mismatch in the linear bypass path. This residual can be tweaked out with the coupler if desired.

### Notes

[1]Electroless silver plating is available from Cool Amp Conducto Lube Co, 15834 Upper Boones Ferry Rd, Lake Oswego, OR 97035.

[2]Finger stock may be ordered from Atlee, 1 Gill St, Woburn, MA 01801, or from Chomerics Inc, 77 Dragon Ct, Woburn, MA 01888.

[3]One source for semirigid miniature coax is Precision Tube Co Inc, Microwave Div, Church Rd and Wissihickon Ave, North Wales, PA 19454.

[4]Copper shielding tape is available from Mouser Electronics.

# A New Way to Tree A Wire

By Stan Gibilisco, W1GV
2301 Collins Ave, #A-632
Miami Beach, FL 33139

Trees are beautiful in many ways. They provide shade and color; they act as air purifiers and air conditioners. They also do a great job of supporting wire antennas.

Rarely are trees used to their full advantage as antenna supports, because it's hard to get wires very high in them. There is a limit to how high you can safely climb a tree. It isn't cost effective for many hams to rent "cherry pickers" to reach up to the tops of most trees. The bow and arrow, with the arrow's point blunted and weighted, is one good way to get wires over treetops. It's also an excellent way to get into trouble with neighbors (broken windows, skewered cats, and such).

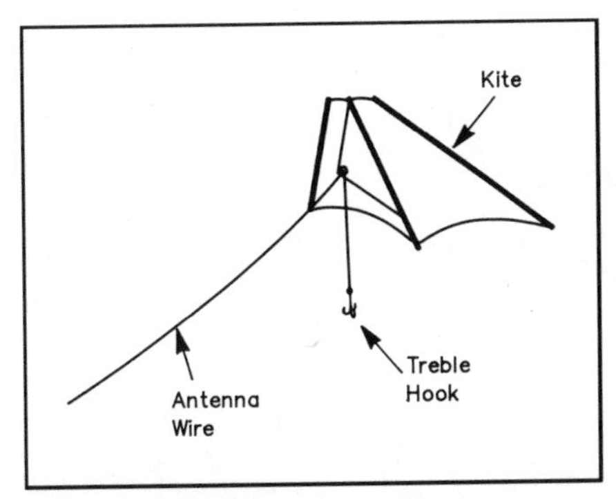


Fig 1—Simple method of treeing a kite for temporary long-wire or random-wire installations. When the kite hits the tree, it'll stay, almost every time.

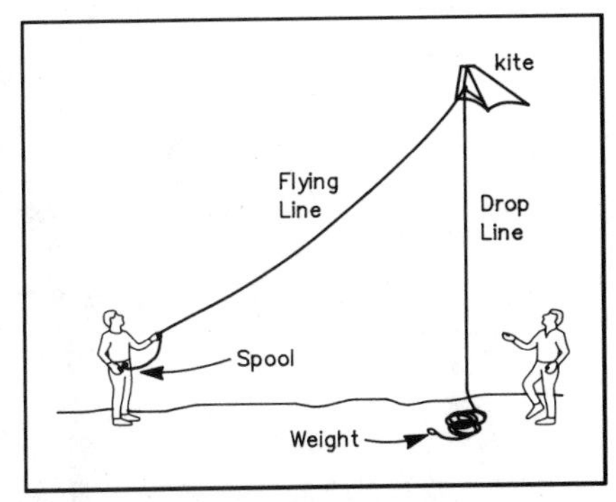


Fig 2—Initial setup for permanent method of putting a line over a tree.

### The Kite

Every March and April we see demonstrations of an elegant, simple, cheap and safe device that's well suited to getting wires just where we want them. Kites, like trees, are pretty to look at. And the two just naturally seem to go together. Kids know that. Kites can be found in trees all year round; sometimes they stay in the branches for months or even years. In anger, the little child, after losing a kite to a tree (again), might exclaim, "May as well just put the thing up there on purpose!" Exactly my idea.

Even a dime-store delta kite flies quite stably in a light breeze, and believe me, if you want to tree it, both the kite and the tree will eagerly accommodate you. All you need is one of those little three-hook things for securely catching fish [properly called a treble hook—*Ed*], available at any tackle shop, placed on a short string dangling from the kite tether point (Fig 1). You can be sure that once your kite is treed, it'll stay treed for as long as the line and the hook and the tree will last, barring major windstorms.

This method doesn't allow for error in measurement of the wire. It also is not very good for stringing heavier gauges of wire, but for low-power and moderate-power amateur transmitters, heavy wire is not necessary anyhow. I have found that no. 18 or no. 20 stranded copper wire is entirely adequate for most installations up to about 500 watts.

The treeing-of-the-kite-on-purpose (TOTKOP) method works best for "random" wires that you simply run out the window to a nearby tree. It is especially convenient for apartment dwellers. No one needs to know that the treed kite has anything to do with your wire antenna. The "stealth factor" can be enhanced by treeing the kite late at night.

### Safe Kite Flying

Use nonconductive line whenever possible, but no matter what type of line you use: Don't fly it where the line can *possibly* contact power lines. *You can be killed!* Also, don't fly a kite in a thundershower, or even when one is forecast or threatening. Use care when flying a kite over parking lots or buildings: You may tangle up the whole neighborhood! Your antenna must never pass over or under utility wires, and it should not be placed where power or phone lines can contact it if a windstorm breaks your antenna wire or a power or phone line. In short, use the same precautions with this method of antenna erection as with any other method.

### For More Permanent Antennas

The TOTKOP method is good for mountain topping, Field Day, and other temporary purposes. For more secure installations, I have developed a safe and elegant way of getting lines over trees and placing the line just where I want it, while allowing adjustment of antenna length and slack.

The initial setup is shown in Fig 2. The best type of kite is a fairly large, stable device, such as the delta or parafoil. The delta flies at a high angle (about 70 degrees with respect to the horizon) in winds of about 10-20 miles per hour. The parafoil flies at a lower angle (about 45 degrees) in winds of 15-30 miles per hour. (A good source of kites is INTO THE WIND, 1408 Pearl Street, Boulder, CO 80302.)

This method does not usually involve loss of the kite to the tree, although trees have a voracious appetite for kites. (You probably learned that as a kid!) It pays to have a spare kite handy, just in case. Con-

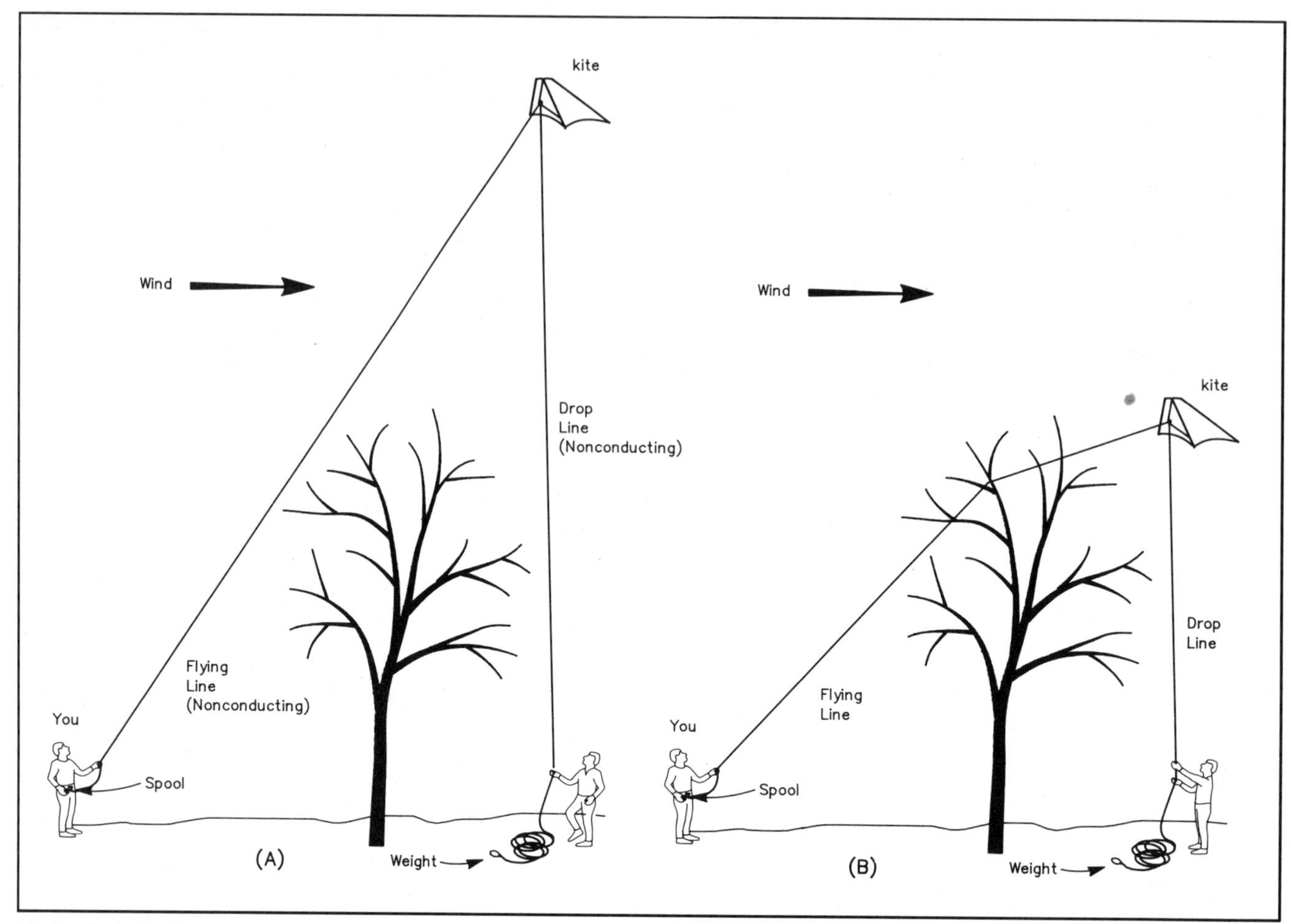


Fig 3—At A, the line is in position for treeing. At B, one person pulls the kite down via the drop line, while the kite flier guides the flying line exactly where wanted.

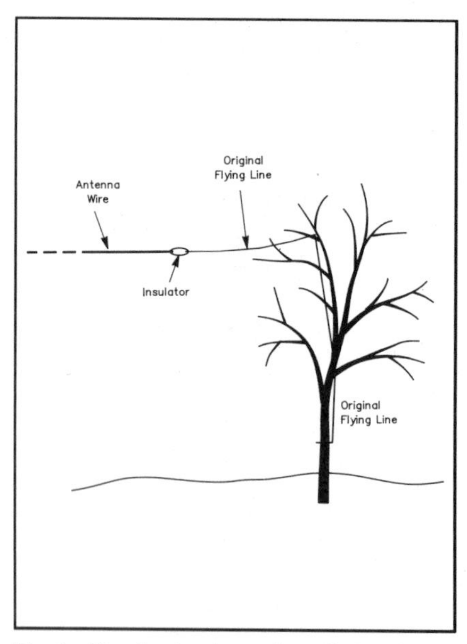


Fig 4—The final antenna installation (one end of the antenna).

ditions are ideal when the wind is steady at about 10-15 miles per hour and you can use a delta with a wingspan of 6 feet or so. The flying line should have a breaking strength sufficient for the antenna you wish to support. The drop line can be a lighter, cheaper string, weighted with a large nut or bolt, and long enough so that you can reach it however high the kite is flown. A good rule is to make the drop line twice as long as the tree is high.

For easiest placing of the line, two people are needed. One flies the kite and the other manages the drop line. The kite can be let out and flown high enough so that the flying line clears the tree you want to run the antenna support over. The person flying the kite (say it's you) should move into position upwind from the tree, while the drop-line person moves around behind the tree (Fig 3A). At this time the kite will be considerably higher than the top of the tree, and the flying line will just clear the treetop.

The drop-line person starts pulling on his line to bring the kite down (Fig 3B) without letting it get caught in the tree. If necessary, you should let out a little more line to ensure that the kite stays away from the tree, since kites and trees display an attraction that varies inversely according to the square of the distance between them (Murphy's Law). Be certain the kite does not get within the radius where the tree claims it forever.

It isn't hard to get the flying line right into the niche you want. Place it as high as possible in the tree, but choose a branch strong enough to support the tension on the wire, once the antenna has been strung. If the branch isn't strong enough, it will break, and the line will settle lower in the tree. Keep this in mind as you select your niche, so that if the branch breaks, the antenna will still be up fairly high and you won't have to restring the line.

### Securing The Line

Have your partner pull the kite all the way down as you let out flying line from the spool. The kite can be retrieved (no small matter if you have spent upwards of $25 for it) and the line will be in a good position for you to attach the antenna wire and pull it up.

You or your partner should attach the line to the tree by tying the line around a protrusion such as a sawed-off branch, or around the tree trunk. You might want to secure the line up 10 or 12 feet from the ground, using a stepladder, so no one can easily get the thrill of untying the line and watching your antenna meet its demise.

Attach an insulator to the other end of the line, knotting the line securely the way you were taught in the Scouts. This means you will have to cut the flying line. With luck and planning, you've bought 500 or 1000 feet of line, so you can string lines over two or more trees as needed, without running out. (You didn't? Murphy strikes again!)

Then attach the antenna wire to the insulator. You can pull the antenna up from the other side of the tree, cutting off the extra line. The result will, when the whole antenna is finally erected, look like Fig 4. You can put a strong door-closing spring in the line just past the insulator, to allow for tree movement in the wind. Since you have attached the line via a rather high point in the tree, this movement may be considerable. The wire will break eventually, of course, no matter what precautions you take. Wire antennas usually don't last long, so work with this fact in mind.

A reminder: Use nonconducting line whenever possible. Ham radio is fun, but it's not worth your life. It makes sense to minimize the chances of electrical disaster by every possible means. Another closing bit of advice: Keep an extra kite handy.

# Front-End Overload, A Worst Case Example

By John Stanley, K4ERO
Box 390 C, Route 2
Rising Fawn, GA 30738

As users of the radio spectrum increase in number, the possibility that your receiver performance will suffer also increases. This article concerns strong off-frequency signals that cause problems with your receiver. Many hams occasionally experience this problem. For some, however, it is a constant problem. Dare I say that some hams have given up entirely because they could not solve an overload problem?

Typical sources of overload problems are the CBer next door, a nearby radio station, or a multi-transmitter contest or Field Day setup. The symptoms are strange sounds in the receiver, signals weaker than one would expect, and extra signals in unpredictable places on the dial. The signs of front-end overload are:

1) The problem goes away when the source of the problem is off the air.

2) The effect is nonlinear. That is, interference decreases at a rate more than the "square law" effect when the distance from the offending source is increased, or its strength is otherwise decreased.

Fig 1—Radio heaven or radio hell? The antenna system of HCJB, near Quito, Ecuador, was located literally in the author's backyard! His house is at the center of the photo, the top building in the row of four.

### You Think You've Got Problems?!

For 13 years, I lived on the transmitter site of HCJB, the shortwave giant in Quito, Ecuador. At distances of from 100 to 1000 yards from my QTH were over 20 antenna systems (Fig 1). Typically, 10 of them at any given time were broadcasting with 10 to 500 kW each. The RF intensity in my yard at ten feet above the ground was about 10 mW/cm$^2$, which at that time was the accepted limit for human exposure to RF! Today, the limit is about 10 times more strict at most high frequencies. It is safe to say that no one is likely to have a worse problem than I lived with. If you do, you should seriously consider moving!

Is it possible to operate a ham station in such an environment? Amazingly enough, the answer is *yes*. After years of trying many different methods of coping with this problem, I was able to operate reliably on all HF bands, to the point of working all states and many foreign stations on each of the then-available bands. Perhaps some of the tricks I tried will solve your problem.

In addition to the effects just mentioned, there were two others that I hope you won't experience. Another ham who worked for HCJB innocently installed a 20-meter quad in his yard, about 20 feet in the air, and connected it to his transceiver. Smoke poured from the rig as the receiver input coils burned up! Warned by his misfortune, I always connected a dummy load to any new antenna and measured the induced voltage first. One 40-meter dipole intercepted enough RF to light a 100-watt bulb to almost full brilliance! On another occasion, the UHF connector on the RG-8 feed line to the same dipole began to arc! No amount of beating, throwing dirt or immersing in water would extinguish it. I finally thought of shorting it, then quickly took down the dipole.

A less serious effect, but also interesting, happened when I used a 1-kW balun with a ferrite core which became noticeably hot when used only for receive. If by now you are convinced that your problem is not quite as bad as mine was, perhaps you will be encouraged to keep trying for a solution.

It may be that the obvious way to handle front-end overload is to buy one of the newer rigs which have super "crunch-proof" front ends. This is an expensive approach, especially if you burn out the receiver front-end anyway. Before you go that route, try some or all of the following suggestions.

**1) Cross polarize!**

Most high-power broadcast-station antennas are either horizontally or vertically polarized. AM-broadcast antennas are always vertically polarized, while shortwave broadcast antennas are usually horizontally polarized. Your antennas should use the opposite polarization. (Commercial FM-broadcast antennas, however, may be circularly polarized.) Since most HF amateur contacts are via ionospheric propagation, which twists the signal polarization anyway, you can use either horizontal or vertical polarization and still make contacts.

If your station is near an AM-broadcast station, avoid verticals like the plague. If you live near a source of high-power horizontally polarized signals like I did, use verticals. In this case, the ground-mounted vertical is best for rejecting nearby signals. If not obstructed in the direction you wish to work it can be very competitive, especially on the low bands. Try to place any obstructions between you and the interfering signal.

**2) Use antenna directivity!**

This step involves more than simple front-to-back ratio. Most antennas have a sharp null in one or more directions. For example, a dipole has a null directly off the ends. A beam has nulls at 90° left and right of its "forward" direction, or off the ends of the elements. You can use this directivity not only to analyze your problem but perhaps also to solve it.

Of course, Murphy's Law guarantees that areas of the world you most want to work will not be at 90° to your interfering source, but remember that the nulls of any antenna are much narrower than the lobes. You may find an antenna that drops the interference into a null, while putting your favorite DX location well up on a lobe.

One example is an array of verticals. It is not difficult to phase two or more verticals for one or more narrow nulls, yet still have nearly full coverage in all other directions. AM-broadcast stations use this technique to allow more stations to operate without mutual interference. A reference book on AM-broadcast arrays has hundreds of patterns from which you can choose.[1]

On the other hand, horizontal long-wire antennas have many different combinations of nulls and lobes, depending on height and the ratio of length to frequency. Try a 2- or 3-λ wire close to the ground, so you can move its far end to null the interfering station. It should still work well in several directions.

One word of warning: Don't assume that the pattern of your antenna will be the same at the interfering frequency as it is at the operating frequency. For example, a 20-meter Yagi with a good F/B ratio might seem a good solution to the presence of a strong signal on 15 MHz. However, the F/B ratio at 15 MHz will not be nearly as good as at 14 MHz. With computer-modeling tools such as NEC or MININEC, you may be able to design an antenna that has a good F/B ratio or pattern null at the interfering frequency, combined with a desirable pattern at the operating frequency.

**3) Filter!**

If the station causing your problem is not on your frequency, then its signal can, in theory at least, be removed by filtering. If the interference is on your frequency, then you must both be hams, and an agreement on operating schedules might be in order!

These two techniques will work whether the interference is actually at the frequency you want to use, or whether it is being produced in your rig through intermodulation. For a filter to be effective, the interfering station must be "clean." Your first job then, is to determine if the problems you experience are due to a signal actually present on the ham bands, or if that signal is being generated in your receiver.

Tune to a moderately strong signal in the ham band and then add attenuation to your receiver input. If the interference drops faster than the desired signal, you have an overload problem. If they both drop at the same rate, the problem is not being produced in your receiver. (The source *may* be in your shack or antenna.)

*Try Another Receiver*

If possible, listen to the same combination of interference and signal on a different receiver. If any combination of antenna, attenuators and receiver allows clear reception on that frequency, then the problem is not with spurious radiation from the offending station. That is good news! It means the solution to the problem is in your hands. The alternate case, that of a real spurious output from the interfering station, is worse, because it involves trying to get someone else to solve the problem, instead of doing it yourself.

Assuming these tests show the interfering signals are not "real" but are being produced in your rig, the next step is to keep them out. This means you must place a filter somewhere between your antenna and the receiver front end. Unless it is quite lossy, a filter will not degrade receiver performance at all on the lower HF bands, and probably very little below 144 MHz. On the other hand, lost transmitter power causes an equal loss of S/N ratio at the other end in all cases.

The alternative to transmitting through your filter is to install it between the TR switch and the receiver. With some transceivers, this may be difficult, unless a separate receive antenna can be connected. Of course, with separate receivers and transmitters it is a simple matter. However, give some thought to transmitting through the filter. If a really strong interfering signal gets into your transmitter, it can cause illegal spurious outputs. You are responsible for what comes out of your rig, even if another station is part of the cause.

This problem is very possible in a multi-transmitter contest or Field Day operation. For example, if 7 MHz gets into the output stage of a transmitter on 14 MHz, it may produce a 21-MHz spur. No amount of filtering on the 21-MHz rig will eliminate this spur, as it is being generated in the 14-MHz rig. The answer is to filter the 14-MHz transmitter at 7 MHz, 21 MHz or both. A band-pass filter on the 14-MHz rig would serve. Another solution is an open coaxial stub cut to ¼ λ at 7 MHz. The stub puts a notch at both 7 and 21 MHz, while having little effect at 14 MHz.

*Where To Install The Filter?*

A filter to be used for both transmit and receive is best located at the antenna. Let us look at that option. Where the interference is on a single frequency, or at most two or three, the notch filter is a good approach. It may allow you to operate on all bands without changing the filter, as would be necessary if band-pass filters were used. A notch can be as simple as a series LC network shorted to ground. It can easily produce a 20- to 40-dB notch at a given frequency, without greatly affecting the desired signal. If you live near an AM-broadcast station and wish to operate 160 meters, try this approach.

In one instance, I wanted to operate on 14 MHz while a nearby 500-kW transmitter operated on 15.250 MHz. I used an open quarter-wave stub cut to 15.250 MHz (Fig 2A). This looked like a large capacitor to ground at 14 MHz and it was paralleled with a strap inductor. The result was a deep notch at 15.250 MHz and less than 0.5 dB loss at 14 MHz. Both the depth of the notch and the loss on 20 meters were a function of the losses in the coaxial stub. I used 1⅞-inch rigid coax. RG-8 type coax is more lossy, so try to use hardline. A lumped-constant equivalent filter is shown schematically in Fig 2B.

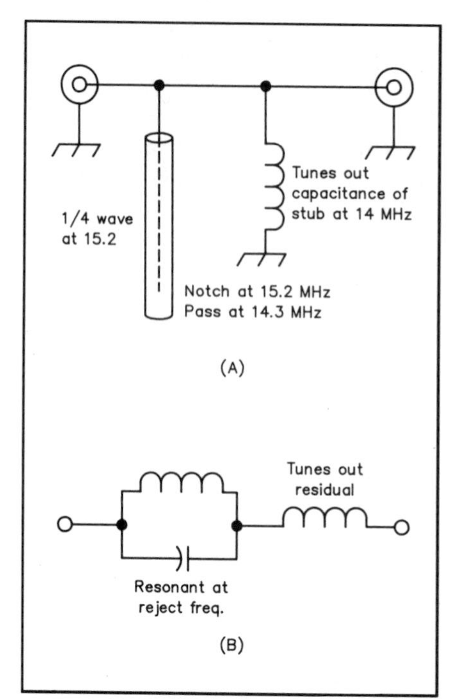


Fig 2—At A, a ¼-wavelength coaxial stub, used as a notch filter. An inductive strap resonates with the stub capacitance at the desired operating frequency. The author recommends using low-loss "hardline" to reduce insertion loss. At B, the lumped-constant equivalent notch filter. Filters made from coaxial line are more likely to withstand transmitting power levels and can be tuned with wire cutters.

*Tunable Filter*

Another use for the series filter is in diagnosing your problem. A broadcast-type variable (about 365-pF maximum capacitance) from a tube-type AM receiver and an appropriate series coil can be fitted with a frequency dial (Fig 3). Tune this circuit through its range while you listen to the desired ham signal. If at some frequency your reception improves, you have determined the frequency of the signal causing the interference. A permanent filter then can be designed.

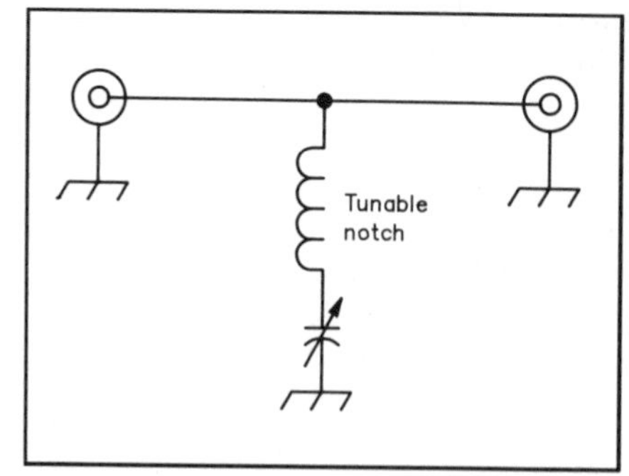


Fig 3—An adjustable notch filter suitable for determining the frequency of an interfering signal. A general-coverage receiver, calibrated signal generator or dip meter is needed to calibrate the capacitor.

If you are trying to reject many different frequencies and pass only one narrow band, such as at a multi-transmitter station, a band-pass filter is the best choice. Just using an antenna tuner may provide all the relief you need. Even though you may not need to improve your match, the tuner will provide some degree of band-pass filtering. Some tuner designs are better than others in this respect, and some can be tuned to improve filtering action with only a small loss of matching effect.

Construction of a filter for just this purpose is not very difficult. From one to four or more poles can be used, but unless a sweep generator is used (sometimes called a scalar network analyzer), the difficulty in tuning your filter increases rapidly as you increase the number of poles. I have made a number of four-pole filters using mica compression-type capacitors and toroidal coils, with good results. These were used only for receiving, not transmitting. A band-pass transmitting filter would likely have only one or two poles, to avoid excessive in-band losses. These might be LC circuits at the lower frequencies or coaxial tanks at the higher frequencies.

**4) Isolate!**

Many problems you will experience in a strong RF environment are caused by energy not picked up by your antenna. It comes in on other "antennas" that connect to your rig. These must be found and eliminated.

Probably the biggest headache is energy coming in on your power lines. One way to find out is to temporarily use battery power. If the problem goes away when you switch to battery power, you know the interference is entering the rig through the power line. One obvious solution is to charge one battery while you use another. On Field Day this is a good choice.

Fortunately, however, effective line filters are not difficult to build. The place to start is with ferrite suppressors. A split toroid that snaps over the power line is a quickly applied remedy. Several turns of the power cord wound through the toroid is most effective. A large toroid can contain 5-10 turns of the power cord. Use a lossy ferrite material optimized for RFI rejection if possible. Otherwise, try whatever your junk box can produce. TV flyback-transformer cores provide an effective split ferrite core.

An alternative approach is a low-capacity power transformer. They are used by AM-broadcast stations to feed power to tower lighting without shorting out the RF. Their big disadvantage is their cost. Surplus outlets may have low-capacity filament transformers, which are sometimes used in grounded-grid power amplifiers. Two of these "back to back" will provide a low-capacity isolation transformer through which common-mode RF cannot be conducted.

A low-capacity transformer can be homemade as well. Acquire two TV-type power transformers and remove all windings except the primaries. Add a low-capacity secondary of a few heavy turns, suspended away from both the primary and the core. A pair of these "back to back" may work.

If you must purchase something, try a commercial bifilar choke, like those used for filament chokes in grounded-grid amplifiers. They should work at 120 volts, and the current rating will be the same as at lower voltages. Be aware that you are exceeding the turn-to-turn voltage rating and will not be able to make a warranty claim if it arcs across. [You can wind your own choke on the ferrite rod from a broadcast radio. These rods are available new from suppliers of ferrite and powdered-iron cores. —*Ed.*]

All these approaches seek to put a high impedance in series with your ac power line. How about a low impedance to ground, to short out the incoming RF? The main problem here is there is usually no good place to shunt this power into. That is, there is no good RF ground point.

Bypassing both sides of the power line to ground with, say, a 0.01-μF capacitor also has a problem in that it will put 55 volts ac on your "ground." While perhaps not deadly, such bypassing can lead to some nasty "bites." If you are going to take such an approach, try to use a ground that has pretty low resistance to the power-line ground. Again, such an approach is not recommended, and I have never found it to be really useful. If you have differential-mode RF on your lines, a capacitor between the two wires, (hot and neutral) will help, but in my experience, most line-conducted signals are common mode.

If all attempts to put a high impedance in series with your ac lines have not solved the problem with line-conducted RF, there is still one more arrow in your quiver.

Having cleaned up your ac line, you may still have common-mode problems with all other cables that connect to your rig. The power line is the place to start, but you may not be able to stop there. Coaxial cables should have toroids installed, and toroids may be needed on any other audio or control cables that connect your rig to the outside world.

Rotator cables are especially important, as they run through high RF zones. In a difficult case, you may wish to use fiber optic cables to connect computers, etc to your rig. With presently available technology, this is not as extreme as it sounds. In the case of computers, fiber optic cables also serve to avoid conducting computer generated hash into your receiver.

Another way to reduce common-mode energy on your antenna leads is to use a "shield breaker," as developed by British hams for use on TV lead-in wires. A similar approach for the lower frequencies is a large toroid with a few turns on one side connected to the antenna, and the same number on the other side going to your receiver. It may introduce some attenuation, but on the lower bands that may actually be an advantage, if the transformer is used only on receive. This approach is essentially an RF version of the ac low-capacity transformer.

**5) Shield!**

What can be very effective, and is in fact the ultimate solution to power-line conducted RF is the use of a shielded room. While this may seem like an extreme solution, it need not be as involved as you think. Apart from purchasing a commercial shielded room, you can build a shielded box that will house only your transceiver and the few accessories that must connect directly to it. One of these accessories need not be the operator.

Amazingly enough, it is possible to leave off one side of such a box and still get 20 dB of ambient RF rejection, at least up through 30 MHz. The operator sits outside and reaches in only to operate the controls. The power line passes through the back wall of the box through a well-designed filter, and the antenna connections, mike and key leads, etc all pass through bulkhead connectors. To avoid floating the shield above ground, use an isolation transformer. Use of a ground-fault interrupter is a wise precaution, although you may then have to isolate the GFI from RF to avoid false trips.

If the shielded room/box approach is used, nearly all common-mode signals should be eliminated. Short of such an extreme solution, there is still a place for shielding. The key to effective shielding, however, is having a good place to bond the outer conductor. This is where the shielded box is so effective. However, the cabinet of a transceiver also provides some shielding. If all conductors in and out of the transceiver are shielded and the shields are connected to the cabinet, the benefit of a shielded box is retained.

For example, if you have an external keyer, the lead(s) to it should be shielded and the shield returned to the cabinet ground of the rig, even though it may not be necessary for proper keying operation. Without shielding, it is an antenna that carries RF through the walls of your transceiver where it can radiate into the circuitry. The same can be said of mike cables, control lines to the amplifier, and so on. The generous use of 0.001-μF bypass capacitors from the center of these leads to ground may also help attenuate some of these signals. The *ARRL Handbook*[2] contains more information on shielding. The same techniques used to keep RF inside a transmitter are useful in keeping unwanted RF out of your receiver or transceiver.

A comprehensive approach to the receiver overload problem may require a number of the above approaches. In an extreme case, you may find yourself "peeling an onion," in which each solution only reveals a lower level of interference. You just keep peeling until you reach the level of operation you desire. With each successful approach, you will be amazed at how much more ham activity has appeared on the bands! You will be able to work 'em, because now you can hear 'em!

**References**

[1]*Directional Antenna Patterns* (Cleveland: Smith Electronics, Inc.)

[2]R. Schetgen, Ed. *The 1993 ARRL Handbook for Radio Amateurs*, 70th ed. (Newington: ARRL, 1992).

# A Remote Field-Strength Metering System

By John Svoboda, W6MIT
2261 Peaceful Garden Way
Rescue, CA 95672

It all started one night while I was walking to the far field to perform another field-strength measurement on my newly designed, yet not quite functional multiband vertical. I didn't mind the curious stares of my neighbors or the long strolls in the dark. The real problem was a matter of too many measurements and too little time!

This, then, is the story of a simple field-strength meter that has grown, in the hands of its creator, into something very . . . *different*. See Fig 1. I don't necessarily recommend that you follow my project design to the letter. Some of the components are unique, as you will see. However, the basic ideas can be adapted to the needs of your particular situation and the size of your junk box!

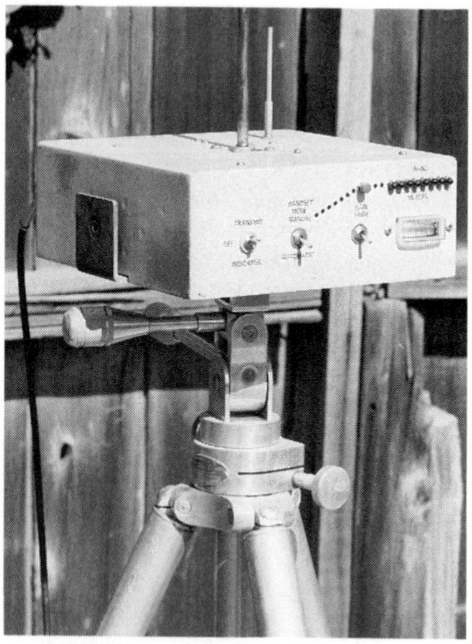

Fig 1—The remote unit mounted on a camera tripod. The HF whip antenna is visible on the top center, along with the shorter UHF telemetry antenna. The switch on the far left is placed in the transmit mode to aid in setting up the UHF link to the shack. Placing the switch in the INDICATOR position enables the HF scanning mode. The row of LEDs above the panel meter indicates which band is active during scanning or manual operation. (Fig 7 shows an interior view.)

### Necessity Is The Mother Of Invention

Short of giving up on my project altogether, getting more and better data was the only alternative. At first I considered a hard-wired remote measurement system. However, stringing out a very long length of cable between the meter and my shack was not acceptable. More radical thinking was required.

In my quest for a solution I turned to my junk box, which had grown substantially after many summer swap meets. Without too much digging I discovered a small cabinet with a big meter and an assortment of goodies, including the electronics from a solid-state TV set. I also found a solid-state UHF oscillator (complete with a TTL input) that had been scavenged from a surplus receiver.

In a stroke of unprecedented luck, the UHF oscillator actually *worked*—and it could be heard at the high end of the UHF TV band (around 800 MHz). Sporting a quarter-wave ground plane, the oscillator produced a good quieting signal at a range of 200 to 300 feet.

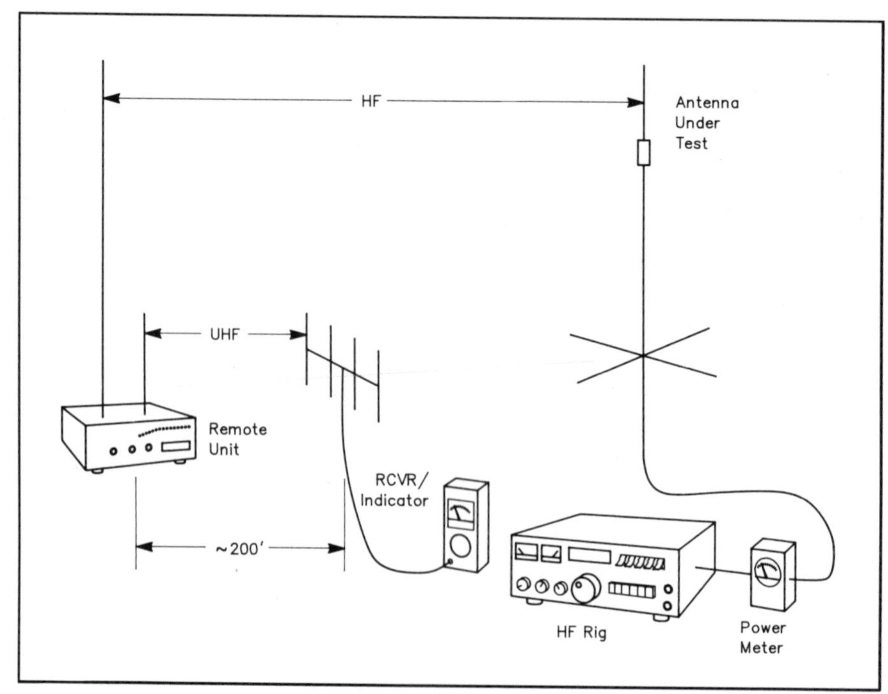


Fig 2—Operational diagram of the remote field-strength metering system.

Fig 3—An old HP-415 has become a fully metered UHF telemetry receiver. Knobs on each side of main tuning dial provide meter-range switching and full-scale adjustment. The small knob to the left of the meter controls audio gain. The switch below the tuning dial selects the CW, MCW or PRR inputs.

Fig 4—Interior view of the UHF telemetry receiver. The VHF tuner is mounted behind the speaker (the UHF tuner is not visible in this photo). Metering circuitry is visible below the chassis.

By examining my assortment of parts, a primitive remote telemetry system began to take shape (Fig 2). A basic field-strength meter detector would drive a voltage-to-frequency converter (an Exar XR-4151 left over from another forgotten project) which, in turn, would pulse modulate the oscillator via the TTL data input. The TV receiver would then detect the signal and convert the pulses to meter readings. Within a short time, a simple, single-band unit was built and tested. To my surprise, it

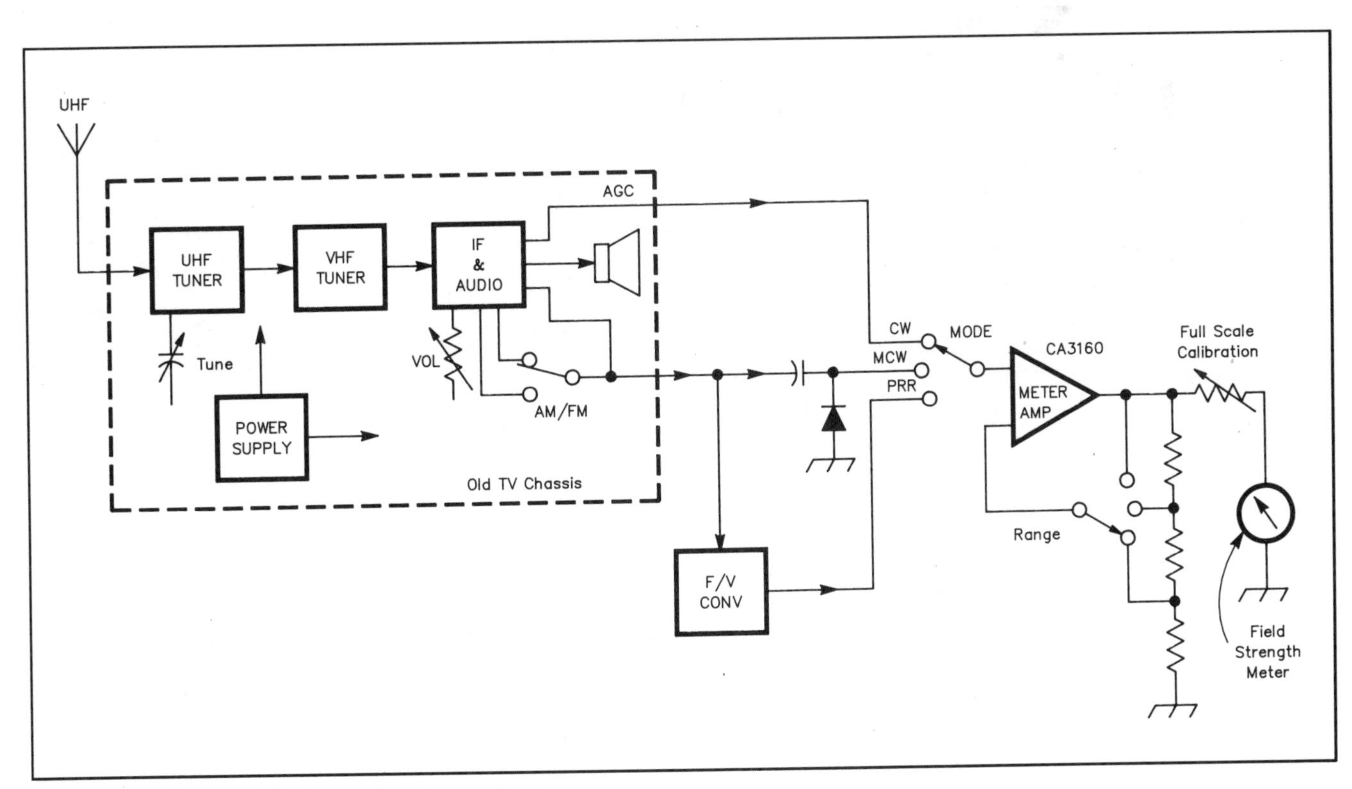


Fig 5—Block diagram of the UHF telemetry receiver.

worked quite well.

The multiband system presented here is an outgrowth of these initial experiments. Circuits were developed rather haphazardly employing standard "ugly" construction practices. Each module was built and tested individually.

### The UHF Telemetry Receiver

From Figs 3 and 4 you will probably recognize an HP cabinet and meter. The cabinet was gutted, retaining only the meter and switches so the TV hardware could be mounted inside (Fig 5). The VHF tuner was mounted inside the box, and the shaft of the UHF tuner (continuous type) was brought out through the front panel and attached to a tuning knob. Modifications to the TV receiver consisted of removing unnecessary circuits such as the traps in the IF amplifiers. The TV audio was retained as an aid in locating the telemetry transmitter signal and as an aural indicator, since the audio pitch rises with increasing signal strength.

The function switch on the front panel selects one of three measurement modes: the signal strength of the UHF telemetry link (CW), telemetry audio (MCW) or the telemetry pulse repetition rate (PRR). With the switch in the MCW position, the meter functions much like an audio VU meter, which tends to fluctuate rapidly. In the PRR mode, however, the variations are smoothed out. This produces a fairly steady field-strength display.

AM sound (for MCW) was taken off just ahead of the 4.5-MHz ratio detector in the TV set. (Note that an AM/FM selector switch has been added to provide monitoring capability for other UHF FM or AM signals if desired.) The input signal for the frequency-to-voltage converter (for PRR) was tapped at the end of the video amplifier string. The F/V circuit was found in the

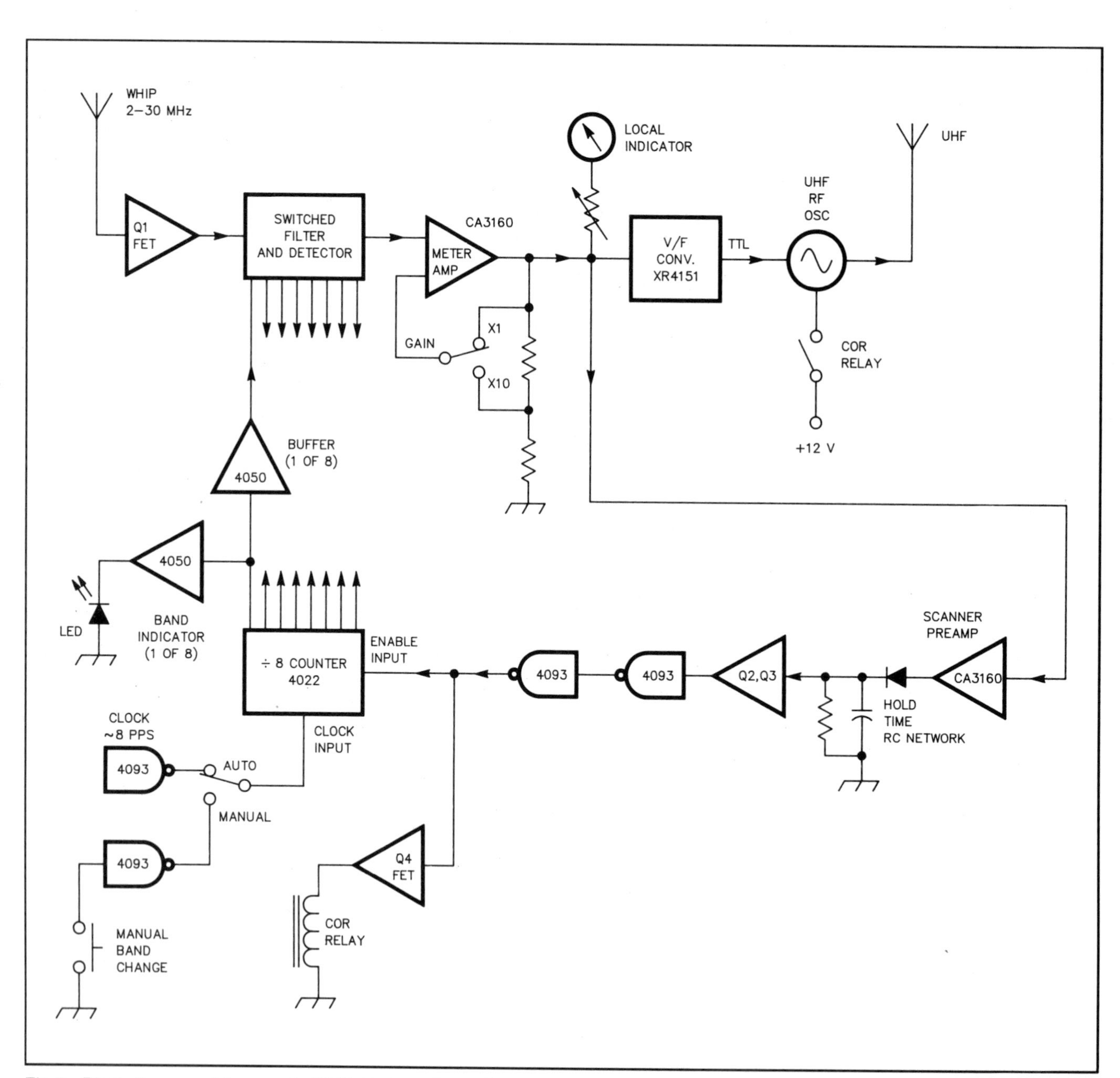


Fig 6—Block diagram of the remote HF field-strength receiver.

*Modern Electronic Circuits Reference Manual.*[1] Another Exar XR-4151 would have worked as well, however. Capacitors were added to the collectors of the video amps to reduce high-frequency response. The CA3160 metering circuit was taken directly from the *RCA Linear Data Manual* with the exception that the range switch and the full-scale pot were not calibrated.

The gain and bandwidth of the modified TV receiver were ideal for use with a simple, low-power oscillator/transmitter. A small UHF Yagi functioned perfectly as the telemetry receiver antenna.

Fig 7—Interior view of the remote unit. At the center is the HF-whip antenna mount and FET follower board. On the left is a portion of the multiband filter/detector circuitry. The digital board (with scanner preamp) occupies the upper area of the photograph with the meter amplifier and V/F converter also visible in the upper right. The UHF oscillator "brick" is located in the spacious area in the lower right. It is coupled to its antenna via a short section of RG-405.

### The Remote HF Field-Strength Receiver

At the heart of the remote field-strength receiver is an 8-band HF scanner (Figs 6 and 7). When a signal of sufficient amplitude is detected, the scanner stops and measures the signal. If the signal disappears, scanning begins again after a delay established by an RC network. (Several seconds of delay are used to permit minor equipment adjustments without unnecessary scanning.)

The circuit design is very straightforward. Q1 provides a high-impedance input for the HF whip antenna and low-impedance output to the filter/detector stage of the scanner. Eight parallel diode switches, one of which is always on, feed eight individually tuned filter circuits, each with its own detector (Fig 8). Separate filter/detector circuits are used for each desired HF band. More complex filters could be used to provide better bandwidth. Only the filter input is switched in an effort to keep the circuits simple. Although the switching diodes see only about 2 mA of switching current, they seem to perform very reliably.

HF band scanning is accomplished through the action of a 4093 quad 2-input NAND Schmitt trigger which provides clock pulses (approximately 8 per second) to a 4022 divide-by-8 counter (a 4017 could also be used if you wanted 10 bands). The 4022 outputs are connected to 4050 buffers that feed the diode switches and band-indicator LEDs. Other sections of the 4093 provide manual band changing and pulse forming for the scanner.

A second CA3160 acts as a scanner preamp. It operates at high gain to detect pulses from the CA3160 meter amp as the receiver scans the bands. When a signal of adequate level is detected, the preamp puts out a pulse which charges the RC network and turns on Q2 and Q3. The output pulse from Q2 and Q3 is shaped by two tandem sections of the 4093. The result is a positive signal at the enable input of the 4022, which stops the counter on the active band and activates the transmit relay (COR) via Q4. (Q4 is a power FET that happened to be available.)

Receiver gain must be adjusted so the scanner will not lock until an adequate HF signal is present. Receiver gain is increased or decreased by selecting one of two available gain-switch positions, or by adjusting the length of the whip antenna. The receiver should be adjusted so a scanner lock does *not* take place until the panel meter indicates a reading greater than the second division above zero.

### The UHF Telemetry Transmitter

A voltage-to-frequency converter is connected to the output of the CA3160 metering amplifier. The TTL output of the V/F converter feeds the data terminals of the UHF oscillator/transmitter. My particular oscillator requires 100 mA at 12 volts at 802 MHz. With its $1/4$-$\lambda$ antenna, it provides adequate receiver quieting at distances up to 300 ft.

The complete remote unit was built into a 10 × 10 × 3.5-inch chassis. It can be mounted on a camera tripod or placed inside a vehicle (using suitable external antennas) that is parked at the desired location. The entire package requires 40 mA on standby and 160 mA on transmit with 12.5 volts. My design utilized an external battery power source.

### Operation

The HF rig should be operated at a low power level (10 watts, for example) that can easily be read on an output power meter. The half-power mark on the output meter (5 watts in this example) should also be easily readable. Before taking any field-strength readings, check to be sure the output power meter is set properly.

When positioning the remote unit, first make sure that the HF signal level will be sufficient at the site. This is accomplished by adjusting the gain of the receiver. A half-scale reading on the panel meter should ensure satisfactory operation. It is important to avoid overdriving the receiver since linearity will suffer as a result.

The next step is establishing the UHF telemetry link to the shack. Switching the remote unit to continuous transmit makes this step easier. (As soon as the link is established, the remote unit must be switched to automatic operation.) Place the field-strength telemetry receiver near your rig and adjust its antenna for maximum signal strength. (I mounted its small Yagi on a photographic light stand and positioned it to receive the telemetry signal.)

When everything is working properly, keying the HF rig will key up the remote unit as well. The telemetry meter can then be adjusted for some convenient indication. Reducing the HF output to the half-power mark should cause a 3-dB change on the telemetry meter. Now, from the comfort of your shack, you can make many field-strength measurements on any HF band you wish!

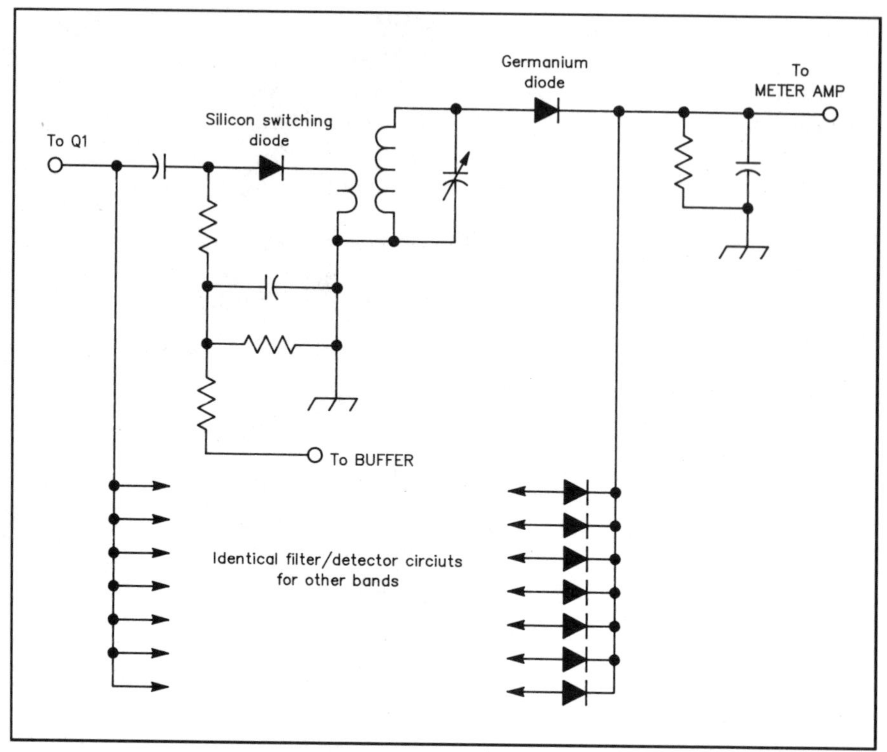


Fig 8—Individually tuned filter/detector circuit. One is required for each HF band.

### Conclusions

This system has performed quite well. Certainly there are many things that need additional work. For example, when the UHF carrier drops the meter pegs when in the PRR position. This is because without a signal present, the telemetry receiver is measuring high-frequency noise.

And why did I include HF band switching in the design? In my case, I have medium and shortwave broadcast stations as well as two airports in my vicinity. In this environment it seemed easier to select what I might *want*—rather than to guess at what I might need to *reject*!

The selection of an operating frequency—accidental in my case—was fortunate. It turned out to be near UHF channel 70, and there were no nearby TV stations. Older style tuners cover up to channel 83, or 890 MHz. It should be fairly simple to slip these tuners up into the 900-MHz region if necessary. Electronic door openers that operate in the microwave region might also be candidates for a short-range transmission system since the models I've seen have separate horns for send and receive. Watch the swap meets!

As in any worthwhile antenna effort, keeping detailed notes and logging test data are the keys to success. It is doubtful that one or two measurements will give any meaningful results. Many measurements and minor adjustments are required to get an overall picture of the performance of your antenna system. Having a remote field-strength metering system makes the job much easier.

### Non-Licensed Operation

Article 15 of the communications code outlines the conditions under which an unlicensed transmitter may be used. A variety of applications are described: telephones, toys, security devices and so on. New devices are being developed all the time. In commercial ventures, these devices must be tested and registered with the FCC before anything can be sold. The FCC has also established a number of bands for unlicensed operation according to the application in question: 46-49 MHz for cordless phones, 902-928 MHz for cordless TV cameras and VCR repeaters, 2450 MHz for microwave ovens and many more. With the exception of microwave ovens, the power from most of these devices must be maintained at very low levels. The remote field-strength metering system is no exception.

How low is low? Two rules of thumb may help: (1) Make sure the range of your device is no greater than that of your cordless telephone and, (2) be aware that where there is no protest as a result of interference to an existing service, or an expressed concern for public safety, unlicensed low-power operation is within the law. If, for whatever reason, you are directed to stop operation by the FCC, of course you must do so immediately.

[1]John Markus, Modern Electronic Circuits Reference Manual, (New York:McGraw-Hill, 1980), Chapter 33.

# The Hybrid Junction Admittance Bridge

By Wilfred N. Caron
2945 Camp Joy Rd
Grants Pass, OR 97526

The following discussion is presented with some hesitation. I have recently learned a great deal about bridges, and I feel that I must share my experiences with the reader for reasons that will soon be evident. I have discovered that most bridges designed around transmission-line type transformers are inherently inaccurate. The problem appears to be centered in the tight coupling between the primary and secondary windings of the transformer and the unbalanced nature of the primary circuit.[1] This condition is illustrated in Fig 1.

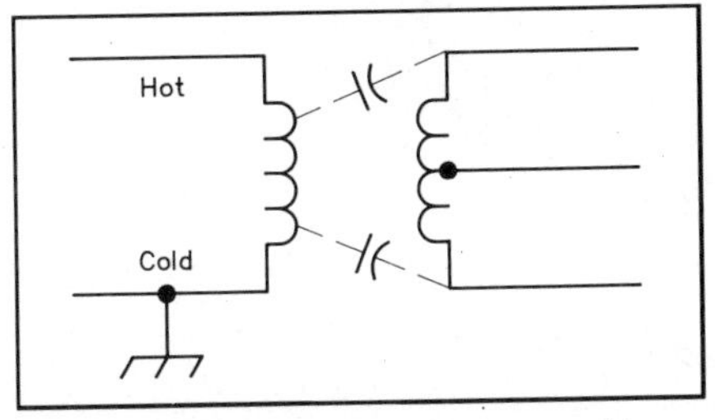


Fig 1—Unbalanced condition caused by capacitive coupling between the "hot" ends of the secondary winding and the "hot" side of the primary winding.

Some bridge designs attempt to correct for this lack of symmetry by using a four-wire transformer.[2] Other designs use a two- or three-wire balun with a three-wire transmission line transformer. These designs are illustrated in Fig 2. These techniques do provide some improvements, but the additional complexity becomes frequency limiting if the balance to ground, the coupling between the coils, and the even division of power between the two arms of the bridge are not completely satisfied. It is particularly difficult to maintain suitable balance for a bridge that must operate across a frequency range that extends four octaves or greater.

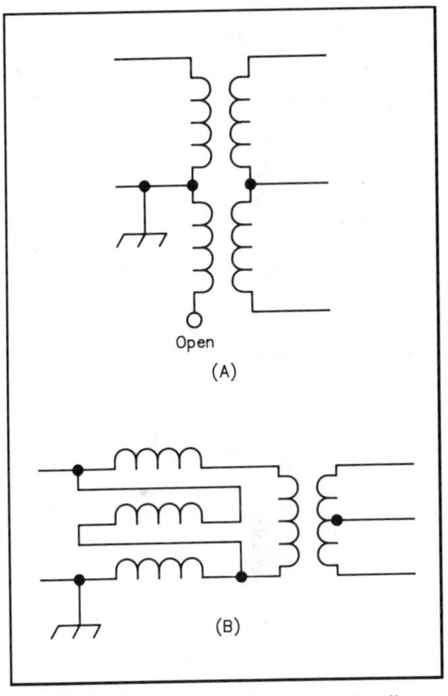


Fig 2—At A, a four-wire transmission line transformer. At B, a three-wire balun used in conjunction with a transmission line transformer.

Another bridge design that is inherently unbalanced is an admittance bridge that incorporates a transmission-line type transformer, a fixed resistance standard and a switchable variable capacitor. Such a bridge is manufactured by Tennatest. The basic bridge circuit is shown in Fig 3. An impedance plot obtained with this bridge is shown in Fig 4. As you can see, the Tennatest bridge cannot be recommended for admittance measurements.

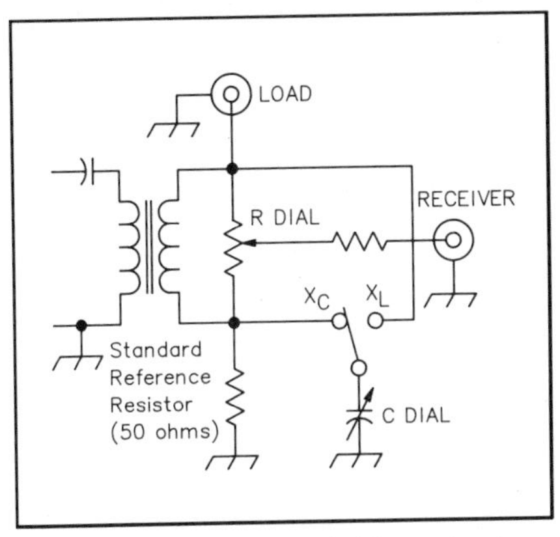


Fig 3—Basic admittance bridge using fixed standard reference resistor and switchable variable capacitor.

Of special interest is the Palomar Engineers R-X noise bridge (see Fig 5). It consists of a transmission-line type transformer and a wide-band noise generator. One arm of the bridge has a calibrated variable capacitor and a calibrated variable potentiometer connected in series. The antenna or other unknown circuit to be measured is connected to the opposite arm of the bridge. A receiver is connected as the detector.

The calibration of the Palomar Engineers R-X noise bridge is extremely coarse. While the instrument is not calibrated as a precision bridge, the potentiometer can be directly calibrated with precision resistors. In any event, measurements performed with the Palomar Engineers bridge leave much to be desired as evidenced by the actual impedance data shown in Fig 6. Again, I cannot recommend the Palomar Engineers bridge for serious impedance measurements.

Regrettably, and much to my chagrin, my discussion must also include the limitations encountered with my *own* bridge design,[3] which is basically a Twin-T admittance bridge circuit with a switchable variable capacitor. The basic Twin-T circuit is shown in Fig 7 and the impedance curve obtained with it is shown in Fig 8.

It appears that an admittance bridge incorporating a fixed resistance standard and

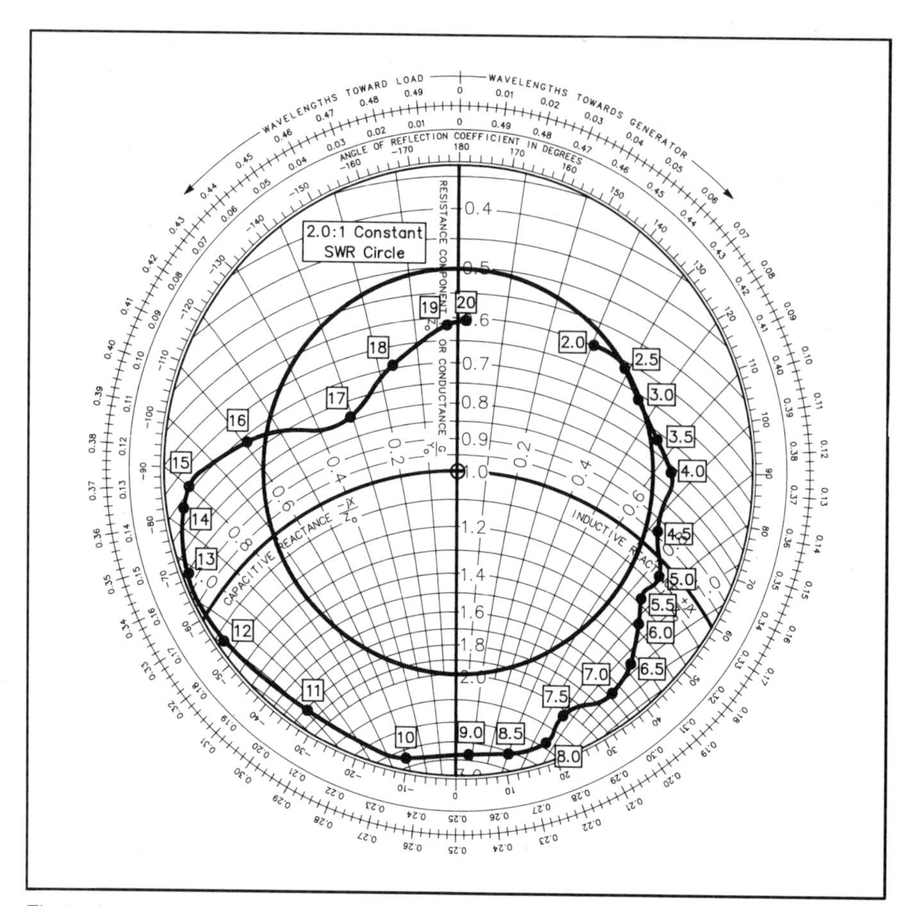


Fig 4—Impedance plot obtained with Tennatest admittance bridge of the type shown in Fig 3. Plot normalized to 52 ohms impedance.

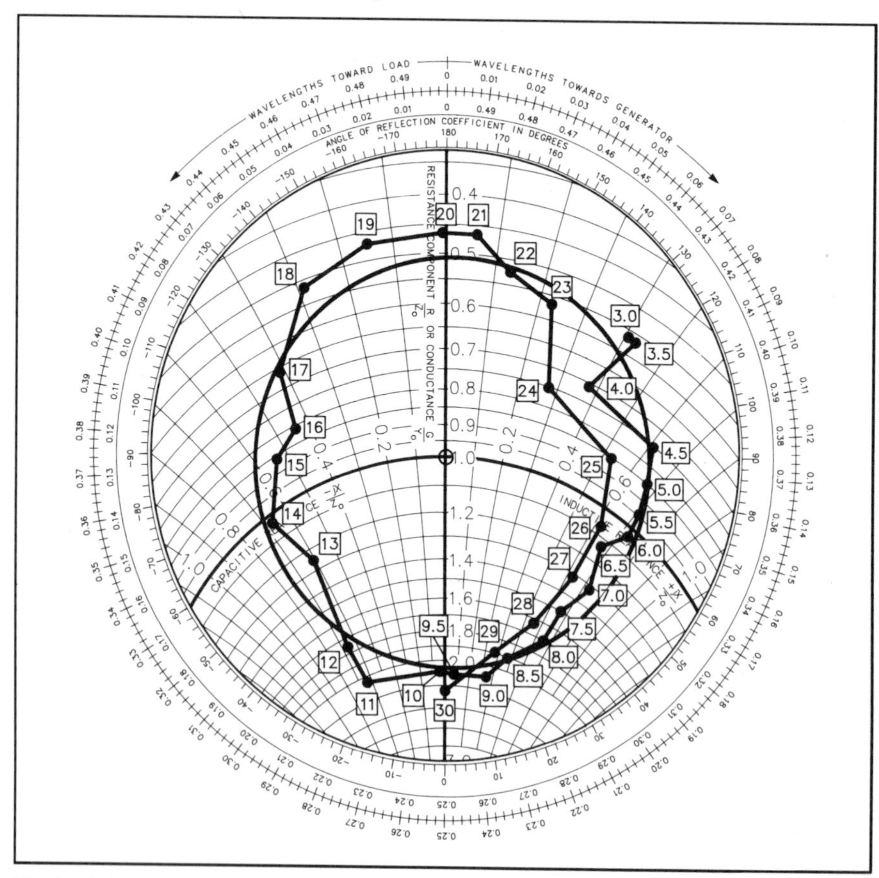


Fig 6—Palomar Engineers R-X noise bridge impedance plot. Plot normalized to 52 ohms.

Fig 5—R-X noise bridge manufactured by Palomar Engineers.

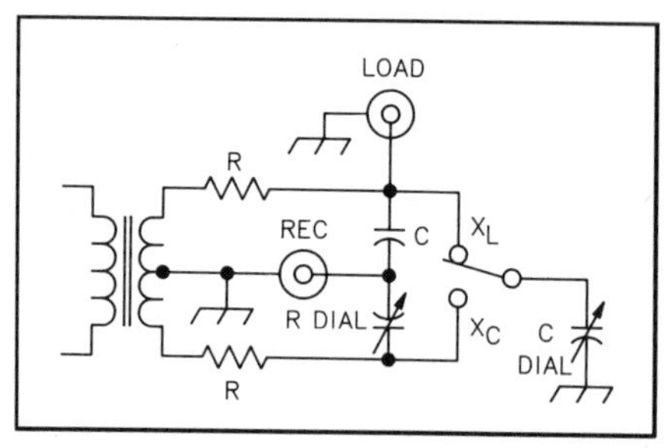


Fig 7—Basic Twin-T admittance bridge using a switchable variable capacitor.

a Twin-T with a switchable variable capacitor is not suited for measuring capacitively reactive loads. Although a null or balanced condition may be obtained with such a bridge, the resulting measurements are not representative of the actual load conditions. The currents flowing in both arms of the bridge are not equal, even when the bridge is nulled. Consequently, the bridge operator is led to wrong conclusions. There is really no remedy for these limitations except to limit the operation of the bridge to inductive reactances and/or resistances. The alternative is to come up with a new bridge design, which is what I have done.

### The Hybrid Junction Admittance Bridge (HJAB)

After exhaustive research and much trial and error testing, a promising bridge design has been developed—much to the delight of my bruised ego. The bridge circuit may appear unusual at first glance. It is based upon a hybrid junction design that provides excellent balance to ground, equal and opposite polarity power division between the two arms of the bridge, and accurate measurements of both inductive and capacitive load reactances as well as load resistances. The development of an improved bridge design, such as the HJAB, is the result of an evolutionary process.

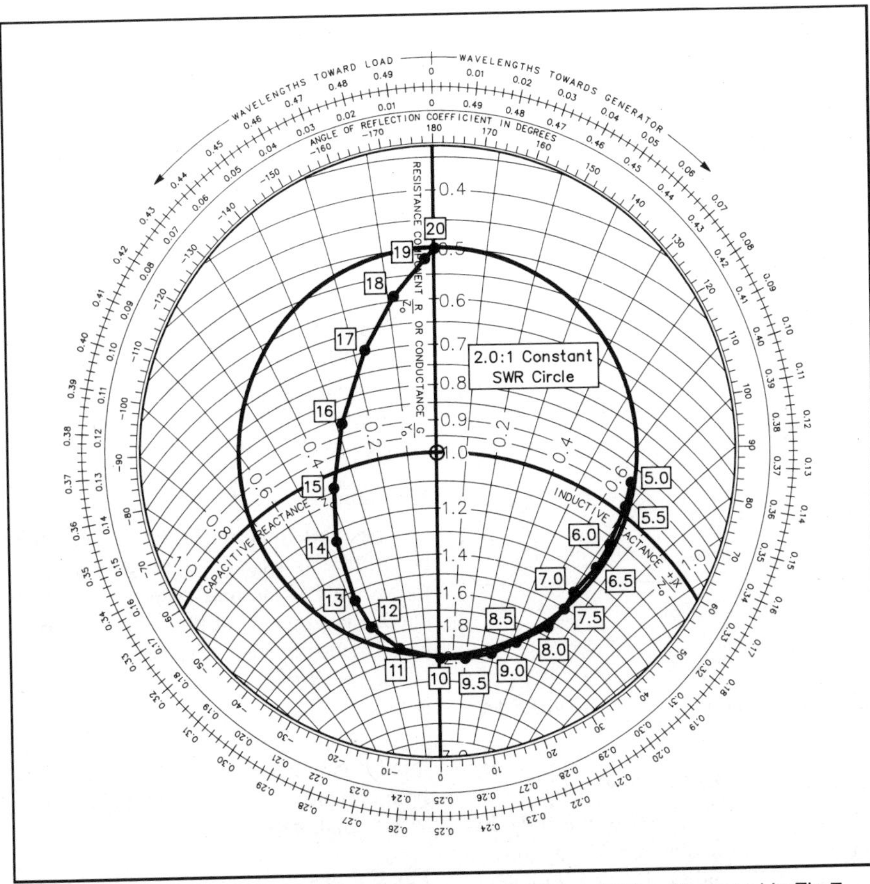

Fig 8—Impedance plot obtained with the Twin-T admittance bridge illustrated in Fig 7.

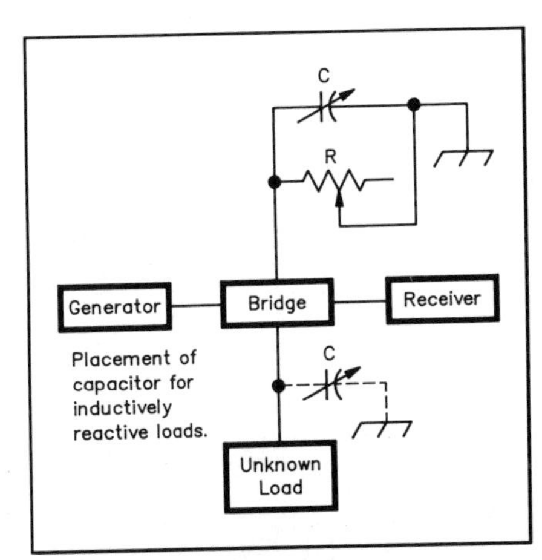

Fig 10—Block diagram of basic measuring system utilizing the practical hybrid junction admittance bridge to measure complex load impedances.

The operation of the practical HJAB is described with reference to Figs 9 and 10. A suitable signal generator is connected to the series branch of the bridge. The parallel branch is connected to a receiver, and side branches (1 and 2) are terminated in 50-ohm loads. The signal passes through the coaxial transmission line that forms coil A, to the small gap C. Opposite potentials are developed across this gap. Under

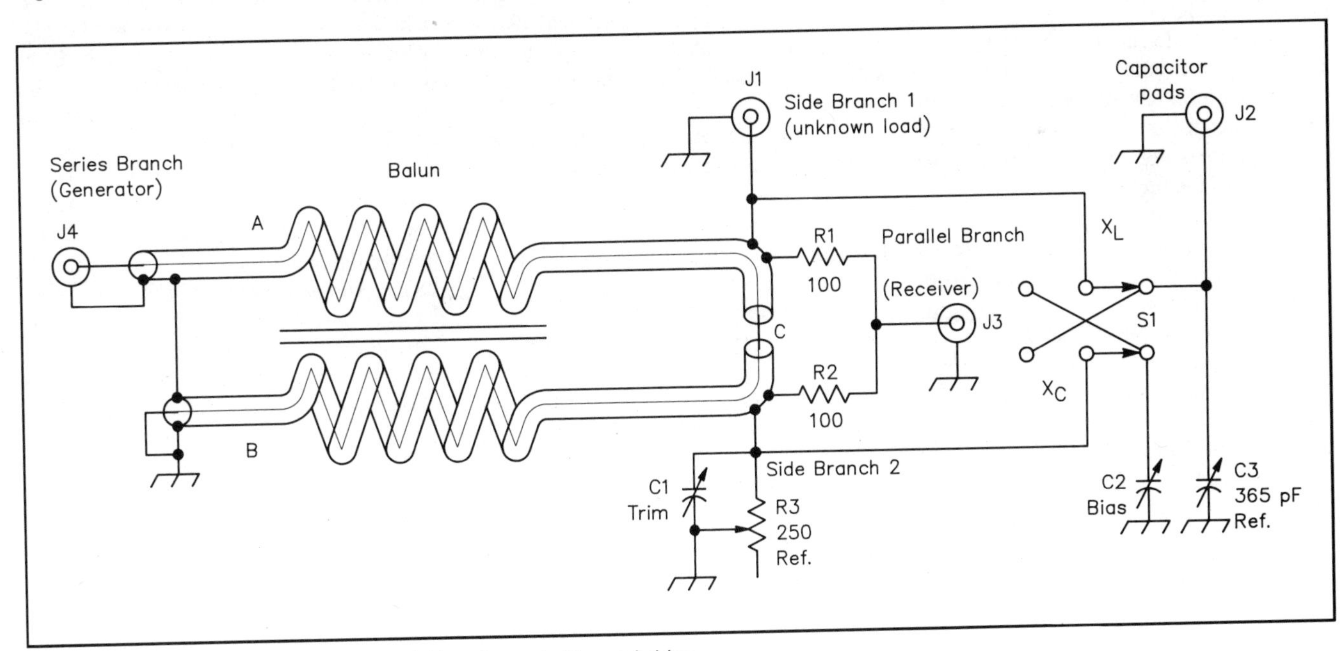


Fig 9—A schematic of the practical hybrid junction admittance bridge.
C1—Trimmer, 0.6-4.5 pF.
C2—Trimmer, 6.0-25 pF.
C3—Air variable, 365 pF.
R1, R2—100-Ω ¼-W metal film.
R3—250-Ω potentiometer, linear, Allen Bradley J style or equiv.
S1—DPDT, center off, Radio Shack 275-664 or equiv.
Balun core—Amidon Associates BLN 43-3312.
Capacitor pads—360 pF, 680 pF, 910 pF, 1200 pF dipped mica.

Misc:
Assorted connectors and hardware.
Cabinet—Bud SC-12100 or equiv.
RG-174/B type miniature coax or shielded microphone cable, Radio Shack 278-752.

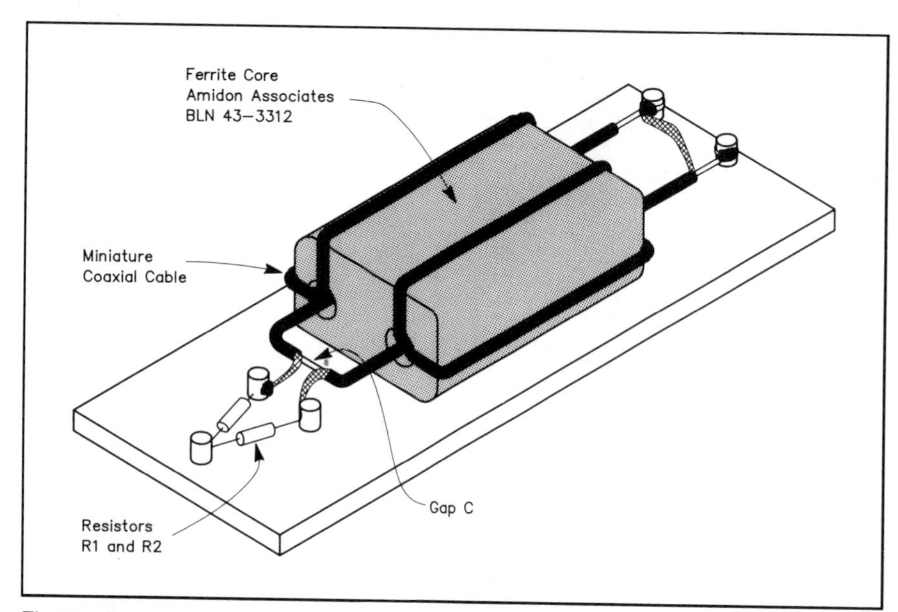


Fig 11—Construction details for ferrite balun.

the given conditions, equal but opposite potentials are applied to the loads connected to side branches 1 and 2. Equal waves of opposite potential travel from gap C, through the resistors and toward the parallel branch, canceling each other at the branch. Therefore, essentially no power is delivered to the receiver connected to the parallel branch.

A second coiled coaxial line (B) is constructed in the same manner as A. It is utilized to preserve mechanical and electrical symmetry, including the balance to ground. It does not enter directly into the operation of the bridge.

If the side branch terminations are replaced by two unequal loads, the waves traveling toward the receiver differ from each other in magnitude or phase (or both) and the waves no longer cancel at the receiver. In the practical HJAB, the reference load is a variable. For instance, if the parallel components of the load impedance are R = 80 ohms and $jX = -20$ ohms, the resistive reference is set to 80 ohms and the reactance reference, a variable capacitor in parallel with the reference resistor, is set to provide –20 ohms at the frequency of interest. (The reference resistance is not frequency dependent.) When the reference load is equal to the unknown load, a deep null is detected at the receiver.

If the reactance of the load is inductive, the reference variable capacitor is simply switched to the load side and adjusted until the capacitive reactance cancels the inductive reactance of the load. With the inductive reactance of the load neutralized by the capacitor, the resultant pure resistance of the load is then equalized by the variable resistor. Again, this creates a deep null at the receiver.

### Limitations of the Practical HJAB

With reference to Fig 9, the only frequency-sensitive element in the bridge design is the shunting effect of conductors A and B at low frequencies. So long as these conductors present an impedance to ground (at gap C) of 50 ohms or greater, the actual impedance measurements will be precise.

The Amidon Associates ferrite core used in the construction (BLN 43-3312) has 3 turns passing through it (see Fig 11). For this core, the $A_L$ value in mH/1000 turns is 5400. To compute for the inductance:

$$L_{mH} = \left(\frac{3\sqrt{5400}}{1000}\right)^2 = 0.0486$$

To compute for the reactance at the lowest design frequency of 2.0 MHz:

$$X_L = 6.28 \times 2.0 \times 10^6 \times 0.0486 \times 10^{-3}$$

$$= 0.610 \times 10^3 = 610 \text{ ohms}$$

It appears that the accuracy of the practical HJAB should be quite good at its lowest design frequency, and its operating range may extend well into the broadcast band. The upper frequency limit of 30 MHz remains to be investigated.

### Stray Capacitances and Inductances

In all high-frequency bridges, stray inductances and stray capacitances should be minimized since they can cause self resonances. Highly unsymmetrical bridges are the result! Specific rules apply when wiring bridge networks: (1) Make all connections as short and as direct as possible. Connections a few inches long, even if large wire is utilized, may cause trouble. (2) The arms of the bridge must be close together. (3) Shielded wire may be used to minimize undesirable coupling. Shielding should be grounded at a common point to prevent ground loops.

A dip meter measurement made of the practical hybrid junction bridge indicated that the HF range was completely free of resonances and that the first resonance detected was at about 100 MHz.

### Reactance of Gap C

The hybrid junction bridge will produce balanced outputs and will act according to theory only if the reactance of the center conductor at gap C in Fig 9 is small compared to the impedance being measured. At HF and VHF, this should not be a problem unless the spacing of conductors A and B is too large. This reactance can be kept small if conductors A and B are extended symmetrically to form a 90-degree bend with a small gap.

### The Balun

The heart of the practical HJAB is the balun. Its assembly is shown in Fig 11. The ferrite core is an Amidon Associates BLN 43-3312. This is a binocular core with the trough holes having an inside diameter of 0.187 inch. This small diameter necessitates that a miniature insulated coaxial cable, such as RG-174/B, be used. About one foot of cable is required. Unfortunately, this cable is usually not available unless you are willing to purchase 100-foot lengths at a price exceeding $50. An alternative which works quite well is a miniature insulated and shielded microphone cable sold by Radio Shack. This microphone cable has an outside diameter of 0.085 inch. Threading this cable through the ferrite core is very easy if the cable is lightly coated with silicon lubricant.

It is very important that the mechanical symmetry for both coils be maintained. Symmetry is required for achieving an electrical balance to ground over the entire operating range. Lack of symmetry causes the null point to shift with frequency, leading to inaccurate measurements.

### Standards of Capacitance

A variable air capacitor is essential as a "standard" for RF bridges. An important requirement of a standard capacitor is that the capacitance remain constant with changes in applied frequency.

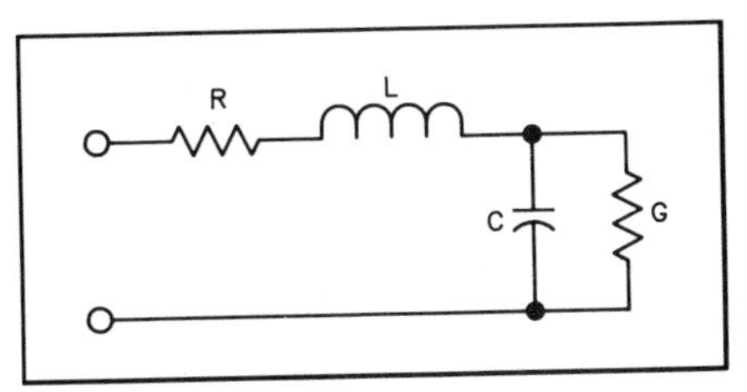


Fig 12—Equivalent circuit of air capacitor.

All capacitors have a small residual inductive reactance. Although it has been customary in some measurements to neglect capacitor lead inductance, such omissions may result in considerable inaccuracy if measurements are performed above 5 MHz. This small residual inductive reactance, when taken into account, gives an equivalent circuit for the variable air capacitor as shown in Fig 12. In this example, C is the static capacitor and L is the inductance of the interconnecting leads *and* the inductance of the frame and shaft connecting the capacitor plates.

The effect of inductance L causes apparent capacitance to be greater than the actual capacitance, as may be seen from the following equation:

$$C_a = \frac{C}{1 - \omega^2 LC} \qquad \text{(Eq 1)}$$

where

$C_a$ = apparent capacitance, farads
C = actual capacitance, farads
L = inductance of capacitor, henries
$\omega = 2\pi f$, hertz

Eq 1 illustrates that the difference between the apparent and the actual capacitances increases rapidly as the frequency increases.

In a properly designed bridge, residual impedances (shunt capacitances and series inductances) are the principle factors that determine the accuracy of the bridge. The solution to this problem involves two possible approaches: (1) Minimize the magnitude of the residual impedances as much as possible, or (2) evaluate the residuals so that corrections can be made for their effects. Option 2 cannot be implemented unless a calibration standard is devised. For this purpose, a calibrated cable terminated into a resistive load will be used.

**Bridge Assembly**

The completed bridge is shown in Fig 13. Note the compact arrangement of the panel components. A smaller resistance dial on the front of the cabinet would have shortened the lead to the potentiometer by about one-half inch. However, the present layout works quite well.

Fig 13—The hybrid junction admittance bridge.

The dial for the variable air capacitor is on top of the cabinet, and the capacitor placement is such that its lug is soldered directly to S1, the L-C switch. The jack to the right of the switch accommodates fixed capacitor pads. It is also connected directly to the variable capacitor lug which eliminates another length of wire. The capacitor pads serve to extend the range of the 365 pF variable capacitor. The L-C switch is a DPDT with a center-off position. The capacitor dial can be calibrated in pF and the resistance dial in ohms. A calibration chart is preferred because it provides greater accuracy.

**Resistance Calibration Procedure**

A number of noninductive resistors between 5 and 200 ohms are required to calibrate the resistance dial and to construct a resistance calibration chart. The initial calibration procedure balances the bridge and is performed as follows.

**Step 1.** With the bridge connected as shown in Fig 10, connect a 50- or 52-ohm noninductive resistor to the load connector.

**Step 2.** Tune the signal generator and the receiver to 30 MHz.

**Step 3.** Set the L-C switch to its center-off position.

**Step 4.** Adjust the potentiometer dial to a null. The null may not be very deep. Adjust trimmer capacitor C1 for a deep null while slowly rocking the potentiometer dial across the null. This procedure tunes out any unbalanced reactance. If the capacitor does not deepen the null, disconnect it and reconnect it across J1. Repeat Step 4.

**Step 5.** Calibrate the resistance dial using precision nonreactive resistors with values between 5 and 200 ohms. Locate the new null position with each resistor and log the dial readings.

**Step 6.** Draw a resistance v dial reading calibration curve from the data obtained in Step 5. This completes the resistance calibration.

**Reactance Calibration Procedure**

A capacitance meter or capacitance bridge is required to calibrate the impedance bridge.

**Step 1.** Set the L-C switch to its center-off position.

**Step 2.** Calibrate the variable capacitor with a capacitance meter connected to J2. Take measurements at the minimum capacitance position and at every ten degrees.

**Step 3.** Set the L-C switch to either the L or C position and set the variable capacitor to its minimum position.

**Step 4.** With a 50- or 52-ohm load connected to the bridge and the signal generator and receiver tuned to 30 MHz, adjust the resistance dial to a null.

**Step 5.** Adjust bias capacitor C2 to obtain the deepest null. This neutralizes the minimum capacitance of the variable capacitor.

**Step 6.** Subtract the minimum capacitance reading obtained in Step 2 from all the capacitance measurements obtained. The minimum effective capacitance is now 0 pF. For a well-constructed bridge, switching from L to OFF to C should not disturb the null depth.

**Step 7.** Draw a capacitance v dial reading calibration curve from the data obtained in Step 6. This completes the capacitance calibration.

**Fixed Capacitor Pads**

The range of the variable capacitor may be inadequate to null the bridge at low frequencies. Therefore, it will be necessary to pad the variable capacitor with additional capacitance. Dipped mica capacitors (connected to phono plugs) provide the necessary pads. Commercially available capacitors with the values of: 360 pF, 680 pF, 910 pF and 1200 pF will provide the necessary padding with adequate overlapping ranges.

**The Calibration Cable Assembly**

To evaluate the accuracy of the bridge, a means must be available to provide predetermined, calculated impedance points. These points are then compared with the measured impedance points. Ideally, the calculated and measured points will have the same $R \pm jX$ values if attenuation, imperfections and other variables are disregarded.

The cable assembly that was used for all

**Table 1**
**Length in Electrical Degrees of a 16-Foot Section of RG-58 Coax**

| $f_{MHz}$ | θ° |
|---|---|
| 2.0 | 17.73 |
| 2.5 | 22.17 |
| 3.0 | 26.60 |
| 3.5 | 31.03 |
| 4.0 | 35.47 |
| 4.5 | 39.90 |
| 5.0 | 44.33 |
| 5.5 | 48.77 |
| 6.0 | 53.20 |
| 6.5 | 57.64 |
| 7.0 | 62.07 |
| 7.5 | 66.50 |
| 8.0 | 70.94 |
| 8.5 | 75.37 |
| 9.0 | 79.80 |
| 9.5 | 84.24 |
| 10.0 | 88.67 |
| 10.15 | 90.00 |
| 11.0 | 97.53 |
| 12.0 | 106.40 |
| 13.0 | 115.27 |
| 14.0 | 124.14 |
| 15.0 | 133.00 |
| 16.0 | 141.87 |
| 17.0 | 150.74 |
| 18.0 | 159.61 |
| 19.0 | 168.47 |
| 20.0 | 177.34 |
| 21.0 | 186.21 |
| 22.0 | 195.07 |
| 23.0 | 203.94 |
| 24.0 | 212.81 |
| 25.0 | 221.67 |
| 26.0 | 230.54 |
| 27.0 | 239.41 |
| 28.0 | 248.28 |
| 29.0 | 257.14 |
| 30.0 | 266.01 |

$$\frac{f_{MHz}}{10.15} \times 90°$$

**Test Data Sheet**
**Test Cable — 16 feet RG-58 terminated into a 25-ohm load.**

| $f_{MHz}$ | $R_{dial}$ | $C_{dial}$ | $R_p$ | $C_p$ | $\pm X_p$ | Normalized $R_s \pm jX_s$ |
|---|---|---|---|---|---|---|
| 2.0 | 52 | 04 + 1020 pF | 36 | 1020 | +78.1 | 0.57 + 0.26 |
| 2.5 | 56 | 86 + 547 pF | 39.5 | 847 | +75.2 | 0.60 + 0.32 |
| 3.0 | 64 | 52 + 547 pF | 45 | 727 | +73 | 0.63 + 0.38 |
| 3.5 | 70 | 20 + 547 pF | 51.5 | 601 | +75.7 | 0.67 + 0.46 |
| 4.0 | 77 | 10 + 475 pF | 58 | 496 | +80.3 | 0.73 + 0.53 |
| 4.5 | 83 | 84 + 107 pF | 63 | 407 | +86.9 | 0.80 + 0.58 |
| 5.0 | 92 | 67 + 107 pF | 70 | 339 | +93.9 | 0.87 + 0.64 |
| 5.5 | 100 | 52 + 107 pF | 77 | 287 | +101 | 0.93 + 0.71 |
| 6.0 | 108 | 66 | 83 | 230 | +115 | 1.05 + 0.76 |
| 6.5 | 116 | 56 | 89 | 194 | +126 | 1.14 + 0.81 |
| 7.0 | 124 | 44 | 96 | 150 | +152 | 1.32 + 0.84 |
| 7.5 | 127 | 36 | 98 | 119 | +178 | 1.44 + 0.80 |
| 8.0 | 130 | 29 | 100 | 90 | +221 | 1.60 + 0.72 |
| 8.5 | 134 | 23 | 104 | 66 | +284 | 1.76 + 0.64 |
| 9.0 | 136 | 17 | 106 | 43 | +412 | 1.91 + 0.49 |
| 9.5 | 137 | 11 | 107 | 24 | +698 | 2.01 + 0.31 |
| 10 | 137.5 | 4 | 107 | 6 | +2654 | 2.06 + 0.09 |
| 11 | 134 | −13.5 | 104 | −33 | −439 | 1.89 − 0.45 |
| 12 | 126 | −21 | 97 | −56 | −237 | 1.60 − 0.65 |
| 13 | 114 | −26.5 | 88 | −80 | −153 | 1.27 − 0.73 |
| 14 | 99 | −31.5 | 76 | −100 | −114 | 1.01 − 0.67 |
| 15 | 84 | −35 | 64 | −115 | −92.3 | 0.83 − 0.58 |
| 16 | 70 | −38 | 53 | −126 | −79 | 0.70 − 0.47 |
| 17 | 57.5 | −38.5 | 41 | −128 | −73.2 | 0.60 − 0.34 |
| 18 | 48 | −35 | 33 | −123 | −71.9 | 0.52 − 0.24 |
| 19 | 43 | −24.5 | 29 | −72 | −116 | 0.52 − 0.13 |
| 20 | 42 | −6 | 28 | −11 | −724 | 0.54 − 0.02 |
| 21 | 43 | +20 | 29 | +55 | +138 | 0.53 + 0.11 |
| 22 | 52 | 26.5 | 36 | 81 | +89.4 | 0.57 + 0.24 |
| 23 | 64 | 28.5 | 46 | 90 | +76.9 | 0.65 + 0.39 |
| 24 | 79 | 27 | 59 | 83 | +79.9 | 0.73 + 0.54 |
| 25 | 93 | 24 | 71 | 70 | +91 | 0.85 + 0.66 |
| 26 | 100 | 20.5 | 84 | 55 | +111 | 1.03 + 0.78 |
| 27 | 120 | 17 | 92 | 43 | +137 | 1.23 + 0.82 |
| 28 | 124 | 13.5 | 100 | 32 | +178 | 1.46 + 0.82 |
| 29 | 134 | 10 | 104 | 21 | +262 | 1.73 + 0.69 |
| 30 | 134 | 6 | 104 | 11 | +483 | 1.91 + 0.41 |

Fig 15—Test data obtained with the practical hybrid junction admittance bridge described.

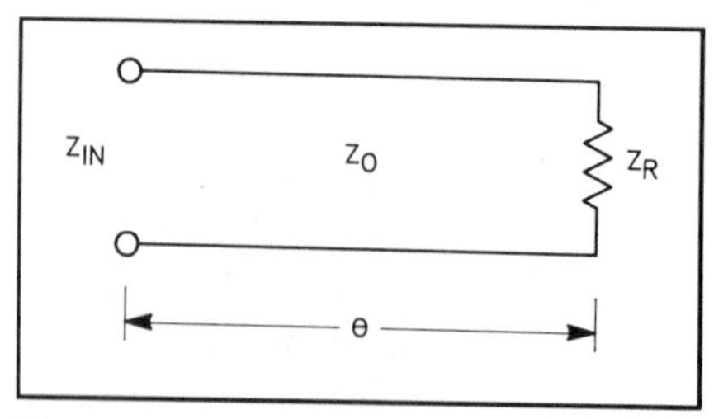

Fig 14—Input impedance of a line.

measurements consisted of a 16-foot length of RG-58 coaxial cable terminated into a 25-ohm load. Actually, any cable length or termination can be used. A longer cable, for instance, will provide impedance points that are further apart. The 25-ohm termination will cause the impedance points to have a constant standing-wave ratio of 2.0 to 1.

We know that when a transmission line with a characteristic impedance ($Z_0$) is terminated by an equal load impedance ($Z_R$), the input impedance of the line will also be $Z_0$, regardless of the length of the line. In any other case, however, the input impedance will depend on the length of the line (θ) and the termination resistance, as illustrated in Fig 14.

The following equation is important for determining the input impedance ($Z_{in}$) of a length of line (measured in electrical degrees, θ), with a characteristic impedance ($Z_0$) terminated by an impedance ($Z_R$).

$$Z_{in} = Z_0\left[\frac{Z_R\cos\theta + jZ_0\sin\theta}{Z_0\cos\theta + jZ_R\sin\theta}\right] \quad \text{(Eq 2)}$$

RG-58 has a velocity of propagation of about 66%. The equivalent electrical length of a 16-ft section is 16/0.66 = 24.24 feet or 7.389 meters. This equals one-quarter wavelength or θ = 90° at 10.150 MHz. The length in degrees v other frequencies is shown in Table 1.

It now remains to compute $Z_{in}$ using the line lengths in electrical degrees from Table 1, the line impedance of 52 ohms for

RG-58 and a termination of 25 ohms. For 2.0 MHz

$$Z_{in} = 52\left[\frac{25\cos 17.73 + j52\sin 17.73}{52\cos 17.73 + j25\sin 17.73}\right]$$

$$= 52\left[\frac{25\,(0.9525) + j52\,(0.3045)}{52\,(0.9525) + j25\,(0.3045)}\right]$$

$$= 52\left[\frac{23.8125 + j15.8340}{49.53 + j7.6125}\right]$$

$$= 52\left[\frac{23.8125 + j15.8340}{49.53 + j7.6125}\right]\left[\frac{49.53 - j7.6125}{49.53 - j7.6125}\right]$$

$$= 52\left[\frac{1179.43 + j784.25 - j181.27 + 120.54}{2453.22 + j377.05 - j377.05 + 57.95}\right]$$

$$Z_{in} = 52\left[\frac{1299.97 + j602.98}{2511.17}\right]$$

$$= 52\,[0.5177 + j0.240]$$

$$= 26.92 + j12.49 \quad \text{Answer}$$

Dispense with the last step when normalizing. Enter R = 0.52 (0.5177 rounded) and $jX$ = 0.24 on the normalized Smith chart. Compared with the measured impedance contained in the Test Data Sheet (Fig 15) for 2.0 MHz, we see a close agreement with the calculated data. Repeat the calculations for all other frequencies of interest and plot as shown in Fig 16.

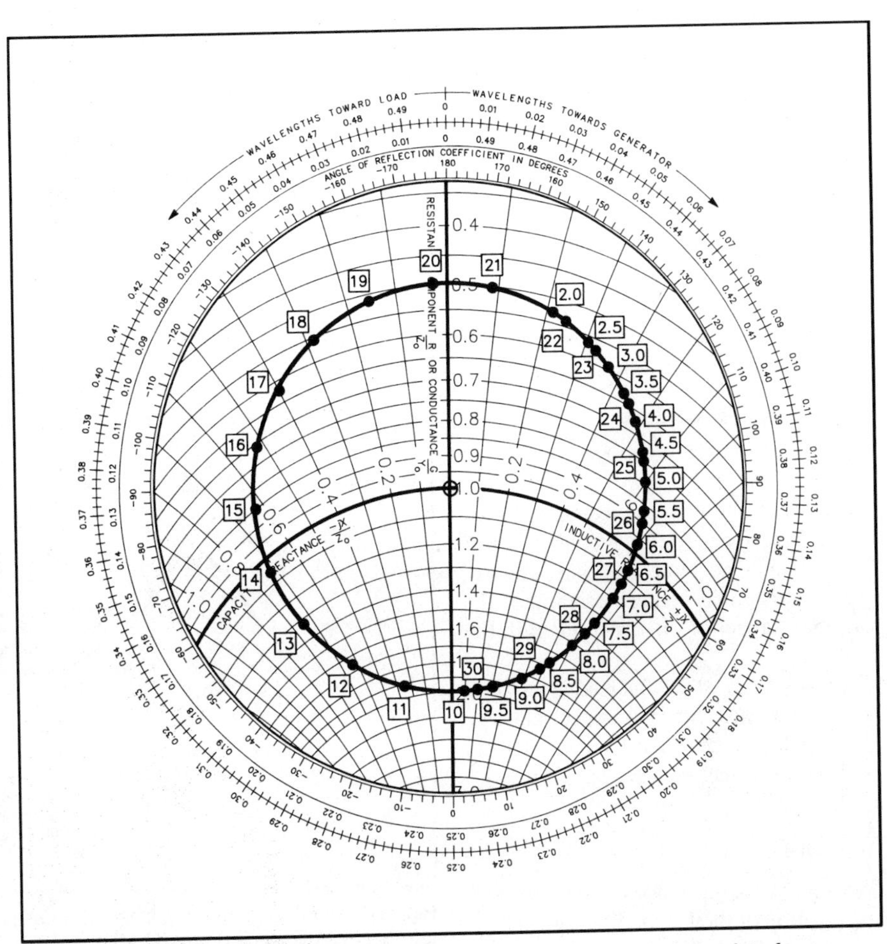


Fig 16—Idealized impedance plot of data obtained from Eq 2 and phase data from Table 1.

## Operating the Bridge

For accurate RF bridge operation, you should exercise considerable care when setting up the testing apparatus. Any direct signal leakage from the signal generator to the receiver will cause errors. Furthermore, if an antenna is being measured, radiation from the generator may be picked up by the antenna. When this signal voltage is subsequently canceled by a voltage of opposite phase in the operation of the bridge, a false null will result. However, with a well-constructed bridge, adequately shielded receiver and soundly constructed interconnecting cables, highly accurate measurements can be obtained.

## Measuring Resistance and Reactance With an Admittance Bridge

The advantage of an admittance bridge is that both the standard variable resistor and standard variable capacitor are "cold" and can be mounted directly to the case. This reduces stray coupling between the components of the bridge, simplifying bridge construction and improving bridge accuracy.

As shown in Fig 17, every impedance can be expressed for any frequency as either a series or parallel combination of resistance and reactance. The relations between the elements of Fig 17 are

$$R_p = \frac{1}{G_p} = \frac{R_s^2 + X_s^2}{R_s} \qquad \text{(Eq 3)}$$

$$X_p = \frac{1}{B_p} = \frac{R_s^2 + X_s^2}{X_s} \qquad \text{(Eq 4)}$$

$$Z = \frac{R_p X_p^2}{R_p^2 + X_p^2} \pm j\frac{R_p^2 X_p}{R_p^2 + X_p^2}$$

$$= R_s \pm jX_s \qquad \text{(Eq 5)}$$

The complex impedance represented by the series combination of resistance R and reactance X is shown in Fig 17A. The complex admittance represented by a parallel combination of conductance G and susceptance B is shown in Fig 17C. In some applications, such as with the Hybrid Junction Admittance Bridge, it is useful to represent impedance by a parallel combination of resistance and reactance as in Fig 17B. The impedance of the circuit in Fig 17B, as of any parallel circuit, is equal to the reciprocal of the sum of the reciprocals

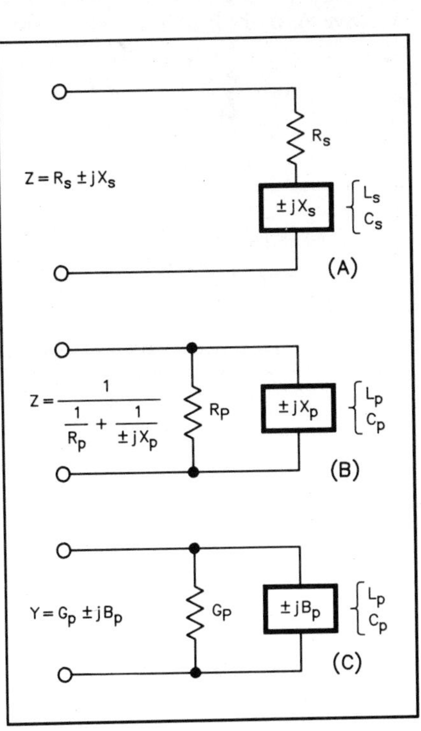


Fig 17—Equivalent complex series and parallel components of impedance Z and complex parallel components of admittance Y.

of the resistive and reactive components. The equation shown can be rationalized to obtain the component parts of the impedance, representing the equivalent series resistance and series reactance as indicated by Eq 5, $R_S \pm jX_S$.

Every impedance can be expressed in terms of either series or parallel equivalence. If both standards of the bridge are adjustable, the balance for the real and imaginary parts of the unknown will be independent of each other.

### Measurement Procedure

**Step 1.** Connect the bridge as shown in Fig 10.

**Step 2.** Following the example shown in Test Data Sheet, Fig 15, record the R and C dial readings. Note that the low frequency range may require padding of the variable capacitor in order to null the bridge.

**Step 3.** Convert the dial readings to their actual resistance and capacitance values.

**Step 4.** Convert the capacitance values to their equivalent reactances.

$$X = \frac{1}{2\pi fC}$$

The sign of the reactance is determined by the position of the L-C switch, +L, –C.

**Step 5.** The above steps provide the parallel components of the load impedance, $R_P$ and $\pm jX_P$. To convert the parallel components of the load impedance to their equivalent series components:

$$R_s = \frac{R_p X_p^2}{R_p^2 + X_p^2} \text{ and } X_s = \frac{R_p^2 X_p}{R_p^2 + X_p^2}$$

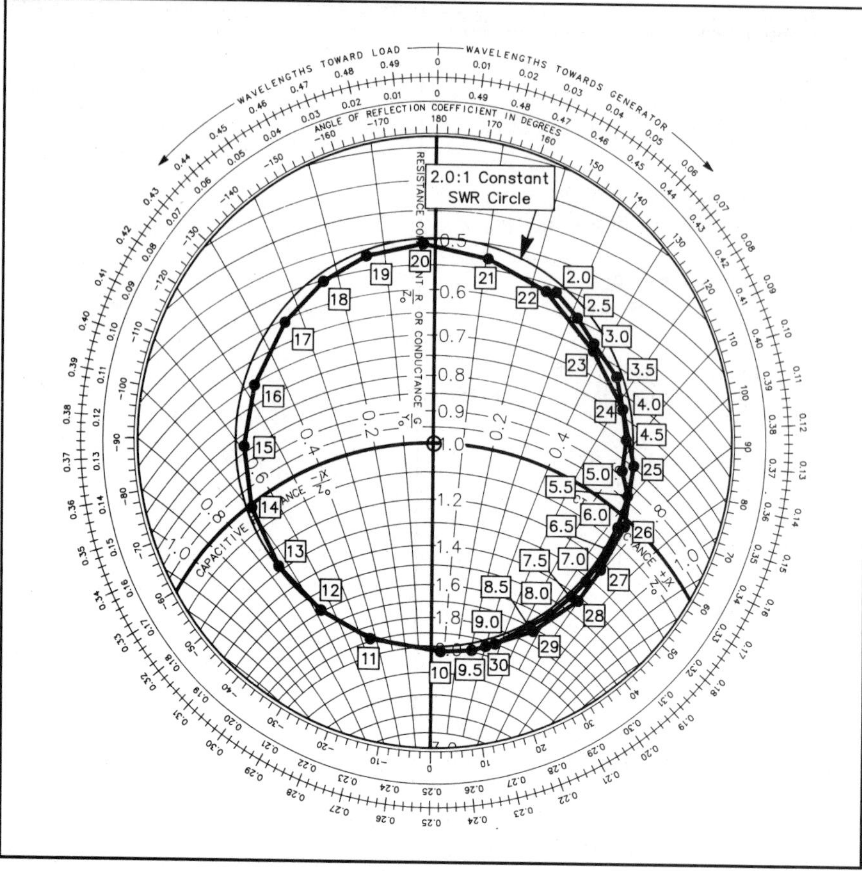


Fig 18—Impedance plot of the test data presented in Fig 15. The frequency range of interest is between 2.0 and 30 MHz.

$$Z = R_s \pm jX_s$$

### References


[1]G. Grammer, "Universal SWR Measurements With a Coaxial Bridge," *QST*, Dec 1950, pp 27-29, 104.

[2]C. Nouel, "How to Build a Cheap and Easy RF Noise Bridge," *CQ*, May 1984, pp 36-37, 41.

[3]W. N. Caron, "An Accurate RF Impedance Bridge," *The ARRL Antenna Book*, 15th Ed (1988), pp 27-19 to 27-23.

# VHF/UHF Ray Tracing With Computer Graphics

By Jack Priedigkeit, W6ZGN
441 Sherwood Way
Menlo Park, CA 94025

My article in November 1986 *QST* generated a bit of interest in VHF/UHF ray tracing.[1] The most interesting letter I received in response to that article was from Dave Palmrose, NY7C. After reading the article, Dave decided to put theory to the test. He watched the weather satellite on TV and waited patiently for nearly solid cloud cover from the Pacific Coast to Hawaii. When that occurred, he went to a 1600-foot elevation site on the Oregon coast, *below* the top of the fog layer, and set up his two-meter rig—just ten watts with a 15-element beam. In less than two hours, he heard and *worked* KH6HMK. Needless to say, Dave became an instant true believer in ray tracing. There have been many transpacific contacts reported, but to my knowledge, Dave is the first amateur to *plan* a transpacific contact using weather information and the concepts of ray tracing.

This article reviews the basic concepts of VHF/UHF radio propagation in a refractive medium. It also presents a menu-driven, hands-on, interactive program for a PC with a Hercules-compatible graphics card.

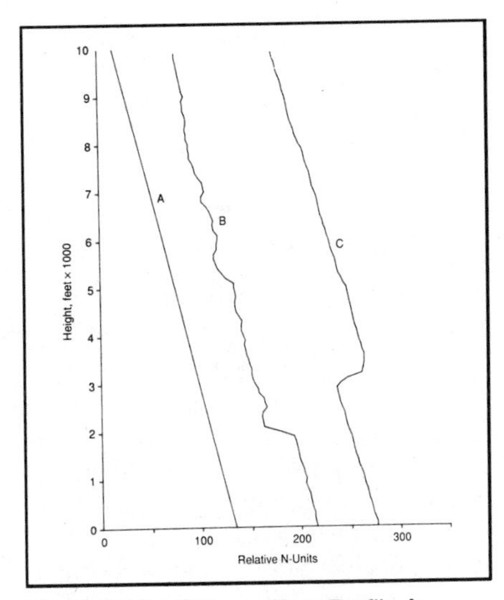


Fig 1—Refractivity profiles. Profile A, linear, and Profile B from West Coast measurements. Profile C, from measurements off the tip of Long Island, New York, does not represent static meteorlogical conditions.

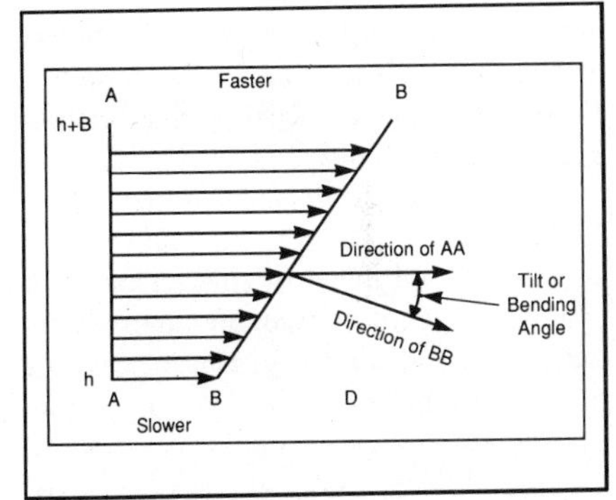


Fig 2—Illustration showing how a radio wavefront is bent or tilted downward by the air at the surface of the earth. This tilt, for a long-term average refractivity profile, is equivalent to the frequently referenced "4/3 earth radius."

### Some Basic Concepts

The propagation path of energy radiated from an antenna depends on the homogeneity of the refractive index in the propagation medium. The path, or ray, is a straight line in a homogeneous medium. However, the ray will bend, due to refraction, in a nonhomogeneous medium.

The absolute refractive index is the ratio of the velocity of propagation in a vacuum to the velocity of propagation in some other medium, in this case air. Refractive index is the square root of the dielectric constant, and this index determines velocity just as the square root of the dielectric constant in an RF transmission line determines the velocity factor.

For most purposes, it is sufficient to assume the dielectric constant and refractive index for air to be 1. However, the refractive index for air is typically 1.00300 for average humidity, temperature, and pressure at sea level. For the convenience of not having to use six-digit decimal numbers, this refractive index is defined as a refractivity of 300 N-units.[2]

The refractivity of air decreases with height to approach zero N-units at great heights (ie, in a vacuum). The long-term average decrease in refractivity is nearly linear below 10,000 feet with a slope, or gradient, of –12 N-units per 1000 feet (see Profile A of Fig 1). Thus a radio signal at height h + Δ travels just a little faster than the same signal at the lower height h. The effect is to tilt the radio wavefront downward, ever so slightly (see Fig 2). This wavefront tilt, or path bending, for this long-term average refractivity profile is equivalent to a 4/3 earth radius, and results in the familiar expression for the distance to the radio horizon:

$$D = \sqrt{2h}$$

where
D = distance to the radio horizon, statute miles
h = antenna height, feet

At any instant, prevailing micro-meteorology will cause the refractivity profile to depart from the long-term average profile and will cause a varying degree of path bending along the signal path. The degree of path bending depends on the gradient of the refractive profile at the signal path height.

Profile B of Fig 1 is a refractivity profile at a particular location and time, measured with an aircraft carrying instruments to record refractive index as a function of altitude.[3,4] It is evident that the average gradient of this profile is similar to that of profile A. However, there is a break, or discontinuity, in the gradient near 2000 feet. This is typical of the West Coast and occurs at the interface between a dense fog layer below 2000 feet and the clear bright sunshine above the fog. The humidity within the fog is 100%, and the temperature is 10° or more below that of the dry air above the fog. Within a 500-foot region, just above the fog, there is both a temperature inversion and a rapid drop in humidity. The refractivity gradient in this region is −160 N-units per 1000 feet, or more than 13 times the normal average downward bending rate. Depending on the antenna height relative to this refractive discontinuity, elevated ducting can occur, as illustrated in Fig 3.

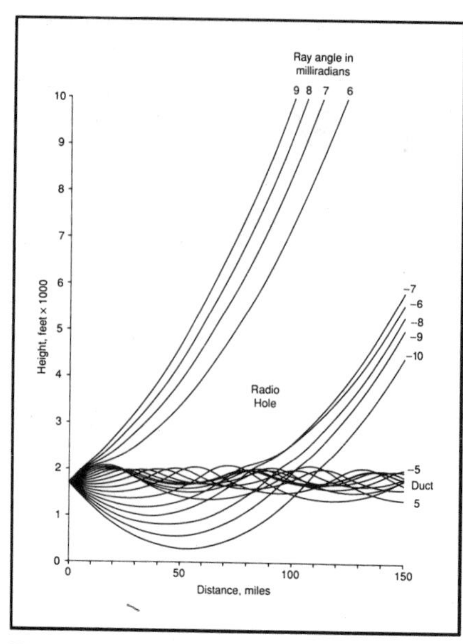

Fig 3—Ray plots for refractivity Profile B of Fig 1. Antenna height is 1700 feet.

Profile C of Fig 1 was measured east of Montauk Point, LI, NY. It should not be considered as typical of this region as it cannot be identified with an almost static meteorological condition, as is the case for the West Coast.

## Ray Tracing

With a computer graphics program, it is possible to explore ray paths for various refractive profiles, antenna heights, and takeoff angles. Fig 3 is a plot of the rays emerging from an antenna at a height of 1700 feet, propagating in a medium with the refractive Profile B of Fig 1. The takeoff angle, in milliradians relative to the horizon at the antenna, is shown on each ray. (A milliradian is 0.0573 degree.) The angles of interest are within 12 milliradians (less than one degree) of the horizon. Thus, in most cases, the vertical directivity of the antenna need not be considered. Fig 3 shows that rays leaving the antenna with takeoff angles between ±4 milliradians are trapped in an elevated duct that is parallel to the surface of the earth. Thus, there is no radio horizon for these rays, and they will continue to propagate as long as the meteorological conditions persist. This is the likely propagation mode for the transpacific VHF/UHF contacts between the West Coast and Hawaii that are reported.

Ducting does not occur with profile B if the antenna height is above the fog. The ray paths are distorted by the refractive discontinuity, but ducting does not occur. Ray tracing using profile C of Fig 1 does not show ducting because the slope of the gradient in the region of the discontinuity is positive, rather than negative, and rays bend up, rather than down.[2,5,6]

## Computer Graphics Program

Program 1 is a listing of a BASIC computer program, RAYPLOT.BAS. The program calculates and displays ray paths on the screen of a computer with Hercules compatible graphics. This is a menu-driven, hands-on, interactive ray-tracing program that lets the user select one of the three refractive profiles shown in Fig 1, specify antenna height, and try various takeoff angles. (The program is available on diskette; see **Diskette Availability** on a page at the front of this book.) Fig 3 represents but one of the many possible ray plots for user-specified conditions that can be displayed on the computer screen.

RAYPLOT.BAS uses Hercules GraphX subroutines to display the graphics. GWBASIC and BASICA do not recognize these Hercules subroutines. Thus, to run RAYPLOT.BAS it is necessary to load Hercules GraphX INT10.COM and HGC FULL or their equivalents into memory before loading GWBASIC, or BASICA. (The INT10.COM and HGC.COM programs are supplied with the Hercules graphics card. Equivalent programs usually accompany compatible cards.)

The program header, lines 5-200, loads the Hercules GraphX binary file BAS2GRPH.BIN (line 40), which must be in the current subdirectory at runtime. An executable version, RAYPLOT.EXE, is available on the companion diskette to this book. This program will run at DOS level after loading INT10.COM and HGC FULL.

Ray tracing for VHF/UHF radio propagation is based on Snell's Law for geometric optics in stratified refractive layers. Numerical integration is used to evaluate the second-order differential equation that defines the ray path in this propagation medium.[7]

The majority of the statements listed in Program 1 are for program and graphics control. Ray paths are calculated and plotted, point by point, with the statements in lines 250-430.

Program lines 500-640 display the program menu. Lines 650-780 query the user for input data. The Y axis is plotted and labeled for ray path height, lines 830-1090. The X axis is plotted and labeled for either relative N-units or statute miles, lines 1440-1850.

Data for Profile A, Fig 1, is calculated in line 1350. Data for Profiles B and C are contained in lines 1870-2030. This is the value of N at 100-foot height intervals. Antenna height and initial ray angle are entered by the user, line 710-770. The gradient of the refractive profile is calculated at the initial height, line 290. Assuming these data to remain constant, a new ray height is calculated for a propagation distance of 0.2 mile, 1056 feet, and this new height is plotted on the monitor screen, lines 380-420. The gradient of the refractive profile at this new height is calculated, line 290 again, and the ray height is again calculated for a propagation distance of 1056 feet. This process is repeated 750 times as the ray propagates from the antenna to a distance of 150 statute miles. Lines 330-360 accomplishes the double integration, summation, to evaluate the differential equation that defines the ray path height versus distance.

The ray is considered as being reflected from the earth when the ray path height is less than 100 feet. At this time, the initial ray angle and ray height are reset to the reflection angle and height at the reflection point. The integrators are set to zero as it is considered to be the start of a new ray path with new initial conditions, lines 300-320. The program pauses at line 440 after each ray plot, and waits for the ENTER key to be pressed by the user. It then clears the screen and returns to the program menu for further instructions from the user, lines 490-640. All ray plots remain in the graphics memory until cleared by the user, and will reappear on screen at the start of a new ray plot. This makes it possible to display a family of rays for a specified initial condition, such as antenna height.

## Notes

[1]J. Priedigkeit, "Ray Tracing and VHF/UHF Radio Propagation," *QST*, Nov 1986, pp 18-21.

[2]B. R. Bean and E. J. Dutton, "Radio Meteorology," *NBS Monograph 92*, Mar 1966.

[3]C. M. Crain, "Apparatus for Recording Fluctuations in Refractive Index of the Atmosphere," *Review of Scientific Instruments*, 21 No. 5, May 1950, pp 456-457.

[4] C. M. Crain, "Survey of Airborne Microwave Refractometer Measurements," *Proc IRE*, 43, No. 10, Oct 1955, pp 1405-1411.

[5] H. R. Reed and C. M. Russell, "Ultra High Frequency Propagation" (Boston: Boston Technical Publishers, 1964).

[6] D. E. Keer, *Propagation of Short Radio Waves* (Boston: Boston Technical Publishers, 1964).

[7] M. S. Wong, "Refraction Anomalies in Airborne Propagation," *Proc IRE*, 46, No. 9, Sep 1958, pp 1628-1639.


---

## Program 1

## Listing for BASIC Program RAYPLOT.BAS

This program is available on a companion diskette; see **Diskette Availability** on an early page of this book. Also included on the companion diskette is RAYPLOT.EXE, a compiled version that runs directly from DOS.

Note: INT10.COM and HGC.COM or their equivalents must be loaded into computer memory before executing either RAYPLOT.BAS or RAYPLOT.EXE.

```
5       ' RAYPLOT.BAS   by Jack Priedigkeit, W6ZGN
10      '       interface header for INT10.COM using BAS2GRPH.BIN
20      '       HERCULES COMPUTER TECHNOLOGY - December 1984
30      DEF SEG = &H3000                'assuming 256k memory
40      BLOAD "bas2grph.bin",0          'interface code from BASIC to INT10
50      GMODE%  = 256*PEEK(1)+PEEK(0)   'call gmode%
60      TMODE%  = 256*PEEK(3)+PEEK(2)   'call tmode%
70      CLRSCR% = 256*PEEK(5)+PEEK(4)   'call clrscr%
80      GPAGE%  = 256*PEEK(7)+PEEK(6)   'call gpage%(page%)
90      LEVEL%  = 256*PEEK(9)+PEEK(8)   'call level%(intens%)
100     DISP%   = 256*PEEK(11)+PEEK(10) 'call disp%(page%)
110     PLOT%   = 256*PEEK(13)+PEEK(12) 'call plot%(x%,y%)
120     GETPT%  = 256*PEEK(15)+PEEK(14) 'call getpt%(x%,y%,intens%)
130     MOVE%   = 256*PEEK(17)+PEEK(16) 'call move%(x%,y%)
140     DLINE%  = 256*PEEK(19)+PEEK(18) 'call dline%(x%,y%)
150     BLKFIL% = 256*PEEK(21)+PEEK(20) 'call blkfil%(x%,y%,w%,h%)
160     TEXTB%  = 256*PEEK(23)+PEEK(22) 'call textb%(x%,y%,msg$)
170     ARC%    = 256*PEEK(25)+PEEK(24) 'call arc%(x%,y%,r%,q%)
180     CIRC%   = 256*PEEK(27)+PEEK(26) 'call circ%(x%,y%,r%)
190     FILL%   = 256*PEEK(29)+PEEK(28) 'call fill%(x%,y%)
200     HRDCPY% = 256*PEEK(31)+PEEK(30) 'call hrdcpy%(optchar%)
210 ON ERROR GOTO 2200
220 DIM N0(110)
230 DEFINT W,X,Y,Z
240 GOTO 2040
250 SUM0=0: SUM1=0: SUM2=0
260 FOR I=0 TO 750
270 H100=INT(HSL*.01)
280 IF H100<0 THEN H100=0
290 NS0=N0(H100+1)-N0(H100): NS0=NS0*10
300 IF HSL<100 THEN 310 ELSE 330
310 TA0=SUM0-TA0
320 SUM0=0: SUM1=0: SUM2=0: H0=HSL
330 SUM0=SUM0+1.056E-06*NS0+.0000505
340 SUM1=SUM1+1056*SUM0
350 SUM2=SUM2+1056*TA0
360 HSL=H0+SUM1+SUM2
370 IF HSL>10000 GOTO 440
380 X%=50+INT(.866*I)
390 Y%=310-INT(HSL*.031)
400 IF I>0 THEN 420
410 CALL MOVE%(X%,Y%)
420 CALL DLINE%(X%,Y%)
430 NEXT I
440 Z0$=INKEY$: IF Z0$<>CHR$(13) THEN 440
450 CALL TMODE%: GOTO 490
```

```
460 CALL GMODE%: Z%=0: CALL GPAGE%(Z%): CALL CLRSCR%
470 Z%=1: CALL GPAGE%(Z%): CALL CLRSCR%: CALL TMODE%
480 FL1=0: FL2=0: FL3=0
490 'MENU
500 CLS: LOCATE 5,5
510 PRINT "             Menu for VHF/UHF Ray Tracing "
520 PRINT
530 PRINT "          1. Select Refractive Index Profile "
540 PRINT "          2. Plot Ray Path "
550 PRINT "          3. Clear Graphics Memory "
560 PRINT "          4. Exit Program "
570 LOCATE 15,20
580 PRINT " Graphics will remain on screen until ENTER is pressed "
590 LOCATE 12,15: INPUT "Select one of the above and ENTER to continue "; SL$
600 IF VAL(SL$)<1 OR VAL(SL$)>4 THEN 490
610 IF SL$="4" GOTO 2210
620 IF SL$="3" THEN 650
630 IF SL$="2" THEN 670
640 IF SL$="1" THEN 790
650 CALL GMODE%: Z%=1: CALL GPAGE%(Z%): CALL CLRSCR%: CALL TMODE%
660 FL3=0: GOTO 490
670 IF FL1=0 THEN 680 ELSE 700
680 CLS: LOCATE 9,20: PRINT " Profile not selected "
690 LOCATE 10,20: INPUT " ENTER to continue "; Q$: GOTO 490
700 CLS: LOCATE 10,15: PRINT "   ENTER to use previous height, or angle "
710 LOCATE 5,10: INPUT "  Antenna height in feet above sea level "; IH$
720 IF IH$="" THEN 740
730 IH=VAL(IH$)
740 LOCATE 6,10
750 INPUT "  Initial angle in milliradians "; IA$
760 IF IA$="" THEN 780
770 A0=VAL(IA$): A0=A0*.001
780 TA0=TAN(A0): HSL=IH: H0=IH
790 CALL GMODE%: Z%=0: CALL GPAGE%(Z%): CALL CLRSCR%
800 Z%=1: CALL GPAGE%(Z%): CALL LEVEL%(Z%): CALL DISP%(Z%)
810 IF SL$="1" THEN FL2=0: GOTO 1120
820 IF SL$="2" THEN FL2=1: GOTO 830
830 'DRAW Y-AXIS
840 IF FL3=1 THEN 250
850 MSG1$="0": MSG2$="1": MSG3$="2": MSG4$="3": MSG5$="4": MSG6$="5"
860 MSG7$="6": MSG8$="7": MSG9$="8": MSG10$="9": MSG11$="10": MSG12$="K"
870 MSG13$="I": MSG14$="L": MSG15$="O": MSG16$="F": MSG17$="E": MSG18$="T"
880 X%=50: Y%=0: CALL MOVE%(X%,Y%)
890 Y%=310: CALL DLINE%(X%,Y%)
900 X%=55
910 FOR I=0 TO 9: Y%=I*31: CALL PLOT%(X%,Y%): NEXT I
920 X%=27: Y%=35: CALL TEXTB%(X%,Y%,MSG10$)
930 Y%=66: CALL TEXTB%(X%,Y%,MSG9$)
940 Y%=97: CALL TEXTB%(X%,Y%,MSG8$)
950 Y%=129: CALL TEXTB%(X%,Y%,MSG7$)
960 Y%=159: CALL TEXTB%(X%,Y%,MSG6$)
970 Y%=190: CALL TEXTB%(X%,Y%,MSG5$)
980 Y%=221: CALL TEXTB%(X%,Y%,MSG4$)
990 Y%=252: CALL TEXTB%(X%,Y%,MSG3$)
1000 Y%=283: CALL TEXTB%(X%,Y%,MSG2$)
1010 Y%=314: CALL TEXTB%(X%,Y%,MSG1$)
1020 X%=10: Y%=90: CALL TEXTB%(X%,Y%,MSG12$)
1030 Y%=106: CALL TEXTB%(X%,Y%,MSG13$)
1040 Y%=122: CALL TEXTB%(X%,Y%,MSG14$)
1050 Y%=138: CALL TEXTB%(X%,Y%,MSG15$)
1060 Y%=170: CALL TEXTB%(X%,Y%,MSG16$)
1070 Y%=186: CALL TEXTB%(X%,Y%,MSG17$)
```

## RAYPLOT.BAS (continued)

```
1080 Y%=202: CALL TEXTB%(X%,Y%,MSG17$)
1090 Y%=218: CALL TEXTB%(X%,Y%,MSG18$)
1100 IF FL2=1 THEN 1740
1110 IF FL2=0 THEN 1440
1120 'PUT #1 PROFILE IN GRAPHICS MEMORY
1125 CALL CLRSCR%
1130 FL3=0
1140 FOR I=1 TO 100
1150 X%=INT(220-I*1.2): Y%=310-INT(I*3.1)
1160 IF I>1 THEN 1180
1170 CALL MOVE%(X%,Y%)
1180 CALL DLINE%(X%,Y%): NEXT I
1190 'PUT #2 PROFILE IN GRAPHICS MEMORY
1200 RESTORE 1870
1210 FOR I=1 TO 100: READ N0(I)
1220 X%=200+N0(I): Y%=310-INT(I*3.1)
1230 IF I>1 THEN 1250
1240 CALL MOVE%(X%,Y%)
1250 CALL DLINE%(X%,Y%): NEXT I
1260 'PUT #3 PROFILE IN GRAPHICS MEMORY
1270 RESTORE 1950
1280 FOR I=1 TO 100: READ N0(I)
1290 X%=200+N0(I): Y%=310-INT(I*3.1)
1300 IF I>1 THEN 1320
1310 CALL MOVE%(X%,Y%)
1320 CALL DLINE%(X%,Y%): NEXT I
1330 GOTO 830
1340 'LOAD N0 ARRAY
1350 FOR I=1 TO 110: N0(I)=INT(140-I*1.2): NEXT I
1360 MSG28$="1": GOTO 1650
1370 RESTORE 1880
1380 FOR  I=1 TO 110: READ N0(I): NEXT I
1390 MSG28$="2": GOTO 1650
1400 RESTORE 1960
1410 FOR I=1 TO 110: READ N0(I): NEXT I
1420 MSG28$="3": GOTO 1650
1430 GOTO 490
1440 'DRAW X AXIS FOR REFRACTIVE PROFILE
1450 MSG19$="100": MSG20$="200": MSG21$="300"
1460 MSG22$="Relative N-Units": MSG23$="Select Profile"
1470 X%=50: Y%=310: CALL MOVE%(X%,Y%)
1480 X%=700: CALL DLINE%(X%,Y%)
1490 Y%=305: FOR I=1 TO 8: X%=50+I*81: CALL PLOT%(X%,Y%): NEXT I
1500 X%=200: Y%=325: CALL TEXTB%(X%,Y%,MSG19$)
1510 X%=363: Y%=325: CALL TEXTB%(X%,Y%,MSG20$)
1520 X%=525: Y%=325: CALL TEXTB%(X%,Y%,MSG21$)
1530 X%=280: Y%=340: CALL TEXTB%(X%,Y%,MSG22$)
1540 X%=535: Y%=340: CALL TEXTB%(X%,Y%,MSG23$)
1550 X%=150: Y%=100: CALL TEXTB%(X%,Y%,MSG2$)
1560 X%=290: Y%=100: CALL TEXTB%(X%,Y%,MSG3$)
1570 X%=475: Y%=100: CALL TEXTB%(X%,Y%,MSG4$)
1580 Z%=2: CALL LEVEL%(Z%)
1590 CRT=0: X%=535: Y%=340: CALL TEXTB%(X%,Y%,MSG23$)
1600 CRT=CRT+1: P$=INKEY$: IF P$="" THEN 1610 ELSE 1620
1610 IF CRT=200 THEN 1590 ELSE 1600
1620 IP=VAL(P$): IF IP<1 THEN 1590
1630 IF IP>3 THEN 1590
1640 ON IP GOTO 1350, 1370, 1400
1650 FL1=1: Z%=1: CALL LEVEL%(Z%)
1660 X%=535: Y%=340: CALL TEXTB%(X%,Y%,MSG23$)
1670 Z%=1: W%=15
1680 ON IP GOTO 1690, 1700, 1710
```

## RAYPLOT.BAS (continued)

```
1690 X%=165: Y%=102: CALL BLKFIL%(X%,Y%,W%,W%): GOTO 1720
1700 X%=305: Y%=102: CALL BLKFIL%(X%,Y%,W%,W%): GOTO 1720
1710 X%=490: Y%=102: CALL BLKFIL%(X%,Y%,W%,W%)
1720 Q$=INKEY$: IF Q$="" THEN 1720:
1730 CALL CLRSCR%: CALL TMODE%: GOTO 490
1740 'DRAW X AXIS FOR RAY PLOT
1750 MSG24$="Statute Miles": MSG25$="50": MSG26$="100": MSG27$="150"
1760 X%=50: Y%=310: CALL MOVE%(X%,Y%)
1770 X%=700: CALL DLINE%(X%,Y%)
1780 Y%=305: FOR I=1 TO 15: X%=50+I*42: CALL PLOT%(X%,Y%): NEXT I
1790 Y%=325: X%=258: CALL TEXTB%(X%,Y%,MSG25$)
1800 X%=46: CALL TEXTB%(X%,Y%,MSG1$)
1810 X%=465: CALL TEXTB%(X%,Y%,MSG26$)
1820 X%=680: CALL TEXTB%(X%,Y%,MSG27$)
1830 X%=320: Y%=340: CALL TEXTB%(X%,Y%,MSG24$)
1840 MSG29$="Profile #": X%=100: Y%=50: CALL TEXTB%(X%,Y%,MSG29$)
1850 X%=190: CALL TEXTB%(X%,Y%,MSG28$)
1860 FL3=1: GOTO 250
1870 'DATA FOR #2 PROFILE
1880 DATA 183,182,181,181,179,178,177,176,174,172,173,171,170,168,166
1890 DATA 165,164,163,162,160,146,129,128,127,128,132,130,129,124,123
1900 DATA 120,118,117,117,115,114,112,111,110,111,109,107,107.108,105
1910 DATA 102,101,101,102,101,100,100,92,88,84,82,80,80,83,83,84,84,80
1920 DATA 79,80,78,76,73,68,68,71,70,69,65,62,61,59,56,56,55,55,53,53
1930 DATA 52,53,53,52,50,50,49,51,49,48,47,46,45,44,43,42,42,41,41,39
1940 DATA 38,37,36,36,37,38,37,36,35
1950 'DATA FOR #3 PROFILE
1960 DATA 308,308,306,304,303,302,300,299,298,296,295,294,292,290,289
1970 DATA 287,286,285,284,283,281,279,278,277,276,274,272,271,270,269
1980 DATA 273,280,294,295,296,296,296,294,293,291,290,289,287,286,285
1990 DATA 284,283,282,281,280,279,276,274,271,270,269,268,266,264,263
2000 DATA 262,260,259,257,256,254,253,252,251,250,248,246,245,244,243
2010 DATA 241,239,238,237,235,234,232,230,229,228,227,225,223,222,221
2020 DATA 220,219,218,216,215,214,212,210,209,208,206,205,204,203,202
2030 DATA 200,199,198,197,198,196,198
2040 CLS: LOCATE 5,25: PRINT " VHF/UHF RAY TRACING  "
2050 PRINT
2060 PRINT "           This is a computer graphics version of my article"
2070 PRINT "      VHF/UHF RAY TRACING, ARRL Antenna Compendium, Volume 3."
2080 PRINT
2090 PRINT "      A Hercules compatable computer system is required to view"
2100 PRINT "      plots like those in my article.  Unlike my article, this "
2110 PRINT "      is a menu driven, hands-on, interective program that lets"
2120 PRINT "      the user select one of three refractive index profiles,"
2130 PRINT "      and specify antenna height and initial ray angle.  The"
2140 PRINT "      plots remain in graphics memory until cleared by the user."
2150 PRINT "      This facilitates the generation of a family of rays for "
2160 PRINT "      specified initial conditions. "
2170 LOCATE 20,25: PRINT "Jack Priedigkeit, W6ZGN          October 1990 "
2180 LOCATE 25,30: INPUT "ENTER to continue ";Q$
2190 GOTO 460
2200 CALL TMODE%: PRINT " Error Code  ";ERR;"     Line No.  ";ERL
2210 END
```

ANTENNA
COMPENDIUM,
VOL 3

PROOF OF
PURCHASE

# FEEDBACK

Please use this form to give us your comments on this book and what you'd like to see in future editions.

Where did you purchase this book? □ From ARRL directly □ From an ARRL dealer

Is there a dealer who carries ARRL publications within: □ 5 miles □ 15 miles □ 30 miles of your location? □ Not sure.

**License class:**

□ Novice □ Technician □ Technician with HF privileges □ General □ Advanced □ Extra

Name ____________________ Call sign ____________________

Address ____________________

City, State/Province, ZIP/Postal Code ____________________

Daytime Phone ( ) ____________________ Age ____________________

If licensed, how long? ____________________ ARRL member? □ Yes □ No

Other hobbies ____________________

____________________

Occupation ____________________

| For ARRL use only | AC V3 |
|---|---|
| Edition | 1 2 3 4 5 6 7 8 9 10 11 12 |
| Printing | 2 3 4 5 6 7 8 9 10 11 12 |

From ______________________________

______________________________

______________________________

Please affix postage. Post Office will not deliver without postage.

EDITOR, ANTENNA COMPENDIUM, VOL 3
AMERICAN RADIO RELAY LEAGUE
225 MAIN ST
NEWINGTON CT 06111-1494

please fold and tape